Birationally Rigid Varieties

Mathematical
Surveys
and
Monographs

Volume 190

Birationally Rigid Varieties

Aleksandr Pukhlikov

AMERICAN MATHEMATICAL SOCIETY · FOUNDED 1888 · ΑΓΕΩΜΕΤΡΗΤΟΣ ΜΗΔΕΙΣ ΕΙΣΙΤΩ

American Mathematical Society
Providence, Rhode Island

2010 *Mathematics Subject Classification*. Primary 14E05, 14E07, 14J45, 14E08, 14E30, 14M22, 14M10, 14M20, 14J30, 14J40.

For additional information and updates on this book, visit
www.ams.org/bookpages/surv-190

Library of Congress Cataloging-in-Publication Data

Pukhlikov, Aleksandr V., 1962–
Birationally rigid varieties / Aleksandr Pukhlikov.
pages cm. – (Mathematical surveys and monographs ; volume 190)
Includes bibliographical references and index.
ISBN 978-0-8218-9476-7 (alk. paper)
1. Geometry, Algebraic. I. American Mathematical Society. II. Title.

QA564.P85 2013
516.3′5–dc23 2013001354

10 9 8 7 6 5 4 3 2 1 18 17 16 15 14 13

Contents

Introduction

0.1. About this book. The purpose of this book is, first, to give a state of the art survey of the theory of birational rigidity and, second, to provide a readable introduction to that theory. Both tasks are not easy and need to be explained. In the past 40 years, starting from the pioneer paper of V.A. Iskovskikh and Yu.I.Manin on the three-dimensional quartics [**IM**], the theory of birational rigidity developed almost from scratch to an important area of birational algebraic geometry. In total, something close to a hundred research papers were published on the subject, some of them well known even beyond the community of experts in birational geometry. The area attracted more and more interest, so the need of a consistent and uniformly written modern survey became obvious. In response to this demand, in the past 15 years extensive survey papers or research papers containing significant survey-type parts, were written with increasing frequency [**IP, CPR, I01, Ch05c, Pukh07b, Pukh10b**]. However, none of them, due to their limited size could possibly cover the entire area, the variety of methods, techniques and ideas. This is what the present book does.

The origins of the theory of birational rigidity go back well into the nineteenth century, to the work of Clebsch, M. Noether and later, Italian geometers. In the 1900s young Gino Fano started his pioneer work on three-dimensional algebraic geometry and published several papers where, in particular, he claimed what is now known to be birational rigidity of several classes of algebraic three-folds (we now say Fano three-folds). However, Fano's arguments were not even an approximation of a rigorous proof and contained serious mistakes, although his intuition and the depth of his insights were amazing. For about 50 years (until 1970) Fano's work and his claims were a very special mysterious subject, not reliable but exciting and potentially promising. In the 1960s, the new foundations of algebraic geometry prompted a review of Fano's work.

The decisive step in the creation of the theory of birational rigidity was made by Manin and Iskovskikh in the above mentioned breakthrough paper (which followed their previous work on birational geometry of algebraic surfaces over non-closed fields [**M66, M67, I67, I70**]. They developed a beautiful scheme of arguments, consisting of several steps and making use of several geometric observations, that provided a rigorous proof of one of Fano's famous claims, that the group of birational self-maps of a non-singular three-dimensional quartic hypersurface in $\mathbb{P}^4$ coincides with the group of its biregular (= projective) automorphisms, being therefore finite, generically trivial. It follows immediately that the quartic hypersurface is non-rational, since the group of birational self-maps is obviously a birational invariant of an algebraic variety and the group of birational transformations of the projective space (the Cremona group) is very large, certainly infinite (it should be noted that non-rationality of the non-singular quartics follows directly from the proof given in [**IM**], without the intermediate reference to the size of the group of birational self-maps, we discuss this point in detail in Chapter 2). Since some non-singular quartic three-folds were known to be unirational, the theorem of Iskovskikh and Manin meant a solution to the famous Lüroth problem ("does unirationality imply rationality?"; the answer is positive for curves and for surfaces over an algebraically closed field; the problem was open for three-folds and higher-dimensional varieties for several decades), which also put the paper [**IM**] on the list of the most remarkable research papers of the twentieth century in algebraic geometry.

0.2. Birational rigidity. It is appropriate now to say a few words about some terminology which is firmly associated with the theory of birational rigidity. First of all, it is the very term "rigidity". Its origin is in the same paper on the three-dimensional quartics: having formulated their main result, the authors remark that the finiteness of the group of birational self-maps demonstrates the extreme "rigidity" of the quartics. However, this was an informal image, when the word was used for the first time. As a rigorous definition, "birational rigidity" appeared first in [**Pukh95**] (and in spoken language, in the author's talk at the conference on algebraic geometry in Warwick in September 1991).

The system of arguments, computational techniques and geometric observations, employed in the theory of birational rigidity, is now known as "the method of maximal singularities", since the concept of a "maximal singularity" of a movable linear system plays the central role in the proofs. To say that a system has a maximal singularity is the same as to say that a certain pair is not canonical; therefore, we can say that the concept of a canonical singularity (of a movable linear system on a non-singular variety) was also first given in [**IM**], much ahead of its time. It should be noted that the method of maximal singularities is about 15 years older than the minimal model program, and it developed independently of the latter, although in the last ten years the influence of ideas and methods of the minimal model program in the theory of birational rigidity was visibly increasing.

Apart from birational (super)rigidity and maximal singularities, there are a few other concepts in this theory which are known in the wider community, such as "maximal subvariety", "K^2-condition", "K-condition", and some general constructions, such as, in the first place, the graph of a sequence of blow ups, the latter seems to have been re-invented several times in different areas of geometry, which is hardly surprising, given its natural definition and essential information accumulated in that graph.

Speaking informally, the phenomenon of birational rigidity is that an algebraic variety that has no non-zero regular differential forms behaves as if it had plenty of them. Indeed, even for a beginner who just started to study algebraic geometry it is obvious why a non-singular hypersurface $V_m \subset \mathbb{P}^d$ is non-rational for $m \geq d+1$ or why if two such hypersurfaces V_{m_1} and V_{m_2} are birational, then $m_1 = m_2$ and they are projectively isomorphic: one has to consider the global differential forms on V_m, V_{m_1} and V_{m_2}. However, it is unexpected and mysterious, why a non-singular hypersurface $V_m \subset \mathbb{P}^d$ for $m \leq d$ should behave in the same way, being rationally connected and for that reason admitting no non-zero differential-geometric birational invariants. A birationally superrigid Fano variety is a Fano variety that behaves in many crucial respects as a variety of general type.

The reasons of this mysterious behaviour are not yet known. However, it is known that birational rigidity is a very typical phenomenon in higher-dimensional algebraic geometry: the most natural constructions of Fano varieties, such as complete intersections in weighted projective space, yield birationally rigid varieties (whenever the study was successfully completed). The underlying reasons are yet to be understood: the area is still full of mysteries!

0.3. The contents of this book. The structure of this book is as follows. Chapter 1, which is aimed at a wide audience, provides an introduction to the subject, tracing the history of birational geometry of rationally connected varieties from its very first steps in the nineteenth century to the present. The chapter is written informally; no special knowledge of algebraic geometry is required to read it through. We show how the straightforward and natural *rationality problem* developed, following its own logic, into the modern problem of describing the structures of a rationally connected fibre space on a given rationally connected variety. Following this course, we mention and sometimes briefly discuss the milestone results that made the area what it is now.

Starting from Chapter 2, the book is a standard mathematical text. To be able to read it, one needs to be acquainted with the basics of algebraic geometry including the intersection theory. R. Hartshorne's "Algebraic Geometry" plus the first three chapters of W. Fulton's "Intersection Theory" are sufficient to read this book without any difficulties, but for an active reader a sufficient background could be reduced to the first volume of I.R. Shafarevich's "Basic Algebraic Geometry". In other words, only some acquaintance with algebraic geometry is assumed, in order to make this book accessible for graduate students on their way into algebraic geometry.

Chapter 2 contains the basics of the method of maximal singularities, illustrated by numerous examples, many of them elementary. A Fano variety V has no global differential-geometric invariants (that is, no non-zero global regular covariant tensors), its canonical class K_V is negative; therefore, the natural idea is to use "the measure of negativity" of K_V instead of the global sections of the sheaves $\mathcal{O}_V(nK_V)$, $n \in \mathbb{Z}_+$, as in the case of varieties of general type. This leads to the classical concept of *termination of canonical adjunction* and eventually, via a natural construction, to the concept of a *maximal singularity* of a movable linear system. We discuss the possible types of maximal singularities and develop techniques to study them. As the first applications, we are able to prove birational (super)rigidity of certain Fano varieties of dimension three and higher. In particular, we give a complete computation of the group of birational self-maps of the complete intersection $V_{2\cdot 3} \subset \mathbb{P}^5$ of quadric and cubic hypersurfaces in $\mathbb{P}^5$, one of the most non-trivial such groups known today.

Chapter 3 is about the technique of hypertangent divisors and linear systems. For Fano varieties of high degree this technique is now the standard and most efficient tool in the proof of their birational (super)rigidity. As a warm-up example, we start with hypersurfaces $V_d \subset \mathbb{P}^d$, $d \geq 5$, providing a series of superrigid Fano varieties in arbitrary dimension. After that we proceed to more sophisticated and technically difficult classes of higher dimensional Fano varieties: complete intersections and cyclic covers. As an additional application, we describe the structures of non-maximal Kodaira dimension on Fano complete intersections; it turns out that all such structures are just pencils of hyperplane sections. However surprising it may look, for any rational map $V \overset{\beta}{\dashrightarrow} S$ onto a positive-dimensional variety S, where V is a generic Fano complete intersection, if β is not a linear projection onto $\mathbb{P}^1$, then the generic fibre of β is of general type!

Chapter 4 is an informal introduction to the method of maximal singularities for the class of *fibre spaces*. By elementary examples we illustrate the new phenomena that occur in the relative case compared to the absolute one (that is, when we study birational geometry of a fibre space $V \to S$ over a positive dimensional variety S rather than that of a primitive Fano variety V with S a point). In particular, we discuss the fibrewise and non-fibrewise birational self-maps and changing the structure of a rationally connected fibre space. After that, we briefly discuss the *Sarkisov Program* — a general theory of factorizing birational maps into elementary modifications (*links*). Finally, we summarize the main results on birational rigidity of Fano fibre spaces, obtained by the method of maximal singularities, and introduce the important K^2-*condition* and K-*condition* that ensure birational (super)rigidity of Fano fibre spaces over $\mathbb{P}^1$.

Chapters 5–8 contain a detailed exposition of the theory outlined in the introductory Chapter 4. Namely, Chapter 5 deals with Fano fibre spaces $V \to \mathbb{P}^1$ with $\dim V \geq 4$. We prove general sufficient conditions of birational superrigidity in terms of numerical geometry of the fibres. This general theory is similar to the methods developed in Chapter 2, but now the relative case requires additional constructions. After that, we develop further the technique of hypertangent divisors, in order to show that generic Fano fibre spaces over $\mathbb{P}^1$ satisfy the needed numerical properties and the general theory applies, giving a proof of their birational rigidity. We consider several classes of Fano fibre spaces, including the two most general ones: fibrations into Fano complete intersections in $\mathbb{P}^n$ and fibrations into Fano cyclic covers. Proofs for the latter two classes are technically the most difficult; they are given in full detail.

Chapter 6 covers three-dimensional fibrations into del Pezzo surfaces. This is by now a very well studied class of varieties, where birational geometry has been exhaustively described even for some of the hardest families. The chapter provides a complete proof of birational rigidity of del Pezzo fibrations of degrees 1, 2 and 3, satisfying the K^2-condition. Furthermore, we explain how the techniques of the proof should be extended (by adding some geometric constructions and observations) in order to make it efficient when the K^2-condition is not satisfied. An "overwhelming majority" of del Pezzo fibrations (for degrees 1 and 2, all but finitely many families) satisfy the K^2-condition, but the extended technique also works in many of the remaining cases. After reading this chapter, one should be able to go through any of the research papers on birational geometry of del Pezzo fibrations.

Chapter 7 introduces and develops the new ideas and constructions of the "linear method", as opposed to the "quadratic method" of the previous chapters. The quadratic constructions were used in the theory of birational rigidity starting from the pioneer paper of Iskovskikh and Manin: if Σ is a movable linear system, then its *self-intersection* $Z = (D_1 \circ D_2)$, where $D_1, D_2 \in \Sigma$ are generic divisors, is an effective cycle of codimension two, and the key principle of the method of maximal singularities was to study the singularities of the system Σ by looking at the induced singularities of the self-intersection Z. It has turned out recently that in certain cases it is much more profitable to study the singularities of the generic divisor $D \in \Sigma$ without taking the self-intersection, for example, by restricting D onto a suitable subvariety and applying the Connectedness Principle of Shokurov and Kollár. Realizing this approach, we prove in this chapter the theorem on Fano direct products and certain other results that give an alternative approach to proving birational rigidity of Fano fibre spaces over $\mathbb{P}^1$.

Finally, Chapter 8 describes birational geometry of Fano double spaces of index two and dimension at least 6. The study of this class of varieties makes the first step outside the realm of birationally rigid ones. However, it comes naturally in this book, since it requires essentially the same techniques as birationally rigid Fano varieties and fibre spaces. Fano double spaces fit perfectly into the general ideology of the method of maximal singularities and demonstrate that the method potentially applies to a much wider world of algebraic varieties than that of birationally rigid ones. The main result of this chapter is that every non-trivial structure of a rationally connected fibre space on a double space V of index 2 is a map $V \dashrightarrow \mathbb{P}^1$ given by an arbitrary pencil in the half-anticanonical linear system of V. This is precisely what one should expect from a Fano variety of index two and makes an evidence of further potential of the method of maximal singularities.

This completes our brief description of the contents of this book. In more details the contents of each chapter is summarized in the introductory section at the beginning of the chapter. Historical remarks tracing the papers where the particular facts were proven and particular ideas introduced and developed, are collected in special sections at the end of Chapters 2–8; for that reason, we give very few references in the body of the text (e.g. when some claim is stated but not proved in this book, the reference is given immediately after the statement), collecting them in the "Notes and references" section at the end of the chapter.

0.4. References and cross-references. Now we explain the system of enumeration and cross-references. All claims (theorems, propositions, lemmas, corollaries) as well as definitions, remarks and examples are numbered independently in each section and independently of each other. A reference to Theorem (Proposition, Lemma, ...) $a.b$ means a reference to a theorem (proposition,...) in the *current* chapter, which is to be found in Section a (under the number $a.b$) and is the b-th theorem (proposition,...) in that section. A references to Theorem (Proposition, Lemma,...) $a.b.c$ means a reference to a theorem (Proposition,...) in Section b of Chapter a, which has the number $b.c$ in that section; such a reference occurs if Chapter a is *not* the current one. The same principle of numeration and references applies to definitions, remarks and examples.

A reference to Section a means a reference to the section number a in the *current* chapter. If we refer to Section b in Chapter a, which is *not* the current one, we write Section $a.b$. Sections are divided into subsections which are numbered consecutively: Section a consists of Subsections $a.1$, $a.2$,.... For references, we use the same principle as above: a reference to Subsection $a.b$ means the b'th subsection of Section a of the current chapter, whereas a reference to Subsection $a.b.c$ means the Subsection $b.c$ in Chapter a, which is different from the current one. With such a system of enumeration and references any misunderstanding is excluded.

Selection of bibliography also needs some explanation. Providing a comprehensive survey of the theory of birational rigidity, this book is not an encyclopedia of the area. The main achievements are all covered, most of the significant results are either covered or at least mentioned and discussed. It was not possible to include every result on birational rigidity and, the more so, every related result, as the book would have grown at least twice longer in that case and become unreadable.

This applies to the bibliography as well. All the main research and survey papers on birational rigidity are there. All or almost all significant research papers on birational rigidity are there, too. However, including every paper on birational rigidity would have made the list of references unnecessarily long. (It is also possible that I overlooked or forgot some items.) As for the papers related to, but not dealing with, birational rigidity, I had to be very selective there. I would explicitly name three very important areas, which are almost completely missing in the bibliography, although the problems they are studying (but not the methods they are employing, which was the decisive argument for me) are quite close to the problems that are solved or discussed in this book: the investigation of the Cremona group of higher (≥ 3) rank, the (biregular) classification of higher-dimensional Fano varieties and the study of rational curves on algebraic varieties. These areas are getting increasingly popular today and there are some indications that their interaction with the theory of birational rigidity will increase in the future.

0.5. Acknowledgements. The task to read the paper [**IM**] and "do the same thing" for the four-dimensional quintics was given to me by Vassily Alekseevich Iskovskikh, my supervisor in the undergraduate years 3–5 and subsequently my PhD supervisor, in October 1982. I completed a proof of birational superrigidity of four-dimensional quintics in December 1983. This was how the main area of my work in algebraic geometry was determined.

This is an appropriate moment to say thank you to mathematicians and research communities for the support, interaction and attention to my work and my life. I am very grateful: to Yurii Ivanovich Manin, whose support was invaluable for me; to Thomas Peternell; to Askol'd Georgievich Khovanskii. My collaboration with Miles Reid and Alessio Corti in 1995–98 was fruitful and important.

Starting from my very first steps in algebraic geometry, I was a member of the research community, centered around the famous I.R. Shafarevich seminar at the Steklov Institute in Moscow, where dozens of talks on birational rigidity were given, including the dozens of mine. All the results presented in this book were discussed at that seminar, together with various other results on birational rigidity and related topics. Even in the difficult 1990s, research life at the Steklov Institute was very active and intense and a group of young mathematicians working on birational rigidity was formed (Mikhail Grinenko, Ivan Cheltsov, Igor' Sobolev, Konstantin Shramov). I am grateful to the Steklov Institute and the research community of algebraic geometers there; my special thanks go to Armen Glebovich Sergeev, Alexey Nikolaevich Parshin and Dmitrii Orlov.

I am grateful to research centres and research groups where I was fortunate to work for some periods of time: to the Max Planck Institute of Mathematics in Bonn; to the research community of algebraic geometers at the University of Bayreuth; to the Mathematics Institute of the University of Warwick.

Since 2003 I have worked at the University of Liverpool. The creative and friendly atmosphere in the research group of algebraic geometers in Liverpool was very important for my research work. I am happy to thank my colleagues: Viacheslav Nikulin, Peter Newstead, Victor Flynn and Thomas Eckl.

CHAPTER 1

The Rationality Problem

Summary

The aim of this chapter is to give an informal introduction to birational geometry of rationally connected varieties. There are no proofs here and no rigorous arguments. We show, how the rationality problem (which is very natural even from the viewpoint of a naive observer without any acquaintance with algebraic geometry) generalizes to the problem of the description of rationally connected structures on a given rationally connected variety; in fact, the rationality problem generates the latter one.

The first section contains some basic definitions and examples and a brief discussion of the famous theorem of Max Noether on the generators of the Cremona group, together with some other related results on rational surfaces. We explain the rationality problem and some related problems and constructions, using only the language of general mathematics and trying to be as elementary as possible. However, some important ideas of the modern birational geometry are introduced there.

The second section covers the period that spans about a century — more or less, the whole twentieth century. We start with a discussion of Fano's work: Fano guessed, introduced and, to some extent (as much as the technique of his time allowed him), developed some basic concepts of the modern higher-dimensional geometry and formulated certain key results, including what we now call birational superrigidity of three-dimensional quartics. After that, we proceed to the beginning of the modern period. The breakthrough paper of Iskovskikh and Manin [**IM**] is of primary interest to us as it initiated the starting point of the theory about which this book is written. We discuss briefly the negative solution of the Lüroth problem in the early 1970s and then the Sarkisov theorem on conic bundles. All those results clearly implied that in higher-dimensional geometry the concept of rationality, keeping its importance, for classification purposes should be generalized to a wider concept, thus preparing the introduction of rationally connected varieties.

Section 3 starts with a discussion of the famous theorem of Graber, Harris and Starr [**GHS**], introducing the language of rationally connected geometry. We then proceed to precise formulation of the problems which we are going to treat in this book (Subsection 3.2) and, finally, of the basic ideas of the method of maximal singularities (Subsection 3.3). We conclude this section with a brief description of the contents of the next two chapters that prove birational (super)rigidity of many classes of Fano varieties of various dimensions.

1. The classical heritage

1.1. The rationality problem. Examples. The rationality problem is one of the oldest problems in algebraic geometry; in fact, it is one of the problems that became a starting point of algebraic geometry. In the first approximation, algebraic geometry can be understood as a branch of mathematics studying geometry of algebraic sets, that is, the sets of solutions of systems of algebraic equations

$$\left\{\begin{array}{c} f_1(x_1,\dots,x_n)=0, \\ \dots \\ f_m(x_1,\dots,x_n)=0, \end{array}\right. \tag{1}$$

where $f_i(x_*)$ are polynomials. If all polynomials f_i are linear, then (provided that the system is compatible) the set of variables $x_1,\dots,x_n$ can be broken into two disjoint subsets: the leading variables x_i, $i\in I$, and the free variables x_i, $i\in J$, where $I\coprod J=\{1,\dots,n\}$. A general solution of the system (1) now looks as follows: the leading variables are linear functions of the free ones, whereas the latter can take arbitrary values. In other words, the set of solutions of the system (1) can be *parametrized*, taking the free variables as natural parameters.

Taking the ideal picture of the linear world, described above, as a model, one can set up a naive question: Is it possible to parametrize the set of solutions of an arbitrary system (1) of polynomial equations? This is the very *rationality problem.* It seems that one of the key points in the first period of algebraic geometry was the understanding that for a general system (1) the answer is negative. The rationality problem is the problem of solving a system of algebraic equations *explicitly.* Its negative solution means that an explicit solution does not exist.

However, one needs to remember the following two points here. First, even in the comparatively simple cases, when the solutions of the system (1) can be parametrized, it is almost never possible to take a set of coordinate functions x_i, $i\in J$, as parameters, similarly to the linear case. And second, it is almost never possible to parametrize the *whole* set of solutions: the best result one can hope for is to parametrize "almost all" solutions, that is, with the exception of a proper subset of a smaller dimension. The closest approximation of this point in the common language of mathematics is a parametrization of a dense open subset of the set of solutions by rational functions of the coordinates x_*.

One more remark. The rationality problem has an important arithmetic aspect; however, in this book we restrict ourselves to the context of complex algebraic geometry. All objects are defined over the field of complex numbers $\mathbb{C}$.

Definition 1.1. A complex algebraic variety V is said to be *rational*, if one of the following equivalent conditions holds:

(1) there exists a non-empty Zariski open subset $U\subset V$, isomorphic to an open algebraic subset $U^+\subset\mathbb{C}^n$, $n=\dim V$;

(2) there is a non-empty Zariski open subset $U\subset V$, admitting a global coordinate system;

(3) there is an algebraic correspondence $\varphi\colon V\dashrightarrow\mathbb{C}^n$, which is one-to-one on dense open subsets.

There is a purely algebraic version of this definition:

(4) the field of rational functions $\mathbb{C}(V)$ of the variety V is isomorphic to the field of rational functions of n independent variables, $\mathbb{C}(V) \cong \mathbb{C}(t_1, \dots, t_n)$.

Example 1.1. (i) The projective space $\mathbb{P}^n$ with the homogeneous coordinates $(x_0 : x_1 : \cdots : x_n)$ is covered by the standard affine charts $U_i = \{x_i \neq 0\} \cong \mathbb{C}^n$ and therefore is rational.

(ii) The Grassmanians $\mathrm{Gr}(k, n)$ are represented as the unions of the standard charts $U_I = \{\det A_I \neq 0\} \cong \mathbb{C}^{k(n-k)}$ and therefore are rational (we mean the realization of $\mathrm{G}(k, n)$ as the quotient space of $\mathrm{Hom}(\mathbb{C}^k, \mathbb{C}^n)$ (identified with the space of $k \times n$-matrices) by the action of the group $\mathrm{Aut}\,\mathbb{C}^k$).

(iii) Toric varieties contain the algebraic torus $(\mathbb{C}^*)^{\times n}$ as a dense open subset, $\mathbb{C}^* = \mathbb{C} \setminus \{0\}$, and therefore are rational.

(iv) A quadric hypersurface $Q \subset \mathbb{P}^n$ is rational: the stereographic projection from a non-singular point $p \in Q$ parametrizes the quadric Q. Indeed, let $P \cong \mathbb{P}^{n-1}$ be an arbitrary hyperplane, not containing the point p. For any point $x \in Q \setminus \{p\}$ let $L_{xp} \subset \mathbb{P}^n$ be the line joining x and p. Set $\pi(x) = L_{xp} \cap P$. Thus we define a map $\pi\colon Q \setminus \{p\} \to P$. It is clear that any point $q \in P$ has exactly one preimage $\pi^{-1}(q)$, provided that the line L_{qp} is not entirely contained in Q.

Let $T \subset \mathbb{P}^n$ be the (unique) hyperplane, that is tangent to Q at the point p. It is easy to see that if a line $L \ni p$ lies entirely in Q, then $L \subset T$ and, conversely, $Q \cap T$ is a quadric cone with the vertex at the point p. Therefore, the stereographic projection gives an isomorphism of open sets $Q \setminus (Q \cap T)$ and $P \setminus (P \cap T)$.

(v) Cubic surfaces give a less trivial example. Let $Q \subset \mathbb{P}^3$ be a non-singular cubic surface. The following fact was one of the achievements of the classic algebraic geometry (and can be found, it seems, in every textbook on algebraic geometry [**Hart, Hul, Mum, Shaf**]): there are precisely 27 lines on the surface Q. The configuration of these lines is known up to the smallest details, however, in order to prove that the surface Q is rational, we need only the existence of two *disjoint* lines $L_+, L_- \subset Q$. Obviously, $\langle L_+, L_- \rangle = \mathbb{P}^3$ (the symbol $\langle \cdot, \cdot \rangle$ stands for the linear span). Let $x_+ \in L_+$ and $x_- \in L_-$ be general points. The line L that joins the points x_+ and x_- intersects Q at three points, two of which we already know: $L \cap Q = \{x_+, x_-, x\}$. The correspondence $(x_+, x_-) \mapsto x$ gives a well-defined map

$$\gamma\colon (L_+ \times L_-) \setminus Y \to Q,$$

where Y consists of the pairs (x_+, x_-) such that the line L lies entirely in Q. There are finitely many such pairs, so that the set Y is finite and γ is a map $(\mathbb{P}^1 \times \mathbb{P}^1) \setminus Y \to Q$. It is easy to see, furthermore, that for a point of general position $x \in \mathbb{P}^3$ there is a unique pair of points $x_\pm \in L_\pm$ such that $x \in L$. In terms of suitable homogeneous coordinates on $\mathbb{P}^3$ the map $\beta\colon x \mapsto (x_+, x_-)$ corresponds to the isomorphism $\mathbb{C}^4 \cong \mathbb{C}^2 \oplus \mathbb{C}^2$. The map β is not defined at the points $x \in L_+ \cup L_-$. Obviously, β and γ are inverse to each other, which proves rationality of the surface Q.

The examples given above are geometrically elementary, but just one step separates the last example from the modern algebraic geometry: increasing the dimension of the cubic by one, we get the following famous result.

Theorem 1.1 (Clemens, Griffiths; 1971). *A smooth cubic hypersurface $Q \subset \mathbb{P}^4$ is non-rational.*

The proof of this theorem employs complicated techniques of modern algebraic geometry; below we will come back to the field that grew from the remarkable paper of Clemens and Griffiths [**CG**].

Making one more step and increasing dimension again, we get an open problem: *Is a generic smooth cubic hypersurface* $Q \subset \mathbb{P}^n$ *rational for* $n \geq 5$*?*

It is believed that the answer is negative for $n \in 2\mathbb{Z}$ even, and for n odd it is negative for a hypersurface of general position. It is easy to see that for $n = 2k+1$ the construction of example (v) gives rationality of the cubic in the case when Q contains a pair of disjoint planes $P_{\pm}$ of dimension k. It is known, however, that for $n \geq 5$ on a general cubic there are no k-planes.

One more reason why the problem of rationality of an even-dimensional cubic is interesting, is that it is connected with the following question: Is it possible to deform smoothly a *smooth* rational variety into a non-rational one? If a general even-dimensional cubic is non-rational, then by what was said above, the answer to this question is positive. There are very few papers dealing with the problem of rationality of a four-dimensional cubic hypersurface. It is not even clear which approach will be successful. For a recent discussion of this problem in the spirit of the transcendent method, see [**Kul06**].

The rationality problem, the formulation of which by its simplicity is within grasp of an advanced schoolboy or schoolgirl, is important for the following reasons:

- rational varieties in an obvious sense are "the simplest" algebraic varieties, and one should be able to distinguish them from other algebraic varieties;
- the rationality problem is a part of tradition that goes back to the beginning of algebraic geometry;
- the rationality problem is usually very hard, provided that it does not admit a trivial solution for the given class of varieties; for instance, an obvious geometric construction (like the stereographic projection) or an obvious birational invariant like differential forms, which we will discuss below; the alternative "either very easy, or very hard" is very typical for the rationality problem and adjacent questions;
- the rationality problem is important from the arithmetic viewpoint, when solutions of a system of polynomial equations are sought in the field of rational numbers $\mathbb{Q}$ or its finite extensions: if the variety can be parametrized, then the parameters can take arbitrary values from a given field;
- the rationality problem, which is important by itself, generates many other deep problems of more general character and independent meaning; in particular, the rationality problem is important from the viewpoint of classification of algebraic varieties.

1.2. Birational geometry. Most of the maps and correspondences and most of the natural geometric constructions emerging in algebraic geometry are defined at the points of general position only (for instance, the stereographic projection in not defined at the point which is the centre of the projection). This leads to the concept of a *rational map*; that is, a map, which is defined on a dense open subset of an algebraic variety. An arbitrary map $\varphi\colon \mathbb{P}^n \dashrightarrow \mathbb{P}^m$ has the form

$$(x_0 : \cdots : x_n) \mapsto (f_0(x_*) : \cdots : f_m(x_*)),$$

where f_* are homogeneous polynomials of the same degree $d \geq 1$ without a non-trivial common divisor. Obviously, φ is not defined at the points of the closed algebraic set $f_0(x_*) = \cdots = f_m(x_*) = 0$, which is of codimension ≥ 2 by what has been said about the set of polynomials f_*. If this set is non-empty, the map φ is called a rational map (otherwise, a regular map). The broken arrow $\dashrightarrow$ means that the map, generally speaking, is not everywhere defined. For the regular maps, the usual arrow $\to$ is used. The most important information about the map is concentrated exactly on the set where it is not defined.

The central object of study in algebraic geometry are projective varieties and the concept of a rational map generalizes naturally to a rational map between arbitrary projective varieties $\varphi\colon X \dashrightarrow Y$. (In the algebraic terminology, rational dominant maps φ correspond to inclusions of the fields of rational functions $\varphi^*\colon \mathbb{C}(Y) \hookrightarrow \mathbb{C}(X)$.) Invertible rational maps are called *birational maps.* (In the algebraic language, birational maps $\varphi\colon X \dashrightarrow Y$ correspond to *isomorphisms* of the fields of rational functions $\mathbb{C}(Y) \leftrightarrow \mathbb{C}(X)$.) Birational classification of algebraic varieties (that is, classification up to birational equivalence) is the most reasonable method of classification.

The main complication in working with (bi)rational maps comes from the fact that every map has its own domain of definition. The main method of working with (bi)rational maps is to regularize them by a sequence of blow ups. The blow ups are the simplest birational maps described in the textbooks [**Hart, Hul, Mum, Shaf**]. The blow up of the origin $o \in \mathbb{C}^n$ is the map $\sigma\colon X \to \mathbb{C}^n$ where $X \subset \mathbb{C}^n \times \mathbb{P}^{n-1}$ is the closure of the set $((z_1, \dots, z_n), (z_1 : \cdots : z_n))$ for $z \neq (0, \dots, 0)$. The map σ determines an isomorphism of the sets $X \setminus \sigma^{-1}(o)$ and $\mathbb{C}^n \setminus \{o\}$ and $\sigma^{-1}(o) = \mathbb{P}^{n-1}$, that is, instead of the point o we paste into $\mathbb{C}^n$ a smooth hypersurface, isomorphic to $\mathbb{P}^{n-1}$. The blow up of a plane $\{z_1 = \cdots = z_n = 0\} = \{o\} \times \mathbb{C}^m \subset \mathbb{C}^{n+m}$ is the map $X \times \mathbb{C}^m \to \mathbb{C}^n \times \mathbb{C}^m = \mathbb{C}^{n+m}$; this gives an idea how the blow up of arbitrary smooth subvariety $Y \subset V$ is defined.

Now, instead of interpreting a rational map $\varphi\colon V \dashrightarrow W$ as a regular map $\varphi\colon U \to W$ of a dense open subset $U \subset V$, we apply to V a sequence of blow ups

$$V_N \stackrel{\sigma_N}{\to} V_{N-1} \to \cdots \to V_{i+1} \stackrel{\sigma_i}{\to} V_i \to \cdots \to V = V_0$$

so that the composite map $\psi = \varphi \circ \sigma_1 \circ \cdots \circ \sigma_N\colon V_N \dashrightarrow W$ is everywhere defined, that is, this map is a regular map (a morphism) $\psi\colon V_N \to W$. Here the varieties V and V_N are birationally equivalent. The procedure described above is called *resolving singularities* (in the classical terminology, *resolving the points of indeterminancy*) of the map φ. The theory of resolution of singularities is a separate branch of birational geometry. Existence of resolution of singularities in arbitrary dimension was first proven by Hironaka [**Hi**]; in the past fifteen years shorter and simpler proofs appeared, and the problem of existence of a resolution and of its realization was considered in many versions; see [**Kol07**] and the references therein.

In the general context of birational classification of algebraic varieties the rationality problem is formulated as the question of birational equivalence of a given variety V and the projective space $\mathbb{P}^n$, $n = \dim V$.

The rationality problem generates numerous natural questions. Let us give some examples.

- For a given variety V, does there exist a rational function $f\colon V \dashrightarrow \mathbb{C}$, a general fibre of which $f^{-1}(a)$, $a \in \mathbb{C}$, is a rational variety? And if such a function exists, does that imply rationality of the variety V itself?
- Generalizing the previous question, does there exist, for a given variety V, a rational dominant (that is, covering a dense open set on the image) map $\varphi\colon V \dashrightarrow S$, the fibre of general position of which $\varphi^{-1}(s)$, $s \in S$, is rational? And again, if the answer is yes, and such a map φ does exist, whereas the variety S is rational, does it imply rationality of V?
- The rationality problem is closely connected with the problem of description of the group of birational self-maps $\operatorname{Bir} V$ of the variety V.
- The rationality problem generalizes to the unirationality problem: For a given variety V, does there exist a rational dominant map $\mathbb{P}^n \dashrightarrow V$ (if it does, then the variety V is said to be *unirational*)? In other words, a multi-valued parametrization of V is meant. (In the algebraic language, the Lüroth problem is formulated in the following way: Is it true that any subfield $F \subset \mathbb{C}(t_1, \dots, t_n)$ of the field of rational functions, $\operatorname{tr.deg} F/\mathbb{C} = n$, is itself a field of rational functions of n independent variables, $F \cong \mathbb{C}(t_1, \dots, t_n)$?) The question, whether unirationality implies rationality, is called *the Lüroth problem*; we will discuss it in more detail below.

These and many other problems are trivial for the "majority" of algebraic varieties, like the rationality problem itself. For a smooth projective variety V let us consider its cotangent bundle T^*V. Let E be a vector bundle on V, obtained from T^*V by applying the following operations: taking the symmetric and skew-symmetric power, tensor product and tensor power (for instance, $(T^*V)^{\otimes n} \otimes (\Lambda^k T^*V)^l$). Essentially, we exclude the dualizing operation only: E should be a strictly covariant tensor object. Let $s \in H^0(V, E)$ be a non-zero global section of the bundle E. Then for any rational dominant map $\beta\colon W \dashrightarrow V$ the pullback of the section s, $\beta^* s$, extends to a non-zero global section of the vector bundle F, obtained from the cotangent bundle T^*W by applying the identical set of tensor operations. For instance, the pullback $\beta^*\omega$ of a regular differential form of degree i, $\omega \in H^0(V, \Omega^i_V)$ (where Ω^i_V is the sheaf of sections of the vector bundle $\Lambda^i T^*V$), is a regular form of degree i on W. This comes from the above mentioned fact: the rational map β is regular on an open set $W \setminus Z$, where $Z \subset W$ is of codimension ≥ 2, so that the regular form $\beta^*\omega$ extends from $W \setminus Z$ to all of W.

Therefore, if on V there is at least one regular covariant tensor, then the variety V is non-rational, since, as it is easy to check, on the projective space there is no such tensors (for instance, all differential forms on $\mathbb{P}^n$ have hypersurfaces of poles).

Example 1.2. It is well known that a smooth cubic curve $C \subset \mathbb{P}^2$ is isomorphic as a complex manifold to a complex torus $\mathbb{C}/\Lambda$, where $\Lambda \cong \mathbb{Z} \oplus \mathbb{Z}$ is a lattice. If z is a complex coordinate on $\mathbb{C}$, then the space $H^0(C, \Omega_C)$ of global regular (that is, holomorphic) forms on C is one-dimensional and is of the form $\{\lambda dz \mid \lambda \in \mathbb{C}\}$. Therefore, the curve C is non-rational.

Example 1.3. According to the classic adjunction formula the space of regular differential forms of maximal degree on a smooth hypersurface $V = V_d \subset \mathbb{P}^N$ for $d \geq N + 1$ is isomorphic to the space of polynomials of degree $(d - N - 1)$ in the homogeneous variables $x_0, \dots, x_N$, whence it follows immediately that V is non-rational. In fact, one can say much more: for $d \geq N + 2$ there are "plenty" of forms of maximal degree on V, the sheaf of sections of the bundle $\Lambda^{N-1}T^*V$ is very ample (the zeros of the sections separate points and tangent directions), so that birational geometry of these varieties is completely reduced to their projective geometry. For instance, two smooth hypersurfaces V_1 and V_2 of degrees $d_1, d_2 \geq N+2$ are birationally equivalent if and only if $d_1 = d_2$ and V_1 is projective equivalent to V_2, that is, can be transformed into V_2 by a projective automorphism, induced by a linear automorphism of the space $\mathbb{C}^{N+1}$.

For this reason, studying of birational geometry of algebraic varieties is most complicated exactly when there are no non-zero global differential forms (and, more generally, global covariant tensors), or, at least, when there are not too many of them. It is for those varieties that the rationality problem is non-trivial (and, as a rule, very hard). For instance, to this class belong hypersurfaces of small degree $V_d \subset \mathbb{P}^N$, $d \geq N$, and more generally, complete intersections $V_{d_1 \dots d_k} = F_{d_1} \cap \dots \cap F_{d_k} \subset \mathbb{P}^N$ of hypersurfaces $F_{d_1}, \dots, F_{d_k}$ of degrees $d_1, \dots, d_k$, satisfying the inequality $d_1 + \dots + d_k \leq N$. An extremal case is given by the projective space $\mathbb{P}^N$ itself; for this variety there is no rationality problem, however, it does not make an exception from the general rule: its birational geometry is most rich and complicated.

1.3. Two classical theorems. In the dimensions 1 and 2 differential geometric invariants completely solve the rationality problem. This was already realized by the classical geometers. A curve C is rational if and only if its genus $g_C = \dim H^0(C, \Omega^1_C)$ is equal to zero. In particular, the concepts of unirationality and rationality for curves coincide. For surfaces the situation is similar (however, in dimension 2 it is already essential that the field of constants is algebraically closed: for non-closed fields geometry is much more complicated, see [**M66, M67**]). Namely, the following theorem holds.

Theorem 1.2 (The Castelnuovo rationality criterion). *Let S be a smooth projective surface, satisfying the conditions*

$$H^0(S, \Omega^1_S) = H^0(S, \omega_S^{\otimes 4}) = H^0(S, \omega_S^{\otimes 6}) = \{0\},$$

where $\omega_S = \Lambda^2 \Omega_S$ is the sheaf of forms of maximal degree. Then S is rational.

Since for a unirational variety all differential geometric invariants are trivial, the Castelnuovo rationality criterion gives a complete solution to the Lüroth problem in dimension two: unirational surfaces are rational. Therefore, non-trivial birational geometry of surfaces is concentrated on the single object: the projective plane $\mathbb{P}^2$. Rational surfaces have very rich birational geometry. Here we consider only a few questions that are important in the sequel.

One of the most remarkable results in two-dimensional birational geometry is the following theorem, due to Max Noether [**N**].

Theorem 1.3. *The group of birational self-maps of the projective plane, that is, the Cremona group* $\mathrm{Bir}\,\mathbb{P}^2 = \{\chi\colon \mathbb{P}^2 \dashrightarrow \mathbb{P}^2\}$ *is generated by the subgroup of projective automorphisms* $\mathrm{Aut}\,\mathbb{P}^2 = PGL(2)$ *and one quadratic transformation:*

$$\tau\colon (x_0 : x_1 : x_2) \mapsto (x_1 x_2 : x_0 x_2 : x_0 x_1).$$

Birational self-maps of the projective space are often called Cremona transformations, and the group $\mathrm{Bir}\,\mathbb{P}^n$ is called "the Cremona group of rank n", so that the fact that was formulated above is known as "the Max Noether theorem on the Cremona group".

Let us describe the general idea of the proof of Theorem 1.3.

Let $\chi\colon \mathbb{P}^2 \dashrightarrow \mathbb{P}^2$ be an arbitrary birational self-map. Let us modify it, taking compositions $\chi \circ \beta_1 \circ \beta_2 \circ \cdots \circ \beta_k = \chi_k$, where β_i is either a linear automorphism, that is, an element of the group $\operatorname{Aut}\mathbb{P}^2 = GL(3,\mathbb{C})/\mathbb{C}^* = PGL(2,\mathbb{C})$, or the involution τ. One should construct such a procedure of modification that at some step $\chi_N \in \operatorname{Aut}\mathbb{P}^2$.

Assume that $\chi \notin \operatorname{Aut}\mathbb{P}^2$ (otherwise there is nothing to modify). By definition, there exists a finite set of points $Y \subset \mathbb{P}^2$, such that $\chi\,|_{\mathbb{P}^2\setminus Y}$ is a regular map to $\mathbb{P}^2$. Let us define the *strict transform* $(\chi^{-1})_*C$ of the curve $C \subset \mathbb{P}^2$ with respect to χ as the closure in $\mathbb{P}^2$ of the set

$$(\chi\,|_{\mathbb{P}^2\setminus Y})^{-1}(C) \subset \mathbb{P}^2 \setminus Y.$$

The procedure of taking closure means adding points from the finite set Y. Now let us consider the *linear system* of lines in $\mathbb{P}^2$, that is, the set of all lines $a_0x_0 + a_1x_1 + a_2x_2 = 0$ with $(a_0, a_1, a_2) \neq (0,0,0)$, and construct its strict transform with respect to χ. We get a linear system of curves $|\chi|$ of some degree $n = n(\chi) \geq 1$, that is, the set of curves given by the equation

$$a_0f_0(x_*) + a_1f_1(x_*) + a_2f_2(x_*) = 0,$$

where f_i is a polynomial of degree n, where for a general set of coefficients a_* the polynomial $\Sigma a_i f_i$ is irreducible. In fact, the map χ is given with respect to the system of homogeneous coordinates x_* by exactly these polynomials:

$$\chi\colon (x_0 : x_1 : x_2) \mapsto (f_0(x_*) : f_1(x_*) : f_2(x_*)).$$

It is easy to see that the polynomials f_i can be arbitrary: by the Bezout theorem, two curves $C_1, C_2 \in |\chi|$ intersect each other at n^2 points, whereas the corresponding lines L_1, L_2 meet transversally at one point. One can assert that the curves C_1, C_2 intersect transversally at one point outside the exceptional set Y. Therefore for $n \geq 2$ (and if $n = 1$, then $\chi \in \operatorname{Aut}\mathbb{P}^2$ is a projective automoprhism) the set Y is non-empty and at every point $y \in Y$ a general polynomial $\Sigma a_i f_i$ has a positive multiplicity $m(y) \geq 1$. Blowing up the finite set Y, we obtain a rational surface S equipped with a regular birational map $\sigma\colon S \to \mathbb{P}^2$, where $\sigma\colon S\ \sigma^{-1}(Y) \to \mathbb{P}^2\setminus Y$ is one-to-one, and into each point $y \in Y$ a smooth rational curve is contracted.

Consider the strict transform $(\sigma^{-1})_*|\chi|$ of the linear system $|\chi|$ on S: it is a linear system of curves on S, a general curve of which is obtained from a general curve $C \in |\chi|$ in two steps, first we take the preimage $\sigma^{-1}(C\setminus Y)$, then we take its closure in S. It can happen that the composite map $\chi \circ \sigma\colon S \dashrightarrow \mathbb{P}^2$ has points where it is not defined, that is, at the points of a finite set $Y_1 \subset S_1 = S$ the map $\chi \circ \sigma$ is non-regular. In that case all curves of the system $(\sigma^{-1})_*|\chi|$ pass through each point of the set Y_1, which is called the *base set* of the given linear system (notation: $\operatorname{Bs}(\sigma^{-1})_*|\chi|$; for instance, $Y = \operatorname{Bs}|\chi|$). If $Y \neq \emptyset$, then the procedure described above should be repeated. As a result we get a sequence of blow ups

$$S_N \stackrel{\sigma_N}{\to} S_{N-1} \to \cdots \to S_i \stackrel{\sigma_i}{\to} S_{i-1} \to \cdots \stackrel{\sigma_1=\sigma}{\to} S_0 = \mathbb{P}^2,$$

resolving singularities of the linear system $|\chi|$, that is, its strict transform has no base points, and the composite map $\chi \circ \sigma_1 \circ \cdots \circ \sigma_N\colon S_N \dashrightarrow \mathbb{P}^2$ is regular. Now two curves C_1, C_2 from the linear system $(\sigma_1 \circ \cdots \circ \sigma_N)_*^{-1}|\chi|$ meet transversally at one point without any reservations.

The map $\sigma_i\colon S_i \to S_{i-1}$ blows up a finite set of points $Y_{i-1} \subset S_{i-1}$, the elements of which for $i \geq 2$ are called *infinitely near points of order* $(i-1)$. To each point $y \in Y_i \subset S_i$ corresponds a positive ineteger $m(y) \geq 1$, its multiplicity with respect to a general curve of the system $(\sigma \circ \cdots \circ \sigma_i)_*^{-1}|\chi|$. Now the condition that two distinct lines intersect transversally at one point gives the equality

$$n^2 = 1 + \sum_{j=1}^{K} m_j^2, \tag{2}$$

where $\{y_j \mid j = 1, \dots K\} \subset \coprod_{i=0}^{N-1} Y_i$ is the set of all base points, including infinitely near, and $m_j = m(y_j)$.

The equality (2) is the first restriction on the linear system $|\chi|$, that is, on the polynomials f_i. We get the second restriction, recalling that the curves $C \in |\chi|$ are all rational: $\chi\colon C \dashrightarrow \mathbb{P}^1$ is a birational map. The strict transform $\widetilde{C}$ of a curve C on S_N is a smooth rational curve. The classical formula for the genus of a curve yields the equality

$$\frac{(n-1)(n-2)}{2} = \sum_{j=1}^{K} \frac{m_j(m_j-1)}{2}. \tag{3}$$

Proof of the following fact is left to the reader as an elementary exercise.

Proposition 1.1 (The Clebsch-Noether inequality). *If $n \geq 2$, then $K \geq 3$ and the following inequality holds:*

$$m_1 + m_2 + m_3 > n \tag{4}$$

(where we assume that the multiplicities m_j are ordered: $m_1 \geq m_2 \geq \cdots \geq m_j \geq m_{j+1} \geq \dots$).

Assume that our case is that of general position: the three maximal multiplicities m_1, m_2, m_3 correspond to three distinct points of the set Y (unfortunately, it is not always so, it could happen that already y_2 is an infinitely near point over y_1). It is easy to understand that the points y_1, y_2, y_3 are not collinear; otherwise by the Clebsch-Noether inequality for a general curve $C \in |\chi|$ and the line $L \ni y_i$ we would obtain

$$n = (C \cdot L) \geq \sum_{j=1}^{3} (C \cdot L)_{y_j} \geq m_1 + m_2 + m_3 > n,$$

which is impossible (take into account that a general curve C is irreducible and so cannot contain the line L as a component). Let $\alpha \in \operatorname{Aut} \mathbb{P}^2$ be a projective transformation mapping the points y_1, y_2, y_3 to (1:0:0),(0:1:0) and (0:0:1), respectively. Consider the composition

$$\chi_1 = \chi \circ (\alpha^{-1} \circ \tau \circ \alpha)\colon \mathbb{P}^2 \dashrightarrow \mathbb{P}^2.$$

Proposition 1.2. *The linear system $|\chi|$ consists of curves of degree $n(\chi_1) = 2n(\chi) - m_1 - m_2 - m_3 < n(\chi)$.*

Proof. By construction of the map χ_1, a general curve $C^+ \in |\chi_1|$ is the strict transform of a general curve $C \in |\chi|$ with respect to the quadratic map $\alpha^{-1} \circ \tau \circ \alpha$:

$$\begin{array}{ccccc} \mathbb{P}^2 & \overset{\alpha^{-1}\tau\alpha}{\dashrightarrow} & \mathbb{P}^2 & \overset{\chi}{\dashrightarrow} & \mathbb{P}^2 \\ \cup & & \cup & & \cup \\ C^+ & \leftarrow\!-\!- & C & \leftarrow\!-\!- & \text{line} \end{array}$$

The degree of the curve C^+ is equal to the intersection number $(C^+ \cdot L)$, where $L \subset \mathbb{P}^2$ is a general line. Obviously, the intersection number $(C^+ \cdot L)$ is equal to the *free intersection* number of the curve $\alpha^{-1}\tau\alpha(C^+) = C$ and the conic $Q = \alpha^{-1}\tau\alpha(L)$, that is, their intersection number outside the points where the map $\alpha^{-1}\tau\alpha$ is not defined (note that the latter map is an involution). Now

$$(C^+ \cdot L) + \sum_{j=1}^{3} (C \cdot Q)_{y_j} = (C \cdot Q) = 2n,$$

which implies our claim.

The procedure described above, that is, the procedure of replacing χ by χ_1 is called *untwisting* the triple (y_1, y_2, y_3). As we have mentioned above, untwisting is well defined, if among the points y_* there are no infinitely near points. In the original Noether's argument [**N**] the possibility of infinitely near points was not taken into consideration. It took a few decades to learn how to untwist sets of points that include infinitely near points. For the modern proofs, see [**AlgS**] or [**I80, R91**]. However, all proofs are based on Noether's idea: taking the composition with a suitable "elementary" birational map and simplify the given map χ. The degree of simplification is measured by integral invariants (in our case, by the degree $n(\chi) \geq 1$) and for this reason cannot be infinite. At some step we get

$$\chi \circ \alpha_1 \circ \tau \circ \alpha_2 \circ \tau \circ \cdots \circ \tau \circ \alpha_N \in \operatorname{Aut} \mathbb{P}^2,$$

which implies the Noether theorem.

There is an immense literature on the Noether theorem and Cremona transformations. Birational self-maps of the plane (Cremona transformations) are both very natural and very non-trivial objects, generating many problems (to describe the involutions in the Cremona group up to conjugation, to describe finite subgroups [**DI09a, DI09b**], to describe the subgroup fixing a certain geometric object, e.g., a curve, etc.). Birational transformations in dimension three are much more complicated and various. Small wonder that on the border of the nineteenth and the twentieth centuries hundreds of papers were devoted to them; see the bibliography in [**Hud**]. Today studying various particular families of birational self-maps of the projective space is a quite popular subject, as well. Among the most spectacular results one can name is the proof that the Cremona group (of the plane) is not simple (the problem remained open for decades); see [**CaL10**] (however, as a topological group it is simple, see [**Bl10**]). On the whole it seems that the interest to various aspects of Cremona transformations has been steadily growing in the recent past, let us mention a selection of recent papers on the topic: [**BaBeau00, Beau08, BeBl04, Bl07, Bl11, dF04, Dol09, Dol10, FL10**]; 20 to 30 years ago only isolated papers like [**Dol83**] or the papers by Iskovskikh and his students (see, for instance, [**IT94, I96b**]) were published on the subject. The number of publications on the Cremona group of higher rank, especially rank 3, is also growing, due to the rapid progress of the minimal model program; also, birational transformations in many papers are now treated from the viewpoint of dynamical systems theory. However, this book is about non-rational varieties. Coming back to the Noether theorem, let us list the main components of his arguments:

- the invariant $n = n(\chi) \geq 1$, measuring the complexity of a birational map (the degree of curves in the linear system $|\chi|$),
- the "maximal" triples of base points of the linear system $|\chi|$ (satisfying the Clebsch-Noether inequality),
- the untwisting procedure, decreasing the invariant n and simplifying the map χ.

Let us make the following important observation: Noether's argument remains valid, if for the linear system $|\chi|$ we take an arbitrary pencil of rational curves, that is, a linear system of the form $a_0 f_0(x_*) + a_1 f_1(x_*) = 0$ (provided that the general curve is rational). Repeating the proof of the Noether theorem almost word for word, we obtain the following important claim.

Theorem 1.4. *Any pencil of rational curves on $\mathbb{P}^2$ can be transformed into a pencil of lines $a_0x_0 + a_1x_1 = 0$ by a suitable Cremona transformation.*

Furthermore, it is clear from the arguments above that relations between the generators of the Cremona group appear when the untwisting procedure is not unique, and in that case only. This is indeed possible. In this way relations in the two-dimensional Cremona group were described by Gizatullin [**Giz**]. Later his proof was radically simplified by Iskovskikh [**I91a**].

Finally, let us mention one more classical fact from the theory of algebraic surfaces.

Theorem 1.5. *If on a smooth projective surface S there is a pencil of rational curves, then S is rational.*

Proof can be found in the standard textbooks; see for instance [**GrH**].

Let us summarize the picture of two-dimensional birational geometry (created mainly by M. Noether and Italian geometers):

- unirationality implies rationality;
- a surface S admits a rational function $f\colon S \dashrightarrow \mathbb{C}$, the general fibre of which $f^{-1}(a)$, $a \in \mathbb{C}$, is a rational curve, if and only if there is a complementary rational function $g\colon S \dashrightarrow \mathbb{C}$ such that the pair $(f, g)\colon S \dashrightarrow \mathbb{C}^2$ defines a birational map;
- the rationality problem for a surface S solves completely in terms of its differetial geometric invariants;
- the Cremona group $\operatorname{Bir} \mathbb{P}^2$ is generated by automorphisms $\operatorname{Aut} \mathbb{P}^2$ and one quadratic transformation.

Higher-dimensional birational geometry started with attempts to extend these results to dimension three.

2. Towards the modern birational geometry

2.1. Fano's work. Gino Fano occupies an intermediate position in the history of algebraic geometry. He was the last outstanding representative of the old Italian school. The main achievement of that school is the theory of algebraic surfaces, however, the main subject of Fano's was algebraic three-folds. Whereas one can work with surfaces on a more or less elementary level, appealing when it is necessary to geometric intuition, in the higher dimensions one needs a systematically developed foundation, claims that look correct from the intuitive viewpoint, may well turn out to be false. Sheaves and cohomology, intersection theory, the general theorem of Riemann-Roch-Hirzebruch-Grothendieck, resolution of singularities — all came to algebraic geometry after Fano. However, in spite of the fact that his papers contain many mistakes and that essentially he failed to prove all of his big theorems, Fano did suggest a series of fruitful ideas and guessed a series of fundamental facts of three-dimensional algebraic geometry [**Fano1, Fano2, Fano3**].

Fano started with an attempt to extend Noether's arguments to dimension three. His first object of study was the three-dimensional quartic $V = V_4 \subset \mathbb{P}^4$. On a smooth three-dimensional quartic there are no non-zero global differential forms and, more generally, no covariant tensors, so that it is impossible to solve the rationality problem for the quartic in terms of differential geometric invariants. Fano tried to describe the group of birational self-maps of the quartic $\operatorname{Bir} V$. Arguing in accordance with Noether's scheme, for a birational automorphism $\chi\colon V \dashrightarrow V$ he considered the strict transform of the linear system of hyperplane sections of V with respect to χ:

$$\begin{array}{ccc} V & \stackrel{\chi}{\dashrightarrow} & V \\ & & \\ \left\{\begin{array}{c}\text{the linear}\\ \text{system of}\\ \text{divisors } |\chi| \text{ on } V\end{array}\right\} & \stackrel{(\chi^{-1})_*}{\longleftarrow} & \left\{\begin{array}{c}\text{the linear system}\\ \text{of hyperplane}\\ \text{sections}\end{array}\right\} \end{array}$$

One obtains a linear system that is cut out on V by hypersurfaces of degree $n = n(\chi) \geq 1$ in $\mathbb{P}^4$. Similar to the two-dimensional case, we get the following alternative: either $n = 1$, and in this case $\chi \in \operatorname{Aut} V$ is a regular automorphism (hyperplane sections are carried to hyperplane sections, so that χ is the restriction $\chi = \hat{\chi}\,|_V$ of some projective automorphism $\hat{\chi} \in \operatorname{Aut}\mathbb{P}^4$, preserving V), or $n \geq 2$, and in this case the map χ is non-regular, in particular, the base set $\operatorname{Bs}|\chi|$ is non-empty.

Fano claimed that the case $n \geq 2$ is impossible. Similar to the two-dimensional case, for $n \geq 2$ in the base set $\operatorname{Bs}|\chi|$ one should find a subset of high multiplicity (an analog of the triple of points y_1, y_2, y_3). Arguing not very formally, Fano came to a conclusion that for $n \geq 2$ one of the following cases holds:

- there is a point $x \in \operatorname{Bs}|\chi|$ where the map is not defined, such that $\operatorname{mult}_x |\chi| > 2n$,
- there is an entire curve $B \subset \operatorname{Bs}|\chi|$ where the map is not defined, such that $\operatorname{mult}_B |\chi| > n$,
- there is a subset of high multiplicity, consisting of infinitely near points.

Let us explain what is meant here: Fano himself did not (presumably, could not, due to the state of foundations of algebraic geometry in his time) give a precise definition, replacing it with examples, one of which we reproduce here. This example is similar to the case when, in the course of the proof of the Noether theorem, there were infinitely near points among y_1, y_2, y_3. In Fano's example there is a point $x \in \operatorname{Bs}|\chi|$ and a whole line B of infinitely near points over x, that is, a line $B \subset E \cong \mathbb{P}^2$ in the exceptional divisor $E \subset \widetilde{V}$ of the blow up $\varphi\colon \widetilde{V} \to V$ of the point x, such that the inequality

$$\operatorname{mult}_x |\chi| + \operatorname{mult}_B |\tilde{\chi}| > 3n$$

holds, where $|\tilde{\chi}|$ is the strict transform of the linear system $|\chi|$ on $\widetilde{V}$.

Let us stress that the latter example just gives some idea of the infinitely near case, not more than that. To embrace all possibilities, one needs a special formalism, which Fano did not have, and which was developed in the pioneer paper of Iskovskikh and Manin [**IM**], discussed below.

Let us come back to Fano's argument. Considering linear systems on the quartic, he came to a conclusion that none of the three possibilities listed above can occur. (Note that some of his arguments were mistaken, but the final conclusion was correct.) This implies that the case $n \geq 2$ is impossible, so that the groups of biregular and birational self-maps of the quartic coincide: $\operatorname{Bir} V = \operatorname{Aut} V$. The latter group is finite, generically trivial, which immediately implies non-rationality of the quartic $V_4 \subset \mathbb{P}^4$.

Fano's conclusion showed that three-dimensional birational geometry is essentially more complicated than geometry of surfaces. He studied three-folds with similar properties to the quartic. Studying a smooth complete intersection of a quadric and a cubic $V = V_{2\cdot 3} \subset \mathbb{P}^5$, Fano discovered, that unlike the situation on the quartic, some linear systems $|\chi|$ on V, cut out by hypersurfaces of degree $n \geq 2$ in $\mathbb{P}^5$, can have a base curve B of multiplicity $\operatorname{mult}_B |\chi| > n$. This can happen if $B = L$ is a line in $\mathbb{P}^5$ (Fano overlooked another possibility, when B is a conic of a special type, see Section 2.5 below). If this case takes place, then similar to the proof of the Noether theorem there exists a procedure of untwisting the line L. Let us describe it briefly for completeness of the picture.

It is easy to see that the projection from the line L defines a rational dominant map $\pi \colon V \dashrightarrow \mathbb{P}^3$. Its degree at the generic point is equal to 2: the fibre $\pi^{-1}(x)$, $x \in \mathbb{P}^3$, is the intersection $V \cap P$ with the plane $P \supset L$, $P \cong \mathbb{P}^2$, corresponding to the point x, *minus the line L itself.* But $V = F_2 \cap F_3$ is a complete intersection of a quadric and a cubic, so that $F_2 \cap P = L \cup L_P$ and $F_3 \cap P = L \cup Q_P$, where L_P and Q_P are a line and a conic, respectively. Thus $\pi^{-1}(x) = L_P \cap Q_P$, as we claimed. Let $\tau_L \in \operatorname{Bir} V$ be the Galois involution, permuting the points in the fibres of the projection π. It is easy to compute that the linear system $|\tau_L|$ is cut out on V by quartics and $\operatorname{mult}_{\mathrm{L}} |\tau_L| = 5$. Taking the composition $\chi \circ \tau_L \colon V \dashrightarrow V$, we obtain the linear system $|\chi \circ \tau_L|$, which is cut out on V by hypersurfaces of degree $4n - 3 \operatorname{mult}_L |\chi| < n$, which defines an inductive procedure of untwisting and eventually makes it possible to describe the group of birational self-maps $\operatorname{Bir} V_{2\cdot 3}$. Untwisting for the complete intersection $V_{2\cdot 3}$ is considered in full detail below, in Section 2.5.

Therefore, Fano found analogs for all main ingredients of the Noether method:

- the invariant $n \geq 1$ measuring the comlexity of a birational map,
- subsets of high multiplicity in the base set $\operatorname{Bs} |\chi|$, including infinitely near points on the level of explicit examples,
- the procedure of untwisting or excluding the subsets of high multiplicity, simplifying the map χ by decreasing the invariant n.

It seems that Fano's main target in birational geometry was the three-dimensional cubic $V = V_3 \subset \mathbb{P}^4$. Description of the group of birational self-maps $\operatorname{Bir} V$ is so far out of reach even for the modern "high-technology" methods, but Fano tried to get it on the level of his explicit geometric constructions and intuitively plausible arguments, which was, of course, unrealistic. His special interest to the cubic came from the point that the cubic is obviously unirational (see Subsection 2.2 below), so that a proof of its non-rationality gives at once a counterexample to the Lüroth problem, which remained one of the main problems in algebraic geometry up to 1971, when a (negative) solution was found.

Note that Fano's geometric intuition always led him in the correct direction, whether his arguments were correct or not.

Studying conic bundles (see Subsection 2.3 below on the Sarkisov theorem), Fano came to the conclusion that a general variety of that type is *non-unirational.* The problem of unirationality of a rationally connected variety is out of reach up to this day: apart from the cases, when unirationality can be shown explicitly by a direct construction, the unirationality problem is open. However, the general opinion is on Fano's side. Recently, new evidence in favour of non-unirationality of some higher-dimensional varieties was discovered [**BeS**]. This example, together with the constructions described above, was much ahead of its time: many of Fano's predictions remain non-proved and non-rejected up to this day.

Outside the framework of this book we leave the "biregular" component of Fano's work, namely, his classification of three-dimensional algebraic varieties with the negative (that is, anti-ample) canonical class (quartics, cubics, complete intersections $V_{2\cdot 3} \subset \mathbb{P}^5$, all belong to that type of variety), which are nowadays called Fano varieties. Similar to the birational direction, Fano's results were incomplete, and their proofs were not flawless, however, by now Fano's program has been completely realized [**IPr**] in the papers of Iskovskikh, Shokurov, Mori, Mukai, Kawamata, Miyaoka. This program started a whole new field in the modern algebraic geometry, the minimal model program (in dimension three it is Mori theory); see, for instance, [**BCHM, ISh, Kol93, Sh93**] and the bibliography in these papers.

2.2. Solution of the Lüroth problem. The modern period in birational geometry started with the papers of Clemens and Griffiths [**CG**] (on non-rationality of a smooth three-dimensional cubic $V_3 \subset \mathbb{P}^4$) and Iskovskikh and Manin [**IM**] (on non-rationality of a smooth three-dimensional quartic $V_4 \subset \mathbb{P}^4$). Since the cubic is obviously unirational and there are known examples of smooth unirational quartics (due to Segre), in the papers [**CG, IM**] the Lüroth problem got a negative solution in dimension three. Soon after that the paper of Artin and Mumford [**ArMum**] appeared, containing examples of non-rational unirational varieties of arbitrary dimension.

Recall the construction of unirationality of a smooth three-dimensional cubic $V = V_3 \subset \mathbb{P}^4$. Let $x \in V$ be a point of general position, $T(x) \subset \mathbb{P}^4$ the tangent hyperplane to V at the point x. A general line $L \subset T(x)$, $L \ni x$ has intersection number with V precisely two at the point x, $(L\cdot V)_x = 2$, so that $(V \circ L) = 2x + p(L)$, where $p(L) \in L$ is the third point of intersection of V and L. Here the symbol $\circ$ means the *scheme-theoretic intersection* of the line L and hypersurface V in $\mathbb{P}^4$, that is, the intersection taking into account the multiplicity of each irreducible component. Therefore, we obtain a rational map

$$p_x \colon \mathbb{T}(x) \dashrightarrow V \cap T(x),$$

where $\mathbb{T}(x) = \mathbb{P}(T_xV) \cong \mathbb{P}^2$ is the set of lines $L \subset T(x)$, containing the point x. In fact, p is birational and shows rationality of the cubic surface $V \cap T(x)$ with the double point x. Now let us consider an arbitrary rational curve $C \subset V$ (for instance, a line). Since the tangent bundle is locally trivial, we get a birational equivalence

$$\bigcup_{x\in C} \mathbb{T}(x) \dashrightarrow C \times \mathbb{P}^2,$$

so that the birational maps p_x constructed above generate a rational dominant map $p\colon C \times \mathbb{P}^2 \dashrightarrow V$, which is what we need since $c \times \mathbb{P}^2$ is birational to $\mathbb{P}^3$.

Recall the construction of a unirational three-dimensional quartic (due to Segre). Let us try to follow the example of the cubic hypersurface described above. Let $V = V_4 \subset \mathbb{P}^4$ be a smooth quartic, $x \in V$ a point of general position. In order for the intersection $L \cap V$, $L \ni x$, to consist of two points, we need L to have a *double* tangency with V at the point x, that is, $(L\cdot V)_x = 3$. In that case $(L \circ V) = 3x + p\,(L)$ where $p\,(L) \in V$ is the residual point and the rational map p is well defined.

Let us consider a system of affine coordinates $z_1, \dots, z_4$ on $\mathbb{P}^4$ with the origin at the point x. The quartic V is given by a non-homogeneous polynomial $f(z_*) = q_1 + q_2 + q_3 + q_4$, where $q_i(z_*)$ is a homogeneous polynomial of degree i. The line $L \ni x$, $L = \{t(a_1, \dots, a_4) \,|\, t \in \mathbb{C}\}$ has at least a double tangency with V at the point x if and only if $q_1(a_*) = q_2(a_*) = 0$. Interpreting (a_*) as homogeneous coordinates on the projectivized tangent space $\mathbb{P}(T_x\mathbb{P}^4) \cong \mathbb{P}^3$, so that the plane $\{q_1 = 0\} \subset \mathbb{P}^3$ is $\mathbb{P}(T_xV)$, we see that the lines we are looking for are parametrized by the conic $Q(x) = \{q_1 = q_2 = 0\} \subset \mathbb{P}(T_xV)$. We obtain a well-defined map $p_x\colon Q(x) \to V$.

It is not too difficult to construct explicit examples of quartics containing a rational surface $S \subset V$ (which plays the same part as the rational curve C for the cubic hypersurface above). A classical construction: let $S = \{G(x_1, \dots, x_4) = 0\} \subset \mathbb{P}^3$ be a rational quartic with a unique triple point $(1 : 0 : 0 : 0)$ (the projection $S \dashrightarrow \mathbb{P}^2$ is a birational equivalence), then a general polynomial $G + x_0H$, where $H(x_0, \dots, x_4)$ is a homogeneous polynomial of degree three, defines a quartic $V \supset S$. We obtain a rational dominant map

$$p : Q_S = (\coprod_{x\in S} Q(x)) \dashrightarrow V.$$

The variety Q_S, equipped with the natural projection $\pi\colon Q_S \to S$ is a *conic bundle* (see below). Unfortunately, a typical conic bundle is non-rational, so that the existence of a map $p : Q_S \to V$ does not mean that V is unirational. It is well known, however, that a conic bundle admitting a section is rational. We leave the elementary proof to the reader as an exercise: up to a fibrewise birational equivalence, the variety Q_S is given by one equation,

$$\sum_{0\le i\le j\le 2} A_{ij}u_iu_j = 0$$

in $S \times \mathbb{P}^2$, where A_{ij} are some rational functions on the surface S, $(u_0 : u_1 : u_2)$ are homogeneous coordinates on $\mathbb{P}^2$; now one can apply the fibrewise stereographic projection from the section.

It is easy to select polynomials G and H is such a way that the fibration $\pi\colon Q_S \to S$ has a section (and the quartic V is smooth) (examples are given in [**IM**]). For those special smooth quartics the construction described above gives unirationality. Note that the unirationality problem is still open for a generic three-dimensional quartic.

Therefore, an arbitrary smooth cubic $V_3 \subset \mathbb{P}^4$ and some special smooth quartics $V_4 \subset \mathbb{P}^4$ are unirational. The hardest point in solving the Lüroth problem, however, is proving non-rationality of these varieties, which is what was done in the papers of Clemens and Griffiths [**CG**] and Iskovskikh and Manin [**IM**].

The negative solution of the Lüroth problem meant that the rationality problem could not be solved in terms of differential geometric invariants (in dimension three and higher). Moreover, the problem of geometric characterization of varieties with no non-trivial differential geometric invariants emerged. It became clear that for a satisfactory classification of higher-dimensional algebraic varieties one needs to introduce new classes of varieties, whereas the rationality problem is an (important) particular case of more general questions.

That is why the breakthrough that was made in [**CG, IM**] was of such great importance.

Note that each of the papers [**CG, IM**] generated a new direction of further research: the so-called transcendental method and the method of maximal singularities, respectively. By means of the transcendental method very precise results on non-rationality of three-dimensional conic bundles were obtained, and the Clemens degeneration techniques made it possible to prove non-rationality of generic (in the complex analytic sense) pencils of cubic surfaces and many classes of Fano three-folds [**Clem74, Clem82, Clem83, Bar, Tyu79**]. Tikhomirov applied the transcendental method to a more delicate problem of existence of conic bundle structures on certain Fano varieties [**Tikh81, Tikh83, Tikh86**]. However, this remarkable method applies to three-dimensional varieties only. In this book we neither discuss nor use the method of Clemens and Griffiths (known also as the method of intermediate Jacobian) and refer the reader to the survey paper [**Tyu79**]. (The list of papers cited above is by no means complete; in particular, we do not even mention the papers dealing specifically with conic bundles and proving their non-rationality in terms of their discriminant curve by the method of intermediate Jacobian.)

The subject of this book is the method of maximal singularities that accumulated ideas of Noether and Fano. Note, slightly in advance, that the results of [**IM**] imply immediately, without any extra arguments, that a smooth three-dimensional quartic $V = V_4 \subset \mathbb{P}^4$ cannot be fibred by a rational map $\varphi\colon V \dashrightarrow Y$ into rational surfaces (the case when $Y = \mathbb{P}^1$) or curves (the case $Y = \mathbb{P}^2$: since in the small dimensions the Lüroth problem has the positive solution, the image of a rational dominant map φ with $\dim Y < \dim V$ can be either a rational curve or a rational map only). Since a linear projection $\mathbb{P}^3 \dashrightarrow \mathbb{P}^k$, $k = 1, 2$, fibres $\mathbb{P}^3$ into planes and lines, respectively, this gives one more proof of non-rationality of a smooth three-dimensional quartic, and moreover, as it soon became clear [**I80**], this argument was better than that of the finiteness of the group of birational self-maps $\operatorname{Bir} V = \operatorname{Aut} V$, since there are other varieties, for which the latter group is infinite, but non-rationality of which still can be shown via the fact that there are no structures of a fibration into rational surfaces or curves.

Thus the following question (or rather a class of questions) acquires the key importance: to describe, for a given variety V, all rational maps $\varphi\colon V \dashrightarrow Y$, the base Y and the fibre of general position $\varphi^{-1}(y)$, $y \in Y$, all of which belong to a given class of varieties. Before we give precise definitions, let us consider one more example, the Sarkisov theorem on (higher-dimensional) conic bundles [**S80, S82**]. The Sarkisov theorem was an especially impressive breakthrough in the early 80s because that fact was so visual and simple (of course we mean the formulation of the result, not the proof). The Sarkisov theorem showed once again that the very rationality problem needs to be modified to develop an adequate higher-dimensional theory, and confirmed the direction, in which this generalization was to be sought.

2.3. The Sarkisov theorem on conic bundles. Let S be a smooth projective variety of dimension $\dim S \geq 2$, $\rho\colon E \to S$ an (algebraic) vector bundle of rank 3, $\rho\colon \mathbb{P}(E) \to S$ its projectivization, that is, a locally trivial $\mathbb{P}^2$-bundle over S. A hypersurface $V \subset \mathbb{P}(E)$, equipped with the natural projection $\pi\colon V \to S$, $\pi = \rho\,|\,_V$, is called a *conic bundle* over S, if every fibre $\pi^{-1}(s) \subset \mathbb{P}^2 = \rho^{-1}(s)$ is a conic in $\mathbb{P}^2$. Let $\mathcal{E}$ be the sheaf of sections of the *dual* bundle $E^* \to S$. Then $\mathbb{P}(E)$ coincides with the projective bundle $\mathbb{P}(\mathcal{E})$ in the sense of Grothendieck (**Proj**, see [**Hart**]). Let $\mathcal{L}$ be the tautological sheaf on $\mathbb{P}(\mathcal{E})$. The hypersurface V is given by a section of the line bundle $\mathcal{L}^{\otimes 2} \otimes \rho^*\mathcal{F}$, where $\mathcal{F}$ is an invertible sheaf on the base S. The simplest example of a conic bundle is a divisor of bidegree $(2, N)$ on the direct product $\mathbb{P}^2 \times \mathbb{P}^M$, $M \geq 2$. Since the vector bundle E is locally trivial, any conic bundle is birational to a conic bundle of the latter example, however, this equivalence does not simplify anything; on the contrary, it makes everything more complicated, since a smooth variety $V \subset \mathbb{P}(E)$ is transformed into a variety $V^\sharp \subset \mathbb{P}^2 \times \mathbb{P}^M$ with, generally speaking, arbitrary singularities, which may contain entire planes $\mathbb{P}^2 \times \{p\}$, $p \in \mathbb{P}^M$. The "correct" object for studying are the *standard conic bundles* $\pi\colon V \to S$, where V is a smooth variety, satisfying the following minimality condition: $\operatorname{Pic} V = \mathbb{Z}K_V \oplus \pi^* \operatorname{Pic} S$. That condition implies, that the fibre space $V\backslash S$ has no sections (in the birational sense, that is, over a Zariski open subset $\mathcal{U} \subset S$): any divisor D on V has an even intersection number with a generic fibre $C_s = \pi^{-1}(s)$, since $(K_V \cdot C_s) = -2$ and $(\pi^* R \cdot C_s) = 0$ for any divisor R on the base S. For this reason, the fibre space V/S does not admit a fibrewise trivialization (in the birational sense!). The Segre construction from Subsection 2.2 gives an opposite example.

With every standard conic bundle $\pi\colon V \to S$ an important object is associated, the *discriminant divisor* $D \subset S$, that consists of the points $x \in S$, the fibre over which $\pi^{-1}(x) \subset \mathbb{P}^2$ is a non-smooth conic: a pair of lines (for a point of general position $x \in D$) or a double line. If $V \subset \mathbb{P}^2 \times \mathbb{P}^M$ is given by the equation

$$\sum_{i,j=0}^{2} A_{ij}(x_*) y_i y_j = 0,$$

where $(y_0 : y_1 : y_2)$ are homogeneous coordinates on $\mathbb{P}^2$, (x_*) are homogeneous coordinates on $\mathbb{P}^M$, $A_{ij} = A_{ji}$ are homogeneous polynomials of degree $N \geq 1$, then the discriminant divisor $D \subset \mathbb{P}^M$ is given by the equation $\det \|A_{ij}\| = 0$ (that is where the name comes from). The Sarkisov theorem asserts that birational geometry of the variety V is determined by the properties of the pair (S, D).

Theorem 2.1. *Let $\pi\colon V \to S$ be a standard conic bundle, satisfying the condition $|\,4K_S + D\,| \neq \emptyset$. Then there are no other structures of a conic bundle on V, apart from π: if $\varepsilon\colon V \dashrightarrow S^+$ is a rational dominant map, the generic fibre of which*

$\varepsilon^{-1}(s)$, $s \in S^+$, is a rational curve, then there exists a birational map $\alpha\colon S \dashrightarrow S^+$, such that $\varepsilon = \alpha\pi$, that is, the diagram

$$\begin{array}{ccc} V & = & V \\ \pi \downarrow & & \downarrow \varepsilon \\ S & \stackrel{\alpha}{\dashrightarrow} & S^+ \end{array}$$

is commutative.

Therefore, in the assumptions of the Sarkisov theorem the variety V can be fibred (by a *rational* map) into rational curves in a unique way. Since the projective space $\mathbb{P}^n$ has infinitely many structures of a conic bundle (to begin with, there is an n-dimensional family of linear projections from points $\mathbb{P}^n \dashrightarrow \mathbb{P}^{n-1}$), we get

Corollary 2.1. *In the assumptions of Theorem 2.1 the variety V is non-rational.*

The breakthrough, realized in Sarkisov's papers [**S80, S82**], was so impressive because it embraced all dimensions (≥ 3). To compare: the rival method of intermediate Jacobian up to this day works in three-dimensional geometry only. One has to admit, however, that the results on non-rationality of conic bundles, obtained by the method of intermediate Jacobian, are more precise (cover a bigger class of varieties).

Note that conic bundles have immense groups of fibrewise birational self-maps $\text{Bir}(V/S)$, that is, self-maps, tranforming every fibre into itself. To see this, let us consider and arbitrary smooth conic $C \subset \mathbb{P}^2$. Let $x \in \mathbb{P}^2 \setminus C$ be a point, $\rho_x\colon \mathbb{P}^2 \dashrightarrow \mathbb{P}^1$ the projection from the point x, $\rho_x\,|\,_C\colon C \to \mathbb{P}^1$ its restriction onto the curve C. Obviously, $\rho_x\,|\,_C$ is a double cover, so that there is a Galois involution $\tau_x \in \text{Aut}\,C$, permuting the points in the fibres. The group $\text{Aut}\,C = \text{Aut}\,\mathbb{P}^1$ is generated by those involutions. The importance of the involutions τ_x comes from the fact that their construction is easy to globalize to a fibrewise birational self-map $\tau_\sigma \in \text{Bir}(v/S)$, associated with an arbitrary rational section $\sigma\colon S \dashrightarrow \mathbb{P}(\mathcal{E})$ of the projective bundle $\rho\colon \mathbb{P}(\mathcal{E}) \to S$ (recall that V is embedded in $\mathbb{P}(\mathcal{E})$, where $C_s = \pi^{-1}(s)$ is embedded in $\mathbb{P}^2 = \rho^{-1}(s)$ as a conic). The map τ_σ is, generally speaking, non-regular, for the following two reasons: first, the section τ is not everywhere defined and second, the self-map τ_σ is not defined on those fibres C_s, $s \in S$, which intersect the image of the section σ, that is, $\sigma(s) \in C_s$. Since the $\mathbb{P}^2$-bundle $\mathbb{P}(\mathcal{E})/S$ is locally trivial (in the algebraic sense), the rational sections correspond to rational maps $S \dashrightarrow \mathbb{P}^2$. This gives an idea of the "size" of the group $\text{Bir}(V/S)$. One of the corollaries from the Sarkisov theorem is that the groups $\text{Bir}\,V$ and $\text{Bir}(V/S)$ coincide (up to a finite subgroup).

Now let us explain the meaning of the Sarkisov condition $|\,4K_S + D| \neq \emptyset$ by a visual example of a conic bundle $V \subset \mathbb{P}^2 \times \mathbb{P}^M$ of bidegree $(2, N)$. Here $S = \mathbb{P}^M$, so that $K_S = -(M+1)H$, where H is the class of a hyperplane in $\mathbb{P}^M$. Thus the Sarkisov condition takes the form $\deg D \geq 4(M+1)$, and since the hypersurface $D \subset \mathbb{P}^M$ is given by the equation $\det \|A_{ij}\| = 0$, we obtain the inequality $3N \geq 4(M+1)$. Therefore, except for a finite set of families, all smooth conic bundles of the given type satisfy the Sarkisov condition, whereas the condition itself can be interpreted as a bound from below for the "twistedness" of the conic bundle over the base $\mathbb{P}^M$ in some intuitively clear sense. The case is similar for conic bundles of generic type $V \subset \mathbb{P}(\mathcal{E})$, where $\mathcal{E}$ is an arbitrary locally free sheaf over a smooth base S. Therefore, the situation when the structure of a conic bundle is unique is by no means an exceptional phenomenon, on the contrary, it is a typical property satisfied by "almost all" conic bundles.

From the viewpoint of birational classification of algebraic varieties, the importance of the Sarkisov theorem comes also from the fact that it gives examples of fibre spaces $\pi\colon V \to S$ of arbitrary dimension, the base S and the fibre of which are rational, but the variety V itself is non-rational. This shows once again that for a satiafactory classification in higher dimensions the concept of rationality should be in some way generalized, to embrace all natural examples. However, before we turn to a description of such a generalization, which was obtained just a few years ago, let us consider, for the picture to be complete, one more class of three-folds, birational geometry of which was studied very recently: the pencils of cubic surfaces (fibrations into del Pezzo surfaces of degree 3).

Let E be an (algebraic) vector bundle of rank 4 over the projective line $\mathbb{P}^1$ and $\rho\colon \mathbb{P}(E) \to \mathbb{P}^1$ its projectivization, a locally trivial $\mathbb{P}^3$-bundle over $\mathbb{P}^1$. It is more convenient to use the presentation of the variety $\mathbb{P}(E)$ as a projective bundle $\mathbb{P}(\mathcal{E})$ in the sense of Grothendieck, where $\mathcal{E}$ is a locally free sheaf of sections of the dual bundle E^*, $\mathcal{E} = \bigoplus_{i=0}^{3} \mathcal{O}_{\mathbb{P}^1}(a_i)$, and we may assume that $a_0 = 0 \le a_1 \le a_2 \le a_3$. A hypersurface $V \subset \mathbb{P}(\mathcal{E})$ equipped with the natural projection $\pi = \rho\,|\,_V\colon V \to \mathbb{P}^1$, is a *fibration into cubic surfaces* over $\mathbb{P}^1$, if every fibre $\pi^{-1}(t) \subset \mathbb{P}^3 = \rho^{-1}(t)$ is a cubic surface. The hypersurface V is given by a section of the invertible sheaf $\mathcal{L}^{\otimes 3} \otimes \rho^* \mathcal{O}_{\mathbb{P}^1}(l)$ for some $l \in \mathbb{Z}$, where $\mathcal{L}$ is the tautological sheaf in the sense of Grothendieck. As in the case of conic bundles, the simplest (and model) example is given by a hypersurface of bidegree $(3, l)$ in $\mathbb{P}^3 \times \mathbb{P}^1$.

Let $V \subset \mathbb{P}(\mathcal{E})$ be a smooth variety. We say that V satisfies the K^2-*condition*, if for any $N \in \mathbb{Z}_+$ the class $NK_V^2 - f$ is non-effective, where $f \in A^2V$ is the class of a line in a fibre of the projection π (that is, of a usual line on the cubic surface $\pi^{-1}(t) \subset \mathbb{P}^3$), K_V is the canonical class of the variety V, A^2V is the Chow group of classes of curves on V modulo numerical equivalence. The K^2-condition can be expressed in a more compact way: $K_V^2 \notin \operatorname{Int} A_+^2 V$, where $A_+^2 V \subset A^2V \otimes \mathbb{R}$ is the closed cone generated by the classes of effective curves on V, the so-called *pseudoeffective cone*. The Picard group of V is

$$\operatorname{Pic} V = \mathbb{Z}K_V \oplus \mathbb{Z}F = \mathbb{Z}L \oplus \mathbb{Z}F,$$

where F is the class of a fibre $\pi^{-1}(t)$, L is the class of the tautological divisor. For $V \subset \mathbb{P}^3 \times \mathbb{P}^1$ this means simply that every surface on V is cut out by a hypersurface of degree (a, b): in terms of the Picard group every class $D \in \operatorname{Pic} V$ is of the form $D = aL + bF$. In this model case $K_V = -L + (l - 2)F$ and it is easy to compute that the class of the curve K_V^2 has bidegree $(3, -5l + 12)$ in $A^3(\mathbb{P}^3 \times \mathbb{P}^1)$, that is, as the class of a curve in $\mathbb{P}^3 \times \mathbb{P}^1$. For $l \ge 3$ the second component is negative, so that the K^2-condition holds. In the general case it is easy to show that a hypersurface V defined by a section of the sheaf $\mathcal{L}^{\otimes 3} \otimes \rho^* \mathcal{O}_{\mathbb{P}^1}(l)$ satisfies the K^2-condition for $l \ge \lambda(\mathcal{E}) = \lambda(a_1, a_2, a_3)$, that is, for a fixed sheaf $\mathcal{E}$ this condition holds for all pencils of cubic surfaces, except for a finite number of families.

Theorem 2.2. *Assume that the fibre space $V \to \mathbb{P}^1$ satisfies the K^2-condition and is given by a sufficiently general section of the sheaf $\mathcal{L}^{\otimes 3} \otimes \rho^* \mathcal{O}_{\mathbb{P}^1}(l)$ on $\mathbb{P}(\mathcal{E})$. Then on V there are no structures of a conic bundle, and the projection $\pi\colon V \to \mathbb{P}^1$ is the unique structure of a fibration into rational surfaces on V. In particular,* *the variety V is non-rational. The group of birational self-maps of the variety V coincides with the group of fibrewise birational self-maps* $\mathrm{Bir}(V/\mathbb{P}^1)$.

By what was said above, in a certain sense almost all fibrations into cubic surfaces over $\mathbb{P}^1$ satisfy the assumptions of Theorem 2.2 (the theorem was proven in [**Pukh98a**]). Note that all pencils of cubic surfaces are unirational. The projection $\pi\colon V \to \mathbb{P}^1$ has many sections (more precisely, for any set of distinct fibres $\pi^{-1}(t_1), \ldots, \pi^{-1}(t_k)$ and smooth points $x_i \in \pi^{-1}(t_i)$ there exists a section $s\colon \mathbb{P}^1 \to V$ such that $s(t_i) = x_i$), let us take one of them, $s\colon \mathbb{P}^1 \to V$. For every point $s(t) \in F_t = \pi^{-1}(t)$ consider a rational cubic curve $C_t = F_t \cap T_{s(t)}F_t$ with a singular point at $s(t)$. Now for every point of the curve C_t let us reproduce the construction of unirationality of the cubic hypersurface (Subsection 2.2), mapping a generic tangent line to F_t to the third point of its intersection with F_t. This gives a construction of unirationality for V. Thus the pencils of cubic surfaces give new counterexamples to the Lüroth problem in dimension three.

Note, furthermore, that the group of fibrewise self-maps $\mathrm{Bir}(V/\mathbb{P}^1)$ is very big (see [**M72**]), the presentation of the group was also reproduced in [**Pukh98a**]. It is generated by involutions corresponding to the sections and bi-sections of the projection π (which form a very large set, a union of algebraic sets of dimensions, increasing to infinity).

The pencils of cubic surfaces give a good illustration of one of the main versions of the rationality problem (Subsection 1.1): on V there exists a unique rational function $f\colon V \dashrightarrow \mathbb{C}$ (up to a fractional-linear transformation of $\mathbb{C}$), the level sets of which $f^{-1}(a)$, $a \in \mathbb{C}$, are irreducible rational surfaces.

Thus, already in dimension three, rationality of the base and of the generic fibre does not imply rationality of the total space of the fibration. For this reason, the concept of rationality needs to be generalized. We discuss such a generalization in the next section.

3. Rationally connected varieties

3.1. The theorem of Graber-Harris-Starr. A precise analog of rational curves and surfaces in higher dimensions are rationally connected varieties. The concept of rational connectedness goes back to Manin's papers on geometry of cubic surfaces (see [**M72**]); Kulikov started to discuss this concept in three-dimensional geometry [**Kul86**]. Active investigation of higher-dimensional rationally connected varieties was started independently by Campana [**Cam92**] and Kollár, Miyaoka and Mori [**KolMM**].

Definition 3.1. A smooth projective variety X is said to be *rationally connected*, if any two points $x_1, x_2 \in X$ of general position can be joined by an irreducible rational curve, that is, there exists a morphism $f\colon \mathbb{P}^1 \to X$ such that $f(t_1) = x_1$ and $f(t_2) = x_2$.

Obviously, the projective space is rationally connected (any two points can be joined by a line), and the quadrics $Q \subset \mathbb{P}^n$ are rationally connected, as well (any two points lie on a conic $Q \cap \mathbb{P}$, where $P = \mathbb{P}^2 \subset \mathbb{P}^n$ is a suitable plane). It is also obvious from the definition, that rational connectedness is invariant with respect to birational maps and that if $\varphi\colon X \dashrightarrow Y$ is a rational dominant map, where X is rationally connected, then Y is rationally connected, too. Indeed, if $y_1, y_2 \in Y$ are generic points (in particular, lying in the image of φ), then we choose points of general position $x_i \in \varphi^{-1}(y_i)$ and obtain a pair of generic points $x_1, x_2 \in X$ which by assumption can be joined by a rational curve $f\colon \mathbb{P}^1 \to X$, but then the rational curve $\varphi \circ f\colon \mathbb{P}^1 \to Y$ joins the points y_1 and y_2.

The following properties of rationally connected varieties are less obvious, but can be easily proved by means of the standard technique of deformation theory (see [**Kol96**, Chapter IV]):

(1) Any k points $x_1, \ldots, x_k$ (that is, without the condition of general position) can be joined by an irreducible rational curve $f\colon \mathbb{P}^1 \to X$, $f(t_i) = x_i$; moreover, for any set of non-zero tangent vectors $v_i \in T_{x_i}X$ one can find a curve f with exactly these tangent vectors at the points x_i;

(2) the variety X is rationally connected if and only if it contains a *very free* rational curve $f\colon \mathbb{P}^1 \to X$, that is, the pullback of the tangent bundle with respect to f is very ample:

$$f^*T_X \cong \mathcal{O}_{\mathbb{P}^1}(a_1) \oplus \cdots \oplus \mathcal{O}_{\mathbb{P}^1}(a_m),$$

$m = \dim X$, $a_i \geq 1$; informally speaking, in a neighborhood of the curve $f(\mathbb{P}^1)$ there are plenty of vector fields, due to the fact that the curve is movable;

(3) for smooth varieties the property of rational connectedness is equivalent to *chain rational connectedness*, when it is required only that for generic points $x_1, x_2 \in X$ there is a (possible reducible) connected curve $C = \bigcup_{i \in I} C_i$ containing these points, $x_i \in C$, where each component $C_i = f_i(\mathbb{P}^1)$ is rational; thus any two points can be joined by a chain of rational curves.

The following example shows that if $f\colon X \to Y$ is a morphism of smooth varieties with irreducible fibres and the fibre of general position is rationally connected, then some special (degenerated) fibres can be not rationally connected. Indeed, take a generic (in particular, smooth) hypersurface of bidegree (3,1) in $\mathbb{P}^3 \times \mathbb{P}^1$, that is, a pencil of cubic surfaces in $\mathbb{P}^3$. A general surface of the pencil is rational, hence rationally connected. However, it is easy to find an equation such that the fibre over some point $t_0 \in \mathbb{P}^1$ is a cone $S_0 \subset \mathbb{P}^3$ over a smooth cubic curve C in $\mathbb{P}^2$. On the cone there are no rational curves, except for the generators, so that S_0 is not rationally connected. However, S_0 is certainly chain rationally connected, since all rational curves on S_0 have a common point, the vertex of the cone. In fact, this is a particular case of a general fact [**Kol96**, Chapter 4]: if the generic fibre is chain rationally connected, then every fibre is chain rationally connected (under the assumption that $f\colon X \to Y$ is a proper equidimensional morphism). It is easy to construct similar examples in arbitrary dimension. They show that the assumption of smoothness is essential for the equivalence of rational connectedness and chain rational connectedness.

Counting the dimension of the set of rational curves $f\colon \mathbb{P}^1 \to \mathbb{P}^M$ of degree $d \geq 1$ meeting at $dm+1$ points a smooth hypersurface $V_m \subset \mathbb{P}^M$ of degree m, $1 \leq m \leq M$, it is easy to show that the Fano hypersurface V is rationally connected and, more generally, Fano complete intersections $V_{m_1, \ldots, m_k} = F_{m_1} \cap F_{m_2} \cap \cdots \cap F_{m_k} \subset \mathbb{P}^M$, where F_{m_i} is a hypersurface of degree m_i, $m_1 + \cdots + m_K \leq M$, are rationally connected. This is a particular case of a general fact; see [**Kol96**].

Theorem 3.1. *A smooth Fano variety V is rationally connected.*

Recall that a smooth variety V is called a *Fano variety*, if the anticanonical class $-K_V$ is ample, that is, its multiple class gives an embedding $V \subset \mathbb{P}^N$. The theorems of Clemens and Griffiths and Iskovskikh and Manin show that rationally connected varieties are more numerous than rational ones. How much more numerous, can be seen from the Sarkisov theorem, since the following claim holds.

Proposition 3.1. *A conic bundle* $\pi\colon V \to S$ *over a rationally connected base* S *is rationally connected.*

Proof. Let $x, y \in V$ be a generic pair of points and $\bar{x} = \pi(x)$, $\bar{y} = \pi(y)$ their images on S. By assumption, the points $\bar{x}, \bar{y}$ lie on an irreducible rational curve $C \subset S$. The surface $\pi^{-1}(C) = Q$ is rational, hence rationally connected, and contains x, y, which is what we need. Q.E.D.

Therefore, the "majority" of conic bundles are non-rational rationally connected varieties. It is easy to show that on rationally connected varieties there are no non-zero differential forms and moreover, covariant tensors. Thus the class of rationally connected varieties is the required generalization of the class of rational varieties, but provided that the following condition holds: provided that this class is closed with respect to the operation of making fibrations, that is, for every fibre space $\pi\colon V \to S$, a general fibre and the base of which are rationally connected, the variety V is rationally connected, too. This fact was conjectured in [**KolMM**] but proved only ten years later in [**GHS**]. Arguing as in the proof of Proposition 3.1, we reduce the claim to the case $S = \mathbb{P}^1$, and the technique of deformation theory reduces it to the following fact, proven by Graber, Harris and Starr.

Theorem 3.2. *Let* $\pi\colon V \to \mathbb{P}^1$ *be a morphism, the generic fibre of which is irreducible and rationally connected. Then the projection* π *has a section* $s\colon \mathbb{P}^1 \to V$.

In fact, [**GHS**] proves existence of the section for a fibre space $\pi\colon V \to B$, where B is an arbitrary, not necessarily rational, curve. However, the case $B = \mathbb{P}^1$ is central in the proof. It is not possible to reproduce here a full proof of Theorem 3.2 (in this book, we discuss very different geometric ideas), let us just comment briefly on this fundamental fact.

In the fibres of the projection π by assumption there are plenty of rational curves. If $C \subset V$ is a curve, then for points of general position $x_i \in C$, $i = 1, \dots, k$, $\pi(x_i) = t_i \in \mathbb{P}^1$ one can consider movable rational curves $C_i \subset \pi^{-1}(t_i)$ in the corresponding fibres of the projection π, passing through the points x_i, $x_i \in C_i$, and construct a *comb* $\Gamma = C \cup \bigcup_{i=1}^{k} C_i$, a connected reducible curve on V. An easy technique of deformation theory shows that mobility of the "attached" rational curves C_i implies mobility of the curve Γ, that is, Γ moves in a family of curves $(\Gamma_\lambda, \lambda \in \Lambda)$, a general curve of which is irreducible. (In particular, for precisely this reason existence of one section of the projection π implies that V is covered by section.) If the curve C is a multi-section (that is, $\deg(\pi\,|\,_C) = d \geq 2$), then the curves Γ_λ are multi-sections of the same degree $d \geq 2$, as well. The main idea of the proof of Graber-Harris-Starr is to compare deformations of the morphism $\mu\colon C \to V$ (that is, roughly speaking, deformations of the curve C inside V) and deformations of the finite cover $\pi \circ \mu\colon C \to \mathbb{P}^1$. If deformations of μ cover deformations of $\pi \circ \mu$, then we immediately get the claim of the theorem, since by the classical theorem of Hurwitz and Clebsch the rough moduli space $\bar{M}_{g,0}(\mathbb{P}^1, d)$ has one irreducible component, so that the curve C can be deformed inside V into a reducible curve, splitting into d sections of the morphism π. Thus the key fact is that the map of the rough moduli spaces

$$\bar{M}_{g,0}(V,[C]) \to \bar{M}_{g,0}(\mathbb{P}^1,d),$$

is dominant, here $[C]$ is the class of the curve $C \subset V$. This fact was proven in [**GHS**] by very delicate arguments using geometry of combs Γ, described above.

One direct implication of Theorem 3.2 is the existence of a section $s\colon \mathbb{P}^1 \to V$, intersecting non-singular fibres $F_t, \dots, F_{t_k}$ at arbitrary prescribed points $x_i \in F_{t_i}$. This claim can be generalized in the following way [**HasTsch**]: for any points $x_i = F_{t_i}$ on non-singular fibres $F_{t_i} = \pi^1(t_i)$ and any set of l_i-jets of non-singular curves transversal to the fibres F_{t_i} at these points, there exists a section $s\colon \mathbb{P}^1 \to V$ with these jets at the points x_i, $(l_1,\dots,l_k) \in \mathbb{Z}_+^k$ is an arbitrary set of non-negative integers. In other words, one can prescribe a section of tangent vectors (transversal to fibres) and more generally jets of finite order at given points.

Theorem 3.2 is a natural generalization of the classical Tsen theorem [**Shaf, Kol96**] on the existence of a section of the projection $\pi\colon V \to \mathbb{P}^1$, where $V \subset \mathbb{P}^m \times \mathbb{P}^1$ is a hypersurface of bidegree (d,N), $d \le m$. (The proof of the Tsen theorem is quite elementary: a section $s\colon \mathbb{P}^1 \to \mathbb{P}^m \times \mathbb{P}^1$ is represented by an arbitrary set of polynomials $(f_0, f_1, \dots, f_m)$ on $\mathbb{P}^1$, the condition $s(\mathbb{P}^1) \subset V$ gives polynomial equations for the coefficients of the polynomials f_i. The number of equations is less than the number of coefficients, which implies what we need.) A non-trivial generalization of the Tsen theorem for fibrations into Fano complete intersections in the products of weighted projective spaces is given in [**CPP**]. Since Fano complete intersections are rationally connected (this is very easy to prove by elementary methods), the latter result also follows from the theorem of Graber-Harris-Starr.

The theorem of [**GHS**] also completed the proof of the following important classification result: for any uniruled variety X there exists a rational dominant map $\pi\colon X \dashrightarrow Z$ such that the fibres are rationally connected and Z is not uniroled (the maximal rationally connected fibration of X); see [**Cam92, KolMM, Kol96**].

Now we can precisely formulate the problems of birational geometry, discussed in this book.

3.2. Rationally connected fibre spaces and Fano fibre spaces. A surjective morphism $\pi\colon V \to S$ of projective varieties is called a *rationally connected fibre space* if the base S and a fibre of general position $\pi^{-1}(s)$, $s \in S$, are rationally connected varieties. In particular, the variety V itself is rationally connected by Theorem 3.2. We exclude from consideration the trivial cases $V = S$ and $S = *$, the point. To emphasize this, we sometimes speak about *non-trivial* fibre spaces.

A *structure of a rationally connected fibre space* on a (rationally connected) variety V can be defined in two equivalent ways:

(1) as a *birational* map $\chi\colon V \dashrightarrow V^+$ onto a variety V^+ with a fixed morphism $\pi^+\colon V^+ \to S^+$, which is a rationally connected fibre space;

(2) as a *rational* dominant map $\varphi\colon V \dashrightarrow S$, the fibre of general position of which is rationally connected (in particular, irreducible).

The definitions (1) and (2) are equivalent by the theorem on resolution of singularities [**Hi, Kol07**]. Two structures $\varphi_1\colon V \dashrightarrow S_1$ and $\varphi_2\colon V \dashrightarrow S_2$ are considered as identical, if there exists a birational map $\alpha\colon S_1 \dashrightarrow S_2$, such that the following diagram of rational maps commutes

$$\begin{array}{ccccc} & X & \stackrel{\mathrm{id}}{\leftrightarrow} & X & \\ \varphi_1 & \downarrow & & \downarrow & \varphi_2 \\ & S_1 & \stackrel{\alpha}{\dashrightarrow} & S_2, & \end{array} \tag{5}$$

that is, $\varphi_2 = \alpha \circ \varphi_1$. In other words, φ_1 and φ_2 have the same fibres. The set of non-trivial structures of a rationally connected fibre space on the variety X (modulo the identification above) is denoted by $RC(X)$.

The utmost importance for birational geometry of studying the structures of a rationally connected fibre space follows from the following clear reason. One of the main methods of classifying algebraic varieties is by reduction to a smaller dimension. If an algebraic variety V admits a morphism $\pi\colon V \to S$ onto a variety S of positive dimension, where $\dim S < \dim V$, then it is natural to consider V as a family of varieties of dimension $\dim V - \dim S$ (that is, the fibres $F_s = \pi^{-1}(s)$, $s \in S$), parametrized by the base S. Thus the description of the variety V reduces to a description of the geometry of a fibre of general position F_s and a characterization of the "twistedness" of the family F_s, $s \in S$, over the base S.

However, fibre spaces $\pi\colon V \to S$ can be considered also from another viewpoint. Fixing a class $\mathcal{S}$ of varieties S and a class $\mathcal{F}$ of varieties F_s, one can construct new varieties of dimension $\dim S + \dim F$ as fibrations with a given base $S \in \mathcal{S}$ and fibre $F_s \in \mathcal{F}$. This gives a source of constructing new algebraic varieties of arbitrary dimension.

Therefore, it is clear that rationally connected varieties with no (non-trivial) structures of a rationally connected fibre space are of exceptional importance in birational geometry: those varieties "do not come from smaller dimensions" and for this reason represent a new classification type. Of special interest are also varieties with just one or at least just finitely many non-trivial structures: for them reduction of dimension is uniquely determined or is realized in finitely many ways.

On the set $RC(V)$ there is a natural relation of partial order: for $\varphi_1, \varphi_2 \in RC(V)$ we have $\varphi_1 \leq \varphi_2$, if there is a rational dominant map $\alpha\colon S_1 \dashrightarrow S_2$ such that the diagram (5) commutes. In other words, the fibres of φ_1 are contained in the fibres of φ_2. For a general point $s \in S_2$ we have

$$\alpha^{-1}(s) = \varphi_1(\varphi_2^{-1}(s)),$$

therefore, $\alpha \in RC(S_1)$ is a structure of a rationally connected fibre space on S_1. It is easy to see that the correspondence $\varphi_2 \mapsto \alpha$ determines a bijection of the sets $\{\psi \in RC(V) \,|\, \psi \geq \varphi_1\}$ and $RC(S_1)$. Therefore from the geometric viewpoint of primary interest are the *minimal* elements of the ordered set $RC(V)$. Denote the set of minimal elements by $RC_{\min}(V)$. Set also $RC_d(V) \subset RC(V)$ to be the set of structures, the generic fibre of which is of dimension d. Obviously, if $d = \min\{e \in \mathbb{Z}_+ | RC_e \neq \emptyset\}$, then $RC_d \subset RC_{\min}$.

For each $d \in \{1, \ldots, \dim V - 1\}$ on the set $RC_d(V)$ there is a natural relation of *fibrewise birational equivalence*: $\varphi_1 \sim \varphi_2$ if there exists a birational transformation $\chi \in \operatorname{Bir} V$ and a birational map $\alpha\colon S_1 \dashrightarrow S_2$ such that the diagram

$$\begin{array}{ccccc} & V & \stackrel{\chi}{\dashrightarrow} & V & \\ \varphi_1 & \downarrow & & \downarrow & \varphi_2 \\ & S_1 & \stackrel{\alpha}{\dashrightarrow} & S_2, & \end{array}$$

commutes, that is, $\varphi_2 \circ \chi = \alpha \circ \varphi_1$. In other words, the birational self-map χ transforms the fibres of φ_1 into the fibres of φ_2. The quotient set $RC_d(V)/\sim$ we denote by the symbol $\overline{RC}_d(V)$.

To illustrate the introduced concepts and constructions, let us consider the following example [**Sob02**]. Let $V \subset \mathbb{P}^3 \times \mathbb{P}^1$ be a general (in particular, smooth) divisor of bidegree $(3, 2)$ and $\pi\colon V \to \mathbb{P}^1$ the projection. Obviously, $V/\mathbb{P}^1$ is a rationally connected fibre space. As we mentioned above, the group $\mathrm{Bir}(V/\mathbb{P}^1)$ of fibrewise birational self-maps is very large. Furthermore, the projection $q\colon V \to \mathbb{P}^3$ is of degree 2 outside the union of 27 smooth rational curves on V that are contracted by q into 27 points on $\mathbb{P}^3$, so that there is a Galois involution $\tau \in \mathrm{Bir}\, V$, permuting points in the fibre of the general position of the morphism q. Using the methods of [**Pukh98a**], the following claim was proved in [**Sob02**].

Theorem 3.3. (i) *There are no structures of a conic bundle on* V, *that is to say,* $RC_1(V) = \emptyset$.

(ii) *Every pencil of rational surfaces on* V *can be transformed by a map* $\chi \in \mathrm{Bir}\, V$ *into the pencil of fibres of the projection* π: $\overline{RC}_2(V) = \{\pi\}$.

(iii) *The group of birational self-maps is the free product,* $\mathrm{Bir}\, V = \mathrm{Bir}(V/\mathbb{P}^1) * \langle\tau\rangle$.

Therefore, modulo the fibrewise birational equivalence there is a unique rationally connected structure on V, the projection π. (Of course, this implies non-rationality of the variety V.)

The set of rationally connected structures on V equipped with all structures on it is a birational invariant of the variety. Thus to describe this set (and to solve other problems of birational geometry as well) it is natural to choose the most convenient model in the class of varieties birational to V (or one of them, if there are plenty of such models). Finding a model with good properties is the subject of the *minimal model program.* Completed in dimension three, this program is now close to completion in higher dimensions. After the famous paper [**BCHM**] appeared, the progress in the minimal model program accelerated, crucial results followed one another, difficult proofs were getting simplified and the whole theory started to become a standard part of algebraic geometry. In particular, it is now proven that any rationally connected variety V is birational to a variety V^+, fibred by a projection $\pi^+\colon V^+ \to S^+$ into Fano varieties, whereas V^+ has singularities of a given bounded type and the relative Picard number one (such fibrations are called Mori fibre spaces or Fano-Mori fibre spaces). We will come back to this theme in Chapter 4 and give more references there. Here we just point out that, unfortunately, already in dimension three the types of singularities and of corresponding families of Fano varieties are very numerous, so numerous that today computer programs are being developed to list them. In this book we will mostly discuss smooth Fano varieties V with the condition $\mathrm{Pic}\, V = \mathbb{Z}K_V$, the so-called *primitive Fano varieties* (a typical example is the smooth quartic $V_4 \subset \mathbb{P}^4$, which started this branch of birational geometry). Where appropriate we will comment on the known results on singular Fano varieties. In Chapters 4–8, devoted to fibre spaces over a non-trivial base, we will consider *standard Fano fibre spaces* $\pi\colon V \to S$, where V, S are smooth projective varieties and $\mathrm{Pic}\, V = \mathbb{Z}K_V \oplus \pi^* \mathrm{Pic}\, S$. Note that, with all the natural problems of working with singularities, it is the primitive Fano varieties and standard Fano fibre spaces that are the hardest to study.

As we pointed out above, of special interest are primitive Fano varieties V with no non-trivial structures of a rationally connected fibre space, $RC(V) = \emptyset$. These varieties make the main subject of the next two chapters. We will show that typically they satisfy the property of *birational rigidity*, which makes it possible to give a complete description of their birational geometry (to compute $RC(V)$, to describe generators and relations of the group of birational self-maps $\operatorname{Bir} V$, and much more).

To make the picture complete, note in addition that for standard Fano fibre spaces $\pi\colon V \to S$ over a fixed non-trivial base S, with the fibres $F_s = \pi^{-1}(s)$, $s \in S$, belonging to a given family $\mathcal{F}$ of Fano varieties, their birational geometry seems to be determined by the following informal principle:

If the fibration π is "sufficiently twisted" over the base S, then birational geometry of the variety V reduces to birational geometry of the fibre, the structure π is the least element of the set $RC(V)$, and birational self-maps of the variety V are fibrewise.

Now, the class of objects and the set of questions being precisely defined, we describe briefly the structure of the contents of the next two chapters.

3.3. The method of maximal singularities. The principal idea of the method, which today is the main instrument of studying birational maps of Fano varieties and Fano fibre spaces, is as follows. On a rationally connected variety there are no differential forms, but one can say a lot more: such varieties satisfy the classical *condition of termination of canonical adjunction*: for any divisor D the linear system $|D + NK_V|$ is empty for $N \gg 0$, since the canonical class is negative on every family of rational curves sweeping out V. Starting with the work of the Italian school, it was understood that of crucial importance is the parameter that measures the moment of termination of adjunction, in the modern terminology, the *threshold of canonical adjunction* (discussed in detail below). To study a birational map

$$\chi\colon V \dashrightarrow V',$$

one should compare the thresholds of canonical adjunction of an arbitrary effective divisor D' on V' and its strict transform D on V. If adjunction of the canonical class terminates on V not later than on V' for every mobile divisor D', then it turns out that the map χ admits a (comparatively) easy description. Thus all work is concentrated on the maps that do not satisfy this assumption, so that on V, V' there are linear systems Σ, Σ' with no fixed components, such that Σ is the strict transform of Σ' and canonical adjunction terminates for Σ' sooner than for Σ. This is the starting point of this work; the main object now is the system Σ and one can forget about Σ'. Now the following steps should be taken:

(1) One should prove that in the base set of the system Σ there are subsets of high multiplicity, the so-called *maximal singularities* (this is where the name of the method comes from); the corresponding condition "to have a high multiplicity" generalizes the Noether inequality for $\mathbb{P}^2$ (Subsection 1.3) and the Fano inequality (Subsection 2.1) and for this reason is called *the Noether-Fano inequality.*

(2) One should find out which maximal singularities can be realized on the given variety V and which cannot; this procedure is called *exclusion* of maximal singularities and makes the hardest part of work.

(3) For potentially realizable maximal singularities (found out in the previous step) one should produce "untwisting" birational maps, simplifying the map χ (or the linear system Σ); this procedure is called *untwisting* maximal singularities and usually makes no big complications, although sometimes it can be quite non-trivial.

From the purely practical viewpoint, the main drawback of the method of maximal singularities is that if at least one type of maximal singularity is neither untwisted nor excluded (that is, we can neither prove that this type does not realize, nor produce an untwisting map), then, however much work has been done, one can say next to nothing about birational geometry of the given variety, whereas all information that was obtained in no way can be presented as a complete result. This feature is to some extent compensated by the fact that if all types of maximal singularities can be excluded or untwisted, then we get an exhaustive description of birational maps for the given variety.

Step (1) on the modern level was realized in the classical paper of Iskovskikh and Manin on three-dimensional quartics [**IM**]. Essentially, it was in [**IM**] that the *canonical thresholds of pairs* were discovered (see Section 2.2). Fano could explain his idea of a maximal singularity by examples only; in [**IM**] a compact formalism was suggested that made it possible to give a precise definition of a maximal singularity, with which one can work.

Step (3) is geometrically the most visual. Maximal singularities were successfully untwisted by Noether, Castelnuovo and Fano, the latter made very significant progress in that direction [**Fano3**]. Modulo step (2) untwisting is the most pleasant part of the work, not too complicated but produces good results.

Almost all complications are accumulated in step (2). One should note that in the pioneer paper of Iskovskikh and Manin step (3) is completely missing: there is nothing to untwist on the three-dimensional quartic, not one of the types of maximal singularities can be realized. The paper [**IM**] is so important because in that paper, for the first time, an effective method of excluding maximal singularities was suggested. The technique developed in [**IM**] is quite complicated and excludes maximal singularities on smooth varieties of a small (at most 4) degree only. The first attempts to apply this method to varieties of higher degree (the complete intersection $V_{2\cdot 3} \subset \mathbb{P}^5$ of a quadric and a cubic [**I80**]) and singular varieties (three-dimensional quartics with a double point) were not completed; see Notes and references for Chapter 2. At the same time, from the geometric point of view, varieties of higher degree are more interesting; there are not too many Fano varieties of small degree. It seemed for a long time that the method developed in [**IM**] makes it possible to obtain isolated results of exceptional type only [**I80**], whereas the majority of Fano varieties (the more so, in higher dimensions) are out of reach for this approach. Such ideas (based on the failures of the 70s–80s) were popular until 1996 (see the introduction to [**I96b**]), when the situation started to change. (The papers [**Pukh87, Pukh89a, Pukh89c**] that were written in the spirit of the techniques of [**IM**], give a good idea of how complications grew exponentially when the degree was slightly increased or at least one non-degenerate double point was permitted.)

In [**Pukh95, Pukh00c**] the *test class technique*, which formed the technical core of the paper [**IM**], was essentially reconsidered and simplified (the paper [**Pukh00c**], published in 2000, was written and became accessible to all who worked in this field since 1996, first as a Warwick preprint, then on the archive). This new approach, combined with a series of new ideas, made it possible to work effectively with three-folds fibred into del Pezzo surfaces over $\mathbb{P}^1$ and certain Fano fibre spaces over $\mathbb{P}^1$ [**Pukh98a**] (see Chapters 4 and 6 for more detailed discussions). Finally, in [**Pukh98b**] the *technique of hypertangent divisors* was developed that made it possible to apply the method to Fano varieties of arbitrary high degree and dimension and to large classes of Fano fibre spaces over $\mathbb{P}^1$ (see Chapters 4–6), and also to Fano direct products (see Chapter 7). The method of maximal singularities became a powerful instrument of studying birational geometry of typical Fano varieties and fibre spaces.

In Chapter 2 the classical theory is concentrated: we treat the condition of termination of adjunction of the canonical class and the thresholds of canonical adjunction, measuring the speed of termination; the definition of birational (super)rigidity and its main geometric implications; the definition and examples of maximal singularities; some elementary examples of exclusion of maximal singularities and the $4n^2$-inequality; the $8n^2$-inequality and its applications; the three-dimensional quartics and the complete intersections $V_{2\cdot 3} \subset \mathbb{P}^5$, the latter class of varieties being used as an example of the untwisting procedure.

Chapter 3 is devoted to the "industrial" methods that make it possible to embrace, simultaneously, big classes of varieties, in the first place, the technique of hypertangent divisors. This technique works for varieties of general position, satisfying special *regularity conditions* (however, a variety of general position can have singularities, which, in turn, should satisfy some regularity conditions of their own). In order for the results, obtained in this way, to be meaningful, it is necessary for each set of regularity conditions to show that a variety of general position satisfies these conditions; such verification is often non-trivial and requires special work. We will explain the methods of checking the regularity conditions.

The main result, a complete proof of which is given in Chapter 3, is birational superrigidity (in particular, the absence of non-trivial structures of a rationally connected fibre space, the coincidence of the groups of birational and biregular self-maps, and, as a very particular corollary, non-rationality) of generic Fano complete intersections $V_{d_1\cdot\dots\cdot d_k} = F_1 \cap \dots \cap F_k \subset \mathbb{P}^N$, $\deg F_i = d_i$, $d_1 + \dots + d_k = N$, of arbitrary dimension (under the restriction $k < \frac{1}{2}\dim V$ for the number of the defining equations), and certain other natural classes of Fano varieties.

CHAPTER 2

The Method of Maximal Singularities

Summary

In this chapter we give a detailed and complete exposition of the method of maximal singularities in the absolute case (that is, for Fano varieties) and up to, but not including, the technique of hypertangent divisors. We start with reviewing the classical concept of termination of canonical adjunction, discuss its implications (the Noether-Fano inequality), define the birational rigidity and superrigidity and explain the general structure of the method (Sec. 1). Namely, to prove (super)rigidity, one should be able either to *exclude* or to *untwist* every geometric discrete valuation as a maximal singularity.

The exclusion part is the main one of the two. It is most difficult and usually takes a lot of effort. The untwisting part is much easier but for some varieties can be quite non-trivial, as we demonstrate in Section 6. Sections 2, 4 and 5 develop various techniques of excluding maximal singularities: the technique of counting multiplicities (leading to the crucial $4n^2$-inequality) is explained in Section 2, the more recent technique based on the connectedness principle of Shokurov and Kollár is developed in Section 4 and various geometric methods making use of the particular projective embedding of the variety under consideration are discussed in Section 5 by the example of the complete intersection $V_{2\cdot 3} \subset \mathbb{P}^5$. As an application, we prove birational superrigidity of several families of Fano varieties. Sections 5 and 6 together present a complete proof of birational rigidity of the complete intersection $V_{2\cdot 3} \subset \mathbb{P}^5$, Section 6 showing how non-trivial the untwisting procedure can be.

For several reasons, technical and historical, three-dimensional quartics $V_4 \subset \mathbb{P}^4$ are of special importance in birational geometry. For that reason, we include a special section (Sec. 3) dealing with the quartics. In that section, we explain, in particular, the *test class* method of Iskovskikh and Manin, which was the first efficient tool for excluding infinitely near maximal singularities.

1. Canonical adjunction

In this section we use the classical approach of considering the adjoint linear systems to define the basic concepts of the theory: maximal singularities and birational (super)rigidity. We then proceed to construct the oriented graph of a sequence of blow ups (one of the crucial tools of the method) and obtain the standard form of the Noether-Fano inequality in terms of the multiplicities at infinitely near subvarieties. Finally, we describe the general structure of the method of maximal singularities and prove the geometric properties of birational rigid varieties.

1.1. The threshold of canonical adjunction and birational rigidity. A smooth projective rationally connected variety X satisfies the classical condition of termination of adjunction of the canonical class: for any effective divisor D the linear system $|D+mK_X|$ is empty for $m \gg 0$, since K_X is negative on every family of rational curves sweeping out X, whereas an effective divisor is non-negative on any such family. As we pointed out above, it is important to fix the moment of termination precisely. To formalize this idea, let us consider the Picard group $A^1X = \operatorname{Pic} X$, set $A^1_{\mathbb{R}}X = A^1X \otimes \mathbb{R}$ and define the cones $A^1_+X \subset A^1_{\mathbb{R}}X$ of *pseudo-effective* classes and $A^1_{\text{mob}}X \subset A^1_{\mathbb{R}}X$ of *mobile* classes as the closed cones (with respect to the standard real topology of $A^1_{\mathbb{R}}X \cong \mathbb{R}^k$), generated by the classes of effective divisors and mobile divisors (that is, divisors in the linear systems with no fixed components), respectively.

Definition 1.1. The *threshold of canonical adjunction* of a divisor D on the variety X is the number $c(D,X) = \sup\{\varepsilon \in \mathbb{Q}_+ | D + \varepsilon K_X \in A^1_+X\}$. If Σ is a non-empty linear system on X, then we set $c(\Sigma, X) = c(D,X)$, where $D \in \Sigma$ is an arbitrary divisor.

Example 1.1. (i) Let X be a *primitive Fano variety*, that is, a smooth projective variety with the ample anticanonical class and $\operatorname{Pic} X = \mathbb{Z}K_X$. For any effective divisor D we have $D \in |-nK_X|$ for some $n \geq 1$, so that $c(D,X) = n$. If we replace the condition $\operatorname{Pic} X = \mathbb{Z}K_X$ by the weaker one $\operatorname{rk}\operatorname{Pic}X = 1$, that is, $K_X = -rH$, where $\operatorname{Pic} X = \mathbb{Z}H$, $r \geq 2$ is the index of the variety X, then for $D \in |nH|$ we get $c(D,X) = \frac{n}{r}$.

This is a trivial example, because the space $A^1_{\mathbb{R}}X \cong \mathbb{R}$ is one-dimensional and $A^1_+X = \mathbb{R}_+$ is the positive ray, $K_X \in \mathbb{R}_-$.

(ii) Let $\pi\colon V \to S$ be a rationally connected fibre space with $\dim V > \dim S \geq 1$, Δ an effective divisor on the base S. Obviously, $c(\pi^*\Delta, V) = 0$. If $\operatorname{Pic} V = \mathbb{Z}K_V \oplus \pi^* \operatorname{Pic} S$, that is, V/S is a standard Fano fibre space, and D is an effective divisor on V, which is not a pullback of a divisor on the base S, then $D \in |-nK_V + \pi^*R|$ for some divisor R on S, where $n \geq 1$. Obviously, $c(D,V) \leq n$, and moreover, if the divisor R is effective, then $c(D,V) = n$. Indeed, K_V is negative on the fibres of the morphism π (in particular, on the dense families of rational curves sweeping out fibres of π), whereas any divisor, pulled back from the base, is trivial on the fibres.

(iii) Let $F_1, \dots, F_K$ be primitive Fano varieties, $V = F_1 \times \cdots \times F_K$ their direct product. Let $H_i = -K_{F_i}$ be the positive generator of the group $\operatorname{Pic} F_i$. Set

$$S_i = \prod_{j \neq i} F_i,$$

so that $V \cong F_i \times S_i$. Let $\rho_i\colon V \to F_i$ and $\pi_i\colon V \to S_i$ be the projections onto the factors. Abusing our notations, we write H_i instead of $\rho_i^* H_i$, so that

$$\operatorname{Pic} V = \bigoplus_{i=1}^{K} \mathbb{Z}H_i$$

and $K_V = -H_1 - \cdots - H_K$. For any effective divisor D on V we get

$$D \in |n_1H_1 + \cdots + n_KH_K|$$

for some non-negative $n_1, \ldots, n_K \in \mathbb{Z}_+$, and obviously

$$c(D, V) = \min\{n_1, \ldots, n_K\}.$$

This example can be reduced to the previous one: assume that $c(D, V) = n_1$ and set $n = n_1$, $\pi = \pi_1$, $F = F_1$, $S = S_1$. We get

$$\Sigma \subset |-nK_V + \pi^* Y|,$$

where $Y = \sum_{i=2}^K (n_i - n) H_i$ is an effective class on the base S of the fibre space $\pi\colon V \to S$. This is the case of Example 1.1 (ii) above.

The threshold of canonical adjunction is easy to compute, but the main disadvantage of this concept is that it is not a birational invariant.

Example 1.2. Let $\pi\colon \mathbb{P}^M \dashrightarrow \mathbb{P}^m$ be a linear projection from a $(M - m - 1)$-dimensional plane $P \subset \mathbb{P}^M$. Consider a mobile linear system Λ of hypersurfaces of degree n in $\mathbb{P}^m$ and let Σ be its pullback via π. Obviously, $c(\Sigma, \mathbb{P}^M) = \frac{n}{M+1}$. However, let us blow up the plane P, say $\sigma\colon \mathbb{P}^+ \to \mathbb{P}^M$, so that the composite map $\pi \circ \sigma\colon \mathbb{P}^+ \to \mathbb{P}^m$ is a $\mathbb{P}^{M-m}$-bundle. Let Σ^+ be the strict transform of Σ on $\mathbb{P}^+$. Since $\pi \circ \sigma$ is a morphism with rationally connected fibres, we get $c(\Sigma^+, \mathbb{P}^+) = 0$. This example can be easily generalized to linear projections of Fano complete intersections $V \subset \mathbb{P}^M$ of index 2 or higher.

To overcome birational non-invariance of the threshold of canonical adjunction, let us give

Definition 1.2. For a mobile linear system Σ on a variety X define the *virtual threshold of canonical adjunction* by the formula

$$c_{\text{virt}}(\Sigma) = \inf_{X^\sharp \to X} \{c(\Sigma^\sharp, X^\sharp)\},$$

where the infimum is taken over all birational morphisms $X^\sharp \to X$, $X^\sharp$ is a smooth projective model of $\mathbb{C}(X)$, $\Sigma^\sharp$ is the strict transform of the system Σ on $X^\sharp$.

The virtual threshold is obviously a birational invariant of the pair (X, Σ): if $\chi\colon X \dashrightarrow X^+$ is a birational map, $\Sigma^+ = \chi_* \Sigma$ is the strict transform of the system Σ with respect to χ^{-1}, then we get $c_{\text{virt}}(\Sigma) = c_{\text{virt}}(\Sigma^+)$.

Proposition 1.1. (i) *Assume that on the variety V there are no mobile linear systems with the zero virtual threshold of canonical adjunction. Then on V there are no structures of a non-trivial fibration into varieties of negative Kodaira dimension, that is, there is no rational dominant map $\rho\colon V \dashrightarrow S$, $\dim S \geq 1$, the generic fibre of which has negative Kodaira dimension.*

(ii) *Let $\pi\colon V \to S$ be a rationally connected fibre space. Assume that every mobile linear system Σ on V with the zero virtual threshold of canonical adjunction, $c_{\text{virt}}(\Sigma) = 0$, is the pullback of a system on the base: $\Sigma = \pi^* \Lambda$, where Λ is some mobile linear system on S. Then any birational map*

$$\begin{array}{ccc} V & \overset{\chi}{\dashrightarrow} & V^\sharp \\ \pi \downarrow & & \downarrow \pi^\sharp \\ S & & S^\sharp, \end{array} \tag{6}$$

where $\pi^\sharp\colon V^\sharp \to S^\sharp$ is a fibration into varieties of negative Kodaira dimension, is fibrewise, that is, there exists a rational dominant map $\rho\colon S \dashrightarrow S^\sharp$, making the diagram (6) commutative, $\pi^\sharp \circ \chi = \rho \circ \pi$. In other words, $\pi^\sharp \geq \pi$ in the sense of the order on the set of rationally connected structures: π is the least element of $RC(V)$.

Thus for certain rationally connected varieties the virtual threshold of canonical adjunction reduces the problem of describing the set $RC(V)$ to the same problem for the base S. This is a crucial step that in many cases leads to an exhaustive description of the set $RC(V)$. But the main disadvantage of the virtual thresholds is that they are extremely hard to compute.

The **proof** is obvious. Q.E.D.

To be precise, the only known way to compute them is by reduction to the ordinary thresholds.

Definition 1.3. (i) The variety V is said to be *birationally superrigid*, if for any mobile linear system Σ on V the following equality holds:

$$c_{\text{virt}}(\Sigma) = c(\Sigma, V).$$

(ii) The variety V (respectively, the Fano fibre space V/S) is said to be *birationally rigid*, if for any mobile linear system Σ on V there exists a birational self-map $\chi \in \operatorname{Bir} V$ (respectively, a fibrewise birational self-map $\chi \in \operatorname{Bir}(V/S)$), providing the equality

$$c_{\text{virt}}(\Sigma) = c(\chi_*\Sigma, V).$$

In the following examples we list the main classes of Fano varieties and Fano fibre spaces, for which birational rigidity or superrigidity is known today.

Example 1.3. (i) A smooth three-dimensional quartic $V = V_4 \subset \mathbb{P}^4$ is birationally superrigid: see Subsection 2.2, Corollary 2.2.

(ii) Generic smooth complete intersections $V_{2\cdot 3} \subset \mathbb{P}^5$ of cubic and quadric hypersurfaces are birationally rigid, but not superrigid. A description of their groups of birational self-maps is discussed below in Section 5 of the present chapter. Proof of birational rigidity is given in Sections 5 and 6.

(iii) A generic complete intersection $V_{d_1\cdot\dots\cdot d_K} \subset \mathbb{P}^{M+k}$ of index one (that is, $d_1+\dots+d_k = M+k$) and dimension $M \geq 4$ is birationally superrigid for $V \geq 2k+1$, as we will prove in Chapter 3; see Section 3.2, Theorem 3.2.1.

(iv) The first examples of birationally superrigid higher-dimensional Fano varieties were the double spaces $\sigma\colon V \stackrel{2:1}{\to} \mathbb{P}^M$, $M \geq 4$, branched over smooth hypersurfaces $W_{2M} \subset \mathbb{P}^M$ of degree $2M$ and the double quadrics $\sigma\colon V \stackrel{2:1}{\to} Q \subset \mathbb{P}^{M+1}$, $M \geq 4$, branched over smooth complete intersections $W = Q \cap W^*_{2(M-1)}$, where $W^*_{2(M-1)} \subset \mathbb{P}^{M+1}$ is a hypersurface of degree $2(M-1)$; see Subsection 2.2.

(v) Generalizing the previous example, let $\sigma\colon V \to Q \subset \mathbb{P}^{M+1}$ be a double cover, where $Q = Q_m \subset \mathbb{P}^{M+1}$ is a smooth hypersurface of degree m, and the branch divisor $W \subset Q$ is cut out on Q by a hypersurface $W^*_{2l} \subset \mathbb{P}^{M+1}$, where $m + l = M + 1$. The Fano variety V is birationally superrigid for general Q, W^* [**Pukh00e**]. Instead of a double cover an arbitrary cyclic cover could be considered, instead of a hypersurface $Q \subset \mathbb{P}^{M+1}$ a smooth complete intersection $Q \subset \mathbb{P}^{M+k}$ of suitable index and codimension $k < \frac{1}{2}M$. A general variety in each of these classes is birationally superrigid; see Subsection 3.2.3. Another example is given by iterated double covers [**Pukh03**]. All varieties of Example 1.3 are realized as Fano complete intersections in weighted projective spaces.

Conjecture 1.1. *A non-singular Fano complete intersection of index one and dimension ≥ 4 in a weighted projective space is birationally rigid, of dimension ≥ 5 birationally superrigid.*

To conclude, let us recall that for a primitive Fano variety birational (super)rigidity immediately implies the absence of non-trivial structures of a rationally connected fibre space.

1.2. Singularities of linear systems. The starting point of a proof of birational (super)rigidity of a rationally connected variety V: fix a mobile linear system Σ satisfying the inequality

$$c_{\text{virt}}(\Sigma) < c(\Sigma) \tag{7}$$

(if there are no such systems, then the variety is superrigid and there is nothing to prove). By definition, this means that there exists a birational morphism $\varphi\colon V^+ \to V$, for the inequality $c(\Sigma^+, V^+) < c(\Sigma)$ holds, where Σ^+ is the strict transform of the system Σ. Note that φ cannot be an isomorphism in codimension one (that is, outside a closed subset $Y_+ \subset V^+$ of codimension ≥ 2); otherwise for any effective divisor D on V and its strict transform D^+ we would have $c(D) = c(D^+)$. Therefore, there exist exceptional divisors $E \subset V^+$, contracted by the morphism φ (in the classical terminology, the exceptional divisor is the union of all irreducible, or prime, divisors, contracted by the morphism φ; for convenience of our exposition, we call every irreducible component an exceptional divisor, as well). Every such divisor determines a discrete valuation $\text{ord}_{\text{E}}(\cdot)$ of the field of rational functions $\mathbb{C}(V)$. This valuation does not depend on the model V^+ in the following sense: if $\varphi^\sharp\colon V^\sharp \to V$ is another birational morphism, where the birational map $(\varphi^\sharp)^{-1} \circ \varphi\colon V^+ \dashrightarrow V^\sharp$ is an isomorphism at the general point of the divisor E, so that $(\varphi^\sharp)^{-1} \circ \varphi(E) = E^\sharp \subset V^\sharp$ is an exceptional divisor of the morphism $\varphi^\sharp$, then $\text{ord}_{\text{E}} = \text{ord}_{\text{E}}^\sharp$. The irreducible subvariety $\varphi(E) \subset V$ is called the *centre* of the discrete valuation ord_{E} and also does not depend on the choice of the model V^+ (notation: $\text{centre}(E, V)$ or just $\text{centre}(E)$, when it is clear which model is meant). The valuations of the field $\mathbb{C}(V)$, realized by exceptional divisors of birational morphisms, are said to be *geometric*.

Since a divisor D on the variety V is given by local equations, applying the valuation ord_{E}, we obtain the *multiplicity* $\nu_E(D) \in \mathbb{Z}_+$ of an effective divisor D with respect to E. Obviously, $\nu_E(D) \geq 1$ if and only if the support of the divisor D contains $\text{centre}(E)$. If $\mathcal{E}$ is the set of exceptional divisors of the birational morphism φ, D^+ the strict transform of D on V^+, then we get

$$\varphi^* D = D^+ + \sum_{E \in \mathcal{E}} \nu_E(D) E. \tag{8}$$

For the canonical class K_{V^+} we get the presentation

$$K_{V^+} = \varphi^* K_V + \sum_{E \in \mathcal{E}} a(E) E, \tag{9}$$

where $a(E) = a(E, V) \geq 1$ is the discrepancy of the geometric valuation E, also independent of the model V^+. Recall that we work in the class of non-singular projective varieties.

By assumption we have $n = c(\Sigma) > 0$.

Definition 1.4. A geometric discrete valuation ord_E of the field $\mathbb{C}(V)$ is called a *maximal singularity* of the linear system Σ, if the *Noether-Fano* inequality

$$\nu_E(\Sigma) > na(E), \tag{10}$$

holds, where $\nu_E(\Sigma) = \nu_E(D)$ for a general divisor $D \in \Sigma$.

There is an especially simple and important class of maximal singularities.

Definition 1.5. An irreducible subvariety $Y \subset V$ of codimension ≥ 2 is called a *maximal subvariety* of the linear system Σ, if the inequality

$$\mathrm{mult}_Y \Sigma > n(\mathrm{codim}\, Y - 1)$$

holds, where $\mathrm{mult}_Y \Sigma = \mathrm{mult}_Y D$ for a general divisor $D \in \Sigma$.

Note that to define correctly the numbers $\nu_E(D)$ and $a(E)$, one needs only the variety V^+ to be non-singular at the generic point of the divisor E. For instance, in the definitions and constructions given above one can assume only that the singularities of the variety V^+ are of codimension ≥ 2 (this is true, if V^+ is normal). Taking this into account, assume that $Y \subset V$ is a maximal subvariety of the linear system Σ. Blow it up: $\varphi\colon V^+ \to V$. Set $E = \varphi^{-1}(Y)$ to be the exceptional divisor. We get $\nu_E(\Sigma) = \mathrm{mult}_Y \Sigma$ and $a(E) = \mathrm{codim}\, E - 1$, so that E realizes a maximal singularity of the system Σ. Therefore, maximal subvarieties represent a (simplest) type of maximal singularities.

Proposition 1.2. *Assume that the inequality* (7) *holds. Then the linear system* Σ *has a maximal singularity.*

Proof. Let $\varphi\colon V^+ \to V$ be a birational morphism (with a non-singular V^+), satisfying the inequality $c(\Sigma^+) < c(\Sigma) = n$, $\mathcal{E}$ the set of prime divisors, contracted by the morphism φ, $D \in \Sigma$ a general divisor, $D^+ \in \Sigma^+$ its strict transform on V^+. From the relations (8) and (9) we get

$$A^1_+ V^+ \not\ni D^+ + nK_{V^+} = \varphi^*(D + nK_V) - \sum_{E \in \mathcal{E}} e(E)E,$$

where $e(E) = \nu_E(D) - na(E)$. Since $D + nK_V \in A^1_+ V$ and the pull back of a pseudoeffective class is pseudoeffective, we obtain, that there exists at least one exceptional divisor E, for which $e(E) > 0$, which is what we need. Q.E.D.

Note that by the remark above, it is sufficient to assume that the variety V^+ is non-singular in codimension one, so that Proposition 1.2 does not depend on the theorem on resolution of singularities and for this reason holds in any characteristic.

With the exceptional divisor $E \subset V^+$ a uniquely determined sequence of blow ups is associated. Let X be a projective (possibly singular) variety, $\psi\colon V^+ \dashrightarrow X$ a birational map, contracting E to a subvariety $B = \psi(E) \subset X$ of codimension ≥ 2, whereas $B \not\subset \mathrm{Sing}\, X$is not contained entirely in the set of singular points. Let $\sigma_B\colon X(B) \to X$ be the blow up of the subvariety B, $E(B) = \sigma_B^{-1}(B)$ the exceptional divisor.

Proposition 1.3. *The following alternative holds: either the composition of birational maps* $\sigma_B^{-1} \circ \psi\colon V^+ \dashrightarrow X$ *is an isomorphism in a neighborhood of the generic point of the divisor* E, *and in this case* $\sigma_B^{-1} \circ \psi(E) = E(B)$, *or* $B^+ = \sigma_B^{-1} \circ \psi(E)$ *is an irreducible subvariety of codimension* ≥ 2 *and, moreover,* $B^+ \not\subset \mathrm{Sing}\, X(B)$, $B^+ \subset E(B)$ *and* $\sigma_B(B^+) = B$.

Proof. This is obvious: we have just listed the possible cases. Note, however, that the variety $X(B)$ is non-singular outside the σ_B-preimage of the set $\operatorname{Sing} X \cup \operatorname{Sing} B$ and $\sigma_B(B^+) = \sigma_B \circ \sigma_B^{-1} \circ \psi(E) = B$. Note also that outside the σ_B-preimage of the set $\operatorname{Sing} X \cup \operatorname{Sing} B$ the morphism $\sigma_B \colon E(B) \to B$ is a locally trivial $\mathbb{P}^{\operatorname{codim} B - 1}$-fibration and the discrepancy of the exceptional divisor $E(B)$ is $\operatorname{codim} B - 1$. Q.E.D.

Iterating the construction of Proposition 1.3, we obtain a sequence of blow ups

$$\begin{array}{rccc} \varphi_{i,i-1}\colon & V_i & \to & V_{i-1} \\ & \cup & & \cup \\ & E_i & \to & B_{i-1}, \end{array} \tag{11}$$

$i = 1, \ldots$, where $V_0 = V$, $B_0 = \operatorname{centre}(E, V)$ and, furthermore, B_j is the centre of E on V_j, $E_i = \varphi_{i,i-1}^{-1}(B_{i-1})$ is the exceptional divisor, B_{i-1} is the centre of the blow up $\varphi_{i,i-1}$. In other words, we succesively blow up the centres of the valuation E. The varieties $V_1, V_2, \ldots$ can, generally speaking, be singular, however, V_j is non-singular at the generic point of the subvariety $B_j \subset E_j$ for each $j = 1, \ldots$. For $i > j$ set

$$\varphi_{i,j} = \varphi_{j+1,j} \circ \cdots \circ \varphi_{i,i-1} \colon V_i \to V_j,$$

$\varphi_{i,i} = \mathrm{id}_{V_i}$. By (iterated) Proposition 1.3, $\varphi_{i,j}(B_i) = B_j$ for $i > j$. For an irreducible subvariety on V_j (say, $Y \subset V_j$) we denote its strict transform on V_i (provided it is well defined, that is, $Y \not\subset B_j$) by adding the upper index i (say, $Y^i \subset V_i$).

The same notation we use for effective algebraic cycles, say $Z^i = \Sigma m_k Z_k^i$ for a cycle $Z = \Sigma m_k Z_k$ on V_j. When we use this notation, it means that the strict transform is well defined.

Proposition 1.4. *The sequence of blow ups* (11) *terminates: for some* $K \geq 1$ *the first case of the alternative of Proposition 2.3 takes place, that is,* $\sigma_{K,0}^{-1} \circ \psi(E) = E_K$.

Proof. As we will see below, the discrepancies of the exceptional divisors E_i with respect to the model V strictly increase; in particular, $a(E_i, V) \geq i$, at the same time $a(E_i, V) \leq a(E, V)$, since the centre of E on V_i is contained in E_i. Q.E.D.

The sequence (11) of blow ups is called the resolution of the discrete valuation ν_E with respect to the model V. On the set of blow ups $\varphi_{i,i-1}$ or, more precisely, on the set of exceptional divisors $\{E_1, \ldots, E_K\}$ we introduce a structure of an oriented graph in the following way. The vertices E_i and E_j are joined by an oriented edge (an *arrow*), which is denoted as $i \to j$, if $i > j$ and $B_{i-1} \subset E_j^{i-1}$. The graph structure formalizes the operation of computing the strict transforms of exceptional divisors:

$$E_j^i = \varphi_{i,j}^* E_j - \sum_{j \leftarrow k \leq i} \varphi_{i,k}^* E_k.$$

In order to compute, the other way round, the pullback in terms of strict transforms, set, for $i > j$, p_{ij} to be the number of paths from E_i to E_j in the oriented graph described above, $p_{ij} \geq 1$. Also, set $p_{ij} = 1$.

Proposition 1.5. *The following decomposition holds:*

$$\varphi^*_{i,j}E_j = \sum_{k=j}^{i} p_{kj}E^i_k. \tag{12}$$

Proof is given by induction on $i \geq j$. If $i = j$, then there is nothing to prove. If $i = j+1$, then $\varphi^*_{j+1,j}E_j = E^{j+1}_j + E_{j+1}$, since $B_j \subset E_j$ and the divisor E_j is non-singular at the generic point of B_j. Now for $i \geq j+2$ we get:

$$\varphi^*_{i,j}E_j = \varphi^*_{i,i-1}(\varphi^*_{i-1,j}E_j) = \varphi^*_{i,i-1}\left(\sum_{k=j}^{i-1} p_{kj}E^{i-1}_k\right)$$

$$= \sum_{k=j}^{i-1} p_{kj}E^i_k + \left(\sum_{\substack{k=j,\\ B_{i-1}\subset E^{i-1}_k}}^{i-1} p_{kj}\right)E_i.$$

It is easy to see that the following equality holds:

$$p_{ij} = \sum_{i\to k} p_{kj}$$

(in each path from i to j mark *the first* vertex of the graph after i: $i \to k \to \cdots \to j$). Q.E.D. for the proposition.

The combinatorial invariants p_{ij} give explicit presentations for multiplicities and discrepancies. Let Σ^j be the strict transform of the linear system Σ on V_j. Set $\nu_j = \operatorname{mult}_{B_{j-1}} \Sigma^{j-1}$ and $\beta_j = \operatorname{codim} B_{j-1} - 1$. We get

$$\nu_{E_K}(\Sigma) = \nu_E(\Sigma) = \sum_{i=1}^{K} p_{K_i}\nu_i, \quad a(E) = \sum_{i=1}^{K} p_{K_i}\beta_i.$$

Setting for convenience of notations $p_i = p_{K_i}$, we obtain the traditional form of the Noether-Fano inequality:

$$\sum_{i=1}^{K} p_i\nu_i > n\sum_{i=1}^{K} p_i\beta_i. \tag{13}$$

Remark 1.1. From the arguments above one can see that our arguments make sense in a more general context. The variety V may be singular, however, first, it is necessary that at least some multiplicity ND of any Weil divisor D on V is given locally by one equation (in order to be able to pull divisors back with respect to morphisms), and second, for any blow ups all discrepancies of exceptional divisors should be positive. Thus the Noether-Fano inequality and the concept of maximal singularity make sense for varieties with $\mathbb{Q}$-factorial terminal singularities.

Remark 1.2. The Noether-Fano inequality can be reformulated in the following way on the language of $\mathbb{Q}$-divisors (that is, linear combinations of prime divisors with rational coefficients). Let $D \in \Sigma$ be a general divisor. Then the log pair $(V, \frac{1}{n}D)$ is not canonical, that is, has a non-canonical singularity $E \subset V^+$, satisfying the inequality $\nu_E(\frac{1}{n}D) > a(E)$. Recall that a log pair (V, Z), where Z is an effective $\mathbb{Q}$-divisor, is *canonical*, if for any geometric discrete valuation ν_E the inequality $\nu_E(Z) \leq a(E)$ holds. The pair (V, Z) is *terminal* (respectively, *log* *terminal* and *log canonical*), if for any ν_E the inequality $\nu_E(Z) < a(E)$ holds (respectively, $\nu_E(Z) < a(E)+1$ and $\nu_E(Z) \leq a(E)+1$). These concepts work in the modern advanced methods of higher-dimensional birational geometry.

As an illustration of these concepts, let us prove

Proposition 1.6. *Let V be a primitive Fano variety, V' a Fano variety with $\mathbb{Q}$-factorial terminal singularities and Picard number one (that is,* $\operatorname{Pic} V' \otimes \mathbb{Q} = \mathbb{Q}K_{V'}$*), $\chi\colon V \dashrightarrow V'$ a birational map.*

(i) *Assume that V is birationally rigid. Then V and V' are (biregularly) isomorphic (although the map χ itself is, generally speaking, not an isomorphism).*

(ii) *Assume that V is birationally superrigid. Then χ is a biregular isomorphism. In particular, the groups of birational and biregular self-maps of the variety V coincide:* $\operatorname{Bir} V = \operatorname{Aut} V$.

Proof. (i) Let $\chi\colon V \dashrightarrow V'$ be a birational map, $\varphi\colon Y \to V$ its Hironaka resolution, so that $\psi = \chi \circ \varphi\colon Y \to V'$ is a birational morphism. The variety Y is non-singular and

$$\operatorname{Pic} Y = \mathbb{Z}\varphi^* K_V \oplus \bigoplus_{i \in I} \mathbb{Z}E_i,$$

where $\{E_i \mid i \in I\}$ is the set of all the φ-exceptional divisors. By assumption,

$$\operatorname{Pic} Y \otimes \mathbb{Q} = \mathbb{Q}\psi^* K_{V'} \oplus \bigoplus_{j \in J} \mathbb{Q}E'_j,$$

where $\{E'_j \mid j \in J\}$ is the set of all the ψ-exceptional divisors. For simplicity of notations set $K = \varphi^* K_V$, $K' = \psi^* K_{V'}$. We get

$$K_Y = K + \sum_{i \in I} a_i E_i = K' + \sum_{j \in J} a'_j E'_j, \tag{14}$$

where $a_i \in \mathbb{Z}$, $a_i \geq 1$, and $a'_j \in \mathbb{Q}$, $a'_j > 0$. Let $\Sigma' = |-mK_{V'}|$, $m \gg 0$, be a very ample system. Obviously, $c(\Sigma', V') = m$. Take its strict transform $\Sigma = \chi_*^{-1}\Sigma' \subset |-nK_V|$; obviously, $c(\Sigma, V) = n$. Twisting by a suitable birational self-map, we may assume that the equality of the virtual and actual thresholds of canonical adjunction already holds for χ. Therefore, $n \leq m$. The strict transform of the linear system Σ on Y coincides with the strict transform of the linear system Σ' with respect to ψ. Therefore, there exist positive integers b_i, $i \in I$, such that

$$-mK' = -nK - \sum_{i \in I} b_i E_i.$$

Dividing by m and substituting into (14), we get

$$\left(1 - \frac{n}{m}\right) K = \sum_{i \in I} \left(\frac{b_i}{m} - a_i\right) E_i + \sum_{j \in J} a'_j E'_j.$$

Since the divisors E_i are φ-exceptional and $a'_j > 0$, we get the equality $n = m$: otherwise we get a contradiction with the ampleness of the class $(-K_V)$. Furthermore, all the divisors E'_j turn out to be φ-exceptional and, moreover, $\{E_i | i \in I\} = \{E'_j | j \in J\}$, otherwise $\operatorname{rk} \operatorname{Pic} V' \geq 2$. Thus χ is an isomorphism in codimension one: set

$$U = V \setminus \bigcup_{i \in I} \varphi(E_i), \quad U' = V' \setminus \bigcup_{j \in J} \psi(E'_j),$$

then $\chi\colon U \to U'$ is an isomorphism. Therefore, $\Sigma = |-nK_V|$ and χ induces an isomorphism of the linear systems Σ and Σ'. Consequently, $\chi\colon V \to V'$ is an isomorphism. (Strictly speaking, we have proven that for an arbitrary birational map $\chi\colon V \dashrightarrow V'$ there exists a birational self-map $\chi^* \in \operatorname{Bir} V$, such that $\chi \circ \chi^*$ is an isomorphism.)

The claim (ii) is now obvious. Q.E.D. for Proposition 1.6.

1.3. The general structure of the method. Now having a rigorous definition of a maximal singularity, we can formulate precisely the main steps of the method of maximal singularities, described in Section 3.3 of Chapter 1. For every geometric discrete valuation $\nu = \nu_E$ of the field $\mathbb{C}(V)$ (realized by an exceptional divisor $E \subset V^+$ on some model $V^+ \to V$) one should answer the question:

Does there exist a mobile linear system Σ with a threshold of canonical adjunction $n = c(\Sigma) > 0$, for which ν_E is a maximal singularity (that is, the inequality (10) *holds)?*

A procedure of obtaining the *negative* answer to this question is called *exclusion* of the given maximal singularity. As we mentioned above, this is the hardest part of the work. The technical methods of exclusion, that are known today, are considered in detail below.

If the answer to the question above is *positive*, one should produce a mobile linear system Σ with the maximal singularity E. The next step in this case is *untwisting* the maximal singularity E. In the simplest form the untwisting procedure is as follows: one constructs a birational self-map $\tau_E \in \operatorname{Bir} V$, such that for any linear system Σ with the maximal singularity E the inequality

$$c((\tau_E^{-1})_*\Sigma) < (\Sigma),$$

holds, where $(\tau_E^{-1})_*\Sigma$ is the strict transform of Σ with respect to τ_E, and E is no longer a maximal singularity of the system $(\tau_E^{-1})_*\Sigma$. Thus τ_E "untwists" (removes) the maximal singularity E, at the same time decreasing the threshold of canonical adjunction.

Let V be a primitive Fano variety. Assume that for V the scheme described above is successfully realized. Let $\mathcal{M}$ be the set of discrete valuations, for which the answer to the question above is positive, that is, any valuation E is realized as a maximal singularity if and only if $E \in \mathcal{M}$. Under these assumptions we get

Theorem 1.1. (i) *The variety V is birationally rigid.*

(ii) *The group of birational self-maps* $\operatorname{Bir} V$ *is generated by the subgroup of biregular automorphisms* $\operatorname{Aut} V$ *and the subgroup $B(V)$, generated by the untwisting transformations τ_E, $E \in \mathcal{M}$.*

(iii) *If $\mathcal{M} = \emptyset$, then V is birationally superrigid.*

Proof. On a primitive Fano variety the values of the threshold of canonical adjunction are positive integers only. For this reason, the untwisting procedure cannot be infinite: for a sequence of maximal singularities $E_1, \dots, E_N$ the birational self-map $\tau = \tau_{E_1} \circ \cdots \circ \tau_{E_N}$ transforms the system Σ into the system $\Sigma = (\tau^{-1})_*\Sigma$ with no maximal singularities and therefore by Proposition 2.2, we get

$$c(\Sigma^\tau) = c_{\mathrm{virt}}(\Sigma^\tau) = c_{\mathrm{virt}}(\Sigma),$$

which means birational rigidity. If $\mathcal{M} = \emptyset$, then we have this coincidence of thresholds from the start, which proves (iii). It remains to show (ii). For an arbitrary birational self-map $\chi \in \operatorname{Bir} V$ consider a very ample linear system $\Sigma^* = |-mK_V|$ and its strict transform $\Sigma^\chi = (\chi^{-1})_* \Sigma^*$ with respect to χ. If $c(\Sigma^\chi) = c_{\text{virt}}(\Sigma^\chi) = c_{\text{virt}}(\Sigma^*) \leq m$, then $\Sigma^\chi = \Sigma^*$ (say, arguing by the dimension $\dim \Sigma^\chi$), so that $\chi \in \operatorname{Aut} V$ is a biregular automorphism. Thus the procedure of untwisting maximal singularities proves (ii) simultaneously with (i). Q.E.D. for the theorem.

Remark 1.3. In all known cases the untwisting birational maps τ_E are involutions, canonically determined by the corresponding maximal singularities, so that, in particular, for any biregular automorphism $\alpha \in \operatorname{Aut} V$ the equality $\alpha^{-1}\tau_E\alpha = \tau_{\alpha(E)}$ holds and the subgroup $B(V)$ is normal, whereas $\operatorname{Bir} V$ is the extension

$$1 \to B(V) \to \operatorname{Bir} V \to \operatorname{Aut} V \to 1.$$

Furthermore, from the description of the untwisting procedure it is clear that *relations* between the generators τ_E, $E \in \mathcal{M}$, can emerge if and only if the untwisting is not uniquely determined, that is, some mobile linear system Σ can have two distinct maximal singularities $E_1 \neq E_2$ simultaneously. If, on the contrary, this is impossible, then the group $B(V)$ is a *free* product of the cyclic groups $\langle \tau_E \rangle$ (in particularly, it is very big). The argument given above presents a somewhat simplified picture. A precise description is given below, the example of computing the group of birational self-maps $\operatorname{Bir} V_{2\cdot 3}$ of the Fano complete intersection $V_{2\cdot 3} \subset \mathbb{P}^5$.

Remark 1.4. The procedure of untwisting maximal singularities, described above, is satisfactory for by no means all varieties. It is a typical situation when a maximal singularity E is realized by a mobile linear system Σ, but cannot be untwisted by a birational self-map: such a self-map τ_E simply does not exist. The experience of studying such singularities in practice has long ago formed a general belief that in this case the singularity E corresponds to *another structure* of a rationally connected fibre space or Fano variety on V, that is, E is untwisted by a birational map $V \dashrightarrow V^\sharp$. These considerations, based on a huge empirical material, make the essence of *Sarkisov program*, the theory of factorization of birational maps into a composition of very special birational maps, *elementary links*, discussed in more detail in Chapter 4. In dimension three Sarkisov's program is a completely proven theorem, the full details and outlines of which were published many times [**Co95, I01, Ch05c**]. The recent progress of the log minimal model program makes it realistic to hope that the higher-dimensional version of Sarkisov program will be completed in the reasonable future as well. However, we would like to emphasize that this is a general theorem of existence, which itself gives no description of birational geometry of a particular variety V. As a hundred years ago, the main problem is to find the realizable maximal singularities and construct the untwisting maps (the latter never makes a difficult problem for an explicitly defined variety). For this reason, in the present book we restrict ourselves by a simple version of the method of maximal singularities, when untwisting is done by birational *self-maps*. So, one should take into account that there are very few completely studied examples of varieties (except for conic bundles) with links onto another models: as a rule, varieties with non-trivial links have a rich birational geometry and the modern technique is not strong enough for them.

Remark 1.5. We mentioned above, certain versions of the concept of a maximal singularity that came from the minimal model program. If a valuation ν_E satisfies the non-strict Noether-Fano inequality, that is, $\nu_E(\Sigma) \geq na(E)$, then the pair $(V, \frac{1}{n}D)$ is not terminal. If we know such singularities E (in fact, we discuss the Noether-Fano *equality* $\nu_E = na(E)$, since otherwise we get the usual concept of a maximal singularity), then we are able to describe *the structures of a K-trivial fibre space* on V, that is, birational maps $\chi\colon V \dashrightarrow V^\sharp$, where $V^\sharp$ is equipped by a morphism $\pi\colon V^\sharp \to S^\sharp$, the fibre of general position of which has the trivial canonical class, and also birational maps $\chi\colon V \dashrightarrow V^\sharp$ onto Fano varieties with canonical singularities. In particular, a simple modification of the proof of Proposition 1.6 (in the end of the argument one should take into account that the exceptional divisors E_i come with *strictly positive coefficients*) gives

Proposition 1.7. *Assume that on a primitive Fano variety V there are no mobile linear systems with maximal singularities, $\mathcal{M} = \emptyset$, whereas any mobile linear system Σ, satisfying the non-strict Noether-Fano inequality for some ν_E, determines a map $V \dashrightarrow S$ onto a variety of a smaller dimension. Then any birational map $\chi\colon V \dashrightarrow V^\sharp$ onto a variety $V^\sharp$ with canonical singularities (with no restrictions on the rank of the Picard group $\rho(V^\sharp)$) is a biregular isomorphism.*

2. Exclusion of maximal singularities

In this section we consider the problem of exclusion of a maximal singularity $E \subset V^+$ of a mobile linear system Σ on a smooth variety V. The problem breaks naturally into exclusion of maximal subvarieties (Subsection 2.1) and exclusion of an infinitely near maximal singularity. In Subsection 2.2 we develop the technique of counting multiplicities and prove the key local fact, on which the procedure of exclusion is based in most cases, that is, the *$4n^2$-inequality*, in Subsection 2.3 we give another proof of this inequality, based on the connectedness principle of Shokurov and Kollár.

2.1. Maximal subvarieties. We use the notations and constructions of Subsection 1.2. A singularity $E \subset V^+$ and its resolution are fixed. We want to prove that E is not a maximal singularity of any mobile linear system Σ with $n = c(\Sigma) > 0$. The methods of proof are essentially different for different types of singularities and different classes of varieties. First of all, let us break the set of singularities into two natural types. In the notations of Section 1.2 the dimensions of the blown up subvarieties $\dim B_i$, $i = 0, 1, \dots, N-1$, do not decrease. Therefore, the following alternative holds: either $\dim B_0 = \dots = \dim B_{N-1}$, or $\dim B_0 < \dim B_{N-1}$. In the latter case we say that E is an *infinitely near* singularity.

Proposition 2.1. *Assume that E is a maximal singularity of the mobile linear system Σ and* $\dim B_0 = \dim B_{N-1}$. *Then the subvariety $B = B_0$ (the centre of the singularity E on V) is a maximal subvariety of the system Σ.*

Proof. The claim follows obviously from the explicit form (13) of the Noether-Fano inequality, the equality of local discrepancies $\beta_1 = \dots = \beta_N = \operatorname{codim} B - 1$ and non-decreasing of multiplicities $\nu_1 \geq \nu_2 \geq \dots \geq \nu_N$. Q.E.D.

Corollary 2.1. *If the centre $B \subset V$ of the maximal singularity E is not a maximal subvariety of the system Σ, then E is an infinitely near maximal singularity.*

It is exclusion of the infinitely near case that causes the biggest problem. Now let us consider the principal methods of excluding maximal subvarieties. The following two methods are most useful.

Assume that $\dim V \geq 4$ and $A^2V = \mathbb{Z}K_V^2$, where A^2V is the group of classes of codimension two cycles modulo *numerical* equivalence. By the Lefschetz theorem, this condition is satisfied, for instance, by all Fano complete intersections of index 1 and dimension ≥ 5 in the projective space (in fact, in the weighted projective space; this class can be extended further and include families of singular varieties). Let $B \sim mK_V^2$ be an irreducible subvariety of codimension two.

Proposition 2.2. *The subvariety B cannot be a maximal subvariety of a mobile linear system $\Sigma \subset |-nK_V|$.*

Proof. Assume the converse. Let $D_1, D_2 \in \Sigma$ be general divisors. The effective cycle $Z = (D_1 \circ D_2)$ of the scheme-theoretic intersection of D_1 and D_2 is well defined, since $\mathrm{codim}(D_1 \cap D_2) = 2$ by generality of these divisors. The cycle Z (depending, of course, on the choice of D_1, D_2) is called the *self-intersection* of the system Σ. We get

$$n^2 K_V^2 \sim Z = \alpha B + Y,$$

where Y is an effective cycle of codimension two and $\alpha > n^2$. We get a contradiction at once. Q.E.D. for the proposition.

If V is an arbitrary primitive Fano variety (without any additional assumptions about A^2V), then the argument given above makes it possible to bound the set of maximal subvarieties of codimension two.

More precisely, let $Y \subset V$ be an irreducible subvariety (of arbitrary codimension). The number

$$\deg Y = (Y \cdot (-K_V)^{\dim Y}) \in \mathbb{Z}_+$$

is called the (anticanonical) degree of Y. In particular, $\deg V = (-K_V)^{\dim V}$ is the degree of the variety V itself. In the notations of the proof of Proposition 2.2, computing $\deg Z = n^2 \deg V \geq \alpha \deg B$, we get the estimate

$$\deg B < \deg V. \tag{15}$$

For instance, on a three-dimensional quartic $V_4 \subset \mathbb{P}^4$ the degree of a maximal curve can take the values 1, 2 and 3; it is not difficult to show that none of these possibilities can be realized. However, the essence of the estimate (15) is in the fact that the class of subvarieties of codimension 2 that can be maximal subvarieties of mobile linear systems becomes bounded and observable and so the possible cases can be inspected by hand.

The most powerful method of excluding maximal singularities is the following *cone method.* Let us describe it on the example of a smooth Fano hypersurface $V = V_M \subset \mathbb{P}^M$.

Lemma 2.1. *For any curve $C \subset V$,*

$$\mathrm{mult}_C\, \Sigma \leq n.$$

Proof. Let us use the following construction.

Proposition 2.3. *Let $C \subset W$, where $W \subset \mathbb{P}^M$ is a smooth hypersurface. Let $x \in \mathbb{P}^M \backslash W$ be a point of general position, $C(x) \subset \mathbb{P}^M$ the cone with the vertex x*

and the base C. Then

$$C(x) \cap W = C \cup R(x),$$

where the residual curve $R(x)$ intersects C at $\deg R(x)$ distinct points.

Proof. On the surface S the curve C is numerically equivalent to a hyperplane section. However, we need to prove that for a general point x all points of intersection of R and C are of multiplicity 1.

Obviously, C meets $R(x)$ at precisely those points where C intersects the ramification divisor $W_x \subset W$ of the projection $\pi\colon W \to \mathbb{P}^{M-1}$ from the point x. The hypersurface W_x is cut out on W by the hypersurface

$$F_x = \sum_{i=0}^{M} \frac{\partial F}{\partial z_i} x_i = 0,$$

where $(z_0 : \cdots : z_M)$ are homogeneous coordinates on $\mathbb{P}^M$, $F(z_0, \ldots, z_M)$ is the equation of W with respect to those coordinates, and $(x_0 : \cdots : x_M)$ is the point x. But the variety W is non-singular, so that the linear system

$$\left| \sum_{i=0}^{M} \lambda_i \frac{\partial F}{\partial z_i} \right|$$

is base point free. Thus for a general point x the intersection $C \cap W_x$ consists of $(\deg W - 1) \deg C = \deg R(x)$ distinct points, which is what we need. Q.E.D. for the proposition.

Let us come back to the proof of Lemma 2.1. Let $a \in \mathbb{P} \setminus V$ be a general point, $R(a)$ the residual curve of the intersection of the cone $C(a)$ and the hypersurface V. Then the restriction of the linear system $|\chi|$ on $R(a)$ is a linear series of degree $n \cdot \deg R(a)$ with $\deg R(a)$ fixed points of multiplicity $\mathrm{mult}_C |\chi|$. Q.E.D. for the lemma.

The cone method makes it possible to estimate multiplicities of subvarieties of higher codimension, not only divisors [**Pukh02b**], and also exclude maximal singularities on different types of Fano hypersurfaces and complete intersections [**CPR, Ch05a, Pukh06b**].

There are also other methods of excluding maximal subvarieties of positive dimension (see, for instance, [**IM, Pukh87, Pukh89a, Pukh00e**]). All of them, in some way, realize the same idea: to find on the variety V a mobile subvariety S of a small dimension with a possibly bigger intersection with the maximal subvariety, and then to restrict the system Σ onto S. One gets an effective divisor on S with "too big" singularities, which gives a contradiction; for particular examples, see the papers listed above.

2.2. The $4n^2$-inequality. Let us consider the infinitely near case. Let $E \subset V^+$ be an infinitely near singularity of the system Σ, where (in the notations of Section 1.2, which we use without special reservations) $\dim B_0 < \dim D_{N-1}$. In particular, $\mathrm{codim}\, B_0 \geq 3$. Set $B = B_0$ and consider the self-intersection $Z = (D_1 \circ D_2)$ of the system Σ. Recall that $n = c(\Sigma) > 0$ is the threshold of canonical adjunction and the Noether-Fano inequality (10) holds (in the explicit traditional form (13)).

Theorem 2.1 (the $4n^2$-inequality). *The following estimate holds:*

$$\mathrm{mult}_B Z > 4n^2. \tag{16}$$

Corollary 2.2. *A smooth three-dimensional quartic $V = V_4 \subset \mathbb{P}^4$ is birationally superrigid.*

Proof. Assume that $\Sigma \subset |nH|$ is a mobile linear system on V with a maximal singularity. By Lemma 2.1, the centre of the maximal singularity is not a curve, hence it is a point $p \in V$. Therefore, for the self-intersection $Z = (D_1 \circ D_2)$ of Σ we have the inequality (16) with $B = p$. But Z is an effective 1-cycle of degree $4n^2$ in $\mathbb{P}^4$, so this is impossible. This contradiction excludes the maximal singularity and proves birational superrigidity. Q.E.D.

As another application, let us consider two infinite series of smooth Fano varieties of index 1:

(1) Fano double space $V \stackrel{2:1}{\to} \mathbb{P}^M$, branched over smooth hypersurfaces $W_{2M} \subset \mathbb{P}^M$ of degree $2M$ (they can be defined as hypersurfaces of degree $2M$ in the weighted projective space $\mathbb{P}(a_0, a_1, \ldots, a_M, a_{M+1}) = \mathbb{P}(1, 1, \ldots, 1, M)$) and

(2) Fano double quadrics $V \stackrel{2:1}{\to} Q = Q_2 \subset \mathbb{P}^{M+1}$ branched over a smooth divisor $W^* = W_{2(M-1)} \cap Q$, where Q is a smooth quadric hypersurface (they can be defined as complete intersections of the type $2 \cdot 2(M-1)$ in the weighted projective space $\mathbb{P}(a_0, \ldots, a_{M+1}, a_{M+2}) = \mathbb{P}(1, \ldots, 1, M-1)$).

For the both classes the anti-canonical linear system $|H| = |-K_V|$ is base point free and determines the double cover. Moreover,

$$\deg V = H^M \leq 4,$$

and for that reason we can apply Theorem 2.1 and obtain

Corollary 2.3. *Fano double spaces for $M \geq 3$ and Fano double quadrics for $M \geq 4$ are birationally superrigid.*

Proof. Given that a mobile linear system $\Sigma \subset |nH|$ cannot have an infinitely near maximal singularity by the $4n^2$-inequality, it remains to show that Σ cannot have a maximal subvariety of codimension 2. This is an easy exercise, left to the reader (or see [**Pukh89a, Pukh00c**]). Q.E.D.

The double quadrics of dimension 3 are marginally more difficult: there are no infinitely near maximal singularities, but maximal curves do exist. It is easy to show that only lines not contained in the ramification divisor of the double cover can occur as maximal subvarieties, and indeed with each of them a non-trivial birational involution is associated. We leave the proof of birational rigidity to the reader (or see [**I80**]).

Now, given the importance of the claim of Theorem 2.1, let us come back to the proof.

Proof of Theorem 2.1. Consider first the following general situation.

Let $B \subset X, B \not\subset \operatorname{Sing} X$ be an irreducible cycle of codimension ≥ 2, let $\sigma_B \colon X(B) \to X$ be its blowing up, and $E(B) = \sigma_B^{-1}(B)$ the exceptional divisor. Let $Z = \sum m_i Z_i$, $Z_i \subset E(B)$ be a k-cycle, $k \geq \dim B$. We define the *degree* of Z setting

$$\deg Z = \sum_i m_i \deg \left(Z_i \bigcap \sigma_B^{-1}(b) \right),$$

where $b \in B$ is a generic point, $\sigma_B^{-1}(b) \cong \mathbb{P}^{\operatorname{codim} B - 1}$ and the right-hand side degree is the ordinary degree in the projective space.

Note that $\deg Z_i = 0$ if and only if $\sigma_B(Z_i)$ is a proper closed subset of B.

Our computations are based on the following statement.

Let D and Q be two different prime Weil divisors on X, D^B and let Q^B be their proper inverse images on $X(B)$. In general, for any subvariety $Y \subset X$, its strict transform on $X(B)$, is denoted by the symbol Y^B.

Lemma 2.2. (i) *Assume that* $\operatorname{codim} B \geq 3$. *Then*

$$D^B \circ Q^B = (D \circ Q)^B + Z,$$

where $\circ$ *stands for the cycle of the scheme-theoretic intersection,* $\operatorname{Supp} Z \subset E(B)$, *and*

$$\operatorname{mult}_B(D \circ Q) = (\operatorname{mult}_B D)(\operatorname{mult}_B Q) + \deg Z.$$

(ii) *Assume that* $\operatorname{codim} B = 2$. *Then*

$$D^B \circ Q^B = Z + Z_1,$$

where $\operatorname{Supp} Z \subset E(B)$, $\operatorname{Supp} \sigma_B(Z_1)$ *does not contain* B, *and*

$$D \circ Q = [(\operatorname{mult}_B D)(\operatorname{mult}_B Q) + \deg Z]\, B + (\sigma_B)_* Z_1.$$

Proof follows easily from the standard intersection theory.

Now let us come back to our discrete valuation ν.

We divide the resolution $\varphi_{i,i-1} : X_i \to X_{i-1}$ into the *lower part*, for the indices $i = 1, \ldots, L \leq K$, corresponding to the blow ups with $\operatorname{codim} B_{i-1} \geq 3$, and the *upper part*, for the indices $i = L+1, \ldots, K$, corresponding to the blow ups with $\operatorname{codim} B_{i-1} = 2$. It may occur that $L = K$ and the upper part is empty.

Let $D_1, D_2 \in \Sigma$ be two different general divisors. We define a sequence of codimension 2 cycles on X_i's setting

$$\begin{aligned} &D_1 \circ D_2 = Z_0, \\ &D_1^1 \circ D_2^1 = Z_0^1 + Z_1, \\ &\vdots \\ &D_1^i \circ D_2^i = (D_1^{i-1} \circ D_2^{i-1})^i + Z_i, \\ &\vdots \end{aligned}$$

where $Z_i \subset E_i$. Thus for any $i \leq L$ we get

$$D_1^i \circ D_2^i = Z_0^i + Z_1^i + \cdots + Z_{i-1}^i + Z_i.$$

For any $j > i, j \leq L$ set

$$m_{i,j} = \operatorname{mult}_{B_{j-1}}(Z_i^{j-1})$$

(the multiplicity of an irreducible subvariety along a smaller subvariety is understood in the usual sense; for an arbitrary cycle we extend the multiplicity by linearity).

Set $d_i = \deg Z_i$.

We get the following system of equalities:

$$\begin{aligned}
\nu_1^2 + d_1 &= m_{0,1}, \\
\nu_2^2 + d_2 &= m_{0,2} + m_{1,2}, \\
&\vdots \\
\nu_i^2 + d_i &= m_{0,i} + \cdots + m_{i-1,i}, \\
&\vdots \\
\nu_L^2 + d_L &= m_{0,L} + \cdots + m_{L-1,L}.
\end{aligned}$$

Now

$$d_L \geq \sum_{i=L+1}^{K} \nu_i^2 \deg(\varphi_{i-1,L})_* B_{i-1} \geq \sum_{i=L+1}^{K} \nu_i^2.$$

Definition 2.1. A function $a : \{1, \ldots, L\} \to \mathbb{R}_+$ is said to be *compatible with the graph structure*, if

$$a(i) \geq \sum_{j \to i} a(j)$$

for any $i = 1, \ldots, L$.

We will actually use only one compatible function, namely $a(i) = p_{Ki} = p_i$.

Proposition 2.4. *Let $a(\cdot)$ be any function, compatible with the graph structure. Then*

$$\sum_{i=1}^{L} a(i) m_{0,i} \geq \sum_{i=1}^{L} a(i) \nu_i^2 + a(L) \sum_{i=L+1}^{K} \nu_i^2.$$

Proof. Multiply the i-th equality by $a(i)$ and put them all together: in the right-hand side for any $i \geq 1$ we get the expression

$$\sum_{j \geq i+1} a(j) m_{i,j}.$$

In the left-hand side for any $i \geq 1$ we get the summand $a(i)d_i$.

Lemma 2.3. *If $m_{i,j} > 0$, then $i \to j$.*

Proof. If $m_{i,j} > 0$, then some component of Z_i^{j-1} contains B_{j-1}. But $Z_i^{j-1} \subset E_i^{j-1}$. Q.E.D.

Now let us compare the multiplicities m_{ij} with the corresponding degrees.

Lemma 2.4. *For any $i \geq 1, j \leq L$ we have $m_{i,j} \leq d_i$.*

Proof. The cycles B_a are non-singular at their generic points. But since the maps $\varphi_{a,b}\colon B_a \to B_b$ are surjective, we can count multiplicities at generic points. Taking into account that the multiplicities are non-increasing with respect to blowing up of a non-singular subvariety, we reduce the claim to the obvious case of a hypersurface in a projective space. Q.E.D. for the lemma.

As a result we get the following estimate:

$$\sum_{j \geq i+1} a(j) m_{i,j} = \sum_{\substack{j \geq i+1 \\ m_{i,j} \neq 0}} a(j) m_{i,j} \leq d_i \sum_{j \to i} a(j) \leq a(i) d_i.$$

By what was said above, we can throw away all the $m_{i,*}, i \geq 1$, from the right-hand side, and all the $d_i, i \geq 1$, from the left-hand side, replacing the equality sign $=$ by the inequality sign $\leq$. Q.E.D. for the proposition.

Corollary 2.4. *Set $m = m_{0,1} = \text{mult}_B(D_1 \circ D_2), D_i \in \Sigma$. Then the following inequality holds:*

$$m\left(\sum_{i=1}^{L} a(i)\right) \geq \sum_{i=1}^{L} a(i)\nu_i^2 + a(L) \sum_{i=L+1}^{K} \nu_i^2.$$

Corollary 2.5. *The following inequality holds:*

$$m\left(\sum_{i=1}^{L} p_i\right) \geq \sum_{i=1}^{K} p_i \nu_i^2.$$

Proof. For $i \geq L+1$ obviously $p_i \leq p_L$. Q.E.D.

Taking into account the Noether-Fano inequality for the valuation ν, we see that the right-hand side here is strictly greater than the value of the quadratic form $\sum_{i=1}^{K} p_i \nu_i^2$ at the point

$$\nu_1 = \cdots = \nu_K = \frac{\sum\limits_{i=1}^{K} p_i \delta_i n}{\sum\limits_{i=1}^{K} p_i}.$$

Now set

$$\Sigma_l = \sum_{\delta_j \geq 2} p_j, \qquad \Sigma_u = \sum_{\delta_j = 1} p_j.$$

In these notations we get:

$$\text{mult}_B Z > \frac{(2\Sigma_l + \Sigma_u)^2}{\Sigma_l(\Sigma_l + \Sigma_u)} n^2.$$

Now easy computations show that the right-hand side is not smaller than $4n^2$.

Q.E.D. for Theorem 2.1.

2.3. Another proof of the $4n^2$-inequality: the connectedness principle. Because of the importance of Theorem 2.1, we give here another proof of that fact. The idea is to reduce Theorem 2.1 to a known fact about the graphs of blow ups on non-singular surfaces. It is sufficient to consider the case when $B = o$ is a point (take a general germ of a smooth subvariety of dimension $\operatorname{codim} B$ at a point of general position $o \in B$, transversal to B, and restrict the system Σ onto this germ).

Let $S \ni o$ be a general germ of a smooth surface on the variety X. Obviously, $\Lambda = \Sigma|_S$ is a germ of a linear system of curves on S with the point o as an isolated base point. Since the pair $(X, \frac{1}{n}\Sigma)$ is not canonical at the point o (this is just a reformulation of the Noether-Fano inequality (10)), according to the inversion of adjunction (which is a direct consequence of the connectedness principle of Shokurov and Kollár), the pair $(S, \frac{1}{n}\Lambda)$ is not *log*-canonical at the point o. In other words, for a certain birational morphism $\varphi\colon \widetilde{S} \to S$ of smooth surfaces there exists a prime divisor $E \subset \widetilde{S}$, satisfying the log-version of the Noether-Fano inequality

$$\nu_E(\Lambda) > n(a(E)+1), \tag{17}$$

where $a(\cdot)$ is the discrepancy, $\nu_E(\cdot)$ is the multiplicity of a general divisor of the system at E. Let $D_1, D_2 \in \Lambda$ be generic curves, $Z = (D_1 \circ D_2)$ a zero-dimensional subscheme on S. One may assume that it is supported at the point o.

Proposition 2.5. *The following inequality holds:*

$$\operatorname{mult}_o Z (= \deg Z) > 4n^2.$$

Since our considerations are local, $\operatorname{mult}_o Z = \deg Z$ is just the degree of the zero-dimensional scheme Z. Since $S \ni o$ is a general germ of a surface, we obtain immediately the claim of Theorem 2.1.

Proof. We give an elementary argument based on explicit computations. Let

$$\begin{array}{rccc} & S_i & \supset & E_i \\ \varphi_{i,i-1} & \downarrow & & \downarrow \\ & S_{i-1} & \ni & x_{i-1} \end{array}$$

be the resolution of the discrete valuation ν_E, $i = 1, \ldots, N$, $x_0 = o$, $x_1, \ldots, x_{N-1}$ points on $S_1, \ldots, S_{N-1}$, respectively, where $x_i \in E_i$ and $\nu_{E_N} = \nu_E$. Set Γ to be the graph of this resolution: $\{E_1, \ldots, E_N\}$ is the set of vertices, and the vertices i and j, $i > j$, are connected by an oriented edge (notation: $i \to j$ always implies that $i > j$), if and only if $x_{i-1} \in E_j^{i-1}$, where E_j^{i-1} is the strict transform of the exceptional line E_j on S_{i-1}. Set also

$$p_j = (\text{the number of paths from } N \text{ to } j)$$

for $j \le N-1$, $p_N = 1$. Set $\nu_i = \operatorname{mult}_{x_{i-1}} \Lambda^{i-1}$, where Λ^{i-1} is the strict transform of the linear system Λ on S_{i-1}. It is easy to see that in terms of the resolution the log-inequality (17) takes the form

$$\sum_{i=1}^{N} p_i \nu_i > n \left(\sum_{i=1}^{N} p_i + 1 \right).$$

Thus, the following estimate is true:

$$\operatorname{mult}_o Z \ge \sum_{i=1}^{N} \nu_i^2.$$

Lemma 2.5. *The following inequality holds:*

$$\operatorname{mult}_o Z > \frac{\left(\sum\limits_{i=1}^{N} p_i + 1 \right)^2}{\sum\limits_{i=1}^{N} p_i^2} n^2.$$

Proof is obtained by elementary computations: the minimum of the quadratic form

$$\sum_{i=1}^{N} s_i^2$$

under the restriction

$$\sum_{i=1}^{N} p_i s_i = c \tag{18}$$

is attained at $s_i = p_i a$, where the common constant a is found from (18). Q.E.D. for the lemma.

In view of this lemma Proposition 2.5 is an implication of the following fact.

Lemma 2.6. *The following inequality holds:*

$$\left(\sum_{i=1}^{N} p_i + 1\right)^2 \geq 4 \sum_{i=1}^{N} p_i^2. \tag{19}$$

Proof. Note first of all that in (19) the equality can be attained, for instance, when $N = 1$. Assume that $N \geq 2$. Set

$$\{2, \ldots, k \leq N\} = \{i \,|\, i \to 1\}.$$

By the definition of the integers p_i we get the equality

$$p_1 = \sum_{i \to 1} p_i = \sum_{i=2}^{k} p_i.$$

Therefore, (19) can be rewritten as

$$\left(2p_1 + \sum_{i=k+1}^{N} p_i + 1\right)^2 \geq 4 \sum_{i=1}^{N} p_i^2,$$

or, after an easy transformation,

$$4\left(\sum_{i=2}^{k} p_i\right)\left(\sum_{i=k+1}^{N} p_i + 1\right) + \left(\sum_{i=k+1}^{N} p_i + 1\right)^2 \geq 4 \sum_{i=2}^{k} p_i^2 + 4 \sum_{i=k+1}^{N} p_i^2.$$

It is easy to see that if $k = N$, then the subgraph of the graph Γ with the vertices $\{2, \ldots, N\}$ is of the form $2 \leftarrow 3 \leftarrow \cdots \leftarrow N$ (since on any surface S_i the curve

$$\bigcup_{j \leq i} E_j^i$$

is by smoothness a normal crossing divisor). Hence $p_2 = \cdots = p_N = 1$ and the inequality (19) holds in an obvious way. So let us assume that $N \geq k+1$. Arguing by induction on the number of vertices of the graph Γ we may assume that the inequality

$$\left(\sum_{i=k+1}^{N} p_i + 1\right)^2 \geq 4 \sum_{i=k+1}^{N} p_i^2$$

is true. Therefore, it is enough to show that the following estimate is true:

$$\left(\sum_{i=2}^{k} p_i\right)\left(\sum_{i=k+1}^{N} p_i + 1\right) \geq \sum_{i=2}^{k} p_i^2. \tag{20}$$

If $k = 2$, then by construction we get

$$p_2 \leq \sum_{i=3}^{N} p_i,$$

which immediately implies the inequality (20). If $k \geq 3$, then the subgraph of the graph Γ with the vertices $\{2, \ldots, \}$ is a chain:

$$2 \leftarrow 3 \leftarrow \cdots \leftarrow k.$$

Since $k \to (k-1)$ and $k \to 1$, the vertices k and i, $i \in \{2, \ldots, k-2\}$, are not joined by an arrow (oriented edge). Consequently, $j \not\to i$ for $j \geq k+1$, $i \in \{2, \ldots, k-2\}$. Thus each path from the vertex N to the vertex $i \in \{2, \ldots, k-2\}$ must go through the vertex $k-1$. Therefore,

$$p_2 = \cdots = p_{k-1} = p_k + \sum_{\substack{i \to k-1 \\ i \geq k+1}} p_i. \tag{21}$$

Lemma 2.7. *For any $i \in \{1, \ldots, N\}$ the following inequality holds:*

$$p_i \leq \sum_{j \geq i+2} p_j + 1 \tag{22}$$

(if the set $\{j \geq i+2\}$ is empty, the sum is assumed to be equal to zero).

Proof. This is a purely combinatorial fact, shown by decreasing induction on i. If $i = N$ or $i = N-1$, then the inequality (22) is true. Now we get

$$\begin{aligned} p_i - \sum_{j \geq i+2} p_j &= \sum_{j \to i} p_j - \sum_{j \geq i+2} p_j \\ &= p_{i+1} - \sum_{\substack{j \geq i+2 \\ j \not\to i}} p_j. \end{aligned}$$

Write the set $\{j \,|\, j \to i\}$ as $\{i+1, \ldots, i+l\}$. If $l = 1$, then applying the induction hypothesis, we obtain (22). If $l \geq 2$, then similarly to (21) we get

$$p_{i+1} = \cdots = p_{i+l-1} = p_{i+l} + \sum_{\substack{j \to i+l-1 \\ j \geq i+l+1}} p_j.$$

Therefore,

$$p_{i+1} - \sum_{\substack{j \geq i+2 \\ j \not\to i}} p_j = p_{i+l-1} - \sum_{j=i+l+1}^{N} p_j.$$

Applying the induction hypothesis to $i+l-1$, we complete the proof of the lemma.

Now let us come back to the proof of Lemma 2.6. We get

$$p_2 = \cdots = p_{k-1} \leq \sum_{i=k+1}^{N} p_i + 1.$$

But $p_k \leq p_{k-1}$, so that

$$\left(\sum_{i=2}^{k} p_i\right)\left(\sum_{i=k+1}^{N} p_i + 1\right) \geq p_{k-1}\sum_{i=2}^{k} p_i \geq \sum_{i=2}^{k} p_i^2,$$

which is what we need. Q.E.D.

Remark 2.1. It is easy to show that the $4n^2$-inequality cannot be improved in the sense that the constant 4 cannot be replaced by any higher number. In order to see this, consider a simple example. Let $\mathbb{C}^M_{z_1,\dots,z_M}$ be the standard coordinate space and set

$$\Lambda_{i-1} = \{z_i = z_{i+1} = \cdots = z_M = 0\} \subset \mathbb{C}^M,$$

$i = 1, \dots, M$. Let us construct a finite sequence of blow ups

$$\varphi_j \colon X_j \to X_{j-1},$$

$i = 1, 2, \dots$, starting with $X_0 = \mathbb{C}^M$, with non-singular centres $B_{j-1} \subset X_{j-1}$ and exceptional divisors $E_j = \varphi_j^{-1}(B_{j-1})$, as follows. We take $B_0 = o = (0, \dots, 0)$, and for the next $k_1 - 1$ blow ups B_{j-1} is the point

$$\Lambda_1^{j-1} \cap E_{j-1},$$

where the upper index a means the strict transform on X_a. For the next k_2 blow ups, B_{j-1} is the smooth curve

$$\Lambda_2^{j-1} \cap E_{j-1},$$

after which we k_3 times blow up smooth surfaces $\Lambda_3^{j-1} \cap E_{j-1}$ and so on. We complete, blowing up k_{M-2} times smooth subvarieties $\Lambda_{M-2}^{j-1} \cap E_{j-1}$ of codimension 3 and then k_{M-1} times smooth subvarieties $\Lambda_{M-1}^{j-1} \cap E_{j-1}$ of codimension 2. Set $E = E_K$ to be the last exceptional divisor, $K = k_1 + \cdots + k_{M-1}$. Obviously,

$$a(E) = (M-1)k_1 + (M-2)k_2 + \cdots + 2k_{M-2} + k_{M-1}.$$

Now set for $i \geq 2$,

$$a_i = \left[\frac{a(E)}{k_1 + \cdots + k_{i-1}} n\right] + 1$$

and $a_1 = a(E)n + 1$, and consider the linear system

$$\Sigma = \{\lambda_1 z_1^{a_1} + \lambda_2 z_2^{a_2} + \cdots + \lambda_M z_M^{a_M}\},$$

where λ_i are arbitrary constants. It is easy to see that $\operatorname{ord}_E \varphi^*\Sigma > na(E)$, where $\varphi = \varphi_1 \circ \cdots \circ \varphi_K$ is the composite map. On the other hand, as $n \to \infty$ and $k_{M-1} \to \infty$ and $k_{M-2}/k_{M-1} \to \infty$ with $k_{M-1}/k_i \to \infty$ for $i \leq M-3$, we have

$$\frac{\operatorname{mult}_o Z}{n^2} \to 4$$

as

$$\frac{((M-1)k_1 + \cdots + 2k_{M-2} + k_{M-1})^2}{(k_1 + \cdots + k_{M-2})(k_1 + \cdots + k_{M-2} + k_{M-1})} \to 4$$

and clearly $\operatorname{mult}_o Z \leq a_{M-1}a_M$ for $Z = (D_1 \circ D_2)$ with D_1, D_2 generic divisors in the linear system Σ.

3. Three-dimensional quartics

Three-dimensional quartics $V_4 \subset \mathbb{P}^4$ occupy a unique position in the theory of birational rigidity, both for purely mathematical and historical reasons. This class of algebraic varieties is sort of a touchstone for various approaches and methods: if a technique does not work for the quartic three-folds, it is useless to apply it to other families of Fano varieties. The section below consists of three parts: first, we give an alternative proof of birational superrigidity of quartics, next, we explain the original test class technique of Iskovskikh and Manin (it will be clear how it relates to the technique of counting multiplicities), finally, we consider singular quartics.

3.1. The test surface method. Here we give an alternative proof of birational superrigidity of a smooth quartic $V = V_4 \subset \mathbb{P}^4$. Assuming the converse, that is, that there exists a mobile linear system $\Sigma \subset |nH|$ with a maximal singularity $E \subset V^+$, where $\varphi\colon V^+ \to V$ is a birational morphism, contracting the exceptional divisor E to a point $o \in V$, we will construct a *test surface* $S \subset V$, such that the restriction $\Sigma_S = \Sigma|_S$ is a mobile system of curves. Computing its self-intersection in two different ways, we get a contradiction. The construction of a test surface is not very complicated but requires some preparational work.

Let us take a generic line $L \subset \mathbb{P}^4$ (in particular, intersecting V transversally at 4 distinct points, none of which is a base point for Σ), blow it up, $\tau\colon X \to \mathbb{P}^4$ and consider the projection $\pi_L\colon X \to \mathbb{P}^2$. Its restriction onto V we denote by $\pi\colon \widetilde{V} \to \mathbb{P}^2$. By the genericity of L, we may assume that $o \notin L \cap V$ and the differential

$$d\pi\colon T_o V \to T_{\pi(o)}\mathbb{P}^2$$

is surjective, in fact, that the whole curve $\pi^{-1}(\pi(o)) \subset \widetilde{V}$ is non-singular.

Lemma 3.1. *There exists a uniquely determined sequence of blow ups of points*

$$\lambda_{i,i-1}\colon Y_i \to Y_{i-1},$$

$i = 1, \dots, N$, $Y_0 = \mathbb{P}^2$, $\lambda_{i,i-1}$ *blows up the point* $x_{i-1} \in Y_{i-1}$, *with the following properties:*

(i) $x_0 = p = \pi(o) \in \mathbb{P}^2$,

(ii) $\lambda_{i,i-1}(x_i) = x_{i-1}$ *for all* $i = 1, \dots, N-1$,

(iii) *the centre of* E *on the variety* $\widetilde{V} \times_{\mathbb{P}^2} Y$, *where* $Y = Y_N$, *is a curve* Λ, *covering the* (-1)*-curve* $T_N = \lambda_{N,N-1}^{-1}(x_{N-1})$, *whereas the centre of* E *on the variety* $V^{(i)} = \widetilde{V} \times_{\mathbb{P}^2} Y_i$, $i = 0, \dots, N-1$, *is a point.*

Proof of this lemma is in its formulation: the centre of E on $\widetilde{V}$ is the point o, so we blow up the point $x_0 = p = \pi(o)$ and look at the (obviously non-singular) variety $\widetilde{V} \times_{\mathbb{P}^2} Y_1$. By the symbol π_V we denote the projection of that fibre product onto $\widetilde{V}$, by π_Y the other projection onto Y_1. Obviously, $\pi_V^{-1}(o)$ is a smooth rational curve and

$$\pi_Y\colon \pi_V^{-1}(o) \to T_1 \cong \mathbb{P}^1$$

is an isomorphism. Therefore, either the centre of E on $V^{(1)}$ is the curve $\Lambda_1 = \pi_V^{-1}(o)$ and we are done ($N = 1$), or the centre is a point $o_1 \in \Lambda_1$. In the latter case we put $x_1 = \pi_Y(o_1) \in T_1$ and continue in the same way. It is not hard to show that this procedure terminates: for instance, we may take a dense family of curves on V^+, intersecting E transversally at generic points, and consider its image on $\mathbb{P}^2$:

after a finite number of blow ups the branches of two generic curves at the point p will get separated. Q.E.D. for the lemma.

We denote by Λ_i the curve $\pi_V^{-1}(o_{i-1}) \subset V^{(i)}$. Abusing notation, we use the same symbols π_V and π_Y for the projections $V^{(i)} \to V^{(i-1)}$ and $V^{(i)} \to Y_i$, respectively. Note that $\pi_Y \colon \Lambda_i \to T_i \cong \mathbb{P}^1$ is an isomorphism, so that the centre $\Lambda = \Lambda_N$ of E on $V^{(N)}$ is a smooth rational curve.

Next, to construct a test surface, we need a dense family of curves C_t, $t \in \Delta$, such that the strict transform $\widetilde{C}_t$ of a generic curve on Y intersects the curve $T = T_N$ transversally at a generic point (there may be several points of intersection, some of them fixed, but at least one should vary on T and be a point of transversal intersection).

Such a family of curves will be called a *T-transversal family.* Consider the graph Γ_Y of the sequence of blow ups $\lambda_{i,i-1}$, $i = 1, \ldots, N$, with $p_{Y,i}$, $i = 1, \ldots, N$, denoting the number of paths in Γ_Y from the top vertex T_N to T_i, where as usual $p_{Y,N} = 1$ by definition.

Proposition 3.1. *There exists a T-transversal family of rational curves of degree*

$$d \leq p_{Y,1} + \cdots + p_{Y,N}.$$

Proposition 3.1 is a very special case of a much more general fact proved below. Now, assuming the proposition, we fix a T-transversal family $(C_t, t \in \Delta)$ of rational curves of degree $d \leq \sum_{i=1}^{N} p_{Y,i}$. Let $(\widetilde{C_t}, t \in \widetilde{\Delta})$ be its strict transform on Y (with the parametrizing set $\widetilde{\Delta} \subset \Delta$ possibly smaller). A generic surface

$$S_t = \pi_Y^{-1}(\widetilde{C}_t) \subset V^{(N)}$$

is called the *test surface.* Fix a test surface $S = S_t$. By the genericity of the line L, none of the fibres of the projection π is a base curve of the system Σ. By the genericity of the curve C_t in its family, the test surface does not contain any base curves of Σ. We conclude that the linear system of curves $\Sigma_S = \Sigma^{(N)}|_S$ is mobile.

Let $\bar{q}_t \in T$ be a point of transversal intersection of $\widetilde{C}_t$ and T, $q_t \in \Lambda_N = \Lambda$ the corresponding point on $V^{(N)}$. Obviously, S meets Λ transversally at $q = q_t$. As $q \in \Lambda$ is a point of general position, we conclude that the maximal singularity E generates a singularity E_S of the system Σ_S on S identically with the same graph Γ. Let

$$\varphi_{i,i-1} \colon S_i \to S_{i-1}$$

be the resolution of E_S, $\varphi_{i,i-1}$ blowing up a point $y_{i-1} \in S_{i-1}$, $E_i = \varphi_{i,i-1}^{-1}(y_{i-1})$ the exceptional curve, $y_0 = q$, $i = 1, \ldots, K$, and $E_K = E_S$. The number of paths from E_K to E_i will be denoted by p_i. Obviously, we have the equality

$$\operatorname{ord}_E \varphi^* \Sigma = \operatorname{ord}_{E_S} \Sigma_S$$

(the birational morphisms $V^{(i)}) \to V^{(i-1)}$ are the blow ups of curves which are not in the base locus of Σ, hence the multiplicities of Σ and its strict transforms at E are the same).

Proposition 3.2. *The point o comes into the scheme-theoretic self-intersection of the system Σ_S with the multiplicity at least*

$$\frac{1}{\left(\sum\limits_{j=1}^{K} p_i^2\right)}(\mathrm{ord}_E(\varphi^*\Sigma))^2. \tag{23}$$

Proof. Setting $\mu_i = \mathrm{mult}_{y_{i-1}} \Sigma_S^{i-1}$, we get that the desired multiplicity is not less than

$$\sum_{i=1}^{K} \mu_i^2,$$

whereas $\mathrm{ord}_E \varphi^*\Sigma = \mathrm{ord}_{E_S} \Sigma_S = \sum_{i=1}^{K} p_i\mu_i$. The minimum of the quadratic form above on that hyperplane is equal to (23). Q.E.D.

The discrepancy $a(E, V)$ can be computed in terms of our construction as follows:

$$a(E, V) = a(T, \mathbb{P}^2) \,\mathrm{ord}_E\, \pi_Y^{-1}(T) + a(E, V^{(N)}),$$

where

$$a(T, \mathbb{P}^2) = \sum_{i=1}^{N} p_{Y,i}, \quad a(E, V^{(N)}) = \sum_{j=1}^{K} p_j \quad \text{and} \quad \mathrm{ord}_E\, \pi_Y^{-1}(T) = p_1.$$

As the total intersection number of two curves in the system Σ_S is $4dn^2$, combining the Noether-Fano inequality, Proposition 3.2 and the presentation of discrepancy given above, we get the inequality

$$4d\sum_{j=1}^{K} p_j^2 > \left(p_1 \sum_{i=1}^{N} p_{Y,i} + \sum_{j=1}^{K} p_j\right)^2.$$

However, by Proposition 3.1 we can assume that $d \leq p_{Y,1} + \cdots + p_{Y,N}$. Since obviously

$$p_1 \sum_{j=1}^{K} p_j \geq \sum_{j=1}^{K} p_j^2,$$

we get the inequality

$$0 > \left(p_1 \sum_{i=1}^{N} p_{Y,i} - \sum_{j=1}^{K} p_j\right)^2,$$

which is impossible. Together with Lemma 2.1 this proves the birational super-rigidity of a smooth three-dimensional quartic.

3.2. Curves with prescribed branches. Here we give a proof of a much more general claim than that of Proposition 3.1. Let X be an algebraic variety, $\varphi\colon \widetilde{X} \to X$ a birational morphism, $Q \subset \widetilde{X}$ a prime Weil divisor such that $Q \not\subset \mathrm{Sing}\, \widetilde{X}$ and $B = \varphi(Q)$ is a subvariety of codimension at least 2, $B \not\subset \mathrm{Sing}\, X$. We define the positive integer

$$s(X, Q) = s(X, \nu_Q),$$

depending only on the discrete valuation $\nu_Q = \mathrm{ord}_Q\, \varphi^*(\cdot)$ of the field of rational functions $\mathbb{C}(X)$, as follows.

Let $\varphi_{i,i-1}\colon X_i \to X_{i-1}$, $i = 1,\dots,N$, be the resolution of Q, that is, $X_0 = X$ and the centre of the blow up $\varphi_{i,i-1}$ is the centre of Q on X_{i-1}, $E_i \subset X_i$ stands for the exceptional divisor and the discrete valuations ν_{E_N} and ν_Q coincide. The graph Γ_Q of the resolution depends on the discrete valuation ν_Q and the model X only. Set $p_{Q,i}$ for $i = 1,\dots,N-1$ to be the number of paths from the top vertex E_N to E_i and $p_{Q,N} = 1$. Now set

$$s(X,Q) = \sum_{i=1}^{N} p_{Q,i}.$$

Let $X = \mathbb{P}^M$ be the projective space. A family of curves $(C_t, t \in T)$ on X is said to be transversal to Q if the strict transform C_t of a general curve in this family has a point of transversal intersection with Q and these points fill an open set of Q as $t \in T$ varies. The curves are assumed to be irreducible.

Theorem 3.1. *There exists a dense transversal family of rational curves of degree at most* $s(\mathbb{P}^M, Q)$.

Obviously, Proposition 3.1 follows from this theorem.

Proof of the theorem. Let us start in the general situation as described at the beginning of this subsection, with X an arbitrary variety. Set $\mathcal{A}_n = K[t]/(t)^n$ and let $\mathcal{M}_n \subset \mathcal{A}_n$ be the maximal ideal. Multiplication by $a \in \mathcal{M}_n^s$ defines a homomorphism of $\mathcal{A}_n$-modules $(\cdot a)\colon \mathcal{A}_n \to \mathcal{M}_n^s$ with the kernel $\operatorname{Ker}(\cdot a) = \mathcal{M}_n^{n-s}$, so that $(\cdot a)\colon \mathcal{A}_{n-s} \to \mathcal{M}_n^s$ is a well-defined isomorphism of $\mathcal{A}_n$-modules. The inverse isomorphism we denote by $(\cdot a^{-1})\colon \mathcal{M}_n^s \to \mathcal{A}_{n-s}^s$.

We say that a linear space $W \subset \mathbb{C}(X)$ is a *space of local parameters* at the non-singular point $x \in X$, if $W \subset \mathcal{M}_{x,X} \subset \mathcal{O}_{x,X}$ and the differential $d_x\colon W \to \mathcal{M}_{x,X}/\mathcal{M}_{x,X}^2$ is an isomorphism. The following proposition is obvious.

Proposition 3.3. *There is a one-to-one correspondence between the set of ring homomorphisms* $\operatorname{Hom}(\mathcal{O}_{x,X}, \mathcal{A}_n)$ *and the set of* $\mathbb{C}$*-linear maps* $\operatorname{Hom}(W, \mathcal{M}_n) = W^* \otimes \mathcal{M}_n$. *This correspondence defines on the set* $\operatorname{Hom}(\mathcal{O}_{x,X}, \mathcal{A}_n)$ *a structure of an affine algebraic variety, which does not depend on the choice of* $W \subset \mathcal{M}_{x,X}$.

The crucial technical claim is given by the following:

Proposition 3.4. *There is an open subset* $\widetilde{Q} \subset Q$ *and for any* $q \in \widetilde{Q}$ *a non-singular subvariety* $S(q) \subset \operatorname{Hom}(\mathcal{O}_{x,X}, \mathcal{A}_{s+1})$, *where* $x = \varphi(q)$, $s = s(X,Q)$, *such that for any algebraic curve* $C \ni y$, $y \notin \operatorname{Sing} C$, *and a morphism* $\mu\colon C \to X$ *the following two conditions are equivalent:*

(i) *the morphism* μ *lifts, possible except for finitely many points, different from* y, *to a morphism* $\tilde{\mu}\colon C \to \widetilde{X}$, *whereas* $\tilde{\mu}(y) = q$ *and* $\widetilde{d\mu}(T_yC) \not\subset T_qQ$;

(ii) $\mu_{s+1}^* \in S(q)$, *where* $\mu_{s+1}^*\colon \mathcal{O}_{x,X} \to \mathcal{O}_{y,C}/\mathcal{M}_{y,C}^{s+1}$ *is the corresponding ring homomorphism and the quotient ring* $\mathcal{O}_{y,C}/\mathcal{M}_{y,C}^{s+1}$ *is identified with* $\mathcal{A}_{s+1}$ *by choosing a local parameter* $t \in \mathcal{M}_{y,C}$.

Proof is given by induction on $s \geq 1$. If $s = 1$, then everything is obvious. Therefore we assume that $s \geq 2$, and for geometric valuations with $s_1 \leq s-1$ the claim is shown.

Let $\sigma = \sigma_B\colon X(B) \to X$ be the blow up of the centre $B = \varphi(Q)$, $E \subset X(B)$ the exceptional divisor. Obviously,

$$s(X,Q) = \nu_Q(E) + s(X(B),Q),$$

because $\nu_Q(E) = p_{Q,1}$ in the notations above. It is clear that for any prime divisor D on X the multiplicity $\nu_Q(D) = \mathrm{ord}_Q(D)$ is equal to 0 if $D \not\supset B$ and to

$$\mathrm{mult}_B\, D \cdot \nu_Q(E) + \nu_Q(\widetilde{D}),$$

if $B \subset D$, where $\widetilde{D} \subset X(B)$ is the strict transform. Therefore, the estimate

$$\nu_Q(D) \leq s(X,Q)\,\mathrm{mult}_B\, D \tag{24}$$

holds.

Now let us apply the induction hypothesis to the variety $X(B)$ and the same valuation ν_Q, as $s(X(B),Q) < s(X,Q)$. Let $\widetilde{Q} \subset Q$ be the open set described in the proposition (which we can always make smaller by removing proper closed subsets in Q), $q \in \widetilde{Q}$, and $x = \varphi(q) \in B\backslash(\mathrm{Sing}\, X \cup \mathrm{Sing}\, B)$. Choose a space of local parameters $W \subset \mathcal{M}_{x,X}$, containing the subspace $W(B)$ of local equations for B at the point x, that is, $W(B) = d_x^{-1}(\mathrm{Ann}\, T_{x,B}) \subset J_{B,x}$, and fix a direct complement $L \oplus W(B) = W$. Assume that a function $g \in W(B)\backslash\{0\}$ is chosen in such a way that $z = \chi(q)$ is not contained in the strict transform of the hypersurface $\{g = 0\}$ on $X(B)$. Identifying $\sigma^{-1}(x)$ with the projectivization of the fibre of the normal bundle $\mathcal{N}_{B/X}$ at the point x, let us consider the hyperplane $W(B,z) = \mathrm{Ann}(z) \subset W(B)$, which obviously does not contain g. By the definition of a blow up, the linear space of functions

$$\widetilde{W} = \langle g \rangle \oplus \frac{1}{g} W(B,z) \oplus L \subset K(X) = K(X(B))$$

is the space of local parameters at the point z, and moreover, $g \in \mathcal{M}_{z,X(B)}$ is a local equation of E at z. In particular, $\beta = \nu(E) \leq s(X(B),\nu) = r$ by (24). Let $S^*(q) \subset \mathrm{Hom}(\mathcal{O}_{z,X(B)}, \mathcal{A}_{r+1})$ be the subvariety from the claim of the proposition. Consider the projection

$$\pi\colon \mathrm{Hom}(\mathcal{O}_{z,X(B)}, \mathcal{A}_{s+1}) \to \mathrm{Hom}(\mathcal{O}_{z,X(B)}, \mathcal{A}_{r+1})$$

and set $\widetilde{S}(q) = \pi^{-1}(S^*(q))$. Now let us define the map

$$\begin{array}{ccc} \sigma_*\colon \mathrm{Hom}(\mathcal{O}_{z,X(B)}, \mathcal{A}_{s+1}) & \to & \mathrm{Hom}(\mathcal{O}_{z,X(B)}, \mathcal{A}_{s+1}) \\ \| & & \| \\ \widetilde{W}^* \otimes \mathcal{M}_{s+1} & \to & W^* \otimes \mathcal{M}_{s+1} \end{array}$$

using the identification above by the relations $\sigma_* h(g) = h(g)$, $\sigma_* h(f) = h$ for $f \in L$ and the relation $\sigma_* h(f) = h(f/g)h(g)$ for $f \in W(B,z)$. Set $S(q) = \sigma_* \widetilde{S}(q)$. Since for any $h \in S(q)$ we have $h(g) \in \mathcal{M}_{s+1}^{\beta}\backslash\mathcal{M}_{s+1}^{\beta+1}$, one can define the map $\omega\colon S(q) \to S^*(q)$ by the following relations (where $\omega(h,f)$ is the value of $\omega(h) \in \widetilde{W}^* \otimes \mathcal{A}_{r+1}$ at $f \in \widetilde{W}$):

$$\begin{aligned} \omega(h,g) &= h(g) \,\mathrm{mod}\, \mathcal{M}_{s+1}^{r+1}, \\ \omega(h,f) &= h(f) \,\mathrm{mod}\, \mathcal{M}_{s+1}^{r+1} \text{ for } f \in L, \\ \omega(h,f/g) &= (\cdot h(g)^{-1})h(f) \in \mathcal{M}_{r+1}. \end{aligned}$$

Lemma 3.2. *The composition $\omega \circ \sigma_*$ coincides with the map π.*

Proof. This is obvious. Q.E.D.

If the condition (ii) holds, then $\mu_{s+1}^*(g) \in \mathcal{M}_{s+1}^*\backslash\mathcal{M}_{s+1}^{\beta+1}$, therefore, $\mu(C) \not\subset B$ and the morphism can be lifted to $\mu_B\colon C \to X(B)$, and moreover, $\mu_{B,r+1}^* =$

$\omega(\mu_{s+1}^*)$, so that μ_B lifts to a morphism $\tilde{\mu}\colon C \to \widetilde{X}$ and the condition (i) holds. Conversely, if the condition (i) holds, then $\mu(C) \not\subset B$, and, using the equivalence of the conditions (i) and (ii) for the discrete valuations on $X(B)$ and the definition of σ_*, we obtain (ii). Q.E.D. for Proposition 3.4.

To complete the proof of Theorem 3.1, it remains to make the following trivial observation: for rational curves $C \ni x$ of degree d in $\mathbb{P}^M$ the ring homomorphism

$$\mu_{d+1}^*\colon \mathcal{O}_{x,\mathbb{P}^M} \to \mathcal{O}_{x,C}/\mathcal{M}_{y,C}^{d+1}$$

can be prescribed: simply speaking, the values of the coefficients $a_{i,j}$ in the presentation

$$z_i = a_{i,1}t + a_{i,2}t^2 + \cdots + a_{i,d}t^d + \dots,$$

where z_* are *linear* coordinates with the origin at a given point $x \in \mathbb{P}^M$ and t is a *linear* parameter on $\mathbb{P}^1$, can be prescribed in an arbitrary way. This completes the proof of Theorem 3.1 and the alternative proof of birational superrigidity of three-dimensional quartics.

3.3. The test class technique. Here we explain the original test class technique of Iskovskikh and Manin, on the basis of which the technique of counting multiplicities was developed. We will work on a non-singular projective three-fold V with $\operatorname{Pic} V = \mathbb{Z}H$, H is the ample generator. At some point we will specify V to be a three-dimensional quartic, but in order to compare the test class technique with that of counting multiplicities, we keep the more general context.

Next, we fix a mobile linear system $\Sigma \subset |nH|$ and a resolution of singularities $\varphi_{i,i-1}\colon V_i \to V_{i-1}$, $i = 1, \dots, K$, so that the strict transform Σ^N on V^N is base point free. The centre $B_{i-1} \subset V_{i-1}$ of the blow up $\varphi_{i,i-1}$ is either a point or non-singular curve, so that all varieties V_i are non-singular. The exceptional divisors of the blow ups $E_i = \varphi_{i,i-1}^{-1}(B_{i-1})$ are either planes $\mathbb{P}^2$ or ruled surfaces. The resolution of singularities is taken to be the best possible, in particular, when B_{i-1} is a curve, the function of the point $p \in B_{i-1}$,

$$\operatorname{mult}_p \Sigma^{i-1} \in \mathbb{Z}_+,$$

is constant, therefore no fibre of the ruled surface E_i/B_{i-1} is a base curve of the system Σ^i. For some positive integers $\nu_1, \dots, \nu_N$ which we may assume to be non-increasing we can write the class of a divisor in the linear system Σ^N as

$$H' = nH - \sum_{i=1}^{N} \nu_i E_i \in \operatorname{Pic} V_N.$$

Obviously, the class (H'^2) of the self-intersection of the system Σ^N in the numerical Chow group $A^2 V_N$ is represented by an effective 1-cycle. Moreover, by the property of the resolution of singularities, mentioned above, we may assume that (H'^2) is represented by an effective 1-cycle, none of the components of which is the strict transform on V_N of a fibre of a ruled surface E_i. Now the idea of the proof of birational superrigidity of a non-singular three-dimensional quartic, given by Iskovskikh and Manin, was to construct a special divisorial class $y \in \operatorname{Pic} V_N$ (the *test class*), satisfying the two properties:

(i) for any irreducible curve $C \subset V_N$, which is not the strict transform of a fibre of a ruled surface E_i,

$$(C \cdot y) \geq 0;$$

(ii) provided that the linear system Σ has an infinitely near maximal singularity,

$$(H'^2 \cdot y) < 0.$$

Obviously, (i) and (ii) contradict each other and exclude the infinitely near maximal singularity. The construction of the test class makes sense for any non-singular three-fold (in fact, for $\mathbb{Q}$-factorial varieties of any dimension), not necessarily for the quartic three-folds. We now proceed to the construction in the general situation.

Let $a \in \{1, \ldots, K\}$ be an index corresponding to a ruled surface E_a/B_{a-1}. Denote by

$$S_a \subset \{1, \ldots, K\}$$

the subset of indices $j \leq a$ such that

$$\varphi_{a,j}(E_a) \subset E_j,$$

where $\varphi_{a,a} = \mathrm{id}_{V_a}$ and for $j < a$

$$\varphi_{a,j} = \varphi_{j+1,j} \circ \cdots \circ \varphi_{a,a-1} \colon V_a \to V_j$$

is the composite map. For $i > j$, $\{i, j\} \subset S_a$, we set $i \to j$ if

$$B_{i-1} \subset E_j^{i-1}$$

and thus define an oriented graph Γ_a with the vertices in S_a. We break the set S_a into a union of two disjoint subsets:

$$S_a = I_0(S_a) \sqcup I_1(S_a),$$

corresponding to the blow ups of points and curves, respectively. Assume that for any $i \in I_1(S_a)$ the image $\varphi_{j,0}(E_j)$ is a point on V.

Definition 3.1. The class

$$y = \left(\sum_{j \in I_0(S_a)} p_{aj} \right) H - \sum_{j \in I_0(S_a)} p_{aj} E_j$$

is the *test class*, associated with the singularity E_a.

Here p_{aj} is the number of paths from the top vertex E_a in Γ_a to E_j.

Proposition 3.5 (the Positivity Lemma). *Let $C \subset V_i$ be an irreducible curve, $i \in \{0, 1, \ldots, K\}$. Assume that for all $j \leq i$, for which $\dim B_{j-1} = 1$ (that is, E_j is a ruled surface over B_{j-1}), the curve $C_j = \varphi_{ij}(C)$ is not a fibre of the ruling E_j/B_{j-1}. Then $(y \cdot C) \geq 0$.*

Proof. If $i \geq a + 1$, then we note that

$$(\varphi_{i,a})_* C = \deg(\varphi_{i,a} \colon C \to C_a) \cdot C_a,$$

so, using the projection formula, it is sufficient to prove the claim for $i \leq a$, so we will assume that $i \leq a$.

If $C_0 = \varphi_{i,0}(C) \subset V$ is a curve, then $(C \cdot E_j) = 0$ for $j \geq i + 1$, $j \in I_0(S_a)$, by the projection formula and for $j \leq i$, $j \in I_0(S_a)$, we have

$$(C \cdot E_j) = \mathrm{mult}_{B_{j-1}}\ C_{j-1} = \mathrm{mult}_{B_{j-1}}\ C_0^{j-1}$$

(by our assumption $\varphi_{i,0}\colon V_i \to V$ is an isomorphism at the generic point of C). As for any point $p \in V_k$, $k \leq a$,

$$\operatorname{mult}_p C_0^k \leq \deg C_0 = (C \cdot H),$$

we obtain the claim of the positivity lemma in this case.

Now assume that C_0 is a point and set $k \in \{1, \dots, N\}$ to be the minimal integer such that $\dim C_k = 1$. Obviously, B_{k-1} is a point, $C_k \subset E_k \cong \mathbb{P}^2$ is a plane curve and

$$(C \cdot E_k) = -\deg(\varphi_{i,k}\colon C \to C_k) \cdot \deg C_k.$$

(Here we use the assumption that C is not a fibre of the ruling E_j/B_{j-1}.) Furthermore, if $j \in I_0(S_a)$ is such an index that $(C \cdot E_j) \neq 0$, then $i \geq j$ and

$$(C \cdot E_j) = \deg(\varphi_{i,j}\colon C \to C_j) \cdot \operatorname{mult}_{B_{j-1}} C_{j-1}.$$

Taking into account the well-known inequality

$$\operatorname{mult}_{B_{j-1}} C_{j-1} \leq \deg(\varphi_{j,k}\colon C_j \to C_k) \cdot \operatorname{mult}_{B_{j-1,k}} C_k$$

and the obvious equality

$$\deg(\varphi_{i,k}\colon C \to C_k) = \deg(\varphi_{i,j}\colon C \to C_j) \deg(\varphi_{j,k}\colon C_j \to C_k)$$

(and, of course, the obvious equality

$$\deg(\varphi_{j,k}\colon C_j \to C_k) = \deg(\varphi_{j-1,k}\colon C_{j-1} \to C_k)$$

and the obvious estimate $\operatorname{mult}_{B_{j-1,k}} C_k) \leq \deg C_k$), we come to the crucial estimate

$$(C \cdot (-E_k - E_j)) \geq 0. \tag{25}$$

Lemma 3.3. *The following estimate holds:*

$$p_{ak} \geq \sum_{j \in I_0(S_a), (C \cdot E_j) \neq 0} p_{aj}.$$

Proof. For any index $j \in I_0(S_a)$, such that $(C \cdot E_j) \neq 0$, let us construct a special path γ_{jk} from E_j to E_k. As $(C \cdot E_j) \neq 0$, the point B_{j-1} lies on the curve C_{j-1}. If the map $\varphi_{j-1,k}\colon V_{j-1} \to V_k$ is an isomorphism at the generic point of C_{j-1}, then obviously

$$B_{j-1} \in E_k^{j-1},$$

so that $j \to k$ and we take for γ_{jk} this one-step path from E_j to E_k. Otherwise, there is at least one exceptional divisor E_l, such that $l > k$, $l \in I_1(S_a)$ and $C_l \subset E_l$. Let us take k^* to be the maximal index with that property. Obviously, $j \to k^*$ and any path from E_{k^*} to E_k goes through the vertices E_l, $l \in I_1(S_a)$ only, as $\dim C_{l-1} = 1$ and $C_{l-1} = B_{l-1}$.

Now fix any path $\gamma_{jk} = j \to k^* \to \cdots \to k$.

Composing the paths from E_a to E_j with the path γ_{jk}, constructed above, we get an injective map from the disjoint union

$$\bigsqcup_{j \in I_0(S_a), (C \cdot E_j) \neq 0} \{\text{paths from } E_a \text{ to } E_j\}$$

to the set of paths from E_a to E_k. Proof of the lemma is complete. Q.E.D.

Writing

$$(y \cdot C) = p_{ak}(-E_k \cdot C) - \sum_{j \in I_0(S_a), (E_j \cdot C) \neq 0} p_{aj}(E_j \cdot C)$$

and combining the estimate (25) and Lemma 3.3, we complete the proof of the Positivity Lemma. Q.E.D.

Now assume that $\nu_a > n$ and E_a is the maximal singularity, satisfying the Noether-Fano inequality, which in terms of the resolution of singularities of the linear system Σ takes the form

$$\sum_{j \in S_a} p_{aj}\nu_j > 2n \sum_{j \in I_0(S_a)} p_{aj} + n \sum_{j \in I_1(S_a)} p_{aj}.$$

Since $\mathrm{mult}_C \Sigma \leq n$ for any curve $C \subset V$, our assumption that for every $j \in I_1(S_a)$ the image $\varphi_{j,0}(E_j)$ is a point (not a curve) on V, holds, and the claim of the Positivity Lemma is true.

Birational superrigidity of the quartic V follows now from

Proposition 3.6. *The following estimate holds:*

$$((H')^2 \cdot y) < 0.$$

Proof. We observe that $(H \cdot E_i \cdot y) = 0$ for all $i = 1, \ldots, K$ and similarly $(E_i \cdot E_j \cdot y) = 0$ for all $i \neq j$, so that

$$((H')^2 \cdot y) = \left(\left(n^2 H^2 + \sum_{i=1}^{k} \nu_i^2 E_i^2 \right) \cdot y \right).$$

Next, if $\dim B_{i-1} = 0$, then $(E_i^2 \cdot y) \neq 0$ if and only if $i \in I_0(S_a)$, and in the latter case

$$((\nu_i^2 E_i^2) \cdot y) = -p_{ai}\nu_i^2.$$

To simplify the notations, set $p_i = p_{ai}$ for $i \in S_a$.

Furthermore, by the projection formula if $\dim B_{i-1} = 1$, then

$$(E_i^2 \cdot y) = -(B_{i-1} \cdot y) \leq 0,$$

therefore we get the inequality

$$((H'^2) \cdot y) \leq \left(\left(n^2 H^2 + \sum_{i \in S_a} \nu_i^2 E_i^2 \right) \cdot y \right).$$

Finally, observing that for any vertex E_i, $i \in I_1(S_a)$, there is a path from E_i to E_k, where $B_{i-1,k} \subset E_k$ is a curve and $k \in I_0(S_a)$,

$$i \to k_1 \to \cdots \to k_l \to k,$$

with all $k_1, \ldots, k_l \in I_1(S_a)$, and repeating word for word the arguments in the proof of the Positivity Lemma, we get the inequality

$$(B_{i-1} \cdot y) \geq p_j,$$

so that putting

$$\Sigma_0 = \sum_{j \in I_0(S)} p_j \quad \text{and} \quad \Sigma_1 = \sum_{j \in I_1(S)} p_j,$$

we can write

$$((H'^2)\cdot y)\leq 4n^2\Sigma_0-\sum_{j\in S_a}p_j\nu_j^2.$$

However, the minimum of the quadratic form $\sum\limits_{j\in S_a}p_j\nu_j^2$ on the hyperplane

$$\sum_{j\in S_a}p_j\nu_j=(2\Sigma_0+\Sigma_1)n$$

is at least $4n^2\Sigma_0$, which proves Proposition 3.6 and completes the (original) proof of birational superrigidity of a three-dimensional quartic. Q.E.D.

It is a very good exercise (which we leave to the reader) to check that the proof of the $4n^2$-inequality given in Section 2 and the argument above are essentially the same.

3.4. Singular quartic hypersurfaces. Let us consider singular quartic hypersurfaces $V\subset\mathbb{P}^4$. Similar to the smooth case, the first family of singular Fano varieties, for which birational rigidity was proven, are also quartics. Assume that the singularities of V are isolated non-degenerate quadratic points (often called nodes) and that V is factorial. This is true, for instance, when V contains precisely one non-degenerate double point.

With every singular point $p\in\operatorname{Sing}V$ we can associate a birational involution $\tau_p\in\operatorname{Bir}V$, acting as follows. Let $\pi_p\colon V\dashrightarrow\mathbb{P}^3$ be the projection from the point p, then $\deg(\pi_p)=2$ and τ_p is the Galois involution associated with this rational map of degree 2. To understand the action of τ_p on divisors, blow up the singular point: $\sigma_p\colon\widetilde{V}\to V$, set $E_p=\sigma_p^{-1}(p)$ to be the exceptional surface $\cong\mathbb{P}^1\times\mathbb{P}^1$. Set $U_p\subset\widetilde{V}$ to be the complement of the closed set $Y_p\subset\widetilde{V}$, which is the union of the strict transforms on $\widetilde{V}$ of the lines on V passing through p. Let us assume that there are finitely many ($\leq 4!=24$) such lines. Obviously, the projection π_p extends to a regular map $\pi_p\colon\widetilde{V}\to\mathbb{P}^3$ and, moreover,

$$\pi_p\colon U_p\to\mathbb{P}^3\backslash\pi_p(Y_p)$$

is a double cover, so that $\tau_p\in\operatorname{Aut}U_p$. As Y_p by assumption is a curve, the action of τ_p on $\operatorname{Pic}\widetilde{V}\cong\mathbb{Z}H\oplus\mathbb{Z}E_p$ is well defined.

Lemma 3.4. *The action of τ_p on $\operatorname{Pic}\widetilde{V}$ is given by the relations*

$$\tau_p^*H=3H-4E_p\quad\text{and}\quad\tau_p^*E_p=2H-3E_p.$$

Proof. Both relations are almost obvious. For example, let $S\subset V$ be a generic hyperplane section, $p\notin S$. Then $H+\tau_p^*H$ is the class of the divisor $\pi_p^{-1}(\pi_p(S))$. As the surface $\pi_p(S)\subset\mathbb{P}^3$ is a quartic surface (isomorphic to S), we get

$$H+\tau_p^*H=4(H-E_p),$$

which gives the first relation. Similarly,

$$E_p+\tau_p^*E_p=2(H-E_p),$$

which gives the second relation. Q.E.D.

Remark 3.1. It is easy to show (see [**Shr08b**]) that if $\dim Y_p = 2$ (such singular points are sometimes called *Eckardt points* by analogy with cubic surfaces in $\mathbb{P}^3$), then the involution τ_p is biregular.

With every line $L \subset V$, containing a unique singular point $p \in \operatorname{Sing} V$, we can associate a birational involution $\tau_L \in \operatorname{Bir} V$, acting as follows. Let $\pi_L \colon V \dashrightarrow \mathbb{P}^2$ be the projection from the line L. Blowing up first the point p, then the strict transform $\widetilde{L}$ of the line L, we get the birational morphism $\sigma_L \colon \widetilde{V} \to V$ with the two exceptional divisors E_p and E_L, corresponding to the point p and line L, respectively. The projection π_L extends to a morphism

$$\pi_L \colon \widetilde{V} \to \mathbb{P}^2.$$

Assume that there are only finitely many lines on V intersecting L (otherwise, L is sometines called an *Eckardt line*). It is easy to check that π_L fibres V into cubic curves and there is a subset $Y_L \subset \widetilde{V}$, consisting of finitely many fibres, such that every fibre $C_t, t \in \mathbb{P}^2$, outside Y_L is an irreducible plane cubic curve, non-singular at the point $C_t \cap E_p$. The general fibre C_t is a smooth elliptic curve.

The involution τ_L preserves the fibres C_t. For a given $t \in \mathbb{P}^2$ (such that $C_t \not\subset Y_L$)

$$\tau_L|_{C_t}(x) = x' \in C_t$$

such that $x+x' \sim 2(C_t \cap E_p)$ on C_t (that is, the reflexion from 0 in terms of the group law on C_t given by setting $0 = C_t \cap E_p$). Since τ_L is well defined outside finitely many fibres of π_L, it acts biregularly on the invariant open subset $U_L = \widetilde{V} \backslash Y_L$ and therefore its action on $\operatorname{Pic} \widetilde{V} = \mathbb{Z}H \oplus \mathbb{Z}E_p \oplus \mathbb{Z}E_L$ is well defined.

Lemma 3.5. *The involution τ_L acts on $\operatorname{Pic} \widetilde{V}$ as follows*:

$$\tau_L^* H = 11H - 6E_p - 12E_L, \quad \tau_L^* E_p = E_p, \quad \tau_L^* E_L = 10H - 6E_p - 11E_L.$$

Proof is elementary and similar to that of Lemma 3.4, so we leave it to the reader as an exercise (the equality $\tau_L^* E_p = E_p$ is obvious; the τ_L^*-invariance of the class $H - E_p - E_L$ is obvious; now one needs to restrict τ_L to the invariant surface $\pi_L^{-1}(R)$, where $R \subset \mathbb{P}^2$ is a generic line, and to the generic fibre $C_t, t \in \mathbb{P}^2$). Q.E.D.

Remark 3.2. It is easy to show (see [**Shr08b**]) that if L is an Eckardt line, then the construction above gives a biregular involution.

Finally, let us consider a line $L \subset V$, containing precisely two singular points of V, say $p, q \in \operatorname{Sing} V$. In this case the corresponding involution $\tau_L \in \operatorname{Bir} V$ is constructed as follows.

Let $\pi_L \colon V \dashrightarrow \mathbb{P}^2$ be the projection from L. Blowing up p, q and the strict transform of L, we obtain a morphism $\pi_L \colon \widetilde{V} \to \mathbb{P}^2$, the fibres of which are plane cubic curves. The fibration $\widetilde{V}/\mathbb{P}$ has three sections: E_p, E_q and E_L (precisely the three exceptional divisors of the blow up $\widetilde{V} \to V$). We define τ_L on a fibre $C_t = \pi_L^{-1}(t)$ of general position as the projective reflexion from the point $E_L \cap C_t$ (that is, the Galois involution, corresponding to the double cover $C_t \to \mathbb{P}^1$, which is given by the projection from the point $C_t \cap E_L$). Its action on the group

$$\operatorname{Pic} \widetilde{V} = \mathbb{Z}H \oplus \mathbb{Z}E_p \oplus \mathbb{Z}E_q \oplus \mathbb{Z}E_L$$

is again well defined and given by the relations

$$\tau_L^* H = 5H - 3E_p - 3E_q - 6E_L,$$
$$\tau_L^* E_L = 4H - 3E_p - 3E_q - 5E_L,$$
$$\tau_L^* E_p = E_q, \quad \tau_L^* E_q = E_p.$$

Denote by $B(V)$ the subgroup in $\operatorname{Bir} V$, generated by the involutions τ_p for all singular points $p \in \operatorname{Sing} V$, which are not Eckardt points, and the involutions τ_L for all lines $L \subset V$ containing either one or two singular points, which are not Eckardt lines.

Recall that we assume V to be factorial and having finitely many non-degenerate quadratic points as singularities.

Theorem 3.2. (i) *The variety V is birationally rigid.*

(ii) *The subgroup $B(V)$ is normal in* $\operatorname{Bir} V$*, has the trivial intersection with* $\operatorname{Aut} V$ *and together with the latter subgroup generates* $\operatorname{Bir} V$*, so that* $\operatorname{Bir} V$ *is the semi-direct product*

$$1 \to B(V) \to \operatorname{Bir} V \to \operatorname{Aut} V \to 1,$$

where the action of $\operatorname{Aut} V$ *on $B(V)$ is determined by the relations*

$$\rho\tau_p\rho^{-1} = \tau_{\rho(p)} \quad \text{and} \quad \rho\tau_L\rho^{-1} = \tau_{\rho(L)}$$

for all $\rho \in \operatorname{Aut} V$.

(iii) *All relations between the generators τ_p, τ_L are implied by the following basic ones:*

$$(\tau_p\tau_q\tau_r)^2 = \mathrm{id}$$

for every triple $\{p, q, r\}$ of collinear singular points $p, q, r \in \operatorname{Sing} V$ and

$$(\tau_p\tau_q\tau_L)^2 = \mathrm{id}$$

for every line L containing precisely two singular points p, q.

Example 3.1. If $V \subset \mathbb{P}^4$ is a quartic hypersurface with just one non-degenerate double point, through which there are precisely 24=4! lines on V, then V is automatically factorial and

$$B(V) = \mathop{*}_{i=0}^{24} \langle\tau_i\rangle$$

is the free product of 25 cyclic groups of order 2, where τ_0 is the reflexion from the double point and $\tau_1, \ldots, \tau_{24}$ correspond to the lines through the double point.

Historically this example was the first case when birational rigidity was proved for a non-smooth Fano variety; Theorem 3.2 was a follow up of that result.

Example 3.2. There are very few papers dealing with quartic three-folds that are outside the assumptions of Theorem 3.2: either they have more complicated singularities [**CoMe**] or are non-factorial [**Grin11**], namely contain a 2-plane in $\mathbb{P}^4$. Only some intermediate results were obtained in those cases. A complete description of the birational type remains an open problem.

We do not give a detailed **proof** of Theorem 3.2 as it involves, with one exception, no ideas that are not present in the proof of birational rigidity of the complete intersection $V_{2\cdot 3} \subset \mathbb{P}^5$ given in Sections 5 and 6. One needs to show that if a mobile linear system $\Sigma \subset |-nK_V|$ has a maximal singularity, then it has either a maximal point $p \in \operatorname{Sing} V$, which is not an Eckardt point, or a maximal line $L \subset V$ which contains either precisely one or precisely two singular points of V and which is not an Eckardt line. This done, the maximal point or line is untwisted by means of the corresponding involution $\tau \in B(V)$ and the claim of Theorem 3.2 follows. As usual, the relations between the generating involutions arise when more than one maximal subvariety exists. The reader can easily reconstruct all details using Section 5 as a model (or looking into the papers [**CoMe, Shr08b**]).

In order to show the existence of a maximal subvariety of the very special sort, as explained above, one needs

- to exclude infinitely near maximal singularities,
- to show that no other curve can be maximal for the system Σ, apart from a line L containing one or two singular points.

The second part is not hard but very tedious and requires a case-by-case study. Again, we refer the reader to [**Shr08b**] or offer this task as an exercise following Section 5. However, in the first part there is a case which we will consider in detail.

Namely, if $V^+ \stackrel{\varphi}{\to} V$ is a birational morphism, E^+ an irreducible prime exceptional divisor which is a maximal singularity for Σ, then $\varphi(E^+)$ is a double point $p \in \operatorname{Sing} V$ (if $p \notin \operatorname{Sing} V$, then the argument given for the smooth quartic works). Let $\sigma\colon \widetilde{V} \to V$ be the blow up of p, with $E = \sigma^{-1}(p) \cong \mathbb{P}^1 \times \mathbb{P}^1$ as the exceptional divisor.

Proposition 3.7. *E is a maximal singularity of the system* $\Sigma : \operatorname{ord}_E \sigma^*\Sigma > n$.

Proof. Assume the converse: $\beta = \operatorname{ord}_E \sigma^*\Sigma \leq n$. In that case, the centre of E^+ on $\widetilde{V}$ is inside E. Setting $\widetilde{\Sigma}$ to be the strict transform of Σ on $\widetilde{V}$, $\widetilde{D} \in \widetilde{\Sigma}$ the strict transform of a generic divisor $D \in \Sigma$, we see that the pair $\left(\widetilde{V}, \frac{1}{n}\widetilde{D}\right)$ is not canonical and therefore the pair $\left(E, \frac{1}{n}\widetilde{D}_E\right)$ is not log canonical, where $\widetilde{D}_E = \widetilde{D}|_E = (\widetilde{D} \circ E)$ is the restriction of $\widetilde{D}$ onto E. Identifying $E \cong \mathbb{P}^1 \times \mathbb{P}^1$, we see that there is an effective 1-cycle C of bidegree (β, β) with $\beta \leq n$ such that $\left(E, \frac{1}{n}C\right)$ is not log canonical. We may assume the worst: $\beta = n$.

As C has no components of multiplicity higher than n, the centre of a non-log canonical singularity must be a point $x \in E$. Let us resolve the singularity

$$\lambda_i\colon X_i \to X_{i-1},$$

$i = 1, \ldots, K$, $X_0 = E, \lambda_i$ blows up a point $x_{i-1} \in X_{i-1}$, $Q_i = \lambda_i^{-1}(x_{i-1})$ is the exceptional line, $x_i \in Q_i$ and Q_K represents the non-log canonical singularity. Let Γ be the graph of the resolution, and set

$$m_i = \operatorname{mult}_{x_{i-1}} C^{i-1}.$$

We get the Noether-Fano inequality

$$\sum_{i=1}^K p_i m_i > n\left(\sum_{i=1}^K p_i + 1\right).$$

Let us show that this is impossible for an effective 1-cycle of bidegree (n, n).

Let A and B be the two lines on E through the point $x = x_0$ (that is, the curves of bidegrees (1,0) and (0,1)). We may assume that if $K \geq 2$, then $x_1 \notin B^1$. We decompose the cycle C,

$$C = \alpha F + \beta B + C_+,$$

where C_+ does not contain A, B, and set

$$\mu_i = \mathrm{mult}_{x_{i-1}} C_+^{i-1}.$$

Also set $k = \max\{i \mid x_{i-1} \in A^{i-1}\} \geq 1$. Note that

$$\min(n-\alpha, n-\beta) \geq \mu_1 \geq \mu_2 \geq \dots$$

and $\mu_1 + \dots + \mu_k \leq n - \beta$. Now if $k = 1$, then $m_1 = \alpha + \beta + \mu_1$ and $m_i = \mu_i$ for $i \geq 2$, so that

$$\sum_{i=1}^{K} p_i m_i = p_1(\alpha + \beta + \mu_1) + \sum_{i=2}^{K} p_i \mu_i$$

$$\leq p_1(\alpha + \mu_1) + \beta \sum_{i=2}^{K} p_i + \sum_{i=2}^{K} p_i \mu_i \leq n \sum_{i=1}^{K} p_i,$$

contrary to the Noether-Fano inequality. If $k \geq 2$, then $m_1 = \alpha + \beta + \mu_1$ and $m_i = \alpha + \mu_i$ for $i \in \{2, \dots, k\}$ and $m_i = \mu_i$ for $i \geq k+1$. Thus, by Lemma 2.7,

$$p_1 = \dots = p_{k-1} \leq \sum_{i=k+1}^{K} p_i + 1.$$

Therefore,

$$\sum_{i=1}^{K} p_i m_i = p_1(\alpha + \beta + \mu_1) + \sum_{i=2}^{K} p_i(\alpha + \mu_i) + \sum_{i=k+1}^{K} p_i \mu_i$$

$$\leq \sum_{i=1}^{k} p_i(\alpha + \mu_1) + \beta \sum_{i=k+1}^{K} p_i + \beta + \sum_{i=k+1}^{K} p_i \mu_i \leq n \sum_{i=1}^{K} p_i + n,$$

which gives a contradiction and proves the proposition. Q.E.D.

Thus we have shown that if there is an infinitely near maximal singularity over a double point p, then the point p itself is a maximal singularity. Now it is a straightforward (but very tedious) work to complete the proof of Theorem 3.2 and leave it to the reader. Q.E.D. for Theorem 3.2.

3.5. The weighted Fano hypersurfaces. Three-dimensional quartics $V_4 \subset \mathbb{P}^4$ can be viewed as a particular case of the larger and more versatile class of *weighted Fano hypersurfaces* $V_d \subset \mathbb{P}(1, a_1, a_2, a_3, a_4)$ with $d = a_1 + a_2 + a_3 + a_4$. (The same is true about the Fano double spaces of index 1: they can be realized either as the double covers $\sigma: V \to \mathbb{P}^3$ branched over a smooth sextic surfaces $S_6 \subset \mathbb{P}^3$ or as the weighted hypersurfaces of degree 6 in $\mathbb{P}(1,1,1,1,3)$.) It is well known that there are 95 families of such hypersurfaces (that is, 95 tuples (a_1, a_2, a_3, a_4)) which are Fano varieties with $\mathbb{Q}$-factorial terminal quotient singularities of type $\frac{1}{r}(1, a, r-a)$. Of these 95 families, two correspond to smooth Fano three-folds (the quartics and the double spaces). In terms of the parameters of the family, the quartics form the biggest one.

Birational rigidity of a general variety in each of the 95 families was shown in [**CPR**]. As usual, the proof included explicit constructions of birational involutions, untwisting maximal subvarieties (points and curves). Those involutions generate the groups of birational self-maps; sometimes there are relations between the generators, which were described in [**ChP06**].

Here we will only comment on the techniques of the proof, emphasizing the new effects that come from the weighted projective space.

As usual, to prove birational rigidity of a weighted hypersurface $V = V_d \subset \mathbb{P}(a_*)$, one needs to fix a mobile linear system $\Sigma \subset |-nK_V|$ with a maximal singularity and either exclude or untwist all potential maximal singularities. Note first of all that the anticanonical degree of V is

$$\deg V = \frac{d}{a_1 a_2 a_3 a_4}$$

which, except for the quartics, is strictly less than 4 and in fact for most families is a small rational number. For that reason, it is not too difficult to exclude infinitely near maximal singularities over smooth points. The only source of trouble is that, to apply the $4n^2$-inequality and exclude a maximal singularity over a smooth point $p \in V$, one needs a linear system of divisors Λ on V without base curves. While for $\mathbb{P}^4$ one can take the system of hyperplanes, containing the point p, and apply the $4n^2$-inequality as it was done in Section 2 to exclude the infinitely near case on the quartic, in a weighted projective space a much higher multiple $|kH|$, where $\mathcal{O}_{\mathbb{P}(a_*)}(H) = \mathcal{O}_{\mathbb{P}(a_*)}(1)$, might be needed, depending on the numbers a_i. In a few cases a lot of tedious work is required. However, in this way all infinitely near maximal singularities centred at smooth points can be excluded.

The new ingredient of the proof is the Kawamata theorem that implies that if $E \subset \widetilde{V}$ is a maximal singularity of the system Σ such that for some quotient terminal point $P \in \operatorname{Sing} V$,

$$P \in \operatorname{centre}(E, V),$$

then P itself is a maximal singularity of Σ (see [**CPR**]). The proof of Kawamata's theorem is short and elementary [**Ka96**]. Thus the terminal quotient singularities turn out to be an easy case (somewhat similar to the non-degenerate quadratic points considered in Subsection 3.4 but much easier).

Finally, it is also not hard (but sometimes again tedious) to show that no curve $C \subset V$ lying entirely in the smooth locus of V can be a maximal subvariety of Σ.

Therefore, to prove birational rigidity of V, it remains to study the singular points: either they can be excluded or untwisted by a suitable birational involution.

The currently available proof involves a huge amount of case by case study and computations and for that reason cannot be reproduced here, we refer the reader to the original work [**CPR**].

4. The connectedness principle

In this section, we give a proof of another local fact, important for the theory of birational rigidity — the $8n^2$-inequality for the self-intersection of a mobile linear system. As an application, we prove birational superrigidity of smooth Fano hypersurfaces $V_M \subset \mathbb{P}^M$ of degrees $M = 5, 6, 7, 8$. The proof of the $8n^2$-inequality makes heavy use of the connectedness principle of Shokurov and Kollár.

4.1. Set up of the problem and start of the proof. Let $o \in X$ be a germ of a smooth variety of dimension $\dim X \geq 4$. Let Σ be a mobile linear system on X, and the effective cycle

$$Z = (D_1 \circ D_2),$$

where $D_1, D_2 \in \Sigma$ are generic divisors, its self-intersection. Blow up the point o:

$$\varphi\colon X^+ \to X,$$

$E = \varphi^{-1}(o) \cong \mathbb{P}^{\dim X - 1}$ is the exceptional divisor. The strict transform of the system Σ and the cycle Z on X^+ we denote by the symbols Σ^+ and Z^+, respectively.

Theorem 4.1 (the $8n^2$-inequality). *Assume that the pair*

$$(X, \frac{1}{n}\Sigma)$$

is not canonical, but canonical outside the point o, where n is some positive number. There exists a linear subspace $P \subset E$ of codimension two (with respect to E), such that the inequality

$$\mathrm{mult}_o Z + \mathrm{mult}_P Z^+ > 8n^2$$

holds. If $\mathrm{mult}_o Z \leq 8n^2$, then the subspace P is uniquely determined by the pair.

Proof. Restricting the system Σ onto a germ of a generic smooth subvariety, containing the point o, we may assume that $\dim X = 4$. Moreover, we may assume that $\nu = \mathrm{mult}_o \Sigma \leq 2\sqrt{2}n < 3n$, since otherwise

$$\mathrm{mult}_o Z \geq \nu^2 > 8n^2$$

and there is nothing to prove.

Lemma 4.1. *The pair*

$$(X^+, \frac{1}{n}\Sigma^+ + \frac{(\nu - 2n)}{n}E) \tag{26}$$

is not log canonical, and the centre of any of its non-log canonical singularities is contained in the exceptional divisor E.

Proof. Let $\lambda\colon \widetilde{X} \to X$ be a resolution of singularities of the pair $(X, \frac{1}{n}\Sigma)$ and $E^* \subset \widetilde{X}$ a prime exceptional divisor, realizing a non-canonical singularity of that pair. Then $\lambda(E^*) = o$ and the Noether-Fano inequality holds:

$$\nu_{E^*}(\Sigma) > na(E^*).$$

For a generic divisor $D \in \Sigma$ we get $\varphi^* D = D^+ + \nu E$, so that

$$\nu_{E^*}(\Sigma) = \nu_{E^*}(\Sigma^+) + \nu \cdot \nu_{E^*}(E)$$

and

$$a(E^*, X) = a(E^*, X^+) + 3\nu_{E^*}(E).$$

From here we get

$$\nu_{E^*}\left(\frac{1}{n}\Sigma^+ + \frac{\nu - 2n}{n}E\right) = \nu_{E^*}\left(\frac{1}{n}\Sigma\right) - 2\nu_{E^*}(E)$$
$$> a(E^*, X^+) + \nu_{E^*}(E) \geq a(E^*, X^+) + 1,$$

which proves the lemma. Q.E.D.

Let $R \ni o$ be a generic three-dimensional germ and $R^+ \subset X^+$ its strict transform on the blow up of the point o. For a small $\varepsilon > 0$ the pair

$$\left(X^+, \frac{1}{1+\varepsilon}\frac{1}{n}\Sigma^+ + \frac{\nu - 2n}{n}E + R^+\right)$$

still satisfies the connectedness principle (with respect to the morphism $\varphi\colon X^+ \to X$), so that the set of centres of non-log canonical singularities of that pair is connected. Since R^+ is a non-log canonical singularity itself, we obtain, that there is a non-log canonical singularity of the pair (26), the centre of which on X^+ is of positive dimension, since it intersects R^+.

Let $Y \subset E$ be a centre of a non-log canonical singularity of the pair (26) that has the maximal dimension.

If $\dim Y = 2$, then consider a generic two-dimensional germ S, intersecting Y transversally at a point of general position. The restriction of the pair (26) onto S is not log canonical at that point, so that, applying Proposition 4.1, which is proven below, we see that

$$\mathrm{mult}_Y(D_1^+ \circ D_2^+) > 4\left(3 - \frac{\nu}{n}\right) n^2,$$

so that

$$\begin{aligned}\mathrm{mult}_o Z &\geq \nu^2 + \mathrm{mult}_Y(D_1^+ \circ D_2^+) \deg Y \\ &> (\nu - 2n)^2 + 8n^2,\end{aligned}$$

which is what we need.

If $\dim Y = 1$, then, since the pair

$$\left(R^+, \frac{1}{1+\varepsilon} \frac{1}{n} \Sigma_R^+ + \frac{\nu - 2n}{n} E_R\right), \tag{27}$$

where $\Sigma_R^+ = \Sigma^+|_{R^+}$ and $E_R = E|_{R^+}$, satisfies the condition of the connectedness principle and R^+ intersects Y at $\deg Y$ distinct points, we conclude that $Y \subset E$ is a line in $\mathbb{P}^3$.

Now we need to distinguish between the following two cases: when $\nu \geq 2n$ and when $\nu < 2n$. The methods of proving the $8n^2$-inequality in these two cases are absolutely different. Consider first the case $\nu \geq 2n$.

Let us choose as $R \ni o$ a generic three-dimensional germ, satisfying the condition $R^+ \supset Y$. Since the pair (26) is effective (recall that $\nu \geq 2n$), one may apply inversion of adjunction (see Chapter 7) and conclude that the pair (27) is not log canonical at Y.

Applying now to the pair (27) (where $R^+ \supset Y$) Proposition 4.1 in the same way as it was done for $\dim Y = 2$, we obtain the inequality

$$\mathrm{mult}_Y(D_1^+|_{R^+} \circ D_2^+|_{R^+}) > 4\left(3 - \frac{\nu}{n}\right) n^2.$$

On the left in brackets we have the self-intersection of the mobile system Σ_R^+, which breaks into two natural components:

$$(D_1^+|_{R^+} \circ D_2^+|_{R^+}) = Z_R^+ + Z_R^{(1)},$$

where Z_R^+ is the strict transform of the cycle $Z_R = Z|_R$ on R^+ and the support of the cycle $Z_R^{(1)}$ is contained in E_R. The line Y is a component of the effective 1-cycle $Z_R^{(1)}$.

On the other hand, for the self-intersection of the mobile linear system Σ^+ we get

$$(D_1^+ \circ D_2^+) = Z^+ + Z_1,$$

where the support of the cycle Z_1 is contained in E. From the genericity of R it follows that outside the line Y the cycles $Z_R^{(1)}$ and $Z_1|_{R^+}$ coincide, whereas for Y we get the equality

$$\mathrm{mult}_Y Z_R^{(1)} = \mathrm{mult}_Y Z^+ + \mathrm{mult}_Y Z_1.$$

However, $\mathrm{mult}_Y Z_1 \leq \deg Z_1$, so that

$$\begin{aligned} &\mathrm{mult}_o Z + \mathrm{mult}_Y Z^+ \\ &= \nu^2 + \deg Z_1 + \mathrm{mult}_Y Z^+ \\ &\geq \nu^2 + \mathrm{mult}_Y Z_R^{(1)} > 8n^2, \end{aligned}$$

which is what we need. This completes the case $\nu \geq 2n$.

Note that the key point in this argument is that the pair (26) is effective. For $\nu < 2n$, inversion of adjunction cannot be applied.

4.2. The case $\nu < 2n$. Consider again the pair (27) for a generic germ $R \ni o$. Let $y = Y \cap R^+$ be the point of (transversal) intersection of the line Y and the variety R^+. Since $a(E_R, R) = 2$, the non-log canonicity of the pair (27) at the point y implies the non-log canonicity of the pair

$$\left(R, \frac{1}{n}\Sigma_R\right)$$

at the point o, whereas the centre of some non-log canonical (that is, log maximal) singularity on R^+ is a point y.

Now the $8n^2$-inequality comes from the following fact.

Lemma 4.2. *The following inequality holds:*

$$\mathrm{mult}_o Z_R + \mathrm{mult}_y Z_R^+ > 8n^2,$$

where Z_R is the self-intersection of a mobile linear system Σ_R and Z_R^+ is its strict transform on R^+.

Proof. Consider the resolution of the maximal singularity of the system Σ_R, the centre of which on R^+ is the point y:

$$\begin{array}{ccc} R_i & \stackrel{\psi_i}{\to} & R_{i-1} \\ \cup & & \cup \\ E_i & & B_{i-1}, \end{array}$$

where B_{i-1} is the centre of the singularity on R_{i-1}, $R_0 = R$, $R_1 = R^+$, $E_i = \psi_i^{-1}(B_{i-1})$ is the exceptional divisor, $B_0 = o$, $B_1 = y \in E_1$, $i = 1, \ldots, N$, where the first L blow ups correspond to points, for $i \geq L+1$ curves are blown up. Since

$$\mathrm{mult}_o \Sigma_R = \mathrm{mult}_o \Sigma < 2n,$$

we get $L < N$, $B_L \subset E_L \cong \mathbb{P}^2$ is a line and for $i \geq L+1$,

$$\deg[\psi_i|_{B_i} : B_i \to B_{i-1}] = 1,$$

that is, $B_i \subset E_i$ is a section of the ruled surface E_i.

Consider the graph of the sequence of blow ups ψ_i.

Lemma 4.3. *The vertices $L+1$ and $L-1$ are not connected by an arrow:*

$$L+1 \nrightarrow L-1.$$

Proof. Assume the converse: $L+1 \to L-1$. This means that

$$B_L = E_L \cap E_{L-1}^L$$

is the exceptional line on the surface E_{L-1}^L and the map

$$E_{L-1}^{L+1} \to E_{L-1}^L$$

is an isomorphism. As usual, set

$$\nu_i = \operatorname{mult}_{B_{i-1}} \Sigma_R^{i-1},$$

$i = 1, \dots, N$. Let us restrict the mobile linear system Σ_R^{L+1} onto the surface E_{L-1}^{L+1} (that is, onto the plane $E_{L-1} \cong \mathbb{P}^2$ with the blown up point B_{L-1}). We obtain a non-empty (but, of course, not necessarily mobile) linear system, which is a subsystem of the complete linear system

$$\left|\nu_{L-1}(-E_{L-1}|_{E_{L-1}}) - (\nu_L + \nu_{L+1})B_L\right|.$$

Since $(-E_{L-1}|_{E_{L-1}})$ is the class of a line on the plane E_{L-1}, this implies that

$$\nu_{L-1} \geq \nu_L + \nu_{L+1} > 2n,$$

such that $\nu_1 = \nu > 2n$. A contradiction. Q.E.D. for Lemma 4.3.

Similarly, one can deduce from the inequality $\nu_1 \leq 2n$ that the upper part of the graph of the sequence of blow ups ψ_i, that is, the part, corresponding to the vertices

$$L+1, \dots, N,$$

is a chain. In other words, the section B_i of the ruled surface E_i, $i \geq L+1$, is different from the section $E_i \cap E_{i-1}^i$. Now set, as usual,

$$m_i = \operatorname{mult}_{B_{i-1}} (Z_R)^{i-1},$$

$i = 1, \dots, L$, so that, in particular,

$$m_1 = \operatorname{mult}_o Z_R \quad \text{and} \quad m_2 = \operatorname{mult}_y Z_R^+.$$

For $1 \leq i \leq L-1$ set $p_i \geq 1$ to be the number of paths in the graph of the sequence of blow ups ψ_i from the vertex L to the vertex i, and

$$p_N = p_{N-1} = \dots = p_L = 1$$

by definition. By the technique of counting multiplicities, developed in Subsection 2.2, we get the inequality

$$\sum_{i=1}^{L} p_i m_i \geq \sum_{i=1}^{N} p_i \nu_i^2. \tag{28}$$

Thus, the inequality

$$\sum_{i=1}^{N} p_i \nu_i > n \left(2 \sum_{i=1}^{L} p_i + \sum_{i=L+1}^{N} p_i \right) \tag{29}$$

holds. (In fact, a somewhat stronger inequality holds, the *log* Noether-Fano inequality, but we do not need that.) The inequality (29) follows from the Noether-Fano inequality for the maximal singularity E_N by the assumption that $\nu_1 \leq 2n$.

Indeed, let p_i^* be the number of paths in the graph of the sequence of blow ups $\{\psi_i\}$, going from the vertex N to the vertex i. By what we have proved,

$$p_N^* = p_{N-1}^* = \dots = p_L^* = p_{L-1}^* = 1$$

and for $i \leq L$ we have $p_i \leq p_i^*$, since p_i is the number of paths from the vertex N to the vertex i, *passing through the vertex* L. The Noether-Fano inequality can be written in the form

$$\sum_{i=L+1}^{N} p_i^*(\nu_i - n) > \sum_{i=1}^{L} p_i^*(2n - \nu_i),$$

where in the right-hand side each summand is non-negative, since $2n \geq \nu_i$ for all $i = 1, \dots, N$. It follows that

$$\sum_{i=L+1}^{N} p_i(\nu_i - n) > \sum_{i=1}^{L} p_i(2n - \nu_i)$$

(recall that $p_i = p_i^*$ for $i = L+1, \dots, N$, so that the left-hand side of the inequality remains unchanged whereas the right-hand side can only get smaller), as we claimed. From the estimates (28, 29), repeating the computations of Subsection 2.2, one gets the inequality

$$\sum_{i=1}^{L} p_i m_i > \frac{(2\Sigma_0 + \Sigma_1)^2}{\Sigma_0 + \Sigma_1} n^2,$$

where $\Sigma_0 = \sum_{i=1}^{L} p_i$ and $\Sigma_1 = \sum_{i=L+1}^{N} p_i = N - L$. Taking into account that for $i \geq 2$ we get

$$m_i \leq m_2$$

and the obvious inequality $(2\Sigma_0 + \Sigma_1)^2 > 4\Sigma_0(\Sigma_0 + \Sigma_1)$, we obtain the following estimate

$$p_1 m_1 + (\Sigma_0 - p_1) m_2 > 4n^2 \Sigma_0.$$

Now assume that the claim of the lemma is false:

$$m_1 + m_2 \leq 8n^2.$$

Lemma 4.4. *The following inequality holds:* $\Sigma_0 \geq 2p_1$.

Proof. By construction, p_1 is the number of paths from the vertex L to the vertex 1. Marking in each path the last but one vertex, we get

$$p_1 = \sum_{L \geq i \to 1} p_i,$$

so that $p_1 \leq \Sigma_0 - p_1$, which is what we need. Q.E.D. for the lemma.

Now, taking into account that $m_2 \leq m_1$, we obtain

$$\begin{aligned} p_1 m_1 + (\Sigma_0 - p_1) m_2 &= p_1(m_1 + m_2) + (\Sigma_0 - 2p_1) m_2 \\ &\leq 8p_1 n^2 + (\Sigma_0 - 2p_1) \cdot 4n^2 = 4n^2 \Sigma_0. \end{aligned}$$

This is a contradiction. Q.E.D. for Lemma 4.2.

Proof of Theorem 4.1 is complete.

Remark 4.1. Let us explain the key point in the proof of $8n^2$-inequality for $\nu_1 \leq 2n$. As it follows from the technique of counting multiplicities, the graph of the sequence of blow ups $\{\psi_i\}$ can be modified in such a way that all applications still hold, namely, one can delete all the arrows going from the vertices

$$L + 1, \dots, N$$

of the upper part of the graph to the vertices

$$1,\dots,L-1$$

of the lower part (and both the Noether-Fano inequality and the estimate for the multiplicities of the self-intersection of the linear system are intact). The graph, modified in this way, satisfies the property of Lemma 4.4.

4.3. A local inequality for a surface. Let $o \in X$ be a germ of a smooth surface, $C \ni o$ a smooth curve and Σ a mobile linear system on X. Let, furthermore, $Z = (D_1 \circ D_2)$ be the self-intersection of the linear system Σ, that is, an effective 0-cycle. Since the situation is local, we may assume that the support of the cycle Z is the point o, that is,

$$\deg Z = (D_1 \cdot D_2)_o.$$

Proposition 4.1. *Assume that for some real number $a < 1$ the pair*

$$\left(X, \frac{1}{n}\Sigma + aC\right) \tag{30}$$

is not log canonical (that is, for a general divisor $D \in \Sigma$ the pair $(X, \frac{1}{n}D + aC))$ is not log canonical, where $n > 0$ is a positive number. Then the following estimate holds:

$$\deg Z > 4(1-a)n^2. \tag{31}$$

Proof. We will show that the inequality (31) follows directly from some well-known facts about infinitely near singularities of curves on a non-singular surface. Assume that the sequence of blow ups

$$\varphi_{i,i-1}\colon X_i \to X_{i-1},$$

$i = 1,\dots,N$, where $X_0 = X$, resolves the non-log canonical singularity of the pair (30). We use the standard notations and conventions: the centre of the blow up $\varphi_{i,i-1}$ is the point $x_{i-1} \in X_{i-1}$, its exceptional line is

$$E_i = \varphi_{i,i-1}^{-1}(x_{i-1}) \subset X_i,$$

the first point to be blown up is $o = x_0$, the blown up points x_i lie over each other: $x_i \in E_i$. The last exceptional line E_N realizes the non-log canonical singularity of the pair (30), that is the log Noether-Fano inequality holds:

$$\sum_{i=1}^{N} \nu_i p_i + an \sum_{x_{i-1}\in C^{i-1}} p_i > n\left(\sum_{i=1}^{N} p_i + 1\right), \tag{32}$$

where $\nu_i = \operatorname{mult}_{x_{i-1}} \Sigma^{i-1}$, the symbols Σ^i and C^i stand for the strict transforms on X_i and p_i is the number of paths in the graph Γ of the constructed sequence of blow ups from the vertex E_N to the vertex E_i. Assume that

$$x_{i-1} \in C^{i-1}$$

for $i = 1,\dots,k \le N$, then the inequality (32) takes the form

$$\sum_{i=1}^{N} \nu_i p_i > n\left(\sum_{i=1}^{k}(1-a)p_i + \sum_{i=k+1}^{N} p_i + 1\right). \tag{33}$$

Lemma 4.5. *The following inequality holds:*

$$\deg Z \geq \sum_{i=1}^{N} \nu_i^2. \tag{34}$$

Proof. This is obvious. Q.E.D.

Lemma 4.6. *For each $i \in \{1, \dots, N-1\}$ the following estimate holds:*

$$\nu_i \geq \sum_{j \to i} \nu_j. \tag{35}$$

Proof. This is a very well-known property of multiplicities of curves at infinitely near points on a non-singular surface. Q.E.D.

Lemma 4.7. *The following estimate is true:*

$$\sum_{i=1}^{N} \nu_i^2 > \frac{\Delta^2}{q} n^2,$$

where

$$\Delta = 1 + (1-a)\sum_{i=1}^{k} p_i + \sum_{i=k+1}^{N} p_i$$

and $q = \sum_{i=1}^{N} p_i^2$ (so that $n\Delta$ is the right-hand side of the inequality (33)*).*

Proof. The minimum of the quadratic form in the right-hand side of the inequality (34) under the restrictions (35) and

$$\sum_{i=1}^{N} \nu_i p_i = \Delta n \tag{36}$$

is attained at $\nu_i = p_i \theta$, where $\theta = \frac{\Delta n}{q}$ is computed from (36). Q.E.D. for the lemma.

Now the claim of Proposition 4.1 follows from a purely combinatorial fact about the graph Γ, which we will now prove.

Lemma 4.8. *Assume that the starting segment of the graph Γ with the vertices $1, \dots, k$ is a chain. Then the following estimate holds:*

$$\Delta^2 \geq 4(1-a)q. \tag{37}$$

Proof will be given by induction on the number N of vertices of the graph Γ. If $N = 1$, then the inequality (37) holds in a trivial way:

$$(2-a)^2 \geq 4(1-a).$$

Consider the inequality (37) as a claim on the non-negativity of a quadratic function of the argument a:

$$a^2 \left(\sum_{i=1}^{k} p_i\right)^2 + 2a\left(2q - \left(\sum_{i=1}^{k} p_i\right)\left(\sum_{i=1}^{N} p_i + 1\right)\right) + \left(\left(\sum_{i=1}^{N} p_i + 1\right)^2 - 4q\right) \geq 0$$

on the interval $a \leq 1$. Since for $a \to \pm\infty$ this function is positive, it is sufficient to check that its minimum is non-negative. Elementary computations show that, up to an inessential positive factor, this minimum is given by the formula

$$\left(\sum_{i=1}^{k} p_i\right)\left(\sum_{i=k+1}^{N} p_i + 1\right) - \sum_{i=1}^{N} p_i^2. \tag{38}$$

We will prove non-negativity of the latter expression by induction on the number of vertices N. Recall that the only assumption, restricting the choice of the number $k \geq 1$, is that there are no arrows $i \to j$ for $i \geq j+2$ and $i \leq k$.

Consider first the case $k = 1$. Assume that $l \geq 1$ vertices are connected by arrows with 1, that is,

$$2 \to 1, \ldots, l+1 \to 1, \quad \text{but} \quad l+2 \nrightarrow 1.$$

In this case $p_1 = p_2 + \cdots + p_{l+1}$ and the subgraph of the graph Γ with the vertices $\{2, \ldots, l+1\}$ either consists of one vertex or is a chain. The expression (38) transforms to the formula

$$\left(\sum_{i=2}^{l+1} p_i\right)\left(\sum_{i=l+2}^{N} p_i + 1\right) - \sum_{i=2}^{N} p_i^2,$$

so that one can apply the induction hypothesis to the subgraph with the vertices $\{2, \ldots, N\}$. This completes the case $k = 1$.

Now let $k \geq 2$. By Lemma 2.7, we have the inequality

$$p_i \leq \sum_{j=i+2}^{N} p_i + 1.$$

Therefore,

$$p_1 = p_2 = \cdots = p_{k-1} \leq \sum_{i=k+1}^{N} p_i + 1.$$

For this reason, for $k \geq 2$, the expression (38) is bounded from below by the number

$$\left(\sum_{i=2}^{k} p_i\right)\left(\sum_{i=k+1}^{N} p_i + 1\right) - \sum_{i=2}^{N} p_i^2.$$

Now, applying the induction hypothesis to the subgraph with the vertices $\{2, \ldots, N\}$ we complete the proof of Lemma 4.8 and Proposition 4.1. Q.E.D.

4.4. The first applications of the $8n^2$-inequality. As an easy consequence of the $8n^2$-inequality, we obtain

Theorem 4.2. *A smooth Fano hypersurface of degree M in $\mathbb{P}^M$ for the values $M \in \{5, 6, 7, 8\}$ is birationally superrigid.*

Proof. Let $\Sigma \subset |nH|$ be a mobile linear system on $V = V_M \subset \mathbb{P}^M$, where $H = -K_V$ is the positive generator of the Picard group, the class of a hyperplane section. Assume that Σ has a maximal singularity $E \subset \widetilde{V}$ for some model $\widetilde{V}$ of V. As $\text{mult}_C \Sigma \leq n$ for any curve $C \subset V$, the centre of E on V is some point o. Let

$$\sigma\colon V^+ \to V$$

be its blow up, $E^+ = \sigma^{-1}(o)$ the exceptional divisor. Applying Theorem 4.1, we conclude that there is a linear subspace $P \subset E^+ \cong \mathbb{P}^{M-2}$ of codimension 2 such that

$$\operatorname{mult}_o Z + \operatorname{mult}_P Z^+ > 8n^2,$$

where $Z = (D_1 \circ D_2)$ is the self-intersection of Σ and Z^+ is the strict transform of Z on V^+. Assume first that $M \in \{6, 7, 8\}$. Let $\Delta \subset \mathbb{P}^M$ be the (unique) linear subspace of codimension 3 such that $P = \mathbb{P}(T_o\Delta)$ is its projectivized tangent cone. In other words, consider the blow up

$$\sigma_{\mathbb{P}}\colon \mathbb{P}^+ \to \mathbb{P}^M$$

of the point o, $E_{\mathbb{P}} = \sigma_{\mathbb{P}}^{-1}(o)$, so that σ is the restriction of $\sigma_{\mathbb{P}}$ to V, then

$$\Delta^+ \cap E_{\mathbb{P}} = P.$$

Applying the Lefschetz theorem, we see that $\Delta \not\subset V$. Therefore, a general hyperplane $\Lambda \supset \Delta$ does not contain any components of the cycle Z and

$$\operatorname{mult}_o(Z \circ \Lambda) \geq \operatorname{mult}_o Z + \operatorname{mult}_P Z^+ > 8n^2,$$

which is impossible as $\deg(Z \circ \Lambda) = Mn^2 \leq 8n^2$.

To complete the proof of the theorem, it remains to consider the case $M = 5$.

Arguing as above, assume that $\Delta \subset V$ (otherwise, the argument above works word for word).

Decompose

$$Z = a\Delta + Z_1,$$

where $a \in \mathbb{Z}_+$ and the effective 2-cycle Z_1 does not contain the plane Δ as a component (the subspace P in this case is a line so Δ is a 2-plane).

Lemma 4.9. *The coefficient a satisfies the inequality*

$$a \leq n^2.$$

Proof. First of all, let us investigate the numerical properties of the plane Δ on V. Note that in this case the numerical group $A^2V \neq \mathbb{Z}H^2$.

Lemma 4.10. *The following equalities hold:*

$$(\Delta \cdot H^2)_V = 1, \quad (\Delta \cdot \Delta)_V = 13.$$

Moreover, the 2-class $(\Delta + 3H^2)$ is non-negative.

Proof. The first equality is obvious. To obtain the second one, let us draw through Δ two generic 3-planes S_1 and S_2. Then

$$S_i \cap V = \Delta \cup Q_i,$$

where Q_i is a quartic surface in $S_i \cong \mathbb{P}^3$. Clearly,

$$Q_1 \cap Q_2 = (Q_1 \cap \Delta) \cap (Q_2 \cap \Delta)$$

is the intersection of two plane quartic curves so we conclude that

$$(Q_1 \cdot Q_2)_V = 16 = (H^2 - \Delta)^2 = 3 + (\Delta \cdot \Delta)$$

and the second equality follows.

Finally, let $G \neq \Delta$ be an irreducible surface on V. Once again, take a generic 3-plane S through Δ, so that $S \cap V = \Delta \cup Q$, the surface Q is the residual quartic surface. It is obvious that the intersection $Q \cap G$ is 0-dimensional. Therefore,

$$(Q \cdot G)_V \leq \deg Q \cdot \deg G,$$

so $((H^2 - \Delta) \cdot G) \leq 4(H^2 \cdot G)$ and we conclude that

$$((3H^2 + \Delta) \cdot G)_V \geq 0$$

for any surface $G \neq \Delta$ on V, but $((3H^2 + \Delta) \cdot \Delta) = 16 > 0$. Q.E.D. for Lemma 4.10.

Now let us complete the proof of Lemma 4.9. As $Z \sim n^2 H^2$, we have

$$16n^2 = (Z \cdot (3H^2 + \Delta)) = a(\Delta \cdot (3H^2 + \Delta)) + (Z_1 \cdot (3H^2 + \Delta)) \geq 16a,$$

as required. Q.E.D.

Now let us complete the case $M = 5$. Obviously, $\deg Z_1 = \deg Z - a = 5n^2 - a$ and $\operatorname{mult}_o Z_1 = \operatorname{mult}_o Z - a$. Furthermore, as $\operatorname{mult}_P \Delta^+ = 1$, we have

$$\operatorname{mult}_P Z_1^+ = \operatorname{mult}_P Z^+ - a,$$

and using the $8n^2$-inequality we conclude that

$$\operatorname{mult}_o Z_1 + \operatorname{mult}_P Z_1^+ > 8n^2 - 2a.$$

Now we argue word for word in the same way as in the case $M \in \{6, 7, 8\}$: take a general hyperplane $\Lambda \supset \Delta$ and consider the effective 1-cycle $(Z_1 \circ \Lambda)$. We obtain the estimate

$$\operatorname{mult}_o(Z_1 \circ \Lambda) > 8n^2 - 2a,$$

however, $\deg(Z_1 \circ \Lambda) = \deg Z_1 = 5n^2 - a$, so that $5n^2 - a > 8n^2 - 2a$, which is impossible for $a \leq n^2$. Q.E.D. for Theorem 4.2.

Remark 4.2. It is possible to prove birational superrigidity of four-dimensional quintics without using the connectedness principle and the $8n^2$-inequality, as it was originally done in [**Pukh87**]. Naturally, it requires a lot more work and case by case study. The following technical claim is crucial for that more elementary (but much harder) proof.

Lemma 4.11. (i) *Let $X \subset \mathbb{P}^M$ be an irreducible variety, $Q \subset X$ an irreducible subvariety, $x \in Q$ a point. Assume that the inequality*

$$\frac{\operatorname{mult}_x Q}{\deg Q} > \frac{\operatorname{mult}_x X}{\operatorname{mult}_x X + 1}$$

holds. The Q is contained in the tangent cone to X at the point x.

(ii) *Let $Q \subset \mathbb{P}^3$ be an irreducible cone of degree m with the vertex o and $C \subset Q$ a curve. Then*

$$\operatorname{mult}_o C \equiv \deg C \bmod m.$$

Proof is elementary and left to the reader. We do not use the lemma anywhere in this book. Q.E.D.

5. Complete intersections $V_{2\cdot 3} \subset \mathbb{P}^5$. I: Untwisting maximal singularities

In this section, we consider in full detail the procedure of untwisting maximal singularities of a mobile linear system, by the example of the most difficult and versatile case known today, that of the three-dimensional complete intersections of a quadric and a cubic in $\mathbb{P}^5$. We also demonstrate how naturally the method of maximal singularities gives not only the generators of the group of birational self-maps, but also the relations between them, so that the final outcome is a complete copresentation of the group. We start with a brief discussion (partly informal) of the untwisting procedure (Subsection 5.1). After that, assuming the crucial fact — the description of maximal singularities of a mobile linear system, given in Theorem 5.2 (which is shown in the next section), we give a copresentation of the group of birational self-maps of the complete intersection $V_{2\cdot 3}$.

5.1. Untwisting birational self-maps. Let Σ be a mobile linear system on the variety V with a maximal singularity E. Assume that (in the terminology and notations of Sec. 1) the singularity E can be untwisted by a birational self-map $\tau_E \in \operatorname{Bir} V$, *depending on the singularity* E *only*. By this assumption of universality, one can hope that

- the linear system $(\tau_E^{-1})_* \,|-nK_V|$, determining the map τ_E itself, has the maximal singularity E,
- the self-map τ_E untwists the singularity E for itself, so that the composition τ_E^2 does not have E as a maximal singularity; moreover, it is natural to suggest that τ_E^2 has no maximal singularities at all.

In all successfully studied cases this is true. All untwisting transformations turn out to be involutions: $\tau_E^2 = \mathrm{id}$. However, this is a purely empirical fact. There are two methods to construct untwisting involutions. The first method is used when the maximal singularity E generates a rational map $V \overset{2:1}{\dashrightarrow} W$ of degree 2. In that case $\tau_e \in \operatorname{Bir} V$ is a Galois involution, permuting points in the general fibre. The image of the singularity E with respect to τ_E is a divisor on V. The second method is used when the singularity E generates a rational map $V \overset{\gamma}{\dashrightarrow} W$, the general fibre of which is an elliptic curve, whereas γ has a (rational) section. An elliptic curve C with a fixed point $o \in C$ is an abelian variety, $C \cong \mathbb{C}/\Lambda$, where the point o corresponds to $o \in \mathbb{C}$, which makes its possible to define the (elliptic) reflexion $z \mapsto -z$ (in the sense of that isomorphism). Now $\tau_E \in \operatorname{Bir} V$ is a γ-fibrewise map, acting on the general fibre $\gamma^{-1}(w)$ as the elliptic reflection.

The simplest (and historically probably the first) example of a Galois involution is given by untwisting a point on a cubic hypersurface $V = V_3 \subset \mathbb{P}^{M+1}$, $M \geq 3$. Let $o \in V$ be a point, $\varphi \colon \widetilde{V} \to V$ its blow up, and $E = \varphi^{-1}(o) \subset \widetilde{V}$ the exceptional divisor. We get

$$\operatorname{Pic} V = \mathbb{Z}H, \quad \operatorname{Pic} \widetilde{V} = \mathbb{Z}H \oplus \mathbb{Z}E, \quad K_{\widetilde{V}} = -(M-1)H + (M-1)E,$$

where $H \in \operatorname{Pic} V$ is the class of a hyperplane section. The discrepancy is $a(E) = M-1$, so that E is a maximal singularity of a mobile linear system $\Sigma \subset |nH|$, if the inequality

$$\nu = \operatorname{mult}_o \Sigma > n \tag{39}$$

holds (for the strict transform $\widetilde{\Sigma}$ on $\widetilde{V}$ we have $\widetilde{\Sigma} \subset |nH - \nu E|$). Do such linear systems exist?

Let $(z_1, \dots, z_{M+1})$ be a system of affine coordinates with the origin at o,

$$f = q_1 + q_2 + q_3, \quad \deg q_i = i,$$

the Taylor decomposition of the equation $f(z_*)$ of the hypersurface V. Obviously, the closed set $Y \subset V$, consisting of all lines on V, passing through the point o, is given by the system of equations $q_1 = q_2 = q_3 = 0$.

Proposition 5.1. *A mobile linear system, satisfying the inequality* (39), *exists if and only if* $\operatorname{codim}_V Y = 2$.

Proof. For any line $L \subset \mathbb{P}^{M+1}$, $o \in L \subset V$, and an arbitrary divisor $D \in \Sigma$ we get $(D \cdot L) = n$, so that the inequality $\operatorname{mult}_o D > n$ implies that $L \subset D$. Thus if $\operatorname{codim}_V Y = 1$, then the linear system Σ cannot be mobile. Conversely, if $\operatorname{codim}_V Y = 2$, then let us consider the linear system Λ, which is cut out on V by the quadrics

$$\lambda(q_1 + q_2) + l_1(z_*)q_1,$$

where $\lambda \in \mathbb{C}$, $l_1(z_*)$ is an arbitrary linear form. Obviously, by construction

$$(q_1 + q_2)\,|\,_V = -q_3\,|\,_V, \quad q_1\,|\,_V = -(q_2 + q_3)\,|\,_V,$$

so that $\operatorname{mult}_o \Lambda \geq 3$. Since $\Lambda \subset |2H|$ and the base set of the system Λ is precisely Y, the system Λ has no fixed components. Q.E.D. for the proposition.

Obviously, for a general point $o \in V$ the set Y is of codimension 2. Assuming this, fix a mobile system Σ with the maximal singularity E. Consider the projection $\pi_o \colon \mathbb{P}^{M+1} \dashrightarrow \mathbb{P}^M$ from the point o and restrict it onto V. Obviously,

$$\pi = \pi_o\,|\,_V \colon V \dashrightarrow \mathbb{P}^M$$

is a rational map of degree 2, so that there is a Galois involution $\tau_E \in \operatorname{Bir} V$. Let $\widetilde{Y} \subset \widetilde{V}$ be the strict transform of the cone Y on $\widetilde{V}$.

Proposition 5.2. *The involution τ_E extends to a biregular automorphism of the open τ_E-invariant set $\widetilde{V} \backslash \widetilde{Y} = U$, $\tau_E \in \operatorname{Aut} U$.*

Proof. This is obvious. Q.E.D.

Since $\widetilde{Y} \subset \widetilde{V}$ is a closed set of codimension two, the action of the involution τ_E on the Picard group $\operatorname{Pic} \widetilde{V}$ is well defined.

Proposition 5.3. *The action of τ_E on the group* $\operatorname{Pic} \widetilde{V} = \mathbb{Z}H \oplus \mathbb{Z}E$ *is given by the formulas*

$$\tau_E^* H = 2H - 3E, \quad \tau_E^* = H - 2E.$$

Proof. This is obvious because the class $H - E$ is τ_E-invariant and the equality $\tau_E(E) = V \cap T_o V$ holds, where $T_o V \subset \mathbb{P}^{M+1}$ is the tangent hyperplane to V at the point o. Q.E.D. for the proposition.

Finally, let $\Sigma_1 = (\tau_E)_* \Sigma$ be the strict transform of the linear system Σ with respect to τ_E, $\widetilde{\Sigma}_1$ the strict transform of the system Σ_1 on $\widetilde{V}$. It is obvious that

$$\widetilde{\Sigma}_1 = \tau_E(\widetilde{\Sigma}) \subset |n\tau_E^* - \nu \tau_E^* E| = |n_1 H - \nu_1 E|,$$

where, by Proposition 5.2, $n_1 = 2n - \nu$ and $\nu_1 = 3n - 2\nu$.

Since by assumption $\nu > n$, for the thresholds of canonical adjunction we get the inequality

$$c(\Sigma_1) = n_1 = 2n - \nu < n = c(\Sigma).$$

Thus, $n_1 - \nu_1 = \nu - n > 0$, so that E is no longer a maximal singularity of the system Σ_1. The untwisting procedure is completed.

The second method of constructing untwisting maps (by means of special elliptic fibrations) is quite similar to the first one, although from the computational viewpoint it is more complicated. Note that if the elliptic curve C is embedded in $\mathbb{P}^2$ as a smooth cubic curve, then the elliptic involution from the point $o \in C \subset \mathbb{P}^2$ (taken as the zero of the group law on C) is the projective reflection of the cubic curve $C \subset \mathbb{P}^2$ from the point $p \in C$, which is uniquely determined by the condition

$$\mathcal{O}_C(2o + p) = \mathcal{O}_{\mathbb{P}^1}(1) \mid_C,$$

so that elliptic involutions can be easily reduced to projective reflections. However, the explicit geometric constructions of the involutions of the Galois type and elliptic involutions are essentially different, as we will see below when we consider the example of the variety $V_{2\cdot 3} \subset \mathbb{P}^5$. That example also illustrates how relations between the untwisting involutions appear, when the untwisting procedure is not uniquely determined.

The technique of untwisting maximal singularities, including description of relations between the generators of the group of birational self-maps, took its modern form in the papers of Yu. I. Manin on surfaces over non-closed fields [**M66, M67, M72**]. Among the completely studied varieties, there are not so many varieties of dimension 3 and higher with a non-trivial group of birational self-maps, that is, requiring a real application of the untwisting procedure. Here are the main types of such Fano varieties with Picard number 1:

- the double quadrics $V_4 \stackrel{2:1}{\to} Q_2 \subset \mathbb{P}^4$ (see [**I80**]);
- the complete intersection of a quadric and a cubic $V_{2\cdot 3} \subset \mathbb{P}^5$, considered below;
- the three-dimensional quartics with a double point $V_4 \subset \mathbb{P}^4$, considered in Subsection 3.4 above;
- 95 families of weighted Fano hypersurfaces ([**CPR**], see also Subsection 3.5 above);
- the Fano hypersurfaces $V_M \subset \mathbb{P}^M$, $M \geq 5$, with a singular point of multiplicity $M-2$ of general position [**Pukh02a**] (the only example in arbitrary dimension).

Also, a non-trivial untwisting procedure was applied to certain families of Fano fibre spaces: the pencils of del Pezzo surfaces and the pencils of higher-dimensional Fano varieties of index 1; see Chapters 4-7.

5.2. The complete intersection of a quadric and a cubic: generators. We study the complete intersection $V = Q \cap F \subset \mathbb{P}^5$, where Q is a quadric and F is a cubic hypersurface. The variety V is assumed to be smooth and, moreover, generic in the sense described below, in particular, $\operatorname{Pic} V = \mathbb{Z}H$, where $H = -K_V$ is the class of a hyperplane section of $V \subset \mathbb{P}^5$.

Let $L \subset V$ be a line in $\mathbb{P}^5$.

Proposition 5.4. *For the normal sheaf $\mathcal{N}_{L/V}$ there are two possible cases:*

(a) *either* $\mathcal{N}_{L/V} \cong \mathcal{O}_L(-1) \oplus \mathcal{O}_L$ *(L is the line of general type),*

(b) *or* $\mathcal{N}_{L/V} \cong \mathcal{O}_L(-2) \oplus \mathcal{O}_L(1)$ *(L is the line of non-general type).*

Moreover, the line L is of non-general type if and only if any of the following two equivalent conditions holds:

- *there exists a plane* $P \subset \mathbb{P}^5$ *such that* $L \subset P$ *and the scheme-theoretic intersection* $V \cap P$ *is not reduced everywhere along* L,
- *let* $\sigma\colon \widetilde{V} \to V$ *be the blow up of* L, $E = \sigma^{-1}(L)$ *the exceptional divisor. Then restricting to* E *the strict transform on* $\widetilde{V}$ *of a generic hyperplane section containing* L, *we get a non-ample divisor on* E.

Proof. The first claim is straightforward and easy. We leave it to the reader as an exercise. Let us show (ii).

The divisor E is a ruled surface of type $\mathbb{F}_N$, where $N = 1$ or 3 in the cases (a) and (b), respectively. Now $\operatorname{Pic} E = \mathbb{Z}s \oplus \mathbb{Z}f$ with s the class of the exceptional section and f the class of a fibre of the projection $\sigma_E\colon E \to L \cong \mathbb{P}^1$, and $s^2 = -N$. Let $h^* \in \operatorname{Pic} E$ be the class of the restriction to E of the strict transform on $\widetilde{V}$ of a general hyperplane section of V, containing the line L. It is easy to see that $(h^*)^2 = 3$, so $h^* = s + 2f$ in the case (a) and $s + 3f$ in the case (b). Obviously, h^* is ample in the case (a) only.

Now let us consider the blow up σ as the restriction of the blow up $\sigma_L\colon X_L \to \mathbb{P}^5$ of the line L in $\mathbb{P}^5$ with the exceptional divisor $E_L = \sigma_L^{-1}(L) \subset X_L$, $E_L \cong \mathbb{P}^1 \times \mathbb{P}^3$. Let $\pi\colon X_L \to \mathbb{P}^3$ be the natural projection. The fibres of π are the strict transforms of 2-planes containing L.

Obviously, the restriction $\pi|_{E_L}$ is the projection onto the second direct factor $\mathbb{P}^1 \times \mathbb{P}^3 \to \mathbb{P}^3$, and $\pi|_{E_L}$ contracts the curves that are the intersections of the strict transforms of 2-planes, containing L, with E_L, and those curves only. However, the linear system $|h^*|$ on E contracts a curve only if L is of the type (b), and, in turn, this can happen if and only if there exists a plane $P \subset \mathbb{P}^5$ such that $L \subset P$ and $\widetilde{P} \cap E_L \subset \widetilde{V}$ (that is, $\widetilde{P} \cap \widetilde{V}$ contains the corresponding fibre of the projection $\pi|_{E_L}$; as usual, $\widetilde{P} \subset X_L$ is the strict transform of P). Proof of the proposition is complete. Q.E.D.

We will consider the general complete intersections $V = Q \cap F$, satisfying the following conditions of general position:

- V does not contain lines of non-general type (it is easy to check by the usual dimension count that this condition is justified, that is, a general complete intersection satisfies it),
- there are no three lines on V lying in one plane and having a common point,
- the quadric Q is non-degenerate.

From now on, we assume that all three conditions above are satisfied. As their immediate consequence, we get

Proposition 5.5. *For any line* $L \subset V$ *and for any plane* $P \supset L$ *the intersection of* V *with a general hypersurface* $H \subset P$ *is a surface, non-singular along* L:

$$L \cap \operatorname{Sing}(H \cap V) = \emptyset.$$

Proof. Let us compare the two closed sets:

$$\bigcup_{p\in L}\{H\in\mathbb{P}^{5*}\,|\,p\in\operatorname{Sing} H\cap V\}\quad\text{and}\quad\{h\in\mathbb{P}^{5*}\,|\,H\supset P\}.$$

Both sets are irreducible surfaces, so it is sufficient to show that they are distinct. Assume the converse. Then for any point $p\in L$ we get

$$P\subset T_pV=T_pQ\cap T_pF.$$

Thus either one of the hypersurfaces Q,F contains P or both curves $Q|_P$ and $F|_P$ are non-reduced along L. We get a contradiction with our assumption. Q.E.D. for the proposition.

Let $L\subset V$ be a line. The projection $\mathbb{P}^5 \dashrightarrow \mathbb{P}^3$ from L defines a rational map $\pi_L\colon V\dashrightarrow\mathbb{P}^3$ of degree two. Set $\alpha_L\in\operatorname{Bir}V$ to be the corresponding Galois involution.

More formally, let $\sigma\colon\widetilde{V}\to V$ be the blow up of L, $E=\sigma^{-1}(L)\subset\widetilde{V}$ the exceptional divisor. The map π_L extends to a morphism $p=\pi_L\circ\sigma\colon\widetilde{V}\to\mathbb{P}^3$.

Lemma 5.1. *The morphism p is a finite morphism of degree 2 outside a closed subset $W\subset\widetilde{V}$ of codimension two, and $p(W)\subset\mathbb{P}^3$ is a finite set of points. The involution α_L extends to a biregular involution of $\widetilde{V}\setminus W$. Its action on $\operatorname{Pic}\widetilde{V}=\mathbb{Z}H\oplus\mathbb{Z}E$ is given by the formulas*

$$\alpha_L^*(H)=4H-5E,\quad \alpha_L^*(E)=3H-4E.$$

Proof. The projection $p\colon\widetilde{V}\to\mathbb{P}^3$ is a finite morphism outside the set $W\subset\widetilde{V}$ that consists of curves that are contracted by the morphism p. We will show there are finitely many of them. Set $H'=nH-\nu E$ and $E'=mH-\mu E$ to be the classes in $\operatorname{Pic}\widetilde{V}$ of the strict transform of a general hyperplane section and the divisor E with respect to α_L. The linear system $|H-E|$ is clearly invariant under α_L. Take a general surface $S\in|H-E|$.

Since $K_S=0$, the birational involution $\alpha_L\,|\,_S$ extends to a biregular involution of this surface. Denote it by α_S, and the restrictions of H and E to S by H_S and E_S, respectively. We get

$$\alpha_S^*H_S=nH_S-\nu E_S,\quad \alpha_S^*E_S=mH_S-\mu E_S$$

and the class H_S-E_S is α_S^*-invariant, whence we get $n=m+1$, $\nu=\mu+1$. Since α_S is an automorphism,

$$(\alpha_S^*H_S\cdot(H_S-E_S))=(H_S\cdot(H_S-E_S))=5$$

and $(\alpha_S^*H_S)^2=(H_S)^2=6$, whence by the obvious equalities $(H_S\cdot E_S)=1$, $(E_S^2)=-2$ we get the following two possibilities for n,m,ν,μ:

- either $H'=4H-5E$, $E'=3H-4E$,
- or $H'=H$, $E'=E$,

the latter being clearly impossible because α_L cannot be extended to a biregular automorphism of V.

By construction, the system $|4H-5E|$ is mobile. However, if a curve C is contracted by the morphism p, then $(C\cdot(H-E))=0$ and therefore $(C\cdot H')<0$. We conclude that there can be only finitely many such curves. Q.E.D.

Now let $P \subset \mathbb{P}^5$ be a 2-plane such that $P \cap V$ is a union of three lines, $P \cap V = L \cup L_1 \cup L_2$. This is possible only if $P \subset Q$. Let $\sigma\colon \widetilde{V} \to V$ be the composition of three blow ups: first, we blow up L, then the strict transform of L_1, then the strict transform of L_2.

We denote the exceptional divisors on $\widetilde{V}$, corresponding to the lines L, L_1, L_2, by the symbols E, E_1, E_2, respectively.

Lemma 5.2. *The involution α_L extends to a biregular involution on $\widetilde{V} \setminus W$, where W is a closed subset of codimension two. The action of α_L on* $\operatorname{Pic}\widetilde{V} = \mathbb{Z}H \oplus \mathbb{Z}E \oplus \mathbb{Z}E_1 \oplus \mathbb{Z}E_2$ *is given by the formulae:*

$$\begin{aligned} \alpha_L^* &= 4H - 5E - 2E_1 - 2E_2, \\ \alpha_L^* E &= 3H - 4E - 2E_1 - 2E_2, \\ \alpha_L^* E_i &= E_j, \end{aligned}$$

where $\{i, j\} = \{1, 2\}$.

Proof is similar to the previous argument. Q.E.D.

Now let us consider conics on the variety V.

It is easy to see that there is a one-dimensional family of irreducible conics $Y \subset V$ such that the plane $P(Y) = \langle Y \rangle$ is contained entirely in the quadric Q. Obviously,

$$P(Y) \cap V = Y \cup L(Y),$$

where $L(Y)$ is the residual line. We will call the conics described above the *special conics.*

Every special conic Y generates the following construction. Set $P = P(Y)$. Consider the projection $\pi_P\colon \mathbb{P}^5 \dashrightarrow \mathbb{P}^2$ from the plane P. The fibres of π_P are 3-planes $S \supset P$, so that $S \cap Q = P \cup P(S)$, where $P(S)$ is the residual plane. Therefore, π_P fibres V over $\mathbb{P}^2$ into elliptic curves $C_S = P(S) \cap F$, that is, plane cubics. A general curve C_S intersects the residual line $L(Y)$ on one point, which is $L(Y) \cap P(S)$. We define the involution $\beta_Y \in \operatorname{Bir} V$ as a fibrewise map, setting $\beta_Y|_{C_S}$ to be the elliptic reflection, where the group law on C_S is defined by the point $L(Y) \cap P(S)$ as the zero.

Let $\sigma\colon \widetilde{V} \to V$ be the composition of the blow up of the conic Y and the blow up of the strict transform of the line $L(Y)$, E and E^+ be the corresponding exceptional divisors. Obviously, $\pi_P \circ \sigma\colon V \to \mathbb{P}^2$ is a morphism, the general fibre of which is an elliptic curve C_t, $t \in \mathbb{P}^2$. The divisor E^+ is a section of this elliptic fibration, $(E^+ \cdot C_t) = 1$.

Lemma 5.3. *The birational involution β_Y extends to a biregular involution on the complement $\widetilde{V} \setminus W$, where W is a closed subset of codimension two, and moreover, $\pi_P \circ \sigma(W) \subset \mathbb{P}^2$ is a finite set. The action of β_Y on* $\operatorname{Pic}\widetilde{V} = \mathbb{Z}H \oplus \mathbb{Z}E \oplus \mathbb{Z}E^+$ *is given by the formulas*:

$$\begin{aligned} \beta_Y^* H &= 13H - 14E - 8E^+, \\ \beta_Y^* E &= 12H - 13E - 8E^+, \\ \beta_Y^* E^+ &= E^+. \end{aligned}$$

Proof is quite similar to the proof of Lemma 5.2. Q.E.D.

As we will show in the next subsection, the involutions α_L and β_Y make it possible to untwist the maximal singularities of any mobile linear system on the variety V.

5.3. The complete intersection of a quadric and a cubic: relations. Let $P \subset \mathbb{P}^5$ be a plane such that $P \subset Q$ and $P \cap F = L_1 \cup L_2 \cup L_3$ is a union of three lines.

Lemma 5.4. *The following relation holds:*

$$(\alpha_{L_1} \circ \alpha_{L_2} \circ \alpha_{L_3})^2 = \mathrm{id}_V .$$

Proof. Obviously, each of the three involutions α_{L_i} preserves the fibres of the projection $\pi_P \colon V \dashrightarrow \mathbb{P}^2$ from the plane P. Recall that a general fibre $\pi_P^{-1}(t)$ is a cubic curve C_t, where $C_t \cap P = \{x_1, x_2, x_3\}$, $x_i = C_t \cap L_i$. Take a point $x \in C_t$; obviously,

$$\alpha_{L_i}(x) + x + x_i \sim x_1 + x_2 + x_3$$

on C_t. Therefore we compute:

$$\begin{array}{rcl} \alpha_{L_3}(x) & \sim & x_1 + x_2 - x, \\ \alpha_{L_2} \circ \alpha_{L_3}(x) & \sim & x_3 - x_2 + x, \\ \alpha_{L_1} \circ \alpha_{L_2} \circ \alpha_{L_3}(x) & \sim & 2x_2 - x, \\ (\alpha_{L_1} \circ \alpha_{L_2} \circ \alpha_{L_3})^2(x) & \sim & x, \end{array}$$

which is what we need. Q.E.D.

After this preparatory work we can formulate the main theorem describing birational geometry of V.

Set $\mathcal{L}$ and $\mathcal{C}$ to be the sets of lines and special conics on V, respectively. Let G^+ be the free group generated by symbols A_L and B_Y for all $L \in \mathcal{L}$ and $Y \in \mathcal{C}$, respectively. Let $R^+ \subset G^+$ be the normal subgroup, generated by the words A_L^2 for all $L \in \mathcal{L}$, B_Y^2 for all $Y \in \mathcal{C}$ and, finally, $(A_{L_1} A_{L_2} A_{L_3})^2$ for all triples of distinct lines $L_1, L_2, L_3 \in \mathcal{L}$ such that $\langle L_1 \cup L_2 \cup L_3 \rangle = \mathbb{P}^2$.

Set $G = G^+/R^+$ to be the quotient group. We construct a semi-direct product $G \operatorname{Aut} V$ using the obvious action of $\operatorname{Aut} V$ on G: for $\rho \in \operatorname{Aut} V$ set

$$\rho A_L \rho^{-1} = A_{\rho(L)}, \quad \rho B_Y \rho^{-1} = B_{\rho(Y)}.$$

Let $\varepsilon \colon \operatorname{Aut} V \to \operatorname{Bir} V$ be the homomorphism, sending A_L to α_L, B_Y to β_Y and identical on $\operatorname{Aut} V$.

Theorem 5.1. *The variety V is birationally rigid and ε is an isomorphism of groups.*

Proof. Set $\mathcal{B} = \mathcal{L} \cup \mathcal{C}$. Take any mobile linear system $\Sigma \subset |nH|$ on V. Obviously, $c(\Sigma, V) = n$. In order to prove that ε is a bijection, we take Σ to be the strict transform of the linear system $|H|$ of hyperplane sections with respect to a fixed birational self-map $\chi \in \operatorname{Bir} V$. Clearly, in that case $n = 1$ if and only if $\chi \in \operatorname{Aut} V$ (and by construction biregular automorphisms are in the image of ε).

We will prove birational rigidity and surjectivity of ε simultaneously, using the following crucial technical fact.

Theorem 5.2. *Assume that $c_{\mathrm{virt}}(\Sigma) < n$. Then there exist a subvariety $B \in \mathcal{B}$ (that is, a line or a special conic) such that $\operatorname{mult}_B \Sigma > n$. Moreover, there are at* *most two subvarieties in $\mathcal{B}$ with that property, and if there are two, say $B_1, B_2 \in \mathcal{B}$, then they are lines, $B_1, B_2 \in \mathcal{L}$, their span $\langle B_1, B_2\rangle$ is a plane $P = \mathbb{P}^2$, and $P \subset Q$.*

Proof will be given below in Section 6.

Lemma 5.5. (i) *Let $L \subset V$ be a line, $\Sigma^+ \subset |n_+H|$ the strict transform of the linear system Σ with respect to α_L. The following equalities hold:*

$$n_+ = 4n - 3\,\mathrm{mult}_L\,\Sigma, \quad \mathrm{mult}_L\,\Sigma^+ = 5n - 4\,\mathrm{mult}_L\,\Sigma.$$

(ii) *Let $Y \in \mathcal{C}$ be a special conic, $L = L(Y) \in \mathcal{L}$ the residual line, and $\Sigma^+ \subset |n_+H|$ the strict transform of the linear system Σ with respect to β_Y. The following equalities hold:*

$$n_+ = 13n - 12\,\mathrm{mult}_Y\,\Sigma, \quad \mathrm{mult}_Y\,\Sigma^+ = 14n - 13\,\mathrm{mult}_Y\,\Sigma,$$
$$\mathrm{mult}_L\,\Sigma^+ = 8n - 8\,\mathrm{mult}_Y\,\Sigma + \mathrm{mult}_L\,\Sigma.$$

(iii) *Let $P \subset \mathbb{P}^5$ be a 2-plane such that $P \cap V = L \cup L_1 \cup L_2$, Σ^+ as in (i) above. Then for $\{i, j\} = \{1, 2\}$ we have*

$$\mathrm{mult}_{L_i}\,\Sigma^+ = 2n - 2\,\mathrm{mult}_L\,\Sigma + \mathrm{mult}_{L_j}\,\Sigma.$$

Proof is a straightforward application of Lemmas 5.1–5.3. Q.E.D.

Corollary 5.1. *An involution $\tau = \alpha_L$ or β_Y satisfies the inequality $n_+ < n$ if and only if L or Y is a maximal curve of the linear system Σ, respectively, where $\Sigma^+ \subset |n_+H|$ is the strict transform of Σ with respect to τ.*

Corollary 5.2. *In the notations of the previous corollary assume that $n_+ = n$. Then $\tau = \alpha_L$ for line $L \in \mathcal{L}$ and there exist lines $L_1, L_2 \in \mathcal{L}$, such that $L \cup L_1 \cup L_2 = P \cap V$, where $P \subset Q$ is a plane.*

Now let us prove birational rigidity of V and surjectivity of ε. Assume that $c_{\mathrm{virt}}(\Sigma) < n$ for a mobile linear system Σ. By Theorem 5.2, there exists a curve $B \in \mathcal{B}$ such that $\mathrm{mult}_B\,\Sigma > n$. Let $\tau \in \varepsilon(G)$ be the corresponding involution (that is, $\tau = \alpha_L$ if $B = L \in \mathcal{L}$ and $\tau = \beta_Y$ if $B = Y \in \mathcal{C}$). By Corollary 5.1, $\Sigma^+ \subset |n_+H|$ with $n_+ < n$, where Σ^+ is the strict transform of Σ with respect to τ. Iterating this procedure, we construct a sequence of involutions $\tau_i \in \varepsilon(G)$ such that the strict transforms $\Sigma^{(i)} \subset |n_iH|$ of the system Σ with respect to the compositions $\tau_i \dots \tau_1$ satisfy the inequalities $n_i < n_{i-1}$. Since $n_i \in \mathbb{Z}_+$, at some step we cannot decrease the threshold $c(\Sigma^{(i)}, V)$ any longer. Therefore, for some $k \geq 1$ we get

$$c(\Sigma^{(k)}, V) = c_{\mathrm{virt}}(\Sigma^{(k)}, V) = c_{\mathrm{virt}}(\Sigma, V),$$

which is birational rigidity. Moreover, if we fix a birational self-map $\chi \in \mathrm{Bir}\,V$ and take Σ to be the strict transform of the system $|H|$ via χ, then the procedure described above gives $n_k = 1$ for some k, that is, $\Sigma^{(k)} \subset |H|$. Comparing dimensions, we get $\Sigma^{(k)} = |H|$, which implies that $\tau_k \dots \tau_1 \chi \in \mathrm{Aut}\,V$ is a biregular map. This proves surjectivity of ε.

The last step in the proof of Theorem 5.1 is to show that ε has the trivial kernel.

For convenience of notations, we denote words in A_L, B_Y, using capital letters and corresponding birational self-maps using small letters, say, $t = \varepsilon(T)$, etc. For a self-map $t \in \operatorname{Bir} V$ we define the integer $n(t) \in \mathbb{Z}_+$ by the formula $\Sigma \subset |n(t)H|$, where Σ is the strict transform of the system $|H|$ via t; obviously, $n(t) = 1$ if and only if $t \in \operatorname{Aut} V$. Theorem 5.1 immediately follows from

Proposition 5.6. *Let $W = T_1 \dots T_l$ be an arbitrary word in the alphabet $\{A_L, B_Y \mid L \in \mathcal{L}, Y \in \mathcal{C}\}$. If $w \in \operatorname{Aut} V$ then using the relations in R^+ one can transform the word W into the empty word.*

Proof. Denote by W_i, $i \leq l(W) = l$, the left segment of the word W of length i, that is, $W_i = T_1 \dots T_i$. Set

$$n^*(W) = \max\{n(w_i) \mid 1 \leq i \leq l(W)\},$$

$$\omega(W) = \sharp\{i \mid n(w_i) = n^*(W), 1 \leq i \leq l(W)\}.$$

Now we associate with every word W the ordered triple

$$(n^*(W), \omega(W), l(W)).$$

We order the set of words, setting $W > W'$, if either $n^*(W) > n^*(W')$, or $n^*(W) = n^*(W')$ and $\omega(W) > \omega(W')$, or $n^*(W = n^*(W')$, $\omega(W) = \omega(W')$ and $l(W) > l(W')$. It is easy to see that every decreasing chain of words $W^{(1)} > W^{(2)} > \dots$ breaks. Therefore, it is sufficient to show that if $w \in \operatorname{Aut} V$, then the word W can be transformed into a word W' such that $W > W'$, $w = w'$.

If the word W contains the subword $A_L A_L$ or $B_Y B_Y$, then, eliminating this subword, we get a smaller word W' (because the image of each left segment of the word W' coincides with the image of some left segment of the word W and the map of the set of left segments of W' into the set of left segments of W is injective).

So we can assume that W does not contain subwords $A_L A_L$ or $B_Y B_Y$.

Since $n(w) = 1$, we can assume that $n^*(W) \geq 2$ (otherwise there is nothing to prove). Let $s = \min\{i \mid n(w_i) = n^*(W)\} \leq l(W) - 1$. Let us consider the two cases $T_s = A_L$ and $T_s = B_Y$ separately.

Case 1. $T_s = B_Y$. In this case $n(w_{s-1}) = n(w_s \beta_Y) < n(w_s)$, by the choice of s. By Corollary 5.1, $\operatorname{mult}_Y \Sigma_s > n(w_s)$, where Σ_s is the strict transform of $|H|$ via w_s. Since by construction $n(w_{s+1}) \leq n(w_s)$, we get $T_{s+1} = T_s = B_Y$. A contradiction to our assumption that W does not contain subwords $A_L A_L$ and $B_Y B_Y$.

Case 2. Let $T_s = A_L$. By the choice of s we get

$$\operatorname{mult}_L \Sigma_s > n(w_s).$$

By assumption, $T_{s+1} \neq T_s$ and $n(w_{s+1}) \leq n(w_s)$. By Corollary 5.2, $T_{s+1} = A_{L'}$, where $L' \subset V$ is a line such that there exists a third line $Z \subset V$,

$$L \cup L' \cup Z = P \cap V$$

for some plane $P \subset Q$.

Lemma 5.6. (i) *Z is a maximal line of the map w_{s-1}, that is, $\operatorname{mult}_Z \Sigma_{s-1} > n(w_{s-1})$. Therefore,*

$$n(w_{s-1}\alpha_Z) < n(w_{s-1}).$$

(ii) *The equality*

$$n(w_{s-1}\alpha_Z) - \operatorname{mult}_{L'} \Sigma' = n(w_s) - \operatorname{mult}_{L'} \Sigma_s \leq 0$$

holds, where Σ' is the strict transform of $|H|$ with respect to $w_{s-1}\alpha_Z$. Therefore,

$$n(w_{s-1}\alpha_Z\alpha_{L'}) \leq n(w_{s-1}\alpha_Z).$$

Proof. Straightforward computations based on Lemma 5.5. We will consider the claim (i) only, leaving (ii) to the reader. Since $w_s = w_{s-1}\alpha_L$, we get $w_{s-1} = w_s\alpha_L$ and by Lemma 5.5,

$$n(w_{s-1}) = n(w_s\alpha_L) = 4n(w_s) - 3\,\mathrm{mult}_L\,\Sigma_s,$$

$$\mathrm{mult}_Z\,\Sigma_{s-1} = 2n(w_s) - 2\,\mathrm{mult}_L\,\Sigma_s + \mathrm{mult}_{L'}\,\Sigma_s.$$

Therefore, $n(w_{s-1}) - \mathrm{mult}_Z\,\Sigma_{s-1} = 2n(w_s) - \mathrm{mult}_L\,\Sigma_s - \mathrm{mult}_{L'}\,\Sigma_s < 0$, which is what we need. For the claim (ii), the arguments are similar. Q.E.D.

Finally, let us complete the proof of Theorem 5.1. Consider first the case when $\mathrm{mult}_{L'}\,\Sigma_s > n(w_s)$. Using the relations $A_Z^2 = e$ and $A_Z A_{L'} A_L = A_L A_{L'} A_Z$, we can replace the subword $A_L A_{L'}$ by the subword $A_Z A_{L'} A_L A_Z$. This operation increases the length. Denote the new word by W^+.

Obviously, $W_i^+ = W_i$ for $i \leq s-1$. Furthermore,

$$w_s^+ = w_{s-1}\alpha_Z, \quad w_{s+1}^+ = w_{s-1}\alpha_Z\alpha_{L'}$$

and $w_{s+2}^+ = w_{s-1}\alpha_Z\alpha_{L'}\alpha_L = w_{s+1}\alpha_Z$, whereas

$$w_{s+i}^+ = w_{s+i-2}$$

for $i \geq 3$. By the lemma above, $n(w_i^+) < n(w_s) = n^*(W)$ for $i = s, s+1, s+2$ (and by construction this is true for the smaller values $i < s$, as well). Therefore, if $\omega(W) \geq 2$, then $n^*(W^+) = n^*(W)$ and $\omega(W^+) = \omega(W) - 1$. If $\omega(W) = 1$, then $n^*(W^+) < n^*(W)$. In any case, $W^+ < W$.

It remains to consider the case $\mathrm{mult}_{L'}\,\Sigma_s = n(w_s)$. In this case $n(w_{s+1}) = n(w_s)$, $\mathrm{mult}_{L'}\,\Sigma_{s+1} = n(w_{s+1})$. Since by assumption there are no subwords $A_{L'}A_{L'}$, we must have $T_{s+2} = A_Z$. Now let us replace the subword

$$T_s T_{s+1} T_{s+2} = A_L A_{L'} A_Z$$

by the subword $A_Z A_{L'} A_L$. Denote the new word by W^+. Now the length is the same, and by Lemma 5.5 we obtain the inequalities $n(w_i^+) < n^*(W)$ for $i = s, s+1, s+2$. Arguing as in the previous case, we complete the proof.

Q.E.D. for Theorem 5.1.

6. Complete intersections $V_{2\cdot 3} \subset \mathbb{P}^5$. II: Excluding maximal singularities

In this section, we prove Theorem 5.2, describing precisely for a mobile system Σ on the complete intersection $V_{2\cdot 3}$, what maximal singularities can it have. This is the third section of the chapter devoted to the methods of excluding maximal singularities. The technique of counting multiplicities works here to its full power, complemented by various geometric constructions, using the projective embedding $V \subset \mathbb{P}^5$.

6.1. The first steps. Maximal curves. Proof of Theorem 5.2 is hard and requires a lot of tedious small work. To begin with, we break it into two major steps, introducing the following intermediary claim.

Theorem 6.1. *In the assumptions of Theorem 5.2 there exists an irreducible curve $B \subset V$ which is a maximal curve for the system $\Sigma : \operatorname{mult}_B \Sigma > n$.*

We need to prove Theorem 6.1 and then deduce Theorem 5.2 from that theorem. The arguments for both parts are both and lengthy and independent of each other. Reversing the order, we start with proving Theorem 5.2, assuming that Theorem 6.1 is shown.

First, as $\deg V = 6$, we can conclude that if $B_1, \dots, B_k$ are distinct irreducible curves on V such that $\operatorname{mult}_{B_i} \Sigma > n$, then $\sum_{i=1}^{k} \deg B_i \leq 5$.

Lemma 6.1. *Let $C \subset V$ be an irreducible curve not contained in any 3-plane. Then* $\operatorname{mult}_C \Sigma \leq n$.

Proof. We may assume that deg $= 4$ or 5. We need to consider the following four cases:

(Case 1) C is a rational normal curve in some 4-plane, $\deg C = 4$;

(Case 2) C is a rational normal curve in $\mathbb{P}^5$;

(Case 3) C is a rational curve of degree 5 in some 4-plane with at most one double point;

(Case 4) C is a smooth elliptic curve of degree 5 in some 4-plane.

First assume that C is smooth. Put $d = \deg C$ and let $m \in \mathbb{Z}_+$ be an integer such that C is cut out by hypersurfaces of degree m in $\mathbb{P}^5$. Let $\sigma \colon \widetilde{V} \to V$ be the blowup of C, $E = \sigma^{-1}(C)$ the exceptional divisor, and set $\nu = \operatorname{mult}_C \Sigma$. The linear system $|mH - E|$ on $\widetilde{V}$ is by definition free, the linear system $|nH - \nu E|$ is free from fixed components.

Therefore,

$$
\begin{gathered}
((nH - \nu E)^2 \cdot (mH - E)) \geq 0 \\
\| \\
6mn^2 - dm\nu^2 - 2nd\nu - \nu^2(2 - 2g - d),
\end{gathered}
$$

where $g \in \{0, 1\}$ is the genus of C. Considering the expression above as a quadratic form in n, ν, we see that if the sum of its coefficients is negative, that is,

$$
(6 - d)m - (d + 2 - 2g) \leq 0,
$$

then $\nu \leq n$; and, indeed, if $d = 4$, we can take $m = 2$, and if $d = 5$, we can take $m = d \leq d + (2 - 2g)$. This completes the non-singular case.

If C is singular, then we are in (Case 3), $d = 5$. Blow up the singular point $p \in C$ first: $\sigma \colon V_1 \to V$, then the strict transform $\widetilde{C} \subset V_1$: $\sigma_2 \colon \widetilde{V} \to V_1$. Let $E_1 = \sigma_1^{-1}(p)$ and $E = \sigma_2^{-1}(\widetilde{C})$ be the exceptional divisors. Projecting the curve C from a general chord $[q_1, q_2] \cong \mathbb{P}^1$, $q_i \in C$, we can see that the linear system $|3H - 2E_1 - E|$ is free. Putting $\nu_1 = \operatorname{mult}_p \Sigma \geq \nu$, we can see that linear system $|nH - \nu_1 E_1 - \nu E|$ is free from fixed components. Therefore, as in the non-singular case above,

$$
\begin{gathered}
((3H - 2E_1 - E) \cdot (nH - \nu_1 E_1 - \nu E)^2) \geq 0, \\
\| \\
18n^2 - 10n\nu - 8\nu^2 - 2(\nu - \nu_1)^2
\end{gathered}
$$

which obviously implies the desired inequality $\nu \leq n$. Q.E.D. for the lemma.

Lemma 6.2. *Assume that $B = B_1 \cup B_2$ is a union of two irreducible curves and*

$$4 \le \dim\langle B\rangle.$$

Assume further that one of the following two cases takes place: either

(i) $\deg B_1 = \deg B_2 = 2$, *or*

(ii) $\deg B_i = i$, $i = 1, 2$.

Then if one of the two numbers $\nu_i = \operatorname{mult}_{B_i} \Sigma$ is higher than n, the other is (strictly) smaller than n.

Proof. We consider the case (i). Let $P = \langle B_1\rangle$ be the plane containing B_1. It is easy to show that the section of V by a general hyperplane $H \supset P$ is a non-singular surface $S = S_H$. Let $x_H \notin P$ be a point of the intersection $H \cap B_2$ *outside* P. Then the restriction $\Sigma_H = \Sigma|_S$ of Σ to S_H is a linear system of curves with the fixed component B_1 of multiplicity ν_1 with the base point x_H of multiplicity ν_2 and, finally, if B_1 is a special conic, then Σ_H may have another fixed component $L = L(B_1) = (F \cap P)\backslash B_1$. Putting $\nu = \operatorname{mult}_L \Sigma_H$ and computing the self-intersection of the mobile part of Σ_H, we obtain the inequality

$$6n^2 - 4n\nu_1 - 2n\nu - 2\nu_1^2 - 2\nu^2 + 4\nu\nu_1 - \nu_2^2 \ge 0$$

(if B_1 is a non-special conic, just put $\nu = 0$). The maximum of this quadratic function of ν is attained at $\nu = \nu_1 - \frac{n}{2}$ and is equal to

$$\frac{13}{2}n^2 - 6n\nu_1 - \nu_2^2 \ge 0.$$

Swapping B_1 and B_2, we get by symmetry that

$$\frac{13}{2}n^2 - 6n\nu_2 - \nu_1^2 \ge 0.$$

Putting the two inequalities together, we get the claim of our lemma in the case (i). The case (ii) is shown in a similar way. Q.E.D.

Lemma 6.3. *Let $P = \mathbb{P}^2$ be a 2-plane and*

$$P \cap V = B = \sum_{i=1}^{k} B_i$$

with irreducible B_i's. Set $d_i = \deg B_i$ and $\mu_i = \operatorname{mult}_{B_i} \Sigma$. Then the following inequality holds:

$$\sum_{i=1}^{k} d_i\mu_i \le n \deg B = n\sum_{i=1}^{k} d_i$$

Proof. Take a general 3-plane $S \supset P$. Two cases are possible.

Case 1. $P \subset Q$, so $\deg B = 3$ is a (possibly reducible) plane cubic curve. We have $S \cap Q = P \cup P(S)$, where $P(S) \subset S$ is another 2-plane and $C_S = P(S) \cap F$ is an irreducible cubic curve. Obviously,

$$C_S \cap P = C_S \cap B$$

consists of 3 distinct points. Restricting Σ to C_S, we get a linear series of degree $3n$ with the base part of degree at least

$$\sum_{i=1}^{k} \sharp(B_i \cap C_S)\mu_i = \sum_{i=1}^{k} d_i\mu_i,$$

which is what we need.

Case 2. $P \not\subset Q$, so $B = Q \cap P$ is a (possibly reducible) conic. In this case $Q_S = Q \cap S$ is an irreducible quadric and $B = P \cap Q_S$ is its plane section. In this case we restrict the system Σ onto the residual curve C_S where

$$C_S \cup B = Q_S \cap F.$$

We get a linear series of degree $n \deg C_S = 4n$ and again the base part of the series is of degree

$$\sum_{i=1}^{k} \sharp(B_i \cap C_S)\mu_i.$$

Since B is an irreducible conic or a pair of intersecting lines, the last expression is not less than

$$\frac{1}{2} \deg C_S \sum_{i=1}^{k} d_i\mu_i.$$

This completes the proof. Q.E.D.

The next claim will be needed in our case by case study of possible configurations of curves.

Lemma 6.4. *Assume that $p \in V$ is a point such that either there are no lines on V through p, or there are no lines on V through p, or there are 6 such lines. Then the inequality $\nu_p = \operatorname{mult}_p \Sigma \leq 2n$ holds.*

Proof. Assume first that there are no lines through p. Blowing p up: $\sigma\colon \widetilde{V} \to V$, $E = \sigma^{-1}(p)$, we claim that the class $(2H - 3E)$ has non-negative intersection with any curve on V. Indeed,

$$2H - 3E = (H - E) + (H - 2E),$$

so that $(2H - 3E)$ could have a negative intersection with a curve $\widetilde{C}$ such that $C \subset V \cap T_pV$ only. It is easy to check explicitly that for every such possible curve,

$$((2H - 3E) \cdot \widetilde{C}) \geq 0.$$

Now we compute

$$((nH - \nu_pE)^2 \cdot (2H - 3E)) = 12n^2 - 3\nu_p^2 \geq 0,$$

as required.

If there are 6 lines through p, then in the above notations consider the linear system $|H - 2E|$ of strict transforms of tangent hyperplanes.

It is easy to see that the restriction

$$|H - 2E|\big|_E$$

is a fixed plane conic in $E = \mathbb{P}^2$. Therefore, there is a hyperplane $H^* \supset T_pV$ such that $\mu = \operatorname{mult}_p H^* \cap V \geq 3$, so that the linear system $|H - \mu E|$ is non-empty.

Computing

$$((nH - \nu_p E) \cdot (H - \mu E) \cdot (H - E)) = 6n - \mu\nu_p \geq 0,$$

we complete the proof. Q.E.D.

6.2. The Plane Section Lemma. The following claim is of crucial importance for proving Theorem 6.1.

Proposition 6.1 (the Plane Section Lemma). *Let $S \subset \mathbb{P}^5$ be an arbitrary 3-plane,*

$$C = V \cap S = \sum_{i=1}^{r} m_i C_i$$

the decomposition of the effective 1-cycle $C = V \cap S$ into irreducible components. Set $\theta_i^* = \mathrm{mult}_{C_i} \Sigma$. Then the following claims hold:

(i) *if $\theta_i^* > n$, then C_i is a line or a special conic;*

(ii) *if $\theta_i^* + \theta_j^* > 2n$ for $i \neq j$, then $C_i \cup C_j$ is a special reducible conic;*

(iii) *$\theta_i^* + \theta_j^* + \theta_k^* \leq 3n$ for all triples of distinct integers $\{i, j, k\} \subset \{1, \ldots, r\}$;*

(iv) *if $S = T_p V$ for some point $p \in V$ and not all the curves $C_i \ni p$ are lines, then*

$$\nu_p = \mathrm{mult}_p \Sigma \leq 2n.$$

The proof of the Plane Section Lemma is hard and tedious. Below we explain a method of proving it which seems to be more or less optimal, minimizing the case by case study, and show how it works in some particular cases. Right now let us prove Theorem 5.2.

By Theorem 6.1, maximal curves do exist. By Lemmas 6.1–6.4 and Proposition 6.1, the maximal curves are either lines or special conics. Moreover, two special conics $Y_1 \neq Y_2$ cannot both satisfy the inequalities $\mathrm{mult}_{Y_i} \Sigma > n$. Indeed, if $\dim\langle Y_1 \cup Y_2 \rangle \geq 4$, then we apply Lemma 6.2; if $\dim\langle Y_1 \cup Y_2 \rangle = 3$, then we apply the Plane Section Lemma. Similarly, if Y is a special conic and L is a line, then we show that both of them cannot be maximal curves for the system Σ: if $\dim\langle Y \cup L \rangle \geq 4$, then by Lemma 6.2; if $\dim\langle Y \cup L \rangle = 3$, then by the Plane Section Lemma, finally, if $L = L(Y)$, then by Lemma 6.3. Since two lines always span at most a 3-plane, the Plane Section Lemma completes the proof of Theorem 5.2.

Now we proceed to explaining the method of proving the Plane Section Lemma. For the full details and the case by case computations, see [**IP**]. The proof consists of several steps.

Step 1. Resolving the singularities. Let Λ_S be the pencil cut out on V by hyperplanes containing the 3-plane S. Obviously, $\mathrm{Bs}\,\Lambda_S = C$. We resolve the singularities of Λ_S in three stages:

- first, we blow up points and get a birational morphism $\varphi_1 \colon V^{(1)} \to V$ such that the base locus of the strict transform $\Lambda_S^{(1)}$ of Λ_S on $V^{(1)}$ is a union of non-singular curves and at every point $p \in V^{(1)}$ at most two of these curves meet transversally; if $S = T_q V$ is the tangent space, then we start with the blow up of q;

- second, we blow up the curves $C_i^{(1)} \subset V^{(1)}$ and get a birational morphism $\varphi_2\colon V^{(2)} \to V^{(1)}$;
- third, we complete the task, blowing up the remaining base curves of the pencil $\Lambda_S^{(2)}$, and get a birational morphism $\varphi_3\colon V^* \to V^{(2)}$ such that the strict transform Λ_S^* of Λ_S is a base point free linear system.

Obviously, for two distinct divisors $D_1^*, D_2^* \in \Lambda_S^*$ we have $(D_1^* \circ D_2^*) = 0$. Let $T_1, \dots, T_N$ be all prime divisors on V^*, exceptional for $\varphi = \varphi_1 \circ \varphi_2 \circ \varphi_3\colon V^* \to V$, where $T_1, \dots, T_r$ correspond to the blow ups of the curves $C_1, \dots, C_r$ (the second stage of our resolution). Consider the lattice

$$\Delta = \bigoplus_{i=1}^{N} \mathbb{Z}T_i,$$

equipped with the integral symmetric bilinear form

$$\langle x, y\rangle = \left(x \cdot y \cdot \left(H - \sum_{i=1}^{N} \lambda_i T_i\right)\right),$$

where $|H - \Sigma\lambda_i T_i| = \Lambda_S^*$, $\lambda_1 = \dots = \lambda_r = 1$.

Step 2. Reduction to the essential sublattice. Assume first that $S \neq T_pV$ for any $p \in V$. Then there is a natural and straightforward procedure of constructing classes

$$T_i^* = T_i + \sum_{j\in\mathcal{P}} a_{ij}T_j, \quad i = 1, \dots, r,$$

where the exceptional divisors T_j, $j \in \mathcal{P}$, correspond to the blow ups of points, carried out at the first stage of the resolution, such that

$$T_i^* \in \left(\bigoplus_{j\in\mathcal{P}} T_j\right)^{\perp}.$$

If $S = T_pV$ for some point $p \in V$, assume that T_{r+1} corresponds to the blow up of p and extend the procedure above to the class T_{r+1}^*, taking $\mathcal{P}$ to be the same as above, *except for* $r + 1 \notin \mathcal{P}$. The coefficients a_{ij} are naturally calculated in terms of the multiplicities of the curves C_i at the corresponding points. They are all non-negative: $a_{ij} \geq 0$. Set

$$\Delta^* = \bigoplus \mathbb{Z}T_i^* \subset \Delta$$

to be *the essential sublattice*. We will consider for simplicity the case $S \neq T_pV$.

The procedure of constructing the classes $T_1^*, \dots, T_r^*$ is in fact just the orthogonal projection

$$p\colon \Lambda \to \left(\bigoplus_{j\in\mathcal{P}} T_j\right)^{\perp}.$$

Combining p with the projection onto the essential sublattice Λ^*, we eventually get the map

$$\pi\colon \Lambda \to \Lambda^*,$$

satisfying the following properties:

(1) $\pi(H) = T_1^* + \cdots + T_r^*$;

(2) $\pi(D^*)$ is non-negative on any non-negative class $a_1T_1^* + \cdots + a_rT_r^*$, $a_i \geq 0$, for any effective divisor D^* on V^*.

If D^* is the strict transform on V^* of a generic divisor $D \in \Sigma$, then it is easy to check that

$$\pi(D^*) = n\pi(H) - \sum_{i=1}^{r} \theta_i T_i^* = \sum_{i=1}^{r}(n - \theta_i)T_i^*.$$

Step 3. Computations in the essential sublattice. Given the above, the claim of the Plane Section Lemma follows from solving a problem of elementary linear algebra: given a class

$$x = \sum_{i=1}^{r} x_i T_i^* \in \Delta^*,$$

we must show that if $\langle x \cdot T_i^* \rangle \geq 0$ for all $i = 1, \ldots, r$, then

(i) $x_i < 0$ can only occur if C_i is a line or a special conic;

(ii) if $x_i + x_j < 0$ for $i \neq j$, then $C_i \cup C_j$ is a special reducible conic;

(iii) $x_i + x_j + x_k \geq 0$ for all triples of distinct integers j, j, k.

(We will comment on the last claim (iv) of Proposition 6.1 later.)

To show (i)–(iii), we have to study the intersection form $\langle \cdot, \cdot \rangle$ on Λ^*. Namely, we show first that the form is non-degenerate. This done, we invert the matrix of the form:

$$\Theta = \|\theta_{ij}\| = \|\langle T_i^* \cdot T_j^* \rangle\|^{-1}_{1\leq i,j\leq r}.$$

For an arbitrary class $x = x_1T_1^* + \cdots + x_rT_r^* \in \Delta^*$, we can retrieve the coefficients x_i from the intersection numbers $a_i = \langle x \cdot T_i^* \rangle$:

$$x_i = \theta_{i1}a_1 + \cdots + \theta_{ir}a_r.$$

Now it is clear that the desired claims (i)–(iii) follow from the three properties of the matrix Θ:

(i) $\theta_{ii} < 0$ implies that C_i is a line or a special conic, all non-diagonal coefficients are non-negative: $\theta_{ij} \geq 0$ for $i \neq j$;

(ii) $\theta_{ii} + \theta_{ij} < 0$ implies that $C_i \cup C_j$ is a special reducible conic;

(iii) for any $\alpha = 1, \ldots, r$ and any three distinct integers i, j, k we have $\theta_{\alpha i} + \theta_{\alpha j} + \theta_{\alpha k} \geq 0$.

Now let us illustrate by examples how this technique works.

Example 6.1. Assume that $Q_S \subset S = \mathbb{P}^3$ is a non-singular quadric. Here $S \cap V = \sum_{i=1}^{r} C_i$, each curve C_i is of bidegree (α_i, β_i). The intersection form on the essential sublattice Δ^* is

$$\langle T_i^* \cdot T_j^* \rangle = \alpha_i\beta_j + \alpha_j\beta_i - 2\delta_{ij}(\alpha_i + \beta_i),$$

where δ_{ij} is the Kronecker symbol.

Lemma 6.5. *The matrix* $\|\langle T_i^* \cdot T_j^* \rangle\|$ *is non-degenerate.*

Proof. Assume that $\varepsilon = \sum_{i=1}^{r} \varepsilon_i T_i^* \in (\Delta^*)^{\perp}$. Set

$$\alpha(\varepsilon) = \sum_{i=1}^{r} \varepsilon_i \alpha_i \quad \text{and} \quad \beta(\varepsilon) = \sum_{i=1}^{r} \varepsilon_i \beta_i.$$

Now $\langle \varepsilon \cdot T_j^* \rangle = \beta_j \alpha(\varepsilon) + \alpha_j \beta(\varepsilon) - 2(\alpha_j + \beta_j)\varepsilon_j = 0$. Summing these up over all $j = 1, \dots, r$, and taking into account that

$$\sum_{i=1}^{r} \alpha_i = \sum_{i=1}^{r} \beta_i = 3,$$

we get $\alpha(\varepsilon) + \beta(\varepsilon) = 0$. Therefore,

$$\varepsilon_j = \frac{\beta_j - \alpha_j}{2(\beta_j + \alpha_j)} \alpha(\varepsilon)$$

and so

$$\alpha(\varepsilon) = \left(\sum_{i=1}^{r} \frac{\alpha_j(\beta_j - \alpha_j)}{2(\beta_j + \alpha_j)} \right) \alpha(\varepsilon).$$

It is easy to transform the last expression to the form

$$\left(\frac{1}{2} - \sum_{i=1}^{r} \frac{\alpha_j^2}{\alpha_j + \beta_j} \right) \alpha(\varepsilon) = 0.$$

Since

$$\frac{\alpha_j^2}{\alpha_j + \beta_j} \geq \frac{\alpha_j}{4},$$

we obtain the estimate

$$\sum_{i=1}^{r} \frac{\alpha_j^2}{\alpha_j + \beta_j} \geq \frac{3}{4},$$

which immediately implies that $\alpha(\varepsilon) = \beta(\varepsilon) = 0$ and $\varepsilon = 0$. Q.E.D. for the lemma.

To compute the inverse matrix, set $d_i = \alpha_i + \beta_i$, $\sigma_i = \alpha_i - \beta_i$ for $i = 1, \dots, r$, and

$$\omega = 1 + \sum_{i=1}^{r} \frac{\sigma_i^2}{4d_i}.$$

The inverse matrix for $\|\langle T_i^* \cdot T_j^* \rangle\|$ is then given by the expression

$$\Theta = \|\theta_{ij}\|, \quad \theta_{ij} = \frac{1}{4} + \frac{\sigma_i \sigma_j}{8\omega d_i d_j} - \frac{2\delta_{ij}}{4d_j}.$$

(the checking, following the same scheme as the proof of Lemma 6.5 above, is left to the reader).

It is easy to check that the properties (i)–(iii) are satisfied. Note that when the quadric Q_S is irreducible, neither of the curves C_i can be a special conic (and the union of any two distinct lines $C_i \cup C_j$ cannot be a special reducible conic as well).

Example 6.2. Q_S is an irreducible cone with the vertex p but $S \neq T_pV$. For each $i = 1, \dots, r$ we set $d_i = \deg C_i$ and $\mu = \operatorname{mult}_p C_i$. The intersection form on the essential sublattice Δ^* is given by

$$\langle T_i^* \cdot T_j^* \rangle = \frac{1}{2}(d_i d_j + \mu_i \mu_j) - (2d_i + \mu_i)\delta_{ij}.$$

A check similar to the proof of Lemma 6.5 shows that the intersection form is non-degenerate. The inverse matrix $\Theta = \|\theta_{ij}\|$ is given by the formula

$$\theta_{ij} = \frac{1}{2}\left(\frac{1}{2} - \frac{\mu_i\mu_j}{\omega(2d_i+\mu_i)(2d_j+\mu_j)} - \frac{2\delta_{ij}}{2d_i+\mu_i}\right),$$

where

$$\omega = \sum_{i=1}^{r} \frac{\mu_i d_i}{2d_i+\mu_i}.$$

A straightforward check shows that the required conditions (i)–(iii) are satisfied.

Example 6.3. Assume that $Q_S = P_1 \cup P_2$ is the union of two planes and $S \cap V$ is the union of six lines $C_1, \dots, C_6$, of which the first 3 lie in P_1 and $C_4, C_5, C_6 \subset P_2$, and moreover, no three lines meet at one point. In this case $S \neq T_pV$ for any $p \in V$. The intersection form $\langle\cdot,\cdot\rangle$ of the essential sublattice is given by the matrix

$$\begin{pmatrix} -2 & 1 & 1 & 1 & 0 & 0 \\ 1 & -2 & 1 & 0 & 1 & 0 \\ 1 & 1 & -2 & 0 & 0 & 1 \\ 1 & 0 & 0 & -2 & 1 & 1 \\ 0 & 1 & 0 & 1 & -2 & 1 \\ 0 & 0 & 1 & 1 & 1 & -2 \end{pmatrix},$$

which is clearly non-degenerate. The inverse matrix is

$$\frac{1}{8}\cdot\begin{pmatrix} -2 & 1 & 1 & 2 & 3 & 3 \\ 1 & -2 & 1 & 3 & 2 & 3 \\ 1 & 1 & -2 & 3 & 3 & 2 \\ 2 & 3 & 3 & -2 & 1 & 1 \\ 3 & 2 & 3 & 1 & -2 & 1 \\ 3 & 3 & 2 & 1 & 1 & -2 \end{pmatrix}.$$

The claim of the Plane Section Lemma is now obvious for the case of this example.

Finally, let us consider briefly the case when $S = T_pV$ for some $p \in V$ (there may be more than one such point, but to prove Proposition 6.1 we need to consider just one of them). As we explained above, in this case we consider a bigger essential sublattice

$$\Delta^* = \bigoplus_{i=1}^{r+1} \mathbb{Z}T_i^*,$$

adding as T_{r+1}^* the exceptional divisor, corresponding to the point p. In this case

$$\pi(H) = T_1^* + \dots + T_r^* + 2T_{r+1}^*,$$

so that

$$\pi(D^*) = \sum_{i=1}^{r}(n-\theta_i)T_i^* + (2n-\theta_{r+1})T_{r+1}^*$$

and to complete the proof of the Plane Section Lemma, we need to show that

$$\theta_{r+1,r+1} \geq 0$$

(together with the other properties of the matrix Θ, discussed above).

Example 6.4. Here $Q_S = P_1 \cup P_2$ is a union of two planes, $S \cap V = C_1 \cup C_2$, where $C_i \subset P_i$ are plane cubic curves with the double point $p \in P_1 \cap P_2$. Here the intersection form is

$$\begin{pmatrix} -2 & 2 & 2 \\ 2 & -2 & 1 \\ 2 & 1 & -2 \end{pmatrix}$$

and the inverse matrix is

$$\frac{1}{6}\begin{pmatrix} 1 & 2 & 2 \\ 2 & 0 & 2 \\ 2 & 2 & 0 \end{pmatrix}.$$

The claim of the Plane Section Lemma follows.

Example 6.5. Let $S = T_pV$, where the curve $S \cap V = \sum\limits_{i=1}^{6} C_i$ is the union of six lines, of which exactly 4, say C_1, C_2, C_4, C_5, contain the point p. The intersection form is given by

$$\begin{pmatrix} -2 & 0 & 1 & 0 & 0 & 0 & 1 \\ 0 & -2 & 1 & 0 & 0 & 0 & 1 \\ 1 & 1 & -2 & 0 & 0 & 1 & 0 \\ 0 & 0 & 0 & -2 & 0 & 1 & 1 \\ 0 & 0 & 0 & 0 & -2 & 1 & 1 \\ 0 & 0 & 1 & 1 & 1 & -2 & 0 \\ 1 & 1 & 0 & 1 & 1 & 0 & -2 \end{pmatrix}.$$

The inverse matrix is

$$\frac{1}{16}\begin{pmatrix} -5 & 3 & 2 & 5 & 5 & 6 & 4 \\ 3 & -5 & 2 & 5 & 5 & 6 & 4 \\ 2 & 2 & -4 & 6 & 6 & 4 & 8 \\ 5 & 5 & 6 & -5 & 3 & 2 & 4 \\ 5 & 5 & 6 & 3 & -5 & 2 & 4 \\ 6 & 6 & 4 & 2 & 2 & -4 & 8 \\ 4 & 4 & 8 & 4 & 4 & 8 & 0 \end{pmatrix}.$$

This proves the claim of the Plane Section Lemma. (Note that $\theta_{77} = 0$ which implies precisely that $\mathrm{mult}_p \Sigma \leq 2n$.)

We gave just (typical) samples, including the most general cases covered by Examples 6.1 and 6.2. The other cases are considered in the same way. Q.E.D. for the Plane Section Lemma.

Remark 6.1. Note that the formal procedure of proving the Plane Section Lemma, described above, is a numerical analog of the most natural geometric approach: to consider a general K3-surface $Q \in \Lambda_S$, restrict the linear system Σ onto Q and study the system of curves Σ_Q with the fixed components $C_1, \ldots, C_r$. In order to work on Σ_Q, however, one needs to resolve the singularities of the surface Q, say $Q^+ \to Q$, and take the strict transform Σ_Q^+ of Σ_Q on Q^+. That is what we are essentially doing above, however, replacing the intersection theory on Q^+ by the intersection form $\langle \cdot, \cdot \rangle$ on some lattice of divisors on the 3-fold V^*. The point is, it seems much more difficult to consider all possible singularities of the surface Q and their resolutions than to resolve the pencil Λ_S and use the formal approach developed above. A lot of case by case study is still to be done. The question is, whether it is possible to do it all in one go? This is the more meaningful question as the situation similar to that of the Plane Section Lemma (requiring essentially the same technique of computing the inverse matrix for the matrix of intersection form) emerges in various problems of birational geometry: for singular three-dimensional quartics (Subsection 3.4), complete intersections of quadrics and quartics in $\mathbb{P}^6$, double spaces of index two (Chapter 8).

6.3. Existence of a maximal curve. To prove Theorem 6.1, we assume the converse,

$$\mathrm{mult}_C\,\Sigma \leq n$$

for every curve $C \subset V$. The aim is to obtain a contradiction. We know that the linear system Σ has a maximal singularity which, due to our assumption and to the fact that

$$\mathrm{mult}_o\,\Sigma \leq 2n \tag{40}$$

for every point $o \in V$ (Lemma 6.4 and the Plane Section Lemma above), has to be an infinitely near one: for some birational morphism $\varphi\colon V^+ \to V$ and an exceptional divisor $E \subset V^+$, the Noether-Fano inequality

$$\mathrm{ord}_E\,\varphi^*\Sigma > na(E)$$

holds. Following the general scheme explained earlier in this chapter (Sec. 2), we resolve the singularity E:

$$\varphi_{i,i-1}\colon V_i \to V_{i-1},$$

$i = 1, \ldots, K$, the centre $B_{i-1} \subset V_{i-1}$ of the blow up $\varphi_{i,i-1}$ being the centre of the maximal singularity E on V_{i-1}, $E_i = \varphi_{i,i-1}^{-1}(B_{i-1}) \subset V_i$ is the exceptional divisor and $E_K \subset V_K$ determines the same discrete valuation as E. We keep the notation L for the lower part of the resolution (so that $B_0, \ldots B_{L-1}$ are points). Due to the inequality (40), the upper part is non-empty: $K \geq L+1$. From the inequality

$$\sum_{i=1}^{K} p_i\nu_i > n \sum_{i=1}^{K} p_i\delta_i$$

it is clear that we may assume that $\nu_K > n$ (otherwise we replace E_K by E_{K^*} with K^* the highest integer $\leq K$ such that $\nu_{K^*} > n$, it is easy to show by the linearity of the Noether-Fano inequality that E_{K^*} is again a maximal singularity of Σ). Therefore, $B_L \subset E_L \cong \mathbb{P}^2$ is a line and for $i \geq L+1$ all maps

$$\varphi_{i,i-1}\colon B_i \to B_{i-1}$$

are isomorphisms. We are therefore in the nice situation when all varieties V_i are non-singular. We will use the special notation $o = B_0$ because of the importance of this point. The next proposition lists the possible options.

Proposition 6.2. *One of the following cases takes place:*

(A) *there is no line on V through o;*

(B) *there is at least one line on V through o but either $L = 1$ or for any line $R \ni o$ on V the next point $B_1 \in V_1$ is not on R, that is, $B_1 \notin R^1$;*

(C) *the lower part contains at least two blow ups, $L \geq 2$, and there is an (obviously unique) line $R \ni o$ on V such that $B_1 \in R^1$, but either $L = 2$ or $B_2 \notin R^2$;*

(D) *the lower part of the resolution contains at least three blow ups, $L \geq 3$, and there is a (unique) line $R \ni o$ on V such that $B_i \in R^i$ for $i = 0, 1, 2$.*

In order to prove Theorem 6.1, we have to show that none of the cases (A)–(D) are realized.

Let us first exclude the easiest case (A). Arguing as in the proof of Lemma 6.4, we see that $(2H - 3E_1)$ is a non-negative class on V_1, that is, every curve $C \subset V$ satisfies the inequality $\mathrm{mult}_o C \leq (2/3) \deg C$. Therefore, for the self-intersection Z of the system Σ we have the inequality

$$\mathrm{mult}_o Z \leq \frac{2}{3} \cdot 6n^2 = 4n^2.$$

On the other hand, by the $4n^2$-inequality

$$\mathrm{mult}_o Z > 4n^2,$$

which is a contradiction. So the case (A) is impossible.

We will deal with the cases (B)–(D) one by one.

Exclusion of the case (B). Assume first that $L = 1$, that is, the centres $B_1, \ldots, B_{K-1}$ are all curves. The Noether-Fano inequality takes the form

$$\nu_1 + \cdots + \nu_K > (K+1)n.$$

In fact, we can be a lot more precise. The proof of the $4n^2$-inequality gives the estimate

$$\mathrm{mult}_o Z > \frac{(K+1)^2}{K} n^2,$$

and the right-hand side is strictly higher than $6n^2$ for $K \geq 4$. Therefore, we have only two options: $K = 2$ or 3.

A simple and straightforward, but a bit tedious check proves

Lemma 6.6. *Let $P \subset \mathbb{P}$ be an arbitrary 2-plane. Then for a general hyperplane $H \supset P$ the surface $S_P = H \cap V$ is non-singular.*

Let us take the unique plane $P \ni o$, the strict transform of which under the blow up of the point o is the line B_1 (here we consider the blow up $\varphi_{1,0}\colon V_1 \to V$ as the restriction to V of the blow up of the point o in $\mathbb{P}^5$). Let $S = S_P$ be a general surface as in Lemma 6.6. Restricting the linear system Σ onto S, we obtain a non-empty linear system of curves. Obviously,

$$\mathrm{mult}_o \Sigma_S = \nu_1 + \nu_2.$$

Considering the ruled surface E_2 of the type $\mathbb{F}_2$, it is easy to show that if $K = 3$, then the strict transform Σ_S^2 on $S^2 \subset V_2$ intersects the curve B_2 at 2 generic points. Now the case (B) for $L = 1$ is excluded by computing the self-intersection of the mobile part of the linear system Σ_S^3 on S^3 for $K = 3$ and of Σ_S^2 on S^2 for $K = 2$, respectively. Obviously,

$$|\,\mathrm{Bs}\,\Sigma_S| \subset P \cap V.$$

Therefore, we need to go through all possible options for the intersection $P \cap V$. We will consider a few typical cases as examples; they are all easy.

Example 6.6. Assume that $\dim P \cap V = 0$. Then the system Σ_S is mobile. By the Noether-Fano inequality, $\nu_1 + \nu_2 > 3n$ if $K = 2$ and $\nu_1 + \nu_2 > \frac{8}{3}n$, if $K = 3$. For the self-intersection Z_S of the mobile system Σ_S we get

$$6n^2 = \deg Z_S \geq \deg(Z_S)_o > \left(\frac{8}{3}n\right)^2,$$

which is a contradiction.

Example 6.7. Assume that $P \cap V = L_1 + L_2$ is a pair of distinct lines. Set $\mu_i = \operatorname{mult}_{L_i} \Sigma$. The self-intersection number of the mobile part Σ_S^{mob} of the system Σ_S is equal to

$$6n^2 - 2n(\mu_1 + \mu_2) - 2\mu_1^2 - 2\mu_2^2 + 2\mu_1\mu_2.$$

Moreover, $\operatorname{mult}_o \Sigma_S^{\mathrm{mob}} = \nu_1 + \nu_2 - \mu_1 - \mu_2$. If $K = 3$, then the strict transform of the mobile system Σ_S^{mob} on S^2 has two more base points of multiplicities at least $\nu_3 - \mu_i$, $i = 1, 2$. Therefore, if $K = 2$, then (since $\nu_1 + \nu_2 > 3n$) we get

$$6n^2 - 2n(\mu_1 + \mu_2) - 2\mu_1^2 - 2\mu_2^2 + 2\mu_1\mu_2 - (3n - \mu_1 - \mu_2)^2 \geq 0.$$

An easy computation shows that for $0 \leq \mu_i \leq n$ this is impossible. If $K = 3$, then $\nu_1 + \nu_2 + \nu_3 > 4n$ and we get an even longer inequality:

$$6n^2 - 2n(\mu_1 + \mu_2) - 2\mu_1^2 - 2\mu_2^2 + 2\mu_1\mu_2$$
$$-(\nu_1 + \nu_2 - \mu_1 - \mu_2)^2 - (\nu_3 - \mu_1)^2 - (\nu_3 - \mu_2^2) \geq 0.$$

The maximum of the left-hand side for $\mu_1 + \mu_2$ and $\nu_1 + \nu_2 + \nu_3$ fixed is attained at $\mu_1 = \mu_2 = \mu$ and $\nu_i = \nu > \frac{4}{3}n$, so the left-hand side simplifies and it is easy to check that it is negative. A contradiction.

Example 6.8. Assume that $P \cap V = C$ is an irreducible cubic curve with $o \in C$ as the singular point. Putting $\mu = \operatorname{mult}_C \Sigma$, we get the inequality

$$6n^2 - 6n\mu - (3n - 2\mu)^2 \geq 0$$

in the case $K = 2$ and

$$6n^2 - 6n\mu - (\nu_1 + \nu_2 - 2\mu)^2 - 2(\nu_3 - \mu)^2 \geq 0$$

in the case $K = 3$. Calculating the maximum of the left-hand side for $0 \leq \mu \leq n$, we see that both cases are impossible.

Arguing as in Examples 6.6–6.8, we exclude all options for the intersection $P \cap V$ and thus show that the case (B) is impossible if $L = 1$.

Now let us consider the same case (B), assuming that $L \geq 2$. Note that by assumption the line in $\mathbb{P}^5$, going through the point o in the direction B_1, does not lie on V. To exclude this case, we need to use the same trick of removing some oriented edges in the graph Γ that was used in Section 4. Namely, we remove the arrows going from the upper vertices $E_{L+2}, \ldots, E_K$ to the lower vertices $E_1, \ldots, E_L$, if there are any such arrows. The modified graph is denoted by Γ^*, the number of paths from E_K to E_i by p_i^*. It is easy to check that the "modified" Noether-Fano inequality

$$\sum_{i=1}^{K} p_i^* \nu_i > n \sum_{i=1}^{K} p_i^* \delta_i$$

still holds (in fact, the difference between the left-hand side and the right-hand side can only increase as $\nu_i \leq 2n$ for the lower indices $i = 1, \ldots, L$). Moreover, putting $m_i = \operatorname{mult}_{B_{i-1}} Z^{i-1}$ for $i = 1, \ldots, L$ (where Z, as usual, stands for the self-intersection of the system Σ) and applying the technique of counting multiplicities, we get the inequality

$$\sum_{i=1}^{L} p_i^* m_i > \frac{(2\Sigma_l^* + \Sigma_u)^2}{\Sigma_l^* + \Sigma_u} n^2,$$

where $\Sigma_l^* = p_1^* + p_2^* + \cdots + p_L^*$ (note that $p_i^* = p_i = 1$ for $i = L, L+1, \ldots, K$). The crucial property of the modified coefficients p_i^* is that

$$p_1^* \leq p_2^* + \cdots + p_L^*$$

as there are no arrows from upper vertices to E_1. Now set

$$Z = aL + Z^*,$$

where Z^* does not contain the line L as a component. Put $m_i^* = \mathrm{mult}_{B_{i-1}}(Z^*)^{i-1}$ for $i = 1, \ldots, L$.

Lemma 6.7. *The following estimate holds:*

$$m_1^* + m_2^* \leq \deg Z^*.$$

Proof. Intersect Z^* with a generic hyperplane, containing the line L^* that goes through the point o in the direction B_1. Q.E.D.

Now we compute:

$$\sum_{i=1}^{L} p_i^* m_i = ap_1^* + \sum_{i=1}^{L} p_i^* m_i^*$$

$$\leq a\left(\frac{1}{2}\Sigma_l^*\right) + p_1^* m_1^* + \left(\sum_{i=2}^{L} p_i^*\right) m_2^*$$

$$\leq \frac{1}{2}\Sigma_l^* \deg Z = 3n^2 \Sigma_l^*,$$

so that

$$3\Sigma_l^*(\Sigma_l^* + \Sigma_u) > (2\Sigma^* + \Sigma_u)^2,$$

which is false. We have proved that the case (B) is impossible.

Exclusion of the case (D). Set $k = \max\{1 \leq i \leq L \,|\, B_{i-1} \in R^{i-1}\}$. By assumption, $k \geq 3$. Let us take the blow up $\sigma\colon V^* \to V$ of the line R with the exceptional divisor $E^* = \sigma^{-1}(R)$. Consider the linear system Λ_R of hyperplane sections of V, containing R, and its strict transform Λ^* on V^*. By the genericity of the complete intersection V, the restriction $\Lambda_E = \Lambda^*|_{E^*}$ is a base point free ample linear system on the surface E^* (see the proof of Proposition 5.4). Therefore, for a generic element $S \subset \Lambda_R$ and any irreducible curve $C \subset E^*$, which is not a fibre of the ruling $\sigma_E\colon E^* \to R$, we have

$$(S^* \cdot C) \geq \deg(\sigma|_C\colon C \to R)$$

and, moreover, S^* meets C transversally at the points of their intersection, the system Λ^* cutting out a base point free system on C.

Now let us compare the self-intersections of the linear system Σ and its strict transform Σ^* on V^*. Let $D_1, D_2 \in \Sigma$ be two generic divisors, so that $D_1^*, D_2^* \in \Sigma^*$ are their strict transforms. We decompose the self-intersections as follows:

$$Z = (D_1 \circ D_2) = Z_{\mathrm{outside}\, R} + dR$$

and

$$Z_* = (D_1^* \circ D_2^*) = Z_{\mathrm{outside}\, E^*} + Z_E + Z_E^f,$$

where $Z_{\text{outside } E^*} = Z^*_{\text{outside } R}$ is the strict transform, Z_E is supported on curves in E^* which are not fibres of the ruling σ_E, Z^f_E is supported on fibres and $Z_{\text{outside } R}$ is supported on curves that are different from R.

The linear system $\Sigma^*|_{S^*}$ is mobile and its self-intersection is bounded from below by the expression

$$\sum_{i=1}^{k}(\nu_i - \nu_R)^2 + \deg(\sigma\colon Z_E \to R).$$

As obviously $d = \nu_R^2 + \deg(\sigma\colon Z_E \to R)$, where $\nu_R = \operatorname{mult}_R \Sigma$, we obtain the estimate

$$6n^2 - 2n\nu_R - \nu_R^2 - \sum_{i=1}^{k}(\nu_i - \nu_R)^2 \geq d. \tag{41}$$

Lemma 6.8. *For any irreducible curve $C \neq R$ on V the following inequality holds:*

$$\sum_{i=1}^{k} \operatorname{mult}_{B_{i-1}} C^{i-1} \leq \deg C.$$

Proof. Taking a generic section $S \subset \Lambda_R$ and counting the intersection number $(S \cdot C) = \deg C$ in terms of the local intersection numbers at the points of $S \cap C$ and infinitely near points over them, which includes $B_0 = o, B_1, \ldots, B_{k-1}$, we get the claim. Q.E.D.

Now set $m_i^+ = \operatorname{mult}_{B_{i-1}} Z^{i-1}_{\text{outside } R}$, so that

$$m_i = m_i^+ + d$$

for $i = 1, \ldots, k$, and for $i \geq k+1$ we have $m_i = m_i^+ \leq m_k^+$. Applying the technique of counting multiplicities, we obtain the estimate

$$d\sum_{i=1}^{k} p_i + \sum_{i=1}^{L} p_i m_i^+ > \frac{(2\Sigma_l + \Sigma_u)^2}{\Sigma_l + \Sigma_u} n^2. \tag{42}$$

Note that by Lemma 6.8,

$$m_1^+ + \cdots + m_k^+ \leq 6n^2 - d. \tag{43}$$

Since the graph Γ of the maximal singularity E_K starts with the chain

$$1 \leftarrow 2 \leftarrow \cdots \leftarrow k,$$

we conclude that $p_1 = p_2 = \cdots = p_{k-1}$, and again as in the case (B) we may (and will) assume that

$$p_{k-1} \leq \sum_{i=k}^{L} p_i.$$

Using this estimate, we see that the maximum of the left-hand side of the inequality (42) on the set (43) is attained at

$$m_1^+ = \cdots = m_k^+ = \cdots = m_L^+ = \frac{6n^2 - d}{k}.$$

Therefore, we can replace (42) by the inequality

$$\left(d+\frac{6n^2-d}{k}\right)\Sigma_l > \frac{(2\Sigma_l+\Sigma_u)^2}{\Sigma_l+\Sigma_u}n^2. \tag{44}$$

Corollary 6.1. *The inequality $d > 3n^2$ holds.*

Proof. Indeed, as $k \geq 3$, if $d \leq 3n^2$, we have

$$d+\frac{6n^2-d}{k} \leq 4n^2,$$

which is impossible as the right-hand side of the inequality (44) is $\geq 4n^2\Sigma_l$. Q.E.D.

The next observation is that for $\nu_1 + \cdots + \nu_{k-1}$ fixed, the maximum of the left-hand side in (41) is attained at $\nu_1 = \cdots = \nu_{k-1}$. Since these multiplicities appear in the Noether-Fano inequality with the same coefficients $p_1 = \cdots = p_{k-1}$, we may (and will) assume that

$$2n \geq \nu_1 = \cdots = \nu_{k-1} = \theta > n.$$

Lemma 6.9. *The inequality $\theta < \frac{3}{2}n$ holds.*

Proof. Assume the converse: $\theta \geq \frac{3}{2}n$.

Computing the maximum of the left-hand side of the inequality (41) in ν_R, we see that $d < 3n^2$, which is impossible by Corollary 6.1. Q.E.D.

Corollary 6.2. *The inequality $\Sigma_u > \Sigma_l$ holds.*

Proof. Replacing in the Noether-Fano inequality all multiplicities by $\frac{3}{2}n$, we get the required claim. Q.E.D.

Corollary 6.3. *The right-hand side of the inequality* (44) *is higher than $\frac{9}{2}\Sigma_l n^2$.*

Proof. Easy arithmetic. Q.E.D.

Lemma 6.10. *The inequality $\theta < \frac{4}{3}n$ holds.*

Proof. Assume the converse: $\theta \geq \frac{4}{3}n$. Computing the maximum of the left-hand side of the inequality (41) in ν_R (set $k = 3$ and $\nu_3 = n$), we obtain the estimate $d < \left(3\frac{2}{9}\right)n^2$. However, this implies that the left-hand side of (44) is smaller than $\left(4\frac{4}{27}\right)\Sigma_l n^2$, whereas by Corollary 6.3 it is higher than $\left(4\frac{1}{2}\right)\Sigma_l n^2$. The contradiction proves the lemma. Q.E.D.

Corollary 6.4. *The inequality $\Sigma_u > 2\Sigma_l$ holds.*

Proof is similar to that of Corollary 6.2. Q.E.D.

Corollary 6.5. *The right-hand side of the inequality* (44) *is higher than $\frac{16}{3}\Sigma_l n^2$.*

Proof. Easy computations. Q.E.D.

Finally, we can complete our exclusion of the case (D). Computing the maximum of the left-hand side of the inequality (41) in ν_R for $k = 3$ and $\nu_1 = \nu_2 = \nu_3 = n$, we see that $d \leq 4n^2$. Therefore, the left-hand side of the inequality (44) is not higher than $\left(4\frac{2}{3}\right)\Sigma_l n^2$. However, by Corollary 6.5 the right-hand side is higher than $\left(5\frac{1}{3}\right)\Sigma_l n^2$, so we get a contradiction.

The case (D) is excluded.

Exclusion of the case (C). We will only sketch the arguments, since in some subcases they repeat the computations in the case (B) and in the other are similar to the case (D).

Assume first that $L = 2$, so that $B_2 \subset E_2 \cong \mathbb{P}^2$ is a line. If $B_2 \cap R^2 = \emptyset$, then we take a generic hyperplane section $S \supset R$ and consider the linear system $\Sigma_S = \Sigma|_S$ of curves on S. Taking into account that every curve B_i, $i = 2, \ldots, K-1$, generates an infinitely near base point of the system Σ_S and estimating the self-intersection of the mobile part of Σ_S as it was done in the case (B), we get the inequality

$$(45)\qquad \begin{aligned} &6n^2 - 2n\nu_R - 2\nu_R^2 - (\nu_1 - \nu_R)^2 - (\nu_2 - \nu_R)^2 - \sum_{i=3}^{K} \nu_i^2 \\ &= 6n^2 + 2(\nu_1 + \nu_2 - n)\nu_R - 4\nu_R^2 - \sum_{i=1}^{K} \nu_i^2 \geq 0. \end{aligned}$$

The graph of the maximal singularity E_K is a chain and the Noether-Fano inequality takes the especially nice form

$$\nu_1 + \nu_2 + \cdots + \nu_K > (K+2)n.$$

Elementary computations show that the left-hand side of the inequality (45) is negative, which gives a contradiction.

Now assume that $L = 2$ but $B_2 \cap R^2 \neq \emptyset$ (it is the point $R^2 \cap E_2$). Here two methods are available. One can either take the unique plane $P \supset R$ in $\mathbb{P}^5$ such that $P^2 \cap E_2 = B_2$ and argue as in the case (B), restricting Σ onto the section of V by a general hyperplane, containing P, or just note that the inequality (41) is valid in this case for $k = 3$ (the point $B_2 \cap R^2$ is an infinitely near base point of multiplicity $(\nu_3 - \nu_R)$ of the mobile part of the linear system Σ_S for a generic hyperplane section $S \supset R$), and so the arguments excluding the case (D) apply.

Finally, let us assume that $L \geq 3$ and $B_2 \not\in R^2$. In this case the arguments are similar to those in the case (D) with a modification which we will now describe. Removing the arrows connecting the upper vertices $E_{L+2}, \ldots, E_K$ with the lower vertices, if there are any, we may assume that the following inequalities hold:

$$p_1 \leq \sum_{i=2}^{L} p_i, \quad p_2 \leq \sum_{i=3}^{L} p_i.$$

We conclude that

$$p_1 + p_2 \leq \frac{3}{4}\Sigma_l.$$

Therefore, while in the present case the inequality (41) is replaced by the following weaker one:

$$(46)\qquad 6n^2 - 2n\nu_R - \nu_R^2 - (\nu_1 - \nu_R)^2 - (\nu_2 - \nu_R)^2 \geq d,$$

the second crucial estimate (44) of the case (D) is replaced by the inequality

$$(47)\qquad \left(\frac{3}{4}d + \frac{6n^2 - d}{2}\right)\Sigma_l = \left(\frac{1}{4}d + 3n^2\right)\Sigma_l > \frac{(2\Sigma_l + \Sigma_u)^2}{\Sigma_l + \Sigma_u} n^2.$$

In particular, if $d \leq 4n^2$, we immediately get a contradiction.

Now we proceed as in the case (D). First of all, the claim of Lemma 6.9 is still valid, as the maximum of the left-hand side of (46) for $\nu_1 = \frac{3}{2}n$ and $\nu_2 = n$ is

$(3\frac{1}{2})n^2$ which is impossible as we have just mentioned. Therefore we get the claims of Corollaries 6.2 and 6.3. But $d \leq 6n^2$ in any case, so the left-hand side of the inequality (47) is bounded from above by $(4\frac{1}{2})\Sigma_l n^2$, which contradicts the claim of Corollary 6.3. This contradiction completes the exclusion of the case (C).

Q.E.D. for Theorem 6.1.

Notes and references

Fano varieties are studied a lot better from the viewpoint of birational rigidity than Fano fibre spaces. Several dozens of papers (counting only significant ones) were published on the subject in the past four decades. Here we only mention the most important points and give the most important references.

(Section 1.) The idea of using adjoint linear systems, adding multiples of the canonical class and fixing the moment when the adjunction terminates, is very old — in fact, the elements of this idea can be found in the work of M. Noether and the old Italian school (for example, see the modern reconstruction of the proof of Castelnuovo's rationality criterion in [**AlgS**]). The canonical adjunction has been extensively used in the modern period of algebraic geometry, including the minimal model program where it is of great importance (see, e.g., [**KMM**]). The concept of the threshold of canonical adjunction is, for some reason, less popular. The virtual threshold of canonical adjunction was introduced by the author in [**Pukh04b**]. This concept is convenient and simplifies and clarifies the language.

The origins of the terms "birationally rigid" and "birationally superrigid" could be traced back to the pioneer paper [**IM**]. In the introduction to [**IM**] we read: "...the absence of non-trivial birational self-maps means the utmost "rigidity" of our unirational varieties". However, the word combination *birational rigidity* has never been used as a rigorous concept of birational geometry until the 90s and for the first time was published in [**Pukh95**] (that first definition was a bit awkward and has been gradually improved to the one used in this book). Note that some authors use different versions of the definition of birational (super)rigidity; see, for instance, [**Co00, CPR, Grin06**], where the main geometric *implication* of birational rigidity, that is, the uniqueness of the structure of a Fano-Mori fibre space is taken for the *definition* of birational rigidity. Cheltsov's version [**Ch05c**] is closer to ours but also different from it. The differences are not very significant, however, sometimes Fano varieties (and fibre spaces) which are birationally rigid in the sense of one definition, do not have that property in the sense of another one. Those who work in this area are already accustomed to these discrepancies and know how to live with them, however, for non-experts they can sometimes be confusing. In any case, the definition used in this book is the oldest one, agrees with the historical tradition and, in the opinion of the author, is definitely the best one. Note that if an algebraic variety V is birationally (super)rigid in the sense of any of the existing definitions, it is birationally (super)rigid in the sense of Definition 1.3. It is also worth mentioning that in all papers on birational rigidity, whichever definition of that concept is used, it is the property defined in Definition 1.3 that is actually proven first. If some other definition of birational rigidity is used, then it is (easily) derived from birational rigidity in our sense.

Birational (super)rigidity of algebraic varieties in the families listed in Example 1.3 was shown, respectively, in (i) [**IM**], (ii) [**I80, Pukh89b**] (see a more detailed comment below), (iii) [**Pukh01**] (see also the comments for Chapter 3), (iv) [**Pukh89a**], (v) [**Pukh00e**] for the double hypersurfaces, [**Pukh09a**] for the cyclic covers, [**Pukh03**] for the iterated double covers. Conjecture 1.1 was formulated in [**Pukh04c**].

The definition of a maximal singularity (Definition 1.4) goes back to Fano [**Fano2**], who considered only particular cases: points and curves on the variety under consideration (in our terminology, maximal subvarieties) and the simplest infinitely near case — a line in the exceptional plane of the blow up of a point. Most probably, he had some idea of a general case but his technique was not sophisticated enough to treat them. The first rigorous definition was given in [**IM**]. Essentially, canonical singularities of pairs were introduced in that paper.

The oriented graph of a sequence of blow ups was introduced by Manin (see [**M67, M72**]) for surfaces and then used in dimension three in [**IM**]. The importance of this construction for the theory of birational rigidity cannot be overestimated. We use it in the simplified setting of the "resolution of a discrete valuation", following [**Pukh00c**], when the graph accumulates only the essential information about the maximal singularity.

The general scheme of exclusion/untwisting, described in Subsection 1.3, also goes back to Max Noether and the old Italian school, Fano in the first place. It is very clearly worded in the introduction of [**IM**] and in many other subsequent papers. Unfortunately, up to this day, it is still an informal program, a "declaration of intentions" rather than a rigorous theory. However, it is very important for the practical work.

(Section 2.) The cone method was introduced in [**Pukh95**] and subsequently developed for subvarieties of higher codimension on Fano hypersurfaces in [**Pukh02b**] and for linear systems on Fano complete intersections in [**Ch05a, Pukh06b**]. Despite its simplicity, it is extremely efficient in excluding maximal subvarieties. The $4n^2$-inequality is one of the crucial facts of the theory. Essentially it was proven in [**IM**] in the language of test class. The technique of counting multiplicities was developed in [**Pukh00c**] on the basis of the proof given in [**IM**]. That re-designing of the techniques simplified and clarified the theory a lot.

It also made it possible to prove birational rigidity of del Pezzo fibrations; see the comments in Chapter 6. The second proof of the $4n^2$-inequality follows the approach suggested in [**Co00**]. For the original proof of Proposition 2.5, see the cited paper. We give another proof in the spirit of traditional arguments based on the combinatorial properties of the graph of a sequence of blow ups, following [**Pukh03**].

(Section 3.) In Subsections 3.1 and 3.2 we reproduce the alternative proof of birational superrigidity of three-dimensional quartics given in [**Pukh95, Pukh95b**]. The approach was developed by the author in the early 90s to prove birational superrigidity of Fano hypersurfaces $V_M \subset \mathbb{P}^M$ for any $M \geq 4$ via degeneration of rational curves. However, the suggested proof of this general fact was not completed.

In Subsection 3.3 we give a brief exposition of the test class technique of Iskovskikh and Manin in its original form. The paper [**IM**] was known to every expert in algebraic geometry from the moment of its publication. Despite that, it seems that very few algebraic geometers understood the proof until the late 90s, and even the experts in birational geometry outside Russia believed the argument to be extremely difficult. While it is difficult indeed, it was originally very well written in [**IM**] with the crucial points highlighted and carefully explained. We reproduce it here to show the origins of the technique of counting multiplicities.

The singular quartic three-folds were one of the most natural objects to investigate after the great success of [**IM**]. As far as the author knows, such an attempt was made in the 70s by a student of Iskovskikh, however, it was unsuccessful and no publication followed. Three-dimensional quartics with one non-degenerate double point (and precisely 24 lines through it) were shown to be birationally rigid in [**Pukh89c**]; see Example 3.1. The double point presented the main problem, the proof was really hard. Later Corti suggested to use the connectedness principle of Shokurov and Kollár for the exclusion of infinitely near maximal singularities over the double point [**Co00**]. In the proof of Proposition 3.7, the fact that for an effective 1-cycle C of bidegree (n,n) on $E \cong \mathbb{P}^1 \times \mathbb{P}^1$ the pair $\left(E, \frac{1}{n}C\right)$ is log canonical was essentially shown in [**Pukh89c**] (in [**Co00**] a different argument was used). The connectedness principle simplified the problem a lot. The general case of an arbitrary $\mathbb{Q}$-factorial nodal quartic was completed in [**Me04, Shr08b**].

(Section 4.) The $8n^2$-inequality was suggested by Cheltsov and (as far as the author knows) first published in [**I01**](in a form, somewhat different from Theorem 4.1). It seems appropriate to give here a brief account of the non-trivial story of this inequality. Corti was the first to apply the connectedness principle to the problem of excluding maximal singularities [**Co00**]; however, that paper was available as a preprint since 1998, and the first version contained mistakes. Inspired by the efficiency of the connectedness principle (for instance, in excluding the infinitely near case over a nodal point), Cheltsov produced in 1998 an argument to show birational superrigidity of any smooth Fano hypersurface $V_M \subset \mathbb{P}^M$, based on a radical improvement of the $4n^2$-inequality, which, in turn, was based on an erroneous claim in the first version of Corti's paper, mentioned above. However, the argument was too good to be true and simple counterexamples (see Remark 2.1) were constructed, by the author of this book and a bit later by Grinenko. Studying the counterexample, it easy to come to the idea of the $8n^2$-inequality (in the notations of Remark 2.1, one should consider the subspace $\{z_{M-1} = z_M = 0\} \cap E_1$). The problem was to find a proof of the fact.

The first argument, published as a proof of the $8n^2$-inequality, can be found in [**I01**], but it was faulty: later it was discovered that in the case $\nu < 2n$ (in the notations of Sec. 4) there was no proof; see [**Pukh08b**]. More sophisticated arguments were presented in [**Ch03, Ch06a, Ch06b**] and some other publications, but they all, like [**I01**], mistreated the case $\nu < 2n$, just one gap in the proof was replaced by another one. The case $\nu \geq 2n$ was shown correctly from the start.

Finally, the case $\nu < 2n$ was sorted out in [**Pukh08b**]. The complete correct proof given in Section 4 follows [**Pukh10**].

Proposition 4.1 was formulated and proved by Corti ([**Co00**]). We give a different argument in the spirit of the technique used in this book. Birational superrigidity of four-dimensional quintics was proved in [**Pukh87**]. Cheltsov simplified the proof in [**Ch00**] and proved Theorem 4.2 (see [**I01**]), using the $8n^2$-inequality, which, as we explained above, at that time was not fully proven. Lemmas 4.9 and 4.10 were shown in [**Pukh87**].

(Sections 5 and 6.) It was Fano who started to investigate the group of birational self-maps of the complete intersection $V_{2\cdot 3} \subset \mathbb{P}^5$ [**Fano3**]. He discovered the involutions, associated with lines but overlooked the the involutions, associated with special conics. The infinitely near case was by far out of reach for his technique. The complete set of involutions was presented in [**I77b, I80**]; the infinitely near case for a point $o \in V_{2\cdot 3}$ through which there are no lines lying on the variety (the case A of the Proposition 6.2) was excluded in [**I80**]. The rest of the work for the infinitely near case was done in [**Pukh89b**] and the complete proof of birational rigidity, together with the copresentation of the group $\mathrm{Bir}\, V_{2\cdot 3}$, was given in [**IP**]. Corti in [**Co00**] suggested a somewhat different approach and sketched a proof, which was much shorter, however, a detailed exposition with every subcase treated carefully would be of a length, similar to Section 6. Our approach seems to be more transparent. Cheltsov and Grinenko went one step further, proving birational rigidity of generic complete intersections $V_{2\cdot 3}$ with a double point [**ChG**].

The rationality problem for the variety $V_{2\cdot 3}$ was studied also by means of the method of intermediate Jacobian, see [**Beau77, Beau11**]; it would be interesting to check, whether the symmetric variety, the non-rationality of which was shown in [**Beau11**], satisfies the genericity conditions of Section 5.

CHAPTER 3

Hypertangent Divisors

In this chapter, we develop the technique of hypertangent divisors and hypertangent linear systems, which is a powerful tool used for excluding infinitely near maximal singularities. This is the third main ingredient of the method of maximal singularities, the first two being the the cone method and the technique of counting multiplicities, developed in Chapter 2. The technique of hypertangent divisors/linear systems makes it possible, at the expense of imposing some conditions of general position (the regularity conditions) on the variety under consideration, to obtain very sharp upper bounds on the multiplicities of subvarieties at singular points. With the technique of counting multiplicities giving a lower bound on the multiplicities, a combination of the two techniques gives a contradiction, excluding the maximal singularity. This is the general idea, in this chapter we explain in detail how it works, proving birational superrigidity of some typical Fano varieties: higher-dimensional Fano hypersurfaces and complete intersections of index one and Fano cyclic covers of index one.

Section 1 is an introductory one. We prove birational superrigidity of generic Fano hypersurfaces, using this example as an illustration of the method. This done, we give a more general and more formal treatment of hypertangent divisors.

Section 2 contains the main examples: Fano complete intersections and cyclic covers of index one. Here the construction of hypertangent divisors and linear systems becomes more sophisticated. The procedure of obtaining the above mentioned upper bounds for the multiplicities is also getting more complicated.

Section 3 deals with the regularity conditions: in order to apply the technique of hypertangent divisors, we need to make sure that a generic variety indeed has very good hypertangent divisors as we need.

Section 4 explains how the technique used for proving birational (super)rigidity makes it possible, as a by-product, to describe the K-trivial structures on Fano varieties. We consider the same class of higher-dimensional varieties as in Section 2, the Fano complete intersections of codimension k in $\mathbb{P}^{M+k}$ with $M \geq 2k+1$, and prove that they admit no other K-trivial structures but the pencils of hyperplane sections.

The technique of hypertangent divisors will be heavily used in this book in Chapters 5 and 7; the most general theorems on birational (super)rigidity cannot be proved today without this technique.

1. Definitions and examples

In this section, we introduce and explain the technique of hypertangent divisors and linear systems by the example of generic Fano hypersurfaces $V = V_M \subset \mathbb{P}^M$,

$M \geq 5$. In Subsection 1.1 we prove that they are birationally superrigid, using hypertangent divisors. In Subsection 1.2 we introduce hypertangent linear systems and describe the procedure of obtaining upper bounds for the multiplicities of subvarieties in the general case. In Subsection 1.3 we treat the regularity conditions for the Fano hypersurfaces $V_M \subset \mathbb{P}^M$.

1.1. Hypertangent divisors. The technique of hypertangent divisors and its modification, the technique of hypertangent linear systems, are the most effective tool for excluding infinitely near maximal singularities. To show how it works, let us prove

Theorem 1.1. *A generic smooth Fano hypersurface $V = V_M \subset \mathbb{P}^M$ of degree M is birationally superrigid.*

Proof. As we have shown in Subsection 2.2.1 (Lemma 2.2.1), a mobile linear system on V cannot have a maximal subvariety. Thus to prove birational superrigidity, it is sufficient to exclude the infinitely near case. By the $4n^2$-inequality (Theorem 2.2.1), an infinitely near maximal singularity is excluded and the proof of Theorem 1.1 is completed by the following claim.

Proposition 1.1. (i) *Let $o \in V$ be a point such that there is at most a finite number of lines $L \subset V$ passing through o. Then for any irreducible subvariety $Y \subset V$ of codimension two the following estimate holds:*

$$\frac{\operatorname{mult}_o Y}{\deg Y} \leq \frac{4}{M}. \tag{48}$$

(ii) *There exists a (Zariski) open set $U \subset \mathbb{P}(H^0(\mathbb{P}^M, \mathcal{O}_{\mathbb{P}^M}(M)))$ of hypersurfaces of degree M, satisfying the finiteness condition (i) at every point.*

To simplify our notations, in the sequel instead of

$$\frac{\operatorname{mult}_o Y}{\deg Y}$$

we write

$$\frac{\operatorname{mult}_o}{\deg} Y$$

(and similarly in other cases), making one joint symbol from the multiplicity and degree.

Proof of Theorem 1.1. Proposition 1.1 immediately excludes an infinitely near singularity of a mobile linear system Σ on V: let $B \subset V$ be the centre of the maximal singularity, $o \in B$ a point of general position, $D_1, D_2 \in \Sigma$ general divisors with $Z = (D_1 \circ D_2)$ the algebraic cycle of their scheme-theoretic intersection. Then $\deg Z = Mn^2$, $\operatorname{mult}_o Z > 4n^2$ where Z is an effective cycle of codimension 2, which contradicts Proposition 1.1. Q.E.D. for Theorem 1.1.

Proof of Proposition 1.1. Let $(z_1, \ldots, z_M)$ be a system of affine coordinates on $\mathbb{P} = \mathbb{P}^M$ with the origin at the point o. Write down the equation of the hypersurface V:

$$f = q_1 + q_2 + \cdots + q_M,$$

where q_i are homogeneous of degree i in z_*. Note that the lines through x on the hypersurface V are given by the system of equations

$$q_1 = q_2 = \cdots = q_M = 0. \tag{49}$$

Therefore the set (49) of common zeros is of dimension at most one. Denote by

$$f_i = q_1 + \cdots + q_i$$

the left segments of the equation f. It is clear that in the affine open set $\mathbb{A} = \mathbb{A}^M_{(z_1,\dots,z_M)} \subset \mathbb{P}$ the algebraic set

$$f_1 = f_2 = \cdots = f_M = 0$$

is the same as (49), therefore it is of dimension at most one. This implies, in turn, that the algebraic set

$$f_1|_{\mathbb{A}\cap V} = f_2|_{\mathbb{A}\cap V} = \cdots = f_{M-1}|_{\mathbb{A}\cap V} = 0$$

on the affine part of the hypersurface V is also of dimension at most one: scheme-theoretically it is the same as (49), supported on the union of lines on V through o.

Let us look at the divisors

$$D_i = \overline{\{f_i|_{\mathbb{A}\cap V} = 0\}},$$

$i = 1, \dots, M-1$. We call them *hypertangent divisors*: if $H \in \operatorname{Pic} V$ is the class of a hyperplane section, then clearly

$$D_i \in |iH|$$

and

$$\operatorname{mult}_o D_i \geq i + 1,$$

since in the affine part of V

$$D_i|_{\mathbb{A}\cap V} = \{(q_{i+1} + \cdots + q_M)|_V = 0\}.$$

Now, by assumption

$$\dim_o(D_1 \cap \cdots \cap D_{M-1}) \leq 1,$$

where $\dim_o$ means the dimension in a neighborhood of the point o.

Now let us construct a sequence of distinct indices $i(1), i(2), \dots, i(M-4)$ and a sequence of irreducible subvarieties

$$Y_0 = Y, Y_1, \dots, Y_{M-4},$$

such that

- $Y_{k+1} \subset Y_k$, $\dim Y_k = M - 3 - k$, $\operatorname{codim}_V Y_k = k + 2$;
- $Y_k \not\subset D_{i(k)}$, so that $(Y_k \circ D_{i(k)})$ is an effective cycle on V and Y_{k+1} is one of its irreducible components;
- the following estimate holds:

$$\frac{\operatorname{mult}_o}{\deg} Y_{k+1} \geq \frac{\operatorname{mult}_o}{\deg} Y_k \cdot \frac{i(k)+1}{i(k)} \tag{50}$$

for each $k = 1, \dots, M-4$. We construct them by induction: for $k = 1$, arguing by dimension, there is a hypertangent divisor $D_{i(1)} \not\subset Y$. In the same way, we choose a divisor $D_{i(k)}$ at each step.

Since $Y_k \not\subset D_{i(k)}$, we get

$$\deg(Y_k \circ D_{i(k)}) = i(k) \deg Y_k,$$

$$\operatorname{mult}_o(Y_k \circ D_{i(k)}) \geq (i(k) + 1) \operatorname{mult}_o Y_k,$$

so that there is an irreducible component Y_{k+1} of the cycle $(Y_k \circ D_{i(k)})$, satisfying the estimate (50).

Now the curve $C = Y_{M-4}$ satisfies the inequality

$$\frac{\text{mult}_o}{\deg} C \geq \frac{\text{mult}_o}{\deg} Y \cdot \prod_{k=1}^{M-4} \frac{i(k)+1}{i(k)}.$$

The minimum of the right-hand side is attained at $i(k) \equiv k+3$ and is equal to $(\text{mult}_o / \deg) Y \cdot (M/4)$. The left-hand side is not greater than 1, which completes the proof of part (i). Proof of part (ii) we put off until Subsection 1.3. Q.E.D. for Proposition 1.1.

Formalizing the given argument, let us consider the following general situation: X is an irreducible projective variety, $o \in X$ a point and, moreover, o is either smooth or an isolated singularity. Let $\varphi\colon X^+ \to X$ be the blow up of the point o, $E = \varphi^{-1}(o)$ the exceptional divisor. We assume that E is reduced and irreducible. Let H be an ample divisor on X, and, moreover, the linear system $|H|$ is free, that is, defines a finite morphism

$$\alpha = \alpha_{|H|}\colon X \to \mathbb{P}^{\dim |H|}.$$

The degrees of subvarieties $Y \subset X$ are understood in the sense of the divisor H, that is, $\deg Y = (Y \cdot H^{\dim Y})$, in particular, the degree of the variety X itself is $\deg X = H^{\dim X} = \deg \alpha \cdot \deg \alpha(X)$. The following fact is well known.

Lemma 1.1. *For any subvariety $Y \subset X$ the inequality* $\text{mult}_y Y \leq \deg Y$ *holds, where $y \in Y$ is an arbitrary point.*

Definition 1.1. An effective divisor D on X is said to be *hypertangent* (with respect to the point o), if $D^+ \in |kH - lE|$, where $l \geq k+1$, D^+ is the strict transform of the divisor D on X^+. The number $\beta(D) = l/k > 1$ is called *the slope* of the divisor D.

Example 1.1. Let $X \subset \mathbb{P}^N$ be a hypersurface of degree $d \geq 2$, $f = q_1 + q_2 + \cdots + q_d$ its equation with respect to a system of affine coordinates $(z_1, \ldots, z_N)$ with the origin at a point o, q_i are homogeneous of degree i. Let D_i be the closure of the set

$$(q_1 + \cdots + q_i) \,|\, _X = 0,$$

$i \leq d-1$, and assume that $q_{i+1} \,|\, _E \not\equiv 0$ (where $(z_1 : \cdots : z_N)$ are considered as homogeneous coordinates on the exceptional divisor of the blow up of the point o in $\mathbb{P}^N$). Then D_i is a hypertangent divisor with the slope $\frac{i+1}{i}$. On the other hand, if $q_{i+1} \equiv \cdots \equiv q_{k-1} \equiv 0$, but $q_k \,|\, _E \not\equiv 0$, then the slope is $\frac{k}{i}$.

Let $\mathcal{D}$ be a finite set of hypertangent divisors, $\sharp \mathcal{D} = N \leq \dim X - 1$. Assume that the equality

$$\text{codim}_o \left(\bigcap_{D \in \mathcal{D}} D \right) = \sharp \mathcal{D}, \tag{51}$$

holds, where codim_o denotes the codimension in an arbitrarily small neighborhood of the point o.

The following fact is true.

Proposition 1.2. *The estimate*

$$\frac{\text{mult}_o}{\deg} Y \leq (\min_{\mathcal{B} \subset \mathcal{D}}{}^{*} \prod_{D \in \mathcal{B}} \beta(D))^{-1}$$

holds, where the symbol min* *means that the minimum is taken over all subsets* $\mathcal{B} \subset \mathcal{D}$ *of cardinality*

$$\sharp\mathcal{B} = \min\{\dim Y - 1, N - \mathrm{codimY}\} = b.$$

Proof. Repeats the argument given above for the Fano hypersurface $V_M \subset \mathbb{P}^M$: we construct a sequence of irreducible subvarieties $Y_0 = Y, Y_1, \dots, Y_b$ and hypertangent divisors $D_1, \dots, D_b$, where $Y_i \not\subset D_{i+1}$ and Y_{i+1} is an irreducible component of the effective cycle $(Y_i \circ D_{i+1})$, satisfying the inequality

$$\frac{\mathrm{mult}_o}{\deg} Y_{i+1} \geq \beta(D_{i+1}) \frac{\mathrm{mult}_o}{\deg} Y_i.$$

We complete the proof, applying Lemma 1.1 to the subvariety Y_b. Q.E.D.

1.2. Hypertangent linear systems. In certain cases it is more convenient to work not with hypertangent divisors, but rather with linear systems, generated by those divisors. Similar to Definition 1.1, let us give

Definition 1.2. A non-empty linear system Σ on X is said to be *hypertangent* (with respect to the point o), if $\Sigma^+ \subset |kH - lE|$, where $l \geq k+1$, Σ^+ is the strict transform of the system Σ on X^+. The number $\beta(\Sigma) = l/k > 1$ is called *the slope* of the system Σ.

The most important characteristic of the hypertangent linear system Σ is the codimension of its base set in a neighborhood of the point o, $\mathrm{codim}_o \mathrm{Bs}\, \Sigma$.

The set of hypertangent divisors $D \in \mathcal{D}$, $D^+ \in |k_D H - l_D E|$, generates a hypertangent system $\Sigma_k = \Sigma_k(\mathcal{D})$ in the following way. Let

$$f_D \in H^0(X, \mathcal{O}_X(k_D H))$$

be a section, defining the divisor D. Set

$$\Sigma_k = \left| \sum_{k_D \leq k} f_D s_D = 0 \right|,$$

where the summation is taken over all hypertangent divisors $D \in \mathcal{D}$ such that $k_D \leq k$, $s_D \in \alpha^* H^0(\mathbb{P}^{\dim|\mathrm{H}|}, \mathcal{O}(k - k_D))$ is the pullback of an arbitrary polynomial of degree $(k - k_D)$, with a zero of order $(k - k_D)$ at the point $\alpha(o)$. It is easy to see that

$$\beta(\Sigma_k) \geq \min_{D \in \mathcal{D}, k_D \leq k} \left\{ \frac{k + l_D - k_D}{k} \right\}.$$

So, since α is a finite morphism, we get the equality

$$\mathrm{codim}_o \mathrm{Bs}\, \Sigma_k = \sharp\{D \in \mathcal{D} \,|\, k_D \leq k\}.$$

From this equality we see, in particular, that the integer-valued function $\mathrm{codim}_o \mathrm{Bs}\, \Sigma_k$ is increasing when $k = k_D$ for some $D \in \mathcal{D}$, and for those values of k only. Define the *ordering function*

$$\chi \colon \{1, \dots, N\} \to \mathcal{K} = \{k_D \,|\, D \in \mathcal{D}\}$$

by the relation

$$\sharp\{D \in \mathcal{D} \,|\, k_D < \chi(i)\} < i \leq \sharp\{D \in \mathcal{D} \,|\, k_D \leq \chi(i)\}. \tag{52}$$

For instance, $\chi(1) = \min\{k_D \,|\, D \in \mathcal{D}\}$, $\chi(N) = \max\{k_D \,|\, D \in \mathcal{D}\}$.

Lemma 1.2. *The following inequality holds:*

$$\operatorname{codim}_o \operatorname{Bs} \Sigma_{\chi(i)} \geq i.$$

Proof. This follows directly from the formula (52) and from the explicit construction of the hypertangent system Σ_j. Q.E.D.

The ordering function is convenient to work with, because for a general set of hypertangent divisors

$$\mathbb{D} = (D_1, \ldots, D_N) \in \prod_{i=1}^{N} \Sigma_{\chi(i)}$$

and an arbitrary subvariety Y of codimension l, containing the point o, we get $Y \not\subset \operatorname{Supp} D_i$ for $i \geq l+1$. In particular, the effective cycle $(Y \circ D_{l+1})$ of codimension $l+1$ is well defined and satisfies the inequality

$$\operatorname{mult}_o(Y \circ D_{l+1}) \geq \beta(\Sigma_{l+1}) \frac{\operatorname{mult}_o}{\deg} Y \deg(Y \circ D_{l+1}).$$

By linearity of this inequality in the cycle $(Y \circ D_{l+1})$ there exists an irreducible component of this cycle, an irreducible subvariety Y_1 of codimension $l+1$, satisfying the inequality

$$\frac{\operatorname{mult}_o}{\deg} Y_1 \geq \beta(\Sigma_{l+1}) \frac{\operatorname{mult}_o}{\deg} Y.$$

Repeating this procedure $b = \min\{\dim Y - 1, N - \operatorname{codim} Y\}$ times, we get an irreducible subvariety Y_b, satisfying the inequality

$$\frac{\operatorname{mult}_o}{\deg} Y_b \geq \frac{\operatorname{mult}_x}{\deg} Y \prod_{i=1}^{b} \beta(\Sigma_{l+i}).$$

Applying Lemma 1.1 to the subvariety Y_b we get

Proposition 1.3. *The following estimate holds:*

$$\frac{\operatorname{mult}_o}{\deg} Y \leq \left(\prod_{i=1}^{b} \beta(\Sigma_{l+i}) \right)^{-1}.$$

The technique of hypertangent divisors and linear systems is applied in most cases to subvarieties of low codimension, mostly to divisors and subvarieties of codimension two, and in the procedure of forming the successive intersections almost all divisors (respectively, linear systems) take part. Moreover, in almost all particular situations for each $D \in \mathcal{D}$ we get $l_D = k_D + 1$, so that all slopes are of the form

$$\beta(D) = \frac{k_D + 1}{k_D},$$

and similarly for hypertangent systems. It is easy to see that for those values of the slopes Propositions 1.2 and 1.3 give the same estimates. However, in the most complicated cases one has to argue in a more delicate way, for instance, it is important that the first intersecting hypertangent divisor could be taken from the system Σ_l, and not Σ_{l+1}, in certain cases this can be justified, using some additional considerations; see Chapters 5, 7 and [**Sob01**]. Thus in more complicated cases the technique of hypertangent linear systems is more efficient and flexible than the technique of hypertangent divisors.

1.3. The regularity conditions. Application of the technique of hypertangent divisors/linear systems is based on the assumption (51). For this reason, the technique works successfully if and only if one can prove that a general variety of a given family satisfies the condition (51) *at every point.* To complete the proof of birational superrigidity of a general Fano hypersurface $V = V_M \subset \mathbb{P} = \mathbb{P}^M$, let us check that indeed (51) holds at every point $o \in V$, provided that the hypersurface V is general.

In fact, a somewhat stronger condition is satisfied. To formulate it, let us consider a system of affine coordinates $(z_1, \dots, z_M)$ with the centre at a point $o \in V$. Let $f = q_1 + q_2 + \dots + q_M$ be the equation of the hypersurface V, where $q_i(z_*)$ is a homogeneous polynomial of degree i.

Definition 1.3. The point $o \in V$ is *regular*, if the polynomials $q_1, \dots, q_{M-1}$ make a regular sequence in $\mathcal{O}_{o,\mathbb{P}}$, that is, the system of equations

$$q_1 = q_2 = \dots = q_{M-1} = 0$$

defines a finite set of lines in $\mathbb{P}$, passing through $o \in \mathbb{P}$.

Obviously, the regularity property does not depend on the choice of a system of coordinates. Furthermore, it is obvious, that if the point $o \in V$ is regular, then there are at most finitely many lines on V passing through the point o. Therefore, birational superrigidity of a general Fano hypersurface follows from

Proposition 1.4. *A general hypersurface $V = V_M \subset \mathbb{P}$ is regular at every point $o \in V$.*

Remark 1.1. The genericity is understood in the sense of Zariski topology on $\mathbb{P}(H^0(\mathbb{P}, \mathcal{O}_{\mathbb{P}}(M)))$.

Proof of Proposition 1.4. Set $\mathcal{P}_{k,K}$ to be the space of homogeneous polynomials of degree k in K variables, $\dim \mathcal{P}_{k,K} = \binom{k+K-1}{K-1}$. Consider the space of pairs

$$S = \{(x, F) \mid x \in \mathbb{P}, F \in H^0(\mathbb{P}, \mathcal{O}_{\mathbb{P}}(M))\}$$

with the natural projections $p\colon S \to \mathbb{P}$ and $q\colon S \to H^0(\mathbb{P}, \mathcal{O}_{\mathbb{P}}(M))$. Let us form inside it the incidence hypersurface $I = \{(x, F) \mid F(x) = 0\}$. Let $Y \subset I$ be the set of all pairs (x, F) such that the point x is not $(M-1)$-regular on the hypersurface $F = 0$. To prove Proposition 1.4, it is sufficient to show that $q(Y)$ is a proper closed subset of the space of homogeneous polynomials $\mathcal{P}_{M,M+1}$. This last statement, in turn, follows from

Lemma 1.3. *The codimension of the set $Y(x) = Y \cap p^{-1}(x)$ with respect to the fibre $p^{-1}(x) = H^0(\mathbb{P}, \mathcal{O}_{\mathbb{P}}(M))$ is equal to $M + 1$.*

Proof of the lemma. Fix an arbitrary point $x \in \mathbb{P}$ and a system of affine coordinates $(z_1, \dots, z_M)$ with the origin at x. Set

$$Y_k(x) = \left\{ (q_1, \dots, q_k) \in \prod_{i=1}^{k} \mathcal{P}_{i,M} \;\middle|\; \operatorname{codim}\{q_1 = \dots = q_k = 0\} < k \right\}.$$

Obviously, $Y(x) = Y_{M-1}(x)$, so that it is sufficient to show that $\operatorname{codim} Y_k(x) = M + 1$ for $k = 1, \dots, M-1$ (the codimension is taken with respect to the fibre $p^{-1}(x)$). We do it by induction on k.

For $k = 1$ the claim is obvious (smooth hypersurfaces do exist!).

Assume that $\operatorname{codim} Y_k(x) = M + 1$ for some $k \in \{1, \ldots, M-2\}$. Let us prove that the same is true for $k+1$, too. Set

$$R_k = \prod_{i=1}^{k} \mathcal{P}_{i,M} \setminus Y_k.$$

Let

$$\pi\colon \prod_{i=1}^{k+1} \mathcal{P}_{i,M} = \prod_{i=1}^{k} \mathcal{P}_{i,M} \times \mathcal{P}_{k+1,M} \to \prod_{i=1}^{k} \mathcal{P}_{i,M}$$

be the natural projection. Since the intersection $I \cap p^{-1}(x)$ has codimension 1 with respect to the fibre $p^{-1}(x)$, it is sufficient to show that

$$\operatorname{codim} Y_{k+1} \cap \pi^{-1}(R_k) \geq M,$$

where the codimension is taken with respect to the space of polynomials $\prod_{i=1}^{k+1} \mathcal{P}_{i,M}$. This inequality, in turn, follows from another estimate:

$$\mu_{k+1} = \operatorname{codim}_{\pi^{-1}(q_*)}(\pi^{-1}(q_*) \cap Y_{k+1}) \geq M,$$

where $q_* = (q_1, \ldots, q_k) \in R_k$ is an arbitrary regular sequence. It is this very inequality that we shall actually prove.

Naturally, $\pi^{-1}(q_*) \cong \mathcal{P}_{k+1,M}$. The condition

$$q_{k+1} \in \pi^{-1}(q_*) \cap Y_{k+1}$$

means that q_{k+1} vanishes along one of the components of the closed algebraic set $\{q_1 = \cdots = q_k = 0\} \subset \mathbb{A}^M$. Since the sequence q_* is regular, all the components of this set are cones of codimension k with the vertices at the point $x = (0, \ldots, 0)$. Projectivizing, we get a closed projective set $Q = \bigcup Q_i \subset \mathbb{P}^{M-1}$, all the components of which have codimension k.

Let $S \subset \mathbb{P}^{M-1}$ be a $(k-1)$-plane of general position, so that the restriction of the corresponding linear projection $\pi_S\colon \mathbb{P}^{M-1} \dashrightarrow \mathbb{P}^{M-1-k}$ onto each component Q_i of the set Q is a finite morphism. Obviously, restricting the space of homogeneous polynomials

$$\pi_S^* H^0(\mathbb{P}^{M-1-k}, \mathcal{O}_{\mathbb{P}^{M-1-k}}(k+1))$$

onto each component Q_i, we get an injective map. It means that we have constructed explicitly a space of homogeneous polynomials $W_{k+1} \subset \mathcal{P}_{k+1,M}$, which intersects Y_{k+1} by zero only. Consequently,

$$\mu_{k+1} \geq \dim W_{k+1} = \binom{M}{k+1} \geq M.$$

Q.E.D. for Lemma 1.3 and Proposition 1.4.

Remark 1.2. (i) Assume that a Fano hypersurface $V = V_M \subset \mathbb{P}^M$ has isolated singular points. The concept of regularity can be extended to singularities so that, similar to Theorem 1.1, a hypersurface, regular at every point, is birationally rigid and, in fact, superrigid if $\operatorname{mult}_o V \leq M-3$ for all $o \in V$. For a generic hypersurface with a singular point of multiplicity $M-2$ one can show birational rigidity and the isomorphism $\operatorname{Bir} V \cong \mathbb{Z}/2\mathbb{Z}$ (the non-trivial birational involution is the Galois involution generated by the projection from the singular point of multiplicity $M-2$, which is clearly a $2:1$ rational map); see [**Pukh02a**] for the proof and details.

(ii) Assume that a Fano hypersurface $V = V_M \subset \mathbb{P}^M$ has at most quadratic singularities of rank at least 5 (not necessarily isolated). The concept of regularity extends naturally to (possibly non-isolated) quadratic singularities and a careful estimate of the codimension of the set of hypersurfaces, non-regular at at least one point makes it possible to prove that the Zariski closure of the set of hypersurfaces which are either not factorial or not birationally superrigid is of codimension at least

$$\frac{(M-3)(M-4)}{2} + 1$$

in the parameter space $\mathbb{P}(H^0(\mathbb{P}^M, \mathcal{O}_{\mathbb{P}^M}(M)))$; see **[EP]** for the proof and details.

2. Fano complete intersections

In this section, we prove birational superrigidity of generic Fano complete intersections $V \subset \mathbb{P}^{M+k}$ of codimension k for $M \geq 2k+1$ (Subsections 2.1 and 2.2) and Fano varieties, realized as K-cyclic covers of Fano hypersurfaces of index ≥ 3, where $K \geq 3$. In the first case of Fano complete intersections the construction of hypertangent divisors is a straightforward generalization of the construction of Section 1. In the second case of Fano cyclic covers a new method is used ("taking the root of the equation of the branch divisor").

2.1. The theorem on birational superrigidity. Consider a Fano complete intersection

$$\mathbb{P} = \mathbb{P}^{M+k} \supset V = V_{d_1\cdot\ldots\cdot d_k} = F_1 \cap F_2 \cap \cdots \cap F_k,$$

where $F_i \subset \mathbb{P}$ are hypersurfaces of degree $d_i \geq 2$, and the integers d_i, $d_1 \leq d_2 \leq \cdots \leq d_k$ satisfy the relation $d_1 + \cdots + d_k = M + k$, $M \geq 4$, $k \geq 2$. The variety V is assumed to be smooth, of dimension M, so that $\operatorname{Pic} V = \mathbb{Z}H$ by the Lefschetz theorem, where H is the class of a hyperplane section, $K_V = -H$, so that V is a Fano variety of index 1.

Theorem 2.1. *For $M \geq 2k+1$ a general (in the sense of Zariski topology) variety V is birationally superrigid.*

In this subsection we formulate precisely the condition of general position for V, in the next one we construct the hypertangent systems and prove birational superrigidity.

Let $(z_1, \ldots, z_{M+k})$ be a system of affine coordinates on $\mathbb{P}$ with the origin at a point $o \in V$. Now write

$$\begin{array}{rll} f_1(z_*) = & q_{1,1} & +q_{1,2} + \cdots + q_{1,d_1}, \\ & \cdots & \\ f_i(z_*) = & q_{i,1} & +q_{i,2} + \cdots + q_{i,d_i}, \\ & \cdots & \end{array}$$

the equations of hypersurfaces $F_1, \ldots, F_k$, decomposed into homogeneous in z_* components $q_{i,j}$, $\deg q_{i,j} = j$.

Definition 2.1. The point $o \in V$ is *regular*, if the set of homogeneous polynomials $\{q_{i,j} \mid 1 \leq i \leq k, 1 \leq j \leq d_i, (i,j) \neq (k, d_k)\}$, consisting of all polynomials $q_{i,j}$, except for q_{k,d_k}, makes a regular sequence in $\mathcal{O}_{o,\mathbb{P}}$, that is, the system of equations $\{q_{i,j} = 0 \mid (i,j) \neq (k, d_k)\}$ defines a finite number of lines in $\mathbb{P}$, passing through the point o.

In particular, there are at most finitely many lines on V, passing through a regular point $o \in V$.

Let

$$\mathcal{H} = \prod_{i=1}^{k} \left(H^0(\mathbb{P}, \mathcal{O}_{\mathbb{P}}(d_i)) \setminus \{0\} \right)$$

be the set of k-uples $(f_1, \ldots, f_k)$ of non-zero homogeneous polynomials of degrees $d_1, \ldots, d_k$, respectively.

Proposition 2.1. *There exists a non-empty Zariski open set $U_{\text{reg}} \subset \mathcal{H}$ such that for $(f_1, \ldots, f_k) \in U_{\text{reg}}$ the complete intersection $V = V(f_1, \ldots, f_k) \subset \mathbb{P}$ is regular at each point.*

Proof is given below, in Section 3.

Proof of the theorem. By Proposition 2.1, it is sufficient to show that a complete intersection V, which is regular at every point, is birationally superrigid. Let us fix such a variety V.

By Theorem 2.2.1, birational superrigidity follows from the inequality

$$\frac{\text{mult}_o}{\deg} Y \leq \frac{4}{\deg V} = \frac{4}{d_1 \ldots d_k}, \tag{53}$$

which should be satisfied for any irreducible subvariety $Y \subset V$ and any point $o \in Y$. Let us fix a pair $o \in Y$ and prove (53). Note that the estimate (53) is optimal: the subvariety $H_1 \cap H_2 \cap V$, where H_i are distinct tangent hyperplanes to V at the point o, has degree $\deg V$ and multiplicity 4 at the point o.

Let $\Lambda_1^{\mathbb{P}} = \overline{\{\lambda_1 q_{1,1} + \cdots + \lambda_k q_{k,1} = 0\}}$ be the linear system of hyperplanes to V at the point o (the overline means the projective closure in $\mathbb{P}$). Let $\Lambda_1 = \Lambda_1^{\mathbb{P}} \mid_V$ be its restriction onto V.

Lemma 2.1. *The following estimate holds:*

$$\text{codim}_V \operatorname{Bs} \Lambda_1 = \text{codim}_{\mathbb{P}} \operatorname{Bs} \Lambda_1^{\mathbb{P}} = k.$$

Proof. It is easy to see that

$$q_{i,1} \mid_V = -(q_{1,2} + \cdots + q_{i,d_i}) \mid_V.$$

Set $\mathbb{T} = T_o V = \{q_{1,1} = q_{2,1} = \cdots = q_{k,1} = 0\} \cong \mathbb{C}^M$. The quadratic components of the functions $q_{i,1} \mid_V$ are $(-q_{i,2}) \mid_{\mathbb{T}}$ (with respect to any system of local parameters at the point o). Since the sequence

$$q_{1,1}, \ldots, q_{k,1}, q_{1,2}, \ldots, q_{k,2}$$

is regular, we conclude that $\operatorname{Bs} \Lambda_1 = \mathbb{T} \cap V$ is a closed subset of codimension precisely k in a neighborhood of the point o. Set $T_i = \{q_{i,1} = 0\}$, so that $\mathbb{T} = T_1 \cap \cdots \cap T_k$.

We claim that for each i, $1 \leq i \leq k$, the closed algebraic set $R_i = T_1 \cap \cdots \cap T_i \cap V$ is an irreducible subvariety of codimension i in V. This is true for $i = 1$ by the Lefschetz theorem, since V is smooth and the group $\operatorname{Pic} V$ is generated by the class of a hyperplane section, so that $T_1 \cap V$ should be irreducible. Assume that our claim is true for each j, $1 \leq j \leq i$, where $i \leq k - 1$. Now $R_i \not\subset T_{i+1}$, because this is true in a neighborhood of the point $x \in R_i$ and R_i is irreducible. Therefore the scheme-theoretic intersection of R_i and T_{i+1} gives an effective cycle $(R_i \circ T_{i+1})$ of codimension $i + 1$. However, by the Lefschetz theorem the group of cycles of codimension $i+1 < M/2$ modulo numerical equivalence is generated exactly by the class $(R_i \circ T_{i+1})$. Therefore $(R_i \circ T_{i+1}) = R_i \cap T_{i+1} = R_{i+1}$ is an irreducible cycle, which is what we need.

Eventually we get that $\mathrm{Bs}\,\Lambda_1$ is an irreducible subvariety of codimension k. Q.E.D. for the lemma.

Corollary 2.1. *Assume that an irreducible subvariety $Y \subset V$ of codimension 2 does not satisfy the inequality* (53). *Then there exists an irreducible subvariety $R \subset V$ of codimension $k+1$, satisfying the inequality*

$$\frac{\mathrm{mult}_o}{\deg} R > 2^{k-1} \frac{\mathrm{mult}_o}{\deg} Y. \tag{54}$$

Proof. Arguing as in Subsection 1.1, we construct a sequence of irreducible subvarieties $Y_0 = Y, Y_1, \ldots, Y_{k-2}$, $\mathrm{codim}\, Y_i = i+2$, satisfying the inequalities

$$\frac{\mathrm{mult}_o}{\deg} Y_i > 2^i \frac{\mathrm{mult}_o}{\deg} Y,$$

Y_{i+1} is an irreducible component of the cycle $(Y_i \circ T_{j(i)})$ for some $j(i) \in \{1, \ldots, k\}$. Such a construction is possible by the previous lemma. Now we get the alternative: either $Y_{k-2} = \mathrm{Bs}\,\Lambda_1$ or $Y_{k-2} \neq \mathrm{Bs}\,\Lambda_1$. In the first case by the regularity condition Y_{k-2} is an irreducible subvariety of degree $\deg V$, which has multiplicity precisely 2^k at the point o (the tangent cone $T_o Y_{k-2}$ is the intersection of quadrics

$$q_{1,2}\,|\,_{\mathbb{T}} = \cdots = q_{k,2}\,|\,_{\mathbb{T}} = 0),$$

so that $(\mathrm{mult}_o / \deg) Y \leq 4/\deg V$ contrary to our assumption. Therefore, the second case takes place, that is, there is a divisor $T_{j(k-2)} \not\supset Y_{k-2}$. Now we take R to be an irreducible component of the effective cycle $(Y_{k-2} \circ T_{j(k-2)})$, satisfying the inequality

$$\frac{\mathrm{mult}_o}{\deg} R \geq 2 \frac{\mathrm{mult}_o}{\deg} Y_{k-2}.$$

Q.E.D. for the corollary.

2.2. Construction of hypertangent divisors. Fix an irreducible subvariety R of codimension $k+1$ satisfying the inequality (54).

Let $\mathbb{A} \subset \mathbb{P}$ be an affine chart with coordinates $(z_1, \ldots, z_{M+k})$, $o = (0, \ldots, 0)$. Set

$$f_{i,\alpha} = q_{i,1} + \cdots + q_{i,\alpha}.$$

The symbol s_a, $a \in \mathbb{Z}_+$, stands for an arbitrary homogeneous polynomial of degree a in (z_*). Consider the linear systems

$$\Lambda^{\mathbb{A}}_{i,j} = \left| \sum_{\alpha=1}^{\min\{j, d_i - 1\}} f_{i,\alpha} s_{j-\alpha} \right|$$

on $\mathbb{A}$. Taking the closure of a general divisor of the system $\Lambda^{\mathbb{A}}_{i,j}$ in $\mathbb{P}$, we get the linear system $\Lambda^{\mathbb{P}}_{i,j}$ on $\mathbb{P}$. Obviously,

$$\mathrm{Bs}\,\Lambda^{\mathbb{A}}_{i,j} = \{q_{i,\alpha} = 0 \,|\, 1 \leq \alpha \leq \min\{j, d_i - 1\}\}, \quad \mathrm{Bs}\,\Lambda^{\mathbb{P}}_{i,j} = \overline{\mathrm{Bs}\,\Lambda^{\mathbb{A}}_{i,j}}.$$

Set $\Lambda^{\mathbb{P}}_j = \sum_{i=1}^{k} \Lambda^{\mathbb{P}}_{i,j}$. For this linear system we get

$$\mathrm{Bs}\,\Lambda^{\mathbb{P}}_j = \overline{\{q_{i,\alpha} = 0 \,|\, 1 \leq \alpha \leq \min\{j, d_i - 1\}\}}$$

and

$$c(j) = \operatorname{codim}_{\mathbb{P}} \operatorname{Bs} \Lambda_j^{\mathbb{P}} = \sum_{i=1}^{k} \min\{j, d_i - 1\}. \tag{55}$$

We get $\operatorname{Bs} \Lambda_{d_k-1}^{\mathbb{P}} = \operatorname{Bs} \Lambda_{d_k}^{\mathbb{P}} = \cdots = \operatorname{Bs}_{\infty}^{\mathbb{P}}$ and $\operatorname{codim}_{\mathbb{P}} \operatorname{Bs}_{\infty}^{\mathbb{P}} = M$.

Definition 2.2. The linear system

$$\Lambda_j = \Lambda_j^{\mathbb{P}}|_V$$

is called the *j-th hypertangent system* at the point o.

For example, Λ_1 is the $(k-1)$-dimensional linear system on V, cut out on V by tangent hyperplanes at o.

Now we will get an estimate for the codimension of the base set of hypertangent systems. A priori one can assert only that $\operatorname{codim}_V \operatorname{Bs} \Lambda_j \geq \operatorname{codim}_{\mathbb{P}} \operatorname{Bs} \Lambda_j^{\mathbb{P}} - k$: when a system is restricted from $\mathbb{P}$ to V, the codimension of the base set may decrease. However, in our case the regularity condition improves the matter.

The system Λ_1 was constructed and used above.

Lemma 2.2. (i) *For* $j \in \{2, \ldots, d_k - 2\}$

$$\operatorname{codim}_V \operatorname{Bs} \Lambda_j = \operatorname{codim}_{\mathbb{P}} \operatorname{Bs} \Lambda_j^{\mathbb{P}} = c(j) \tag{56}$$

in a neighborhood of the point o.

(ii) *For each* j *we have the estimate*

$$\operatorname{codim}_V \operatorname{Bs} \Lambda_j \geq \operatorname{codim}_{\mathbb{P}} \operatorname{Bs} \Lambda_j^{\mathbb{P}} - 1.$$

Proof. It is easy to see that for $\alpha \leq d_i - 1$,

$$f_{i,\alpha}|_{V \cap \mathbb{A}} = -(q_{i,\alpha+1} + \cdots + q_{i,d_i})|_{V \cap \mathbb{A}}. \tag{57}$$

Recall that $\mathbb{T} = T_o V \cong \mathbb{C}^M$ is the tangent space. Let us introduce the auxiliary linear systems

$$\Lambda_{i,j}^{\mathbb{T}} = \left| \sum_{\alpha=1}^{\min\{j, d_i-1\}} q_{i,\alpha+1} s_{j-\alpha} \right|_{\mathbb{T}}, \quad \Lambda_j^{\mathbb{T}} = \sum_{i=1}^{k} \Lambda_{i,j}^{\mathbb{T}}.$$

These are linear systems of cones of degree $j+1$ on the tangent space $\mathbb{T}$.

Replacing a general divisor of a linear system by its tangent cone at a given point, we get the corresponding tangent linear system. This simple operation always makes sense. However, on the level of explicit equations there is a point of possible trouble: replacing an equation by its first homogeneous component, one should take care that it is non-zero.

We claim that

$$\Lambda_j^{\mathbb{T}} = T_x \Lambda_j. \tag{58}$$

Indeed, by (57) the equation of the tangent cone to the divisor $(f_{i,\alpha}|_{V\cap\mathbb{A}})$ at the point o is $q_{i,\alpha+1}|_{\mathbb{T}}$, *provided the latter is non-zero.* Therefore in the following diagram of equalities

$$\begin{array}{ccc} T_x\Lambda_j & \stackrel{?}{=} & \Lambda_j^{\mathbb{T}} \\ \| & & \| \\ T_x(\Lambda_j^{\mathbb{P}}|_V) & \stackrel{?}{=} & (T_x\Lambda_j^{\mathbb{P}})|_{\mathbb{T}} \end{array}$$

we can erase the ? signs and thus obtain the desired equality (58), provided that the system $\Lambda_j^{\mathbb{T}}$ is non-zero. But this is true by the regularity condition. (The inequality $\Lambda_j^{\mathbb{T}} \neq 0$ is a much weaker fact than the regularity condition; it is just below and later in Subsection 2.2 that we use this condition in its full strength.) Obviously,

$$\operatorname{Bs}\Lambda_j^{\mathbb{T}} = \{q_{i,\alpha+1}|_{\mathbb{T}} = 0, 1 \leq \alpha \leq \min\{j, d_i - 1\}\} \subset \mathbb{T}.$$

Again by the regularity condition we get

$$\operatorname{codim}_{\mathbb{T}} \operatorname{Bs}\Lambda_j^{\mathbb{T}} = \sum_{i=1}^{k} \min\{j, d_i - 1\} = \operatorname{codim}_{\mathbb{P}} \operatorname{Bs}\Lambda_j^{\mathbb{P}}$$

for $j \in \{1, \ldots, d_k - 2\}$. The case $j = d_k - 1$ is excluded because q_{k,d_k} is excluded from the regularity condition. Since, obviously, for any linear system Σ we have

$$T_o \operatorname{Bs}\Sigma \subset \operatorname{Bs} T_o\Sigma,$$

we obtain the claim (i).

Let us prove (ii). The base set of the system $\Lambda_j^{\mathbb{P}}$ is given by a system of $c(j)$ equations and has codimension $c(j)$; see (55). The base set of the system Λ_j is given by the same equations, restricted to V, and $c(j) \leq \dim V$. This implies that $\operatorname{codim}_V \operatorname{Bs}\Lambda_j - \operatorname{codim}_{\mathbb{P}} \operatorname{Bs}\Lambda_j^{\mathbb{P}}$ is a non-increasing function of the parameter j. (For $a \geq b$ the codimension $\operatorname{codim}_V \operatorname{Bs}\Lambda_a$ is higher than $\operatorname{codim}_V \operatorname{Bs}\Lambda_b$ by, at most, the number of the new equations, that is, $c(a) - c(b)$.) Thus it is sufficient to prove (iii) for $j = \infty$: $\dim(\operatorname{Bs}_{\infty}^{\mathbb{P}})|_V \leq 1$.

Indeed, the base subscheme $(\operatorname{Bs}_{\infty}^{\mathbb{P}})|_V$ is given on V by the system of equations

$$\{q_{i,\alpha} = 0 \,|\, 1 \leq \alpha \leq d_i - 1\}.$$

Therefore, as a subscheme of the projective space $\mathbb{P}$ the subscheme $(\operatorname{Bs}_{\infty}^{\mathbb{P}})|_V$ is given by the system

$$\{q_{i,\alpha} = 0 \,|\, 1 \leq \alpha \leq d_i\}$$

of $M + k$ equations. By the regularity condition $\dim(\operatorname{Bs}_{\infty}^{\mathbb{P}})|_V \leq 1$, which is what we need. Q.E.D. for Lemma 2.2.

For $j \geq 1$ set

$$w^+(j) = \sharp\{i, 1 \leq i \leq k \,|\, j \leq d_i - 1\}.$$

Furthermore, set $w(j) = w^+(j) - 1$ for $j = 1, 2$ and $d_k - 1$, and $w(j) = w^+(j)$ for the remaining values of j. Since $d_i \geq 2$ for each $i = 1, \ldots, k$, we get $w^+(1) = k$. Furthermore,

$$c(j) = \sharp\{(i, \alpha) \,|\, 1 \leq i \leq k,\ 1 \leq \alpha \leq \min\{j, d_i - 1\}\} = \sum_{\alpha=1}^{j} w^+(\alpha)$$

for each $j \geq 1$. For $1 \leq j \leq d_k - 1$ we introduce the spaces

$$\Delta_j = \underbrace{\Lambda_j \times \cdots \times \Lambda_j}_{w(j)},$$

parametrizing $w(j)$-uples $(D_{j,1}, \dots, D_{j,w(j)}) \in \Delta_j$ of divisors from the linear systems Λ_j. Set

$$\Delta = \prod_{j=2}^{d_k-1} \Delta_j.$$

Lemma 2.3. *For a general set $(D_{*,*}) \in \Delta$ the set-theoretic intersection*

$$R^* = R \bigcap \left(\bigcap_{j=2}^{d_k-1} \bigcap_{a=1}^{w(j)} D_{j,a} \right)$$

is one-dimensional in a neighborhood of the point o.

Proof. First, note that we are taking the intersection of

$$w(2) + \cdots + w(d_k - 1) = c(d_k - 1) - k - 2 = M - k - 2$$

divisors with R, so that $\dim R^* \geq 1$. For $b \in \{2, \dots, d_k - 1\}$ set

$$R_b = R \bigcap \left(\bigcap_{j=2}^{b} \bigcap_{a=1}^{w(j)} D_{j,a} \right).$$

Let us prove by (ascending) induction on b that for a general set $(D_{*,*}) \in \prod_{j=2}^{b} \Delta_j$ we have

$$\operatorname{codim}_o R_b = 2 + \sum_{j=2}^{b} w(j) \tag{59}$$

where codim_o stands for the codimension in a neighborhood of the point o. Set for $e \in \{0, 1, \dots, w(b)\}$:

$$R_{b,e} = R \bigcap \left(\bigcap_{j=2}^{b-1} \bigcap_{a=1}^{w(j)} D_{j,a} \right) \bigcap \left(\bigcap_{a=1}^{e} D_{b,a} \right).$$

In particular, $R_{b,w(b)} = R_{b+1,0} = R_b$. Let us prove by induction on e, that for a general set $(D_{*,*})$ the following equality holds:

$$\operatorname{codim}_o R_{b,e} = k + 1 + \sum_{j=2}^{b-1} w(j) + e. \tag{60}$$

For $b = 2$, $e = 0$ (60) is true by assumption, since $R_{2,0} = R$. Take $e \leq w(b) - 1$, and assume that (60) is true. By (56),

$$\operatorname{codim}_V \operatorname{Bs} \Lambda_b = c(b) = \sum_{j=1}^{b} w^+(j) = k + 1 + \sum_{j=2}^{b-1} w(j) + w(b),$$

so that a generic divisor $D_{b,e+1} \in \Lambda_b$ contains none of the components of the set $R_{b,e}$, which contain the point o. This proves (60).

The equality (60) implies (59) automatically. Q.E.D. for the lemma.

Now let us use the sequence of hypertangent divisors constructed above to estimate the multiplicity $\mathrm{mult}_o R$.

Lemma 2.4. *In each of the closed algebraic sets $R_{b,e}$ there is an irreducible component $Y_{b,e}$ such that the sequence of irreducible subvarieties*

$$\{Y_{b,e} \mid 2 \leq b \leq d_k - 1, 0 \leq e \leq w(b)\}$$

satisfies the following properties:

(i) $Y_{2,0} = R$,

(ii) $Y_{b,e} \ni o$ *and, in particular,*

$$\mathrm{codim}_V Y_{b,e} = k + 1 + \sum_{j=2}^{b-1} w(j) + e,$$

(iii) *for* $e \leq w(b) - 1$,

$$\frac{\mathrm{mult}_o}{\deg} Y_{b,e+1} \geq \frac{b+1}{b} \cdot \frac{\mathrm{mult}_o}{\deg} Y_{b,e}. \tag{61}$$

Proof. Repeats the arguments of Subsection 1.2.

Corollary 2.2. *The following estimate holds*:

$$\frac{\mathrm{mult}_o}{\deg} R \leq \frac{3d_k}{2d_k - 2} \cdot 2^{w^+(2)} \cdot \left(\prod_{d_i \geq 3} d_i\right)^{-1}.$$

Proof. We argue as in Subsection 1.2.

The last variety $Y_{b,e}$ in the sequence constructed above is an irreducible curve. Its multiplicity at the point o does not exceed its degree. Applying part (iii) of Lemma 2.4, we get the estimate

$$\frac{\mathrm{mult}_x}{\deg} R \cdot \prod_{b=2}^{d_k - 1} \left(\frac{b+1}{b}\right)^{w(b)} \leq 1.$$

Recalling the definition of the integers $w(b)$ and $w^+(b)$, we get

$$\prod_{b=2}^{d_k-1} \left(\frac{b+1}{b}\right)^{w(b)} = \frac{2}{3} \cdot \frac{d_k - 1}{d_k} \cdot \prod_{b=2}^{d_k-1} \left(\frac{b+1}{b}\right)^{w^+(b)}$$

$$= \frac{2d_k - 2}{3d_k} \cdot 2^{-w^+(2)} \cdot \left(\prod_{b=3}^{d_k} b^{w^+(b-1) - w^+(b)}\right).$$

However, $w^+(b-1) - w^+(b) = \sharp\{i \mid 1 \leq i \leq k, d_i = b\}$. Consequently, the product in brackets is equal to

$$\prod_{d_i \geq 3} d_i,$$

which is what we need. Q.E.D. for the corollary.

Now, recalling that the subvariety R satisfies the inequality (54), we get the estimate

$$\frac{\mathrm{mult}_o}{\deg} C \leq \frac{3d_k}{4d_k - 4} \cdot \frac{4}{\deg V}.$$

The inequality $M \geq 2k+1$ implies easily that $d_k \geq 4$. This proves the inequality (53) and birational superrigidity of the variety V.

Remark 2.1. (i) Using the $8n^2$-inequality (Theorem 2.4.1) instead of the $4n^2$-inequality and slightly sharpening the regularity conditions, one can prove birational superrigidity of generic Fano complete intersections with the defining parameters k, M, $d_1, \ldots, d_k$ satisfying the conditions

- $\max\limits_{i=1,\ldots,k} \{d_i\} \geq 4$,
- $d_1 + \cdots + d_k \geq \max\{2k+3, k+7\}$,

so that $M \geq 7$ and $M \geq k+3$. This is a much stronger fact than Theorem 2.1. See [**Pukh11**] for the proof and details.

(ii) Using the method of [**Pukh11**] and considerably sharpening the regularity conditions, one can prove birational superrigidity of generic complete intersections of k_1 quadrics and k_2 cubics in $\mathbb{P}^{2k_1+3k_2}$ for $k_1+2k_2 \geq 12$ and also for the families, corresponding to the following values of (k_1, k_2):

$$(5,3), \quad (3,4), \quad (1,5), \quad (2,4), \quad (0,5).$$

See [**Pukh12c, Pukh12a**] for the proof and details.

2.3. Fano cyclic covers. Now let us consider a class of Fano complete intersections in a *weighted* projective space, the class of Fano cyclic covers. Their construction is as follows. Set $\mathbb{P} = \mathbb{P}^{M+1}$ to be the complex projective space, $M \geq 5$, $G = G_m \subset \mathbb{P}$ to be an irreducible hypersurface of degree $m \geq 2$, and $\sigma\colon F \to G$ to be the cyclic cover of degree $K \geq 2$, branched over an irreducible divisor $W \cap G$, where $W = W_{Kl} \subset \mathbb{P}$ is a hypersurface of degree Kl. Introducing a new coordinate u of weight l, one can realize F as a complete intersection of the type $m \cdot Kl$ in the weighted projective space

$$\mathbb{P}(\underbrace{1, \ldots, 1}_{M+2}, l).$$

Namely, F is given by the system of equations

$$\left\{ \begin{array}{l} f(x_0, \ldots, x_{M+1}) = 0, \\ u^K = g(x_0, \ldots, x_{M+1}), \end{array} \right. \tag{62}$$

where $f(x_*)$ and $g(x_*)$ are homogeneous polynomials of degrees m and Kl, respectively.

If the integers m, l and K satisfy the relation $m + (K-1)l = M+1$, then F is a primitive Fano variety of dimension M, that is, $\operatorname{Pic} F = \mathbb{Z}K_F$ and $(-K_F)$ is ample. Similarly to Theorem 2.1, we have

Theorem 2.2. *A sufficiently general (in the sense of Zariski topology) variety F is birationally superrigid.*

Proof. From this moment on, until the end of this section, we assume that the integers m, K, l are fixed and consider the open set

$$\mathcal{F} \subset \mathbb{P}(H^0(\mathbb{P}, \mathcal{O}_{\mathbb{P}}(m))) \times \mathbb{P}(H^0(\mathbb{P}, \mathcal{O}_{\mathbb{P}}(Kl))),$$

defined by the following condition: for a pair of polynomials $(f, g) \in \mathcal{F}$ the corresponding Fano cyclic cover F is irreducible. The set $\mathcal{F}$ makes a natural parameter space for Fano cyclic covers. The set of smooth cyclic covers $F \in \mathcal{F}$ we denote by the symbol $\mathcal{F}_{\text{sm}}$, its complement $\mathcal{F}\setminus\mathcal{F}_{\text{sm}}$ (that is, the set of singular cyclic covers) by the symbol $\mathcal{F}_{\text{sing}}$.

Let $F\sim(f,g)\in\mathcal{F}$ be a variety, $o\in F$ an arbitrary point, $p=\sigma(o)\in G=\{f=0\}$ its image on $\mathbb{P}$. Choose a system of affine coordinates $z_1,\dots,z_{M+1}$ with the origin at the point p. Without loss of generality we may assume that $z_i=x_i/x_0$. Set $y=u/x_0^l$. Now the standard affine set

$$\mathbb{A}^{M+2}_{(z_1,\dots,z_{M+1},y)}$$

is a chart for $\mathbb{P}(1,\dots,1,l)$. Abusing our notations, we use for the non-homogeneous polynomials, corresponding to f and g, the same symbols:

$$f=q_1+\dots+q_m,\qquad g=w_0+w_1+\dots+w_{Kl},$$

where q_i, w_j are homogeneous components of degree i,j in the variables z_*, respectively, so that in the affine chart $\mathbb{A}^{M+2}_{(z_*,y)}$, introduced above, the variety F is given by the pair of equations $f=0$, $y^K=g$ (replacing the system (62)). If the point $o\in F$ does not lie on the ramification divisor of the morphism σ, then we always assume that $w_0=1$. If the point $p\in G$ is non-singular, then without loss of generality we assume that $q_1\equiv z_{M+1}$. In the latter case we set

$$\bar{q}_i=q_i\,|_{\{z_{M+1}=0\}}=q_i(z_1,\dots,z_M,0)$$

and $\bar{w}_j=w_j\,|_{\,z_{M+1}=0}=w_j(z_1,\dots,z_M,0)$ for $i,j\geq 2$.

Now let us formulate the regularity condition for a point $o\in F$. We consider separately two cases, depending on whether the point $o\in F$ lies on the ramification divisor or not.

(R1.1) **The regularity condition for a point outside the ramification divisor.** Here $w_0=1$ and we may assume that $y(o)=1$. Set

$$g^{1/K}=(1+w_1+\dots+w_{Kl})^{1/K}=1+\sum_{i=1}^{\infty}\gamma_i(w_1+\dots+w_{Kl})^i$$

$$=1+\sum_{i=1}^{\infty}\Phi_i(w_1,\dots,w_{Kl}),$$

where $\gamma_i\in\mathbb{Q}$ are defined by the Taylor series of the function $(1+s)^{1/K}$ at zero, s is a variable:

$$(1+s)^{1/K}=1+\sum_{i=1}^{\infty}\gamma_i s^i$$

and $\Phi_i(w_1(z_*),\dots,w_{Kl}(z_*))$ are homogeneous polynomials of degree $i\geq 1$ in the variables z_*. It is easy to see that for $i\in\{1,\dots,Kl\}$ we get

$$\Phi_i(w_*(z_*))=\frac{1}{K}w_i+\Phi_i^\sharp(w_1,\dots,w_{i-1}).$$

In these notations the regularity condition is formulated in the following way: for $m\leq Kl$ the sequence

$$q_1,\dots,q_m,\Phi_{l+1}(w_*(z_*)),\dots,\Phi_{Kl-1}(w_*(z_*)) \tag{63}$$

should be regular in $\mathcal{O}_{o,\mathbb{C}^{M+1}}$; if $m \geq Kl+1$, then we require the regularity of the following sequence:

$$q_1, \dots, q_{m-1}, \Phi_{l+1}(w_*(z_*)), \dots, \Phi_{Kl}(w_*(z_*)).$$

(R2.1) **The regularity condition for a point on the ramification divisor.** Here $w_0 = 0$. We require that the sequence

$$q_1, \dots, q_m, w_1, \dots, w_K$$

should be regular in $\mathcal{O}_{o,\mathbb{C}^{M+1}}$.

Definition 2.3. A Fano cyclic cover $F \in \mathcal{F}$ is said to be *regular*, if every point $o \in F$ satisfies the corresponding regularity condition (Rα.1).

The set of smooth regular cyclic covers will be denoted by the symbol $\mathcal{F}_{\mathrm{reg}}$. Obviously, $\mathcal{F}_{\mathrm{reg}} \subset \mathcal{F}$ is an open set.

Proposition 2.2. *The set $\mathcal{F}_{\mathrm{reg}}$ is non-empty and, moreover, the following estimate holds:*

$$\mathrm{codim}_{\mathcal{F}}(\mathcal{F}_{\mathrm{sm}} \setminus \mathcal{F}_{\mathrm{reg}}) \geq 2.$$

Proof will be given in Subsection 3.3.

Assuming Proposition 2.2, we can prove Theorem 2.2 in the following form: any regular cyclic cover $V \in \mathcal{F}_{\mathrm{reg}}$ is birationally superrigid. Setting $H = -K_V$ to be the σ-pullback of a hyperplane section of G, we get $\mathrm{Pic}\, F = \mathbb{Z}H$, $A^2F = \mathbb{Z}H^2$. In particular, the claim of Theorem 2.2 follows from the estimate

$$\frac{\mathrm{mult}_o}{\deg} Y \leq \frac{4}{\deg V} = \frac{4}{mK}, \tag{64}$$

which should be true for any irreducible subvariety $Y \subset V$ of codimension two and any point $o \in Y$. Let us check it. First assume that $o \in F$ lies outside the ramification divisor. In the notations of the condition (R1.1) set

$$D_i = \sigma^* \overline{\{(q_1 + \dots + q_i)\,|\,_G = 0\}},$$

$i = 1, \dots, m-1$, where the bar means the closure in $\mathbb{P}$, and

$$L_j = \overline{\{(y - 1 - \sum_{i=1}^{j} \Phi_i(w_1, \dots, w_j))\,|\,_F = 0\}},$$

$j = l, l+1, \dots, Kl-1$. Consider first the case $m \leq Kl$. In this case set

$$\mathcal{D} = \{D_i \,|\, i = 1, \dots, m-1\} \cup \{L_j \,|\, j = l, \dots, Kl-2\}.$$

By the regularity condition (R1.1) the codimension of the closed set $\bigcap_{D \in \mathcal{D}} D$ in a neighborhood of the point o is $\sharp\mathcal{D} = M-1$. Furthermore, by construction

$$\mathrm{mult}_o\, D_i = i+1, \quad \mathrm{mult}_o\, L_j = j+1,$$

so that D_i, L_j are hypertangent divisors with the slopes $\beta(D_i) = \frac{i+1}{i}$ and $\beta(L_j) = \frac{j+1}{j}$, respectively. Applying Proposition 1.2, we get the estimate

$$\frac{\mathrm{mult}_o}{\deg} Y \leq \left(\frac{1}{2} \cdot \frac{2}{3} \cdot \prod_{i=1}^{m-1} \frac{i+1}{i} \cdot \prod_{j=l}^{Kl-2} \frac{j+1}{j} \right)^{-1}$$

$$\|$$

$$\frac{3l}{m(Kl-1)},$$

but, since $\dfrac{3l}{m(Kl-1)} < \dfrac{4}{mK}$, this proves the inequality (64). If $m \geq Kl+1$, then set

$$\mathcal{D} = \{D_i \mid i = 1, \ldots, m-2\} \cup \{L_j \mid j = l, \ldots, Kl-1\}.$$

Applying the condition (R1.1) and Proposition 1.2 again, we get the estimate

$$\frac{\mathrm{mult}_o}{\deg} Y \leq \left(\frac{1}{2} \cdot \frac{2}{3} \cdot \prod_{i=1}^{m-2} \frac{i+1}{i} \cdot \prod_{j=l}^{Kl-1} \frac{j+1}{j} \right)^{-1}$$

$$\|$$

$$\frac{3}{(m-1)K}.$$

Again we get $\dfrac{3}{(m-1)K} < \dfrac{4}{mK}$, which proves that the inequality (64) holds, if the point $o \in F$ lies outside the branch divisor.

Now assume that the smooth point $o \in F$ lies on the ramification divisor. In this case set

$$\Delta_j = \sigma^* \overline{\{(w_1 + \cdots + w_j) \mid_G = 0\}}, \quad j = 1, \ldots, K-1,$$

in the notations of the regularity condition (R2.1).

Lemma 2.5. *The equality* $\mathrm{mult}_o \Delta_j = j + 1$ *holds.*

Proof. This is obvious: the polynomial $(y^K - w_1 - \cdots - w_{Kl})$ vanishes on F and $y(o) = 0$, $j \leq K-1$. Q.E.D. for the lemma.

Now applying Proposition 1.2 to the set of hypertangent divisors

$$\mathcal{D} = \{D_i \mid i = 1, \ldots, m-1\} \cup \{\Delta_j \mid j = l, \ldots, K-1\},$$

by the condition (R2.1) we get the estimate

$$\frac{\mathrm{mult}_o}{\deg} Y \leq \left(\prod_{i=2}^{m-1} \frac{i+1}{i} \cdot \prod_{j=2}^{K-1} \frac{j+1}{j} \right)^{-1} = \frac{4}{mK},$$

which is what we need. Q.E.D. for Theorem 2.2.

3. Regular Fano varieties

In Subsection 1.3 above it was proven that a general Fano hypersurface satisfies the regularity conditions. However, the method given in Subsection 1.3 does not work for Fano complete intersections. Below we develop another method that makes it possible to prove the regularity conditions for complete intersections and cyclic covers.

3.1. A method of estimating the codimension. Let $z_1, \dots, z_{N+1}$ be a set of variables. The symbol $\mathcal{P}_a$ stands for the space of homogeneous polynomials of degree a in the variables z_*. Set

$$\mathcal{L} = \prod_{i=1}^{l+1} \mathcal{P}_{m_i} = \{(p_1, \dots, p_{l+1})\}$$

to be the set of all $(l+1)$-uples of homogeneous polynomials in the variables z_*, $0 \le l \le N-1$. With each $(l+1)$-uple $(p_*) \in \mathcal{L}$ we associate the projectivized set of its zeros

$$Z(p_*) = \{p_1 = \dots = p_{l+1} = 0\} \subset \mathbb{P}^N = \mathbb{P}(\mathbf{C}^{N+1}) = X.$$

Here we write $Z(p_*)$ and not $V(p_*)$ in order to make this notation different from our complete intersection $V = V(f_*)$. Let

$$Y = \{(p_*) \in \mathcal{L} | \operatorname{codim}_X Z(p_*) \le l\}$$

be the set of "irregular" $(l+1)$-uples. We need an estimate for the codimension of Y. The case $l = N-1$, when the "correct" dimension of $Z(p_*)$ is zero, is especially important for applications to Fano varieties. However, for technical reasons, it is more convenient to consider the general case of an arbitrary $l \in \{0, \dots, N-1\}$. Set $I = \{1, \dots, l+1\}$ and

$$\mu_j = \min_{S \subset I, \sharp S = j} \left\{ \sum_{i \in S} m_i \right\},$$

$j = 1, \dots, l+1$. Assume that $m = \mu_1 = \min\{m_1, \dots, m_{l+1}\} \ge 2$.

Proposition 3.1. *For any $l \in \{1, \dots, N-1\}$ the following estimate holds:*

$$\operatorname{codim}_{\mathcal{L}} Y \ge \min_{j \in \{0, \dots, l\}} \{(\mu_{j+1} - j)(N - j) + 1\}. \tag{65}$$

Remark 3.1. For $l = 0$ we get the trivial estimate $\operatorname{codim}_{\mathcal{L}} Y = \dim \mathcal{P}_m$.

Proof of Proposition 3.1. Set

$$\mathcal{L}_a = \prod_{i=1}^{a} \mathcal{P}_{m_i} = \{(p_1, \dots, p_a)\}.$$

For each irregular $(l+1)$-uple $(p_1, \dots, p_{l+1})$ fix the first (counting from the left to the right) moment when the codimension of the set of zeros $p_1 = \dots = p_a = 0$ fails to take the correct value. Consider the sets

$$Y_a = \{(p_*) \in \mathcal{L}_a \mid \operatorname{codim}_X Z(p_1, \dots, p_a) = \operatorname{codim}_X Z(p_1, \dots, p_{a-1}) = a-1\}.$$

Obviously,

$$Y = \coprod_{a=1}^{l+1} \left(Y_a \times \prod_{i=a+1}^{l+1} \mathcal{P}_{m_i} \right).$$

In particular, $\operatorname{codim}_{\mathcal{L}} Y = \min\{\operatorname{codim}_{\mathcal{L}_a} Y_a \mid 1 \le a \le l+1\}$. Set $I_a = \{1, \dots, a\} \subset I$ and

$$\mu_{a,j} = \min_{S \subset I_a, \sharp S = j} \left\{ \sum_{i \in S} m_i \right\},$$

$j = 1, \dots, a$. Obviously, $\mu_{a,j} \ge \mu_j$. Therefore, it is sufficient to prove the estimate

$$\operatorname{codim}_{\mathcal{L}_a} Y_a \ge \min_{j \in \{0, \dots, a-1\}} \{(\mu_{a,j+1} - j)(N - j) + 1\} \tag{66}$$

for each $a = 2, \dots, l+1$. We omit the trivial case $a = 1$, because in this case $\operatorname{codim}_{\mathcal{L}_1} Y_1 = \dim \mathcal{P}_{m_1} \ge \dim \mathcal{P}_m$, which is certainly higher than the right-hand side of (65).

The space $\mathcal{L}_a$, the set Y_a and the inequality (66) do not depend on l. Thus we may simplify our notations, setting $a = l+1$ and $\mu_{a,j} = \mu_j$. In other words, we prove the inequality (65) for Y_{l+1} instead of Y.

Denote Y_{l+1} by Y^*. We have reduced our original problem to a simpler task of estimating codimension of Y^* in $\mathcal{L}$, where Y^* consists of all such $(l+1)$-uples of polynomials $(p_1, \dots, p_{l+1})$ that the set $p_1 = \dots = p_l = 0$ has the correct dimension and there exists an irreducible component $B \subset Z(p_1, \dots, p_l)$, on which p_{l+1} vanishes. Let $\langle B \rangle$ be the linear span of B, and set $b = \operatorname{codim}\langle B \rangle \le l$.

Now set $Y^*(b)$ to be the set of all those $(l+1)$-uples $(p_*) \in Y^*$, for which there exists a component $B \subset Z(p_1, \dots, p_l)$ such that

$$\operatorname{codim}_X \langle B \rangle = b, \quad p_{l+1}|_B \equiv 0.$$

Obviously,

$$Y^* = \bigcup_{b=0}^{l} Y^*(b).$$

Thus it is sufficient to prove that

$$\operatorname{codim}_{\mathcal{L}} Y^*(b) \ge (\mu_{b+1} - b)(N - b) + 1 \tag{67}$$

for each $b = 0, \dots, l$. Let us prove (67).

The case $\mathbf{b = 0}$. Here $\langle B \rangle = \mathbb{P}^N$ and therefore each non-zero monomial of degree $m_{l+1} \ge m$ in the linear forms in $z_1, \dots, z_{N+1}$ does not vanish on B. The space of such monomials

$$\left\{ \prod_{i=1}^{m_{l+1}} (a_{i,1} z_1 + \dots + a_{i,N+1} z_{N+1}) \right\} \subset \mathcal{P}_{m_{l+1}}$$

is closed. Its dimension is equal to $m_{k+1} N + 1 \ge \mu_1 N + 1$. On the other hand, the set of polynomials $p_{l+1} \in \mathcal{P}_{m_{l+1}}$ vanishing on B is closed. These two closed sets intersect each other at zero only. Therefore the codimension of $Y(0)$ in $\mathcal{L}$ is no smaller than $\mu_1 N + 1$. This gives the estimate (67) for $b = 0$.

The case $\mathbf{b \ge 1}$. Here $\langle B \rangle = \mathbb{P}^{N-b}$. Our strategy is to reduce this case to the previous one ($b = 0$), restricting the polynomials p_i onto the linear span $P = \langle B \rangle$. Although our arguments are rather simple, they are not straightforward and require some extra work.

Definition 3.1. Let $g_1, \dots, g_e$ be homogeneous polynomials on the projective space P, $e \le \dim P - 1$, $\deg g_i \ge 2$ for $i = 1, \dots, e$. An irreducible subvariety

$C \subset P$ such that $\langle C\rangle = P$ and $\operatorname{codim}_P C = e$ is called an *associated subvariety of the sequence* (g_*), if there exists a chain of irreducible subvarieties $R_j \subset P$, $j = 0, \dots, e$, satisfying the following properties:

- $R_0 = P$;
- for each $j = 0, \dots, e-1$ the subvariety R_{j+1} is an irreducible component of the closed algebraic set
$$\{g_{j+1} = 0\} \cap R_j,$$
where $g_{j+1}|_{R_j} \not\equiv 0$, so that $\operatorname{codim}_P R_j = j$ for all j;
- $R_e = C$.

If the sequence (g_*) has an associated subvariety, this sequence is said to be *good*.

Lemma 3.1. (i) *The property of being good is an open property.*
(ii) *A good sequence* (g_*) *can have at most*

$$\left[\frac{1}{e+1}\prod_{j=1}^{e} \deg g_j\right]$$

associated subvarieties.

Proof is easily obtained by induction on e. For g_1 we have the condition $g_1 \not\equiv 0$, which is clearly an open one. Furthermore, at least one irreducible component of the hypersurface $g_1 = 0$ must be of degree at least two, which is also an open condition. There can be at most $[\deg g_1/2]$ such components.

Assume that Lemma 3.1 is true for each $e = 1, \dots, j$, where $j \leq \dim P - 2$. Denote by G_j the open set of good sequences of length j. By (ii), for each $(g_*) \in G_j$ there exist at most

$$\left[\frac{1}{j+1}\prod_{\alpha=1}^{j} \deg g_\alpha\right]$$

associated subvarieties. The polynomial g_{j+1} should be non-zero on at least one of them, say R_j, and moreover, the intersection

$$\{g_{j+1} = 0\} \cap R_j$$

should contain an irreducible component, the linear span of which is P. Obviously, this determines an open set in

$$G_j \times H^0(P, \mathcal{O}_P(\deg g_{j+1})).$$

Each associated subvariety has codimension $j+1$ and does not lie in a hyperplane; therefore, its degree is not smaller than $j+2$. Q.E.D. for the lemma.

Now let us come back to the polynomials p_* and assume that $l > b$. We claim that we can find $(l-b)$ polynomials among them — after renumbering we may assume that they are $p_1, \dots, p_{l-b}$ — such that the sequence

$$p_1|_P, \dots, p_{l-b}|_P \tag{68}$$

is good and B is one of its associated subvarieties.

Proof. Assume that we have already found j polynomials — let them be $p_1, \dots, p_j$ — such that the sequence $(p_1|_P, \dots, p_j|_P)$ is good and one of its associated subvarieties, say R_j, contains B. If $j < l - b$, then $R_j \neq B$ and there exists a polynomial p_α, $\alpha \in \{j+1, \ldots, l\}$, such that

$$p_\alpha|_{R_j} \not\equiv 0.$$

Otherwise, $R_j \subset Z(p_1, \ldots, p_l)$ and we get a contradiction, since $\dim R_j > \dim B$. After renumbering, we may assume that $\alpha = j+1$. Now $p_{j+1}|_B \equiv 0$, so that there exists an irreducible component R_{j+1} of the set $\{p_{j+1} = 0\} \cap R_j$, such that $R_{j+1} \supset B$. Proceeding in this way, we obtain our claim.

Now fix a projective subspace $P \subset \mathbb{P}^N$ of codimension b. Let $Y^*(P)$ be the set of all $(l+1)$-uples $(p_1, \ldots, p_{l+1}) \in Y^*$ such that there exists a component $B \subset Z(p_1, \ldots, p_l)$, whose linear span is $\langle B \rangle = P$ and $p_{l+1}|_B \equiv 0$.

By Lemma 3.1, good sequences form an open set. Thus we may estimate the codimension of $Y^*(P)$ in $\mathcal{L}$, assuming that $(p_1|_P, \ldots, p_{l-b}|_P)$ make a good sequence. Let

$$B_1, \ldots, B_K$$

be all its associated subvarieties, whose linear span is P. If $(p_1, \ldots, p_{l+1}) \in Y^*(P)$, then the polynomials

$$p_{l-b+1}|_P, \ldots, p_{l+1}|_P$$

must all vanish on one of these subvarieties B_i. Now arguing as in the case $b = 0$, we get

$$N \sum_{j=l-b+1}^{l+1} \deg p_j + b + 1 \geq \mu_{b+1}(N-b) + b + 1$$

independent conditions on $p_{l-b+1}, \ldots, p_{l+1}$. Taking into account that the Grassmanian has dimension $\dim G(N+1-b, N+1) = b(N+1-b)$, we finally get

$$\begin{array}{rcl} \operatorname{codim}_{\mathcal{L}} Y^*(b) & \geq & \mu_{b+1}(N-b) + b + 1 - b(N+1-b) \\ & & \| \\ & & (\mu_{b+1} - b)(N-b) + 1, \end{array}$$

which is what we need.

In our arguments above we assumed that $l > b$. If $l = b$, then $B \subset \mathbb{P}^N$ is a line, $l = N-1$ and the inequality (67) can be obtained by an easy dimension count: for a fixed line B the condition $p|_B \equiv 0$ for a polynomial p of degree $e \geq 1$ defines a closed algebraic set of polynomials of codimension $e+1$ in $\mathcal{P}_e$. Therefore,

$$\operatorname{codim}_{\mathcal{L}} Y^*(N-1) \geq \sum_{i=1}^{N} (m_i + 1) - 2(N-1) = \mu_{l+1} - l + 1,$$

since $\mu_{l+1} = m_1 + \cdots + m_N$.

Q.E.D. for Proposition 3.1.

Corollary 3.1. *In the notations of Proposition 3.1 for $l \leq N-2$ the following estimate holds:*

$$\operatorname{codim}_{\mathcal{L}} Y \geq mN + 1, \tag{69}$$

whereas for $l = N-1$ the following estimate holds:

$$\operatorname{codim}_{\mathcal{L}} Y \geq \min\{mN + 1, \mu_{l+1} - l + 1\}. \tag{70}$$

Proof. Obviously $\mu_j \geq jm$ for each $j = 1, \ldots, l+1$. Thus

$$(\mu_{j+1} - j)(N - j) + 1 \geq \varepsilon(j) + mN + 1,$$

where the polynomial $\varepsilon(t) = -(m-1)t^2 + (Nm - N - m)t$ has two roots, $t = 0$ and $t = N - 1 - \frac{1}{m-1}$. Thus $\varepsilon(0) = 0$ and $\varepsilon(j) \geq 0$ for $j = 1, \ldots, N-2$. Therefore we may omit in (65) the values $j = 1, \ldots, l-1$. If $l \leq N-2$, then we can omit the value $j = l$, as well. Q.E.D. for the corollary.

3.2. Proof of Proposition 2.1. Recall that $\mathcal{H}$ stands for the set of all k-uples $(f_1, \ldots, f_k)$ of non-zero homogeneous polynomials on $\mathbb{P}$ of degrees $d_1, \ldots, d_k$, respectively.

To begin with, there is a Zariski open set $U_{\rm sm} \subset \mathcal{H}$ such that for $(f_1, \ldots, f_k) \in U_{\rm sm}$ the complete intersection $V(f_*)$ is a smooth subvariety of codimension k in $\mathbb{P}$. Furthermore, let us consider the subset

$$Y = \{(o, (f_*)) \,|\, o \in V(f_*) \text{ is a non-regular point}\}$$

in $\mathbb{P} \times U_{\rm sm}$. We claim that the closure

$$\overline{\pi(Y)} \subset U_{\rm sm}$$

is a proper closed subset of positive codimension in $U_{\rm sm}$. Here $\pi\colon \mathbb{P} \times U_{\rm sm} \to U_{\rm sm}$ is the projection onto the second factor. Indeed, set

$$Y(o) = Y \cap (\{o\} \times U_{\rm sm}) \subset \mathbb{P} \times U_{\rm sm},$$

$$I = \{(o, (f_*)) \,|\, o \in V(f_*)\} \subset \mathbb{P} \times U_{\rm sm},$$

$$I(o) = I \cap (\{o\} \times U_{\rm sm}) \subset \mathbb{P} \times U_{\rm sm}.$$

Identifying $\{o\} \times U_{\rm sm} \cong U_{\rm sm}$, we can naturally look at $Y(o)$ and $I(o)$ as subsets in the open set $U_{\rm sm}$.

Lemma 3.2. *The following estimate holds:*

$$\operatorname{codim}_{I(o)} Y(o) \geq M + k + 1. \tag{71}$$

Proof. Indeed, the system of equations $q_{1,1} = \cdots = q_{1,k} = 0$ defines the tangent space $T_oV(f_*)$, which is of dimension exactly M. Projectivizing, we get the set of polynomials

$$\mathcal{Q} = \{q_{i,j} \,|\, 1 \leq i \leq k, 2 \leq j \leq d_i, (i,j) \neq (k, d_k)\}$$

of degree ≥ 2 on $\mathbb{P}^N$, $N = M - 1$. Obviously, $\sharp\mathcal{Q} = N$. Applying Proposition 3.1 to the set $\mathcal{Q}$ with $q_{1,1}, \ldots, q_{1,k}$ fixed, we see that

$$\begin{array}{rcl} \operatorname{codim}_{I(o)} Y(o) & \geq & \min\{2M - 1, \displaystyle\sum_{i=1}^{k-1}\sum_{j=2}^{d_i}(j-1) + \sum_{j=2}^{d_k - 1}(j-1) + 2\} \\ & & \qquad\qquad\qquad \| \\ & & \displaystyle\frac{1}{2}\sum_{i=1}^{k-1} d_i(d_i - 1) + \frac{1}{2}(d_k - 1)(d_k - 2) + 2. \end{array}$$

Denote the last expression by $\xi = \xi(d_1, \ldots, d_k)$. Since $M \geq 2k+1$, the inequality (71) follows from the estimate

$$\xi(d_1, \ldots, d_k) \geq M + k + 1.$$

Let us compute the minimum of the function ξ in the real arguments $(a_1, \ldots, a_k) \in \mathbb{R}^k_+$, satisfying the relation $a_1 + \cdots + a_k = M + k$. Obviously, the minimum is attained at $a_1 = \cdots = a_{k-1} = s \geq 0$, $a_k = M + k - (k-1)s \geq 0$. It is easy to compute that

$$\xi(s, \ldots, s, M + k - (k-1)s)$$
$$= \frac{1}{2}(k-1)(ks^2 - 2(M+k-1)s) + \frac{1}{2}(M+k-1)(M+k-2) + 2 = \zeta(s).$$

The function $\zeta(s)$ attains its minimum at $s_* = (M+k-1)/k$. This value satisfies all the restrictions. We get

$$\zeta(s_*) = \frac{(M+k-1)(M-1)}{2k} + 2.$$

Since $M \geq 2k+1$, we get $\zeta(s_*) \geq M + k + 1$, which is what we need. Q.E.D. for the lemma.

Since obviously $\operatorname{codim}_{U_{\mathrm{sm}}} I(o) = k$, we get the estimate

$$\operatorname{codim}_{U_{\mathrm{sm}}} Y(o) \geq M + k + (k+1).$$

But the set $\pi(Y)$ is covered by the subsets $Y(o)$ for all $o \in \mathbb{P}$, from which we get that

$$\operatorname{codim}_{U_{\mathrm{sm}}} \overline{\pi(Y)} \geq k + 1,$$

so that $\overline{\pi(Y)}$ is a proper subset of U_{sm}. In particular, the open set

$$U_{\mathrm{reg}} = U_{\mathrm{sm}} \setminus \overline{\pi(Y)}$$

is non-empty. Q.E.D. for Proposition 2.1.

3.3. Regular cyclic covers. Using the techniques developed above and in Subsection 1.3, let us prove Proposition 2.2. We need to show that the set of cyclic covers V with at least one non-regular smooth point is of codimension at least two in $\mathcal{F}$. In turn, this claim reduces to a simpler one. Let $p \in \mathbb{P}$ be a fixed point. Consider the set $\mathcal{F}(p) = \{F \in \mathcal{F} \,|\, G \ni p\} \subset \mathcal{F}$. Since the cover σ is cyclic, all the points $o \in \sigma^{-1}(p)$ for $F \in \mathcal{F}(p)$ are either all regular, or all non-regular. Set

$$\mathcal{F}_{\mathrm{reg}}(p) \subset \mathcal{F}(p)$$

to be the set of coverings F such that each point $o \in \sigma^{-1}(p)$ is non-singular and regular. We have

Proposition 3.2. *The following estimate holds:*

$$\operatorname{codim}_{\mathcal{F}(p)}(\mathcal{F}(p) \setminus \mathcal{F}_{\mathrm{reg}}(p)) \geq M + 2. \tag{72}$$

Since $p \in \mathbb{P}^{M+1}$ is an arbitrary point, and $\mathcal{F}(p) \subset \mathcal{F}$ is a divisor, by what was said above Proposition 3.2 implies Proposition 2.2.

Proof of Proposition 3.2. We work in a fixed system of affine coordinates $z_1, \ldots, z_{M+1}$ with the origin at the point p, that is, we realize F as a complete intersection of codimension two in $\mathbb{A}^{M+2}_{(z_1,\ldots,z_{M+1},y)}$. Let us consider separately the two cases, when the conditions (R1.1) and (R2.1) are violated.

Case 1. The point p does not lie on the branch divisor, that is, $w_0 = 1$. We assume that $m \leq Kl$ (if $m \geq Kl+1$, then the arguments are absolutely similar). Non-regularity of the sequence (63), consisting of M polynomials in $M+1$ variables z_*, means that for some irreducible cone $B \ni p$ of dimension ≥ 2 all the polynomials (63) vanish identically on B.

Let us construct by induction a sequence of polynomials $\Phi_i^+(w_1,\dots,w_l)$, setting $\Phi_{l+1}^+ = \Phi_{l+1}^\sharp$ and

$$\Phi_i^+ = \Phi_i^\sharp(w_1,\dots,w_l,-K\Phi_{l+1}^+,\dots,-K\Phi_{i-1}^+).$$

Lemma 3.3. *The sequence* (63) *is regular if and only if the sequence*

$$q_1,\dots,q_m,\ w_{l+1}+K\Phi_{l+1}^+(w_*(z_*)),\ \dots,\ w_{Kl-1}+K\Phi_{Kl-1}^+(w_*(z_*)) \tag{73}$$

is regular.

Proof. This is almost obvious. The sets of common zeros are the same for both sequences (63) and (73). This can be shown by induction using the obvious equality

$$\Phi_i(w_*(z_*))\,|\,_B \equiv 0 \Leftrightarrow w_i\,|\,_B \equiv -K\Phi_i^\sharp(w_*(z_*))\,|\,_B$$

for an arbitrary closed irreducible set B. Q.E.D. for the lemma.

The advantage of the sequence (73) compared to (63) is that the polynomials $w_i(z_*)$ are shifted by polynomials $\Phi_i^+(w_1(z_*),\dots,w_l(z_*))$, which depend on $w_1,\dots,w_l$ only (but not on $w_{l+1},\dots,w_{i-1}$). The set of polynomials w_i, $i \in \{1,\dots,l\}$ can be assumed to be a fixed set of general position. Thus in (73) each of the homogeneous polynomials w_i, $i \in \{l+1,\dots,Kl-1\}$, is shifted by a fixed homogeneous polynomial of degree i.

Let Π be the space of polynomials $q_1,\dots,q_m$, $w_1,\dots,w_{Kl-1}$. Consider an irreducible component $X \subset \Pi$, corresponding to non-regular sequences (63) or (73). For a fixed set of homogeneous polynomials $u_1,\dots,u_l$, $\deg u_i = i$, let $\Pi(u_1,\dots,u_l) = \{w_i = u_i\,|\,i = 1,\dots,l\} \subset \Pi$ be the corresponding affine subspace with the fixed $w_1,\dots,w_l$, and set $X(u_1,\dots,u_l) = X \cap \Pi(u_1,\dots u_l)$. Obviously, $\Pi(u_1,\dots u_l)$ identifies with the space of polynomials $q_1,\dots,q_m$, $w_{l+1},\dots,w_{Kl-1}$, which we denote by the symbol Π^+. Thus we consider $X(u_1,\dots u_l)$ to be embedded in the *linear* space Π^+. For a general set $(u_1,\dots u_l)$ we have

$$\operatorname{codim}_\Pi X = \operatorname{codim}_{\Pi^+} X(u_1,\dots,u_l).$$

Lemma 3.4. $X(0,\dots,0) \neq \emptyset$.

Proof. It is easy to see that the polynomials Φ_i, $\Phi_i^\sharp$ and Φ_i^+ are quasi-homogeneous in w_*, where $\operatorname{wt} w_i = i$. Therefore, for $\lambda \neq 0$,

$$(q_1,\dots,q_m,\lambda^{l+1}w_{l+1},\dots,\lambda^{Kl-1}w_{Kl-1}) \in X(\lambda u_1,\lambda^2 u_2,\dots,\lambda^l u_l)$$

if and only if

$$(q_1,\dots,q_m,w_{l+1},\dots,w_{Kl-1}) \in X(u_1,\dots,u_l)$$

(it is easy to see that the operation $w_i \mapsto \lambda^i w_i$ leaves a non-regular sequence of polynomials in its irreducible component X; so, this operation is invertible). Setting $\lambda = 0$, we obtain the claim of the lemma. Q.E.D.

Corollary 3.2. $\operatorname{codim}_\Pi X \geq \operatorname{codim}_{\Pi^+} X(0,\dots 0)$.

Now it remains to estimate the codimension of the closed set of non-regular sequences $q_1, \ldots, q_m, w_{l+1}, \ldots, w_{Kl-1}$. Let us order these polynomials by increasing degrees: $p_1 = q_1 \not\equiv 0$, $p_2, \ldots, p_M = w_{Kl-1}$. Let Y_a, $a \geq 2$ be the set of the sets $(p_1, \ldots, p_a)$ such that

$$\operatorname{codim}\{p_1 = \cdots = p_a = 0\} = \operatorname{codim}\{p_1 = \cdots = p_{a-1} = 0\} = a - 1. \tag{74}$$

If the sequence $p_1, \ldots, p_M$ is non-regular, then there is $a \in \{2, \ldots, M\}$, satisfying the equality (74). Therefore, it is sufficient to estimate from below the codimension $\operatorname{codim} Y_a$ in the space of all sets $\{p_1, \ldots, p_a\}$. This can be done using one of the two methods developed in this chapter.

Method 1 (see Subsection 1.3). Let $\mathbb{P}^M = \mathbb{P}(\mathbb{C}^{M+1})$ be the projective space with homogeneous coordinates $(z_1 : \cdots : z_{M+1})$, $B \subset \mathbb{P}^M$ an irreducible component of the closed set $\{p_1 = \cdots = p_{a-1} = 0\}$. Let

$$\gamma \colon \mathbb{P}^M \dashrightarrow \mathbb{P}^{M+1-a}$$

be a generic linear projection, $\gamma_B \colon B \to \mathbb{P}^{M+1-a}$ the corresponding finite morphism, $\gamma_B = \gamma\,|\,_B$. For any homogeneous polynomial h on $\mathbb{P}^{M+1-a}$ we have $\gamma_B^* h \not\equiv 0$, that is, $\gamma^* h\,|\,_B \not\equiv 0$. This implies the inequality

$$\operatorname{codim} Y_a \geq \binom{M+1-a+\deg p_a}{M+1-a}. \tag{75}$$

It is easy to check that for $a \in \{2, \ldots, M-1\}$ the right-hand side in (75) is strictly higher than $M+2$, so the estimate (75) is strong enough to prove the inequality (72) provided that regularity is violated at any place except for the last one. If $a = M$, then we have to argue in a different way.

Method 2 (Subsection 3.2 above). Again let $B \subset \mathbb{P}^M$ be an irreducible component of the closed set $\{p_1 = \cdots = p_{M-1} = 0\}$, which has the correct codimension, that is, $B \subset \mathbb{P}^M$ is an irreducible curve. Set $\langle B \rangle$ to be the linear span of the set B, $\operatorname{codim}\langle B \rangle \in \{0, 1, \ldots, M-1\}$. Take the subset $Y_M(b) \subset Y_M$, consisting of the sets $(p_1, \ldots, p_M)$, such that $p_M\,|\,_B \equiv 0$ for some irreducible component B with $\operatorname{codim}\langle B \rangle = b$. Now arguing as in the proof of Proposition 3.1, we get the estimate

$$\begin{aligned} &\operatorname{codim} Y_M(b) \\ &\geq (M-b)\left(\deg p_M + \min_{\substack{I \subset \{1, \ldots, M-1\},\\ \sharp I = b}} \{\sum_{i \in I} \deg p_i\}\right) + b + 1. \end{aligned} \tag{76}$$

It is easy to check that for any $b \in \{0, \ldots, M-1\}$ the expression (76) is strictly higher than $M+2$. Since

$$Y_M = \bigcup_{b=0}^{M-1} Y_M(b),$$

this completes the proof of Proposition 2.2. Q.E.D.

Remark 3.2. In this chapter we have given the principal elements of the technique of hypertangent divisors. In the hardest problem (accessible today) one has to combine it with other methods (for instance, the connectedness principle of Shokurov and Kollár) or to apply it not only to the variety V itself, but also to auxiliary varieties (for instance, to the exceptional divisor of the blow up of the singular point $o \in V$). How to make the techniques of hypertangent divisors stronger, will be considered in the subsequent chapters on the Fano fibre spaces. Here we just remark that in the last ten years there were a few attempts to suggest an alternative technique that would make it possible to exclude infinitely near maximal singularities on Fano varieties of arbitrary degree. However, by its efficiency the method of hypertangent divisors exceeds by far the other constructions. In particular, none of the alternative approaches makes it possible to prove birational superrigidity of Fano complete intersections (and the more so, does not work in the relative case, for Fano fibre spaces). This fact, it seems, has a reason. The method of hypertangent divisors is based on the geometry of lines on the given Fano variety, whereas the geometry of rational curves of small (anticanonical) degree is of exclusive importance for birational geometry of the variety itself.

4. K-trivial structures

In this section, somewhat deviating from the main topic of this book, we consider a new type of problems, closely related, but not identical, to the problem of birational rigidity. Namely, we go outside the class of rationally connected varieties when fibering a given *rationally connected* variety V. Let $\mathcal{F}$ be a class of algebraic varieties. For a particular rationally connected variety V we consider the problem:

> *describe all fibrations $V \dashrightarrow S$, the generic fibre of which is (birational to) a variety in the class $\mathcal{F}$.*

As in the purely rationally connected case, this problem can be reformulated in terms of birational maps:

> *for all fibre spaces $\beta\colon W \to S$, generic fibres of which are in the class $\mathcal{F}$, describe the possible birational maps $V \dashrightarrow W$.*

Below we will consider the latter setting.

It was observed long ago, that if it is possible to describe all rationally connected structures on a given variety V, then a slight modification of the arguments yields a complete description of all structures of Kodaira dimension zero (or even of non-maximal Kodaira dimension). That is to say, if V is shown to be birationally (super)rigid, then it is not hard to extend the techniques in such a way that the case when $\mathcal{F}$ is the class of varieties of Kodaira dimension zero (respectively, the class of varieties Y with $\kappa(Y) < \dim Y$) is covered. This informal principle will be demonstrated below for the Fano complete intersections of index one.

4.1. The non-strict Noether-Fano inequality. By the *relative Kodaira dimension* of the fibre space $\beta\colon W \to S$ we mean the Kodaira dimension of a fibre of general position $\beta^{-1}(s)$, $s \in S$. Notation: $\kappa(W/S)$.

Proposition 4.1. *Let V be an arbitrary primitive Fano variety and $\chi\colon V \dashrightarrow W$ a structure of a fibre space, that is, a birational map. Assume that the inequality*

$$\dim S + \kappa(W/S) < \dim W$$

holds (that is, the fibre of the fibre space W/S is not a variety of general type). Then for any mobile linear system Σ_S on S the pair $(V, \frac{1}{n}(\chi^{-1})_*\beta^*\Sigma_S)$ is not terminal, *where* $\Sigma = (\chi^{-1})_* \beta^* \Sigma_S \subset |-nK_V|$ *is the strict transform of the system* $\beta^* \Sigma_S$ *on* V.

Proof is almost word for word the same as the proof of existence of a maximal singularity in the case of Kodaira dimension $-\infty$ (Proposition 2.1.2). Assume the converse: the pair $(V, \frac{1}{n}\Sigma)$ is terminal. Let $\varphi\colon V^+ \to V$ be a resolution of singularities of the map χ, $\psi = \chi \circ \varphi$ the composite map, a birational morphism. By assumption, for each exceptional divisor $E \subset V^+$ of the morphism φ we have $\nu_E(\Sigma) < na(E)$. Let $D \in \Sigma$ be a general divisor. Since there are finitely many exceptional divisors, for some $n_+ \in \mathbb{Q}_+$, $n_+ < n$, we get in $A^1_{\mathbb{Q}} V^+$,

$$D^+ + n_+ K^+ = (n - n_+)\varphi^*(-K_V) + D^\sharp,$$

where D^+ is the strict transform of D, K^+ is the canonical class of V^+, $D^\sharp$ is an effective divisor. Therefore, for $N \gg 0$ the linear system

$$\Sigma^+ = |N!(D^+ + n_+ K^+)|$$

defines a birational map $\gamma\colon V^+ \dashrightarrow V^\sharp \subset \mathbb{P}^{\dim \Sigma^+}$. Let $W(s) = \beta^{-1}(s)$ be a fibre of general position, $Y(s) \subset V^+$ its strict transform on V^+. Obviously, the linear system $\Sigma^+|_{Y(s)}$ also defines a birational map, so that $\psi_* \Sigma^+|_{W(s)}$ defines a birational map, as well. However,

$$\psi_* \Sigma^+|_{W(s)} = |N! n_+ K_{W(s)}|$$

is a subsystem of the pluricanonical system of the fibre $W(s)$, whereas by assumption $\kappa(W(s)) < \dim W(s)$. A contradiction. Q.E.D. for the proposition.

In other words, any structure of a non-maximal relative Kodaira dimension on V generates an *almost maximal singularity* $E \subset V^+$ of a mobile system $\Sigma \subset |-nK_V|$, where $\varphi\colon V^+ \to V$ is a birational morphism, that is, the inequality $\nu_E(\Sigma) > na(E)$ is satisfied. If the inequality is strict, we are in the well studied standard case of a maximal singularity. Therefore, the only new situation is when we get no maximal singularities but the equality $\nu_E(\Sigma) = na(E)$ is satisfied. If this case can be dealt with, then a complete description of structures of non-maximal relative Kodaira dimension becomes immediately available. Now assume that $V \subset \mathbb{P} = \mathbb{P}^{M+k}$ is a generic Fano complete intersection studied in Section 2.

We fix $k \geq 2$, $M \geq 2k+1$ and a set of integers $(d_1, \dots, d_k) \in \mathbb{Z}^k_+$ satisfying the conditions $d_k \geq \dots \geq d_1 \geq 2$ and

$$d_1 + \dots + d_k = M + k.$$

We assume in addition that if $M = 2k+2$, then $k \geq 4$, and if $M = 2k+1$, then $k \geq 5$ and $(d_1, \dots, d_k) \neq (2, \dots, 2, k+3)$.

The main result of this section is

Theorem 4.1. *Let* $\chi\colon V \dashrightarrow W$ *be a structure of a fibre space of non-maximal relative Kodaira dimension on a sufficiently general complete intersection* $V \subset \mathbb{P}$, *that is, the inequality*

$$\kappa(W/S) + \dim S < \dim W = M$$

holds. Then $\kappa(W/S) = 0$, $S = \mathbb{P}^1$ *and there exists a uniquely determined linear subspace* $\Lambda \subset \mathbb{P}$ *of codimension two such that the following diagram of maps is commutative*

$$\begin{array}{ccccc} & V & \stackrel{\chi}{\dashrightarrow} & W & \\ \pi & \downarrow & & \downarrow & \beta \\ & \mathbb{P}^1 & = & S, & \end{array}$$

where $\pi = \pi_\Lambda|_V\colon V \dashrightarrow \mathbb{P}^1$ *is the restriction onto* V *of the linear projection* $\pi_\Lambda\colon \mathbb{P} \dashrightarrow \mathbb{P}^1$ *from the subspace* Λ.

Obviously, the restriction onto V of the projection π_Λ from a subspace of codimension two is a K-trivial structure.

The proof breaks into several steps. The first task is to show that S is a curve (therefore, $S = \mathbb{P}^1$) and the linear system $\Sigma = (\chi^{-1})_*\beta^*\Sigma_S \subset |-nK_V|$ is composed from a pencil.

4.2. The base S is a curve. We assume that $V \in U_{\mathrm{reg}}$ is a regular complete intersection.

Proposition 4.2. *We have* $\dim S = 1$, *so that* $S = \mathbb{P}^1$.

Proof. The pair $(V, \frac{1}{n}\Sigma)$ by Proposition 4.1 is not terminal. However, by Section 2, it is canonical. Therefore, for some birational morphism $\varphi\colon V^+ \to V$ there is an exceptional divisor $E \subset V^+$, satisfying the equality $\nu_E(\Sigma) = na(E)$. Set $B = \varphi(E) \subset V$ to be the centre of E on V. The following observation is crucial: If

- either $\operatorname{codim} B \geq 4$
- or $\operatorname{codim} B = 3$ and the inequality $\operatorname{mult}_B \Sigma < 2n$ holds,

then the self-intersection $Z = (D_1 \circ D_2)$ of the linear system Σ (where $D_i \in \Sigma$ are generic divisors) still satisfies the $4n^2$-inequality

$$\operatorname{mult}_B Z > 4n^2. \tag{77}$$

Indeed, the first proof of Theorem 2.2.1 works word for word as either the upper part of the resolution is non-empty (and then $\Sigma_u > 0$) or all discrepancies δ_j are at least 2 and at least one of them is at least 3. Alternatively, one can use the second proof of Theorem 2.2.1 and observe that after one restriction to a generic hyperplane by the inversion of adjunction the linear system becomes not *log* canonical, so that on a generic 2-dimensional germ we get a non-log canonical pair again and the proof works.

However, we have shown in Section 2 that the inequality (77) is impossible. Therefore, either $\operatorname{codim} B = 3$ and $\operatorname{mult}_B \Sigma = 2n$, or $\operatorname{codim} B = 2$ and

$$\operatorname{mult}_B \Sigma = n. \tag{78}$$

In any case the blow up of the subvariety B realizes a non-terminal singularity of the system Σ. If $\operatorname{codim} B = 2$, then $Z = n^2 B$, that is, the self-intersection of the system Σ has no mobile part. Thus the system Σ is composed from a pencil, $\dim S = 1$, as we claimed.

To complete the proof, it remains to exclude the first case when $\operatorname{codim} B = 3$. It could be done, using the technique developed in Section 2 again. For instance, if $d_k = \max\{d_i\} \geq 5$, then the equality $\operatorname{mult}_B Z = 4n^2$ implies that for every point $x \in B$ each component of the effective cycle Z is of the form $T_1 \cap T_2$, where $T_1 \neq T_2$ are sections of F by hyperplanes, tangent to F at x. This is, of course, impossible (a section of F by any plane $P \subset \mathbb{P}$ of codimension two has at most a curve of singular points). Now we will show a better way to exclude the case $\operatorname{codim} B = 3$. Assume that it takes place.

Proposition 4.3. *For any irreducible subvariety $Y \subset V$ of dimension k the inequality* $\operatorname{mult}_Y \Sigma \leq n$ *holds.*

Proof. Let $x \in \mathbb{P} \setminus V$ be a point of general position, $C(x) \subset \mathbb{P}$ the cone with the vertex at x and the base Y. It is easy to see that $C(x) \cap V = Y \cup R(x)$, where $R(x) \subset V$ is the residual curve. For a sufficiently general point x the curve $R(x)$ is irreducible. The family of residual curves $R(x)$ sweeps out V.

Lemma 4.1. *If the point $x \in \mathbb{P}$ is sufficiently general, then*

(i) *the cone $C(x)$ is singular at a point $z \neq x$ if and only if z lies on a line $L_{x,y}$ connecting x and y, where $y \in \operatorname{Sing} Y$ is a singularity of Y,*

(ii) *the curve $R(x)$ meets the subvariety Y at smooth points of Y only: $R(x) \cap \operatorname{Sing} Y = \emptyset$.*

Proof is given below.

Let us carry on with the proof of Proposition 4.3. We know that the intersection $R(x) \cap Y$ consists of points which are smooth both on Y and $C(x)$. Therefore, the intersection index $(R(x) \cdot Y)_{C(x)}$ is well defined.

Lemma 4.2. *The following equality holds:* $(R(x) \cdot Y)_{C(x)} = \deg R(x)$.

Proof. It is easy to see that on the natural desingularization of the cone $C(x)$, obtained first by blowing up the vertex x and then by taking the fibre product of $\widetilde{C}(x)$ with Y^+ over Y, where $Y^+ \to Y$ is a desingularization of Y, the strict transform of Y is linearly equivalent to the pullback of a hyperplane section of the cone. The rest is obvious. Q.E.D.

Now assume that $\operatorname{mult}_Y \Sigma > n$, that is, for a general divisor $D \in \Sigma$ the estimate $\operatorname{mult}_Y D > n$ holds.

Lemma 4.3. *The following inequality holds:*

$$(D \cdot R(x))_F > n \deg R(x). \tag{79}$$

Proof (which is obtained by combining Lemma 4.2 with the inequality $\operatorname{mult}_Y D > n$) is given below.

Since $D \in |-nK_F|$, we get $(D \cdot R(x))_F = n \deg R(x)$, which contradicts the inequality (79). Therefore, the inequality $\operatorname{mult}_Y \Sigma > n$ is impossible. Q.E.D. for Proposition 4.3.

Proof of Proposition 4.2 is complete as well. Q.E.D.

Proof of Lemma 4.1. We obtain both claims by a trivial dimension count taking into account the inequality $2k+1 \leq M$. Part (i) follows from the fact that for a general point x the projection from this point,

$$\pi_x \colon \mathbb{P} \setminus \{x\} \to \mathbb{P}^{M+k-1}$$

is one-to-one on Y and, moreover, for any point $y \in Y$ the line $L_{x,y}$ is not tangent to Y at the point y.

Part (ii) follows from the fact that at the points of intersection $y \in R(x) \cap Y$ the line $L_{x,y}$, connecting x and y, is tangent to F, whereas the set

$$\overline{\bigcup_{y \in \operatorname{Sing} Y} T_y F} \subset \mathbb{P}$$

is a proper closed subset of the projective space (since $\dim \operatorname{Sing} Y \leq k-1$). Q.E.D.

Proof of Lemma 4.3. There is a well-known explicit construction for the intersection index of a curve and a divisor, using the "sum over all infinitely near points". Let us apply this construction to the intersection number $(R(x) \cdot Y)_{C(x)}$. We construct by induction a sequence of birational morphisms

$$\varphi_i \colon \mathbb{P}^{(i)} \to \mathbb{P}^{(i-1)}, \quad i = 1, \dots, N,$$

where $\mathbb{P}^{(0)} = \mathbb{P}$, φ_i is the blow up of the finite set of points $S_{i-1} = R^{(i-1)} \cap Y^{(i-1)}$, and $R^{(i)} \subset \mathbb{P}^{(i)}$ and $Y^{(i)} \subset \mathbb{P}^{(i)}$ are strict transforms of $R(x)$ and Y on $\mathbb{P}^{(i)}$, respectively, $R^{(N)} \cap Y^{(N)} = \emptyset$. We emphasize that $S_j = R^{(j)} \cap Y^{(j)}$ is the set-theoretic (not the scheme-theoretic) intersection, so that all varieties $\mathbb{P}^{(j)}$ are non-singular. Now we get

$$(R(x) \cdot Y) = \sum_{i=0}^{N-1} \left(\sum_{y \in S_i} \operatorname{mult}_y R^{(i)} \cdot \operatorname{mult}_y Y^{(i)} \right).$$

To simplify the notations, set

$$R(x) \wedge Y = \coprod_{i=0}^{N-1} S_i$$

to be the set of all points of intersection of $R(x)$ and Y, *including the infinitely near ones*. Now

$$(R(x) \cdot Y) = \sum_{y \in R(x) \wedge Y} \operatorname{mult}_y R \cdot \operatorname{mult}_y Y,$$

where $\operatorname{mult}_y R = \operatorname{mult}_y R^{(i)}$ for $y \in S_i$, and similarly for Y. Taking into account that Y is non-singular at the points of intersection $R(x) \cap Y$, we finally get

$$(R(x) \cdot Y) = \sum_{y \in R(x) \wedge Y} \operatorname{mult}_y R.$$

Let $D \subset F$ be an irreducible divisor, satisfying the inequality $\operatorname{mult}_Y D > n$ and not containing the curve $R(x)$ entirely. For the intersection index $(R(x) \cdot D)_F$ we get:

$$(R(x) \cdot D) = \sum_{y \in R(x) \wedge D} \operatorname{mult}_y R(x) \cdot \operatorname{mult}_y D$$

$$\geq \sum_{y \in R(x) \wedge Y} \operatorname{mult}_y R(x) \cdot \operatorname{mult}_y D > n \sum_{y \in R(x) \wedge Y} \operatorname{mult}_y R(x) = n \deg R(x)$$

(since obviously $\operatorname{mult}_y D^{(i)} \geq \operatorname{mult}_{Y^{(i)}} D^{(i)} > n$ for each $i = 0, \dots, N-1$). Q.E.D.

4.3. Linear projections and cones. Let us describe the structure of our further arguments. Apart from the standard regularity conditions of Section 2, we require some additional genericity properties.

Proposition 4.4. *A sufficiently general complete intersection V satisfies the following condition: for any linear subspace $\Lambda \subset \mathbb{P}$ of codimension $k+1$ the intersection $V \cap \Lambda$ is an irreducible reduced variety of dimension $M-k-1$.*

Proof is given below. By the Lefschetz theorem the claim of the proposition holds for $M \geq 2k+3$ for any smooth complete intersection V. Only two cases need to be considered: $M = 2k+2$ and $M = 2k+1$.

We assume that the complete intersection V satisfies the property, described in Proposition 4.4.

Let $\chi\colon V \dashrightarrow W$ be a structure of a fibre space of non-maximal relative Kodaira dimension, Σ the strict transform on V with respect to χ of the pullback on W of some mobile linear system on the base S. Then $\Sigma \subset |-nK_V|$, $n \geq 1$, the pair $(V, \frac{1}{n}\Sigma)$ is not terminal, the system Σ is composed from a pencil and there is an irreducible subvariety $B \subset V$ of codimension two, such that the equality (78) holds. Let $b \in B$ be a point of general position, $\Delta = T_b B \subset \mathbb{P}$ the tangent space.

Proposition 4.5. *Let $\mu_\Delta\colon \mathbb{P} \dashrightarrow \mathbb{P}^{k+1}$ be the linear projection from Δ and*

$$\mu = \mu_\Delta|_V\colon V \dashrightarrow \mathbb{P}^{k+1}$$

its restriction onto V. Then the fibres of the rational map μ are irreducible and reduced, whereas the linear system Σ is the pullback via μ of a mobile linear system Γ on $\mathbb{P}^{k+1}$.

Proof is given below.

Now let us complete the **proof of Theorem 4.1.** By Proposition 4.4, the set $V \cap \Delta$ has codimension $k+2$ in V. Considering the dimensions, we conclude that $\mu(B) = \bar{B}$ is an irreducible subvariety of codimension two ($\bar{B}$ cannot be a divisor, since in that case the linear system Γ, and therefore also Σ, would have had a fixed component), so that $B = \mu^{-1}(\bar{B})$. Therefore,

$$\operatorname{mult}_{\bar{B}} \Gamma = n,$$

where Γ is a linear system of hypersurfaces of degree n. This is possible in one case only, when $\bar{B} \subset \mathbb{P}^{k+1}$ is a linear subspace of codimension two, and the system Γ is composed from the pencil of hyperplanes, containing $\bar{B}$. But then $B = \Lambda \cap V$, where $\Lambda = \mu_\Delta^{-1}(\bar{B})$ is a linear subspace of codimension two in $\mathbb{P}$, and the system Σ is composed from the pencil of sections of V by hyperplanes containing Λ. Q.E.D. for Theorem 4.1.

Proof of Proposition 4.5. Proposition 4.4 implies that the fibres of μ are irreducible and reduced. To prove the main claim that $\Sigma = \mu^*\Gamma$, let us consider a point of general position $p \in V$. Let $D \in \Sigma$ be the divisor, containing that point. Now we get

Proposition 4.6. *The following inclusion holds:*

$$T_p \mu^{-1}(\mu(p)) \subset T_p D.$$

Proposition 4.5 follows immediately from Proposition 4.6: as the point p is generic, we get that $\mu(D) \subset \mathbb{P}^{k+1}$ is a divisor (that is, $\mu(D) \neq \mathbb{P}^{k+1}$), so that $\Sigma = \mu^*\Gamma$ for some linear system Γ, which is what we need. Q.E.D.

Proof of Proposition 4.6. Set $P = \langle \Delta, p \rangle$ to be the fibre of the linear projection μ_Δ, so that $\mu^{-1}(\mu(p)) = P \cap V$. Since the point p is generic and the fibre $\mu^{-1}(\mu(p))$ is non-singular at this point, we get

$$\mathrm{codim}_{\mathbb{P}}(P \cap T_pV) = \mathrm{codim}_{\mathbb{P}}\, P + k,$$

that is, the linear subspaces P and $T_pV \subset \mathbb{P}$ are in general position.

Assume now that the point of general position $b \in B$ was chosen in the following way: on the variety B we considered an arbitrary family of irreducible k-dimensional subvarieties $\{Y_u, u \in U\}$, sweeping out B, in that family we chose a variety $Y = Y_u$ of general position, and the point b is a point of general position on Y. In particular, T_bY is a generic linear subspace of dimension k in $\Delta = T_bB \subset \mathbb{P}$. In particular,

$$\dim(\langle T_bY, p\rangle \cap T_pV) = 1, \tag{80}$$

that is, the linear subspaces $\langle T_bY, p\rangle$ and T_pV are in general position.

By the symbol $[b,p]$ we denote the line in $\mathbb{P}$, joining these two points, by the symbol (b,p) the set $[b,p] \setminus \{b,p\}$. Take a point $x \in (b,p)$. Set $C(Y,x)$ to be the cone with the vertex x and the base Y.

Proposition 4.7. *For sufficiently general Y, b, p, x the following claims are true:*

(i) *the point $z \in C(Y,x)$ is a singularity of that cone, if either $z = x$, or $z \in [y,x]$, where $y \in \mathrm{Sing}\, Y$,*

(ii) *the closed algebraic set $R(Y,x)$, which is the union of all irreducible components of the intersection $C(Y,x) \cap V$, containing the point p, is an irreducible curve, non-singular at the point p,*

(iii) *the curve $R(Y,x)$ intersects the subvariety Y outside the closed subset $\mathrm{Sing}\, Y$ of singular points of this variety.*

Proof of the claim (i) was given above in Subsection 4.2 (in addition to the arguments, given above, one needs to note that V cannot be contained in the variety of secant lines $\mathrm{Sec}\, Y$ of the variety Y; even if $M = 2k+1$, the variety V is not covered by lines). Furthermore, $[b,p] \cap Y = \{b\}$, so that the point $p \in C(Y,x)$ is non-singular. Obviously,

$$T_pC(Y,x) = \langle T_bY, p\rangle,$$

whence, taking into account (80), it follows that the varieties $C(Y,x)$ and V intersect transversally at the point p, which proves (ii). Finally, the claim (iii) is proved by the arguments of Subsection 4.2 as well, taking into account that when Y, $b \in Y$ and $p \in V$ vary, the points $x \in (b,p)$ fill out an open subset of the projective space $\mathbb{P}$ (for instance, by the surjectivity of the map μ). Q.E.D. for Proposition 4.7.

Now we argue as in Subsection 4.2: the curve $R(Y,x)$ meets Y at the points, which are non-singular both on Y and on the cone $C(Y,x)$. Therefore, we get that the intersection number $(R(Y,x) \cdot Y)_{C(Y,x)}$ is well defined.

Now recall that by Lemma 4.2 the following equality holds:

$$(R(Y,x) \cdot Y)_{C(Y,x)} = \deg R(Y,x).$$

Let us come back to the divisor $D \in \Sigma$, containing the point p.

Lemma 4.4. *The following inclusion holds: $R(Y,x) \subset D$.*

Proof. Obviously, $(D \cdot R(Y,x)) = n \deg R(Y,x)$. On the other hand, as it was shown in the proof of Lemma 4.3 above,

$$\sum_{y \in D \cap R(Y,x) \cap Y} (D \cdot R(Y,x))_y \geq (R(Y,x) \cdot Y)_{C(Y,x)} \operatorname{mult}_Y D.$$

Taking into account that $\operatorname{mult}_Y D = \operatorname{mult}_B D = n$ and the equality of Lemma 4.2, we obtain the claim of Lemma 4.4, since

$$p \in D \cap R(Y,x)$$

and $p \notin Y$. Q.E.D. for the lemma.

Lemma 4.4 implies the inclusion

$$\langle T_b Y, p \rangle \cap T_p V \subset T_p D,$$

whence, since the subspace $T_b Y \subset \Delta$ is generic, we get the inclusion

$$\langle \Delta, p \rangle \cap T_p V \subset T_p D,$$

which completes the proof of Proposition 4.5.

Finally, let us prove Proposition 4.4. As we noted above, we have to consider the two cases: when $M = 2k+2$ and $M = 2k+1$. Consider first the following general problem. Let $X \subset \mathbb{P}^N$ be an irreducible subvariety of dimension $l \geq 2$. By the symbol $\mathcal{P}_d = \mathcal{P}_{d,N}$ we denote the space of homogeneous polynomials of degree d on $\mathbb{P}^N$. Let $U_d(X) \subset \mathcal{P}_d$ be the open set, consisting of such polynomials $f \in \mathcal{P}_d$, that $\{f|_X = 0\}$ is an irreducible reduced subvariety of dimension $(l-1)$. Respectively, let $R_d(X) = \mathcal{P}_d \setminus U_d(X)$ be the set of "incorrect" polynomials. The problem is to estimate from below the codimension of the closed set $R_d(X)$ in the space $\mathcal{P}_d$.

Lemma 4.5. *The following estimate holds:*

$$\operatorname{codim}(R_d(X) \subset \mathcal{P}_d) \geq \binom{d+l-2}{d} - l + 1.$$

Proof. Let $\gamma\colon \mathbb{P}^N \dashrightarrow \mathbb{P}^{l-1}$ be the linear projection from a $(N-l)$-plane of general position, $\gamma_X = \gamma|_X\colon X \dashrightarrow \mathbb{P}^{l-1}$ its restriction onto X (the set of points where γ_X is not defined is zero-dimensional). Obviously, the map γ_X is surjective, all its fibres are one-dimensional and the fibre $\gamma_X^{-1}(z)$ over a point of general position $z \in \mathbb{P}^{l-1}$ is an irreducible curve. The set $\Delta \subset \mathbb{P}^{l-1}$, consisting of such points z, that the fibre $\gamma_X^{-1}(z)$ is reducible or non-reduced, is a proper closed subset of $\mathbb{P}^{l-1}$ (at most a divisor). Therefore, for any irreducible divisor D on $\mathbb{P}^{l-1}$, such that $D \not\subset \Delta$, its inverse image

$$\gamma_X^{-1}(D) = \gamma^{-1}(D) \cap X$$

is irreducible and reduced. In other words, for any irreducible polynomial f on $\mathbb{P}^{l-1}$, such that $\{f=0\} \not\subset \Delta$, we get $\gamma^* f \in U_d(X)$. The set $\gamma^* \mathcal{P}_{d,l-1}$ is a linear subspace of the space $\mathcal{P}_d$ (the same polynomials considered as polynomials in a larger number of variables). Let $R_{d,l-1} \subset \mathcal{P}_{d,l-1}$ be the closed subset of reducible polynomials. From what was said, it follows that

$$\operatorname{codim}(R_d(X) \subset \mathcal{P}_d) \geq \operatorname{codim}(R_{d,l-1} \subset \mathcal{P}_{d,l-1}).$$

The last codimension is easy to compute: the irreducible component of maximal dimension of the set $R_{d,l-1}$ consists of the polynomials of the form $f = f^\sharp h$, where $h \in \mathcal{P}_{1,l-1}$ is a linear form. Q.E.D. for Lemma 4.5.

Let us come back to the proof of Proposition 4.4. Assume that $M = 2k+2$. Assume also that the complete intersection V is generic in the following sense: the variety

$$V^\sharp = \{f_1 = \cdots = f_{k-1} = 0\} \subset \mathbb{P} \tag{81}$$

is smooth. Obviously, $\dim V^\sharp = 2k+3$, so that by the Lefschetz theorem the intersection $V^\sharp \cap \Lambda$ is irreducible, reduced and has dimension $k+2$ for any linear subspace $\Lambda \subset \mathbb{P}$ of codimension $k+1$. Fix such a subspace. Obviously,

$$V \cap \Lambda = \{f_k|_{V^\sharp \cap \Lambda} = 0\}.$$

Let $R_\Lambda \subset \mathcal{P}_{d_k, M+k}$ be the closed subset of polynomials f_k of degree d_k, for which $V \cap \Lambda$ is not an irreducible reduced subvariety of dimension $k+1$. By Lemma 4.5,

$$\operatorname{codim} R_\Lambda \geq \binom{d_k + k}{d_k} - k - 1. \tag{82}$$

The equality $d_1 + \cdots + d_k = M + k$ implies that $d_k \geq 4$. Elementary computations show that the right-hand side of the inequality (82) is strictly higher than the dimension of the projective Grassmanian of $(2k+1)$-planes in $\mathbb{P}$ for $k \geq 4$. This proves Proposition 4.4 for $M = 2k+2$.

Assume now that $M = 2k+1$. In this case the arguments are completely similar to those given above, but we need two steps. First, we consider the complete intersection

$$V^+ = \{f_1 = \cdots = f_{k-2} = 0\} \subset \mathbb{P},$$

which is assumed to be smooth. By the Lefschetz theorem the intersection $V^+ \cap \Lambda$ is irreducible and reduced. Now the arguments similar to those given above for $M = 2k+2$, show that for a generic polynomial f_{k-1} of degree $d_{k-1} \geq 3$ the closed set

$$\{f_{k-1}|_{V^+ \cap \Lambda} = 0\}$$

is irreducible and reduced for any subspace Λ. Now we consider the variety $V^\sharp$, defined by the formula (81), and argue as in the case $M = 2k+2$ and complete the proof. We omit the details of elementary computations.

Q.E.D. for Proposition 4.4.

Remark 4.1. The additional (compared to Section 2) restrictions for the parameters k, $d_1, \ldots, d_k$ are needed precisely for the reason that for the excluded values the proof of Proposition 4.4 does not work. However, there are no doubts that both the claim of Proposition 4.4 and, the more so, Theorem 4.1 are true for those values as well.

Notes and references

The technique of hypertangent divisors was first developed in [**Pukh98b**]. In Subsection 1.1 we follow a slightly modified version given in [**Pukh04c**] as it slightly improves the result (the genericity conditions are a bit relaxed). Proof of the regularity conditions being satisfied on a generic hypersurface in Subsection 1.3 follows [**Pukh98b**]. Section 2 follows [**Pukh01**] in Subsections 2.1 and 2.2 and [**Pukh09a**] in Subsection 2.3. Section 3 follows the two above cited papers as well.

The story of K-trivial structures, considered briefly in Section 4, is much longer and needs some overview. It seems that the problem of description of the structures of fibrations into varieties with the trivial canonical class is as old as that for the structures of rationally connected fibre spaces, at least in the most elementary setting: soon after Max Noether's study of pencils of rational curves [**N**], Halphen started a similar work for pencils of elliptic curves [**Hal**]. In the 1960s, a century later, Dolgachev's paper on the pencils of elliptic curves on rational surfaces [**Dol66**] appeared just after the modern reconstruction of the Noether theorem on pencils of rational curves was done in [**AlgS**], at the same time as the papers of Manin [**M66, M67**] and Iskovskikh [**I67, I70**].

In higher dimensions the connection between the two problems, to describe the rationally connected structures and to describe the K-trivial (or elliptic) structures on a given rationally connected variety is even more straightforward as the non-terminal and non-canonical singularities are very close classes. For many families of Fano varieties and fibre spaces one needs to add just a bit more to a "rationally connected" proof to get a complete solution in the K-trivial case. This is demonstrated in [**Ch00b, Ch04a, Ch05b**]. For some other classes, the amount of additional "K-trivial" work is significant; see [**Ch07b, ChK, ChP09**] and Section 4 of this chapter. Proof of Proposition 4.2 was sketched in [**Pukh01**]. In [**Ch05a**] a simpler argument was suggested to exclude subvarieties of codimension 3, but the proof was not complete. Proposition 4.2 was shown in full in [**Pukh06b**], but the question of whether every pencil with K-trivial fibres on a Fano complete intersection is a pencil of hyperplane sections remained open until [**Pukh12b**]. Our proof of Theorem 4.1 follows that paper.

CHAPTER 4

Rationally Connected Fibre Spaces

Summary

This chapter provides an introduction to the second part of the book, that on birationally rigid fibre spaces over a base of positive dimension. In the relative setting the method of maximal singularities works very well but becomes more complicated and requires some modifications. Naturally, the complications and modifications come from taking into account the structure of the fibre space.

In the study of birational geometry of fibre spaces, three main differences from the absolute case of primitive Fano varieties are to be taken into account:

- the Picard group has rank at least two,
- there is the important subclass of birational correspondences, mapping fibres to fibres, which affects the definitions,
- the maximal singularities can cover the base of the fibre space (generating maximal singularities on the fibres) or be concentrated over the preimage of a proper closed subset of the base.

Whereas the technique of the method of maximal singularities for fibre spaces will be developed in full detail in the subsequent chapters, here we introduce the context, explain some key points and give an overview of the most spectacular results. We start with examples of Fano fibre spaces and fibrewise and non-fibrewise birational correspondences between them (Section 1). This makes it possible for us to explain the general structure of the method for fibre spaces.

Then we compare the approach used by the method of maximal singularities with that of the Sarkisov program (Section 2). There we formulate the problems of birational classification of rationally connected fibre spaces and of factorization of birational correspondences between them, from the viewpoint of the minimal model program. The discussions of Section 2 and the results mentioned there are never used in this book; however, they need to be there to show the correct context of the problem of birational rigidity.

Finally, Section 3 contains the key definitions of the K-condition and the K^2-condition that in many cases imply birational (super)rigidity, outlines the two main approaches to proving that property (the quadratic and linear methods) and lists some of the main results of the relative theory (the proof will be given in Chapters 5–8).

1. Fano fibre spaces

In this section, we introduce the main object of investigation in the theory of birationally rigid fibre spaces, the standard Fano fibre spaces, and give some examples: fibrations into Fano complete intersections and Fano direct products. After that, we explain the general structure of the method of maximal singularities for varieties fibred over a positive-dimensional base. The general concepts are then illustrated by the key examples of fibrewise (for del Pezzo fibrations over $\mathbb{P}^1$ and conic bundles) and non-fibrewise (for a particular family of fibrations into Fano hypersurfaces of index one) birational maps.

1.1. Fano fibre space: definitions and examples. First, recall the well-known

Definition 1.1. A surjective morphism of smooth projective varieties $\pi\colon V \to S$ with connected fibres of the same dimension is called a *Fano fibre space*, if the anticanonical class $(-K_V)$ is relatively ample, that is, ample on the fibres of the projection π. A Fano fibre space is said to be *standard*, if

$$\operatorname{Pic} V = \mathbb{Z}K_V \oplus \pi^* \operatorname{Pic} S, \tag{83}$$

in particular, the relative Picard number is $\rho(V/S) = 1$.

In this book we consider only Fano fibre spaces over a rationally connected base S; in that case the variety V is automatically rationally connected. If the variety V has $\mathbb{Q}$-factorial terminal singularities, S is normal and the condition (83) is replaced by the equality $\rho(V/S) = 1$, then this more general object is often called a *Mori fibre space*; see Section 2. However, the term "a Fano fibre space with singularities" is also justified.

The fibres of a Fano fibre space $\pi\colon V \to S$ are varieties from some family $\mathcal{F}$, the general element of which is a smooth Fano variety, so that the projection π generates a map $S \to \mathcal{F}$, associating to a point $s \in S$ the corresponding fibre. In a sufficiently typical situation the sheaf $\pi_*(-K_V)$ is locally free and allows one to give an explicit construction of a Fano fibre space, which is shown by the following two examples.

Example 1.1 (fibrations into Fano complete intersections). Let S be a smooth projective rationally connected variety of positive dimension, $\mathcal{E}$ a locally free sheaf on S of rank $M+k+1$, where $M, k \geq 1$ are positive integers, $X = \mathbb{P}(\mathcal{E})$ the corresponding projective bundle in the sense of Grothendieck (X is the projectivisation of the vector bundle, the sheaf of sections of which is $\mathcal{E}^*$, the sheaf, dual to $\mathcal{E}$). The projection $\pi\colon X \to S$ is a locally trivial $\mathbb{P}^{M+k}$-fibration. If $L_X \in \operatorname{Pic} X$ is the class of the tautological sheaf $\mathcal{O}_{\mathbb{P}(\mathcal{E})}(1)$, then

$$\operatorname{Pic} X = \mathbb{Z}L_X \oplus \pi^* \operatorname{Pic} S,$$

so that X/S is a Fano fibre space. For the canonical class of the variety X there is a well-known formula

$$K_X = -(M+k+1)L_X + \pi^*(\det \mathcal{E} + K_S).$$

Let $(d_1, \ldots, d_k) \in \mathbb{Z}_+^k$ be a k-uple of integers, satisfying the conditions

$$d_k \geq \cdots \geq d_1 \geq 2 \quad \text{and} \quad d_1 + \cdots + d_k = M + k.$$

Consider, furthermore, the set of classes $A_i \in \operatorname{Pic} S$, $i = 1, \ldots, k$, and assume that there are irreducible divisors

$$G_i \sim d_i L_X + \pi^* A_i, \quad i = 1, \ldots, k,$$

such that the complete intersection

$$V = G_1 \cap \cdots \cap G_k \subset X$$

is a smooth variety, the fibres of which $V \cap \pi^{-1}(s)$, $s \in S$, are of the same dimension M. We denote the restriction of the projection π onto V by the same symbol $\pi\colon V \to S$. The restriction of the tautological class $L_X|_V$ is denoted by the symbol L_V. Obviously,

$$K_V = -L_V + \pi^*\Delta$$

for some $\Delta \in \operatorname{Pic} S$, so that V/S is a Fano fibre space. Moreover, for a point of general position $s \in S$ the fibre $\pi^{-1}(s) \subset V$ is a smooth Fano complete intersection of the type $d_1 \cdot \cdots \cdot d_k$ in $\mathbb{P}^{M+k}$. For $M \geq 3$ this implies that

$$\operatorname{Pic} V = \mathbb{Z}L_V \oplus \pi^* \operatorname{Pic} S = \mathbb{Z}K_V \oplus \pi^* \operatorname{Pic} S, \tag{84}$$

that is, V/S is a standard Fano fibre space. If $M = 2$, then V/S is a fibration into del Pezzo surfaces of degree 3 ($k = 1$), that is, cubic surfaces in $\mathbb{P}^3$, or degree 4 ($k = 2$, $d_1 = d_2 = 2$), that is, complete intersections of the type $2 \cdot 2$ in $\mathbb{P}^4$. If $M = 1$, then V/S is a conic bundle: in that case we require, in addition, that $\dim S \geq 2$. For $M \in \{1, 2\}$ the condition (84) does not hold automatically and we should require it as an extra. However, if the divisors $G_i, 1 \leq i \leq k \in \{1, 2\}$, are ample, then (84) holds by the Lefschetz theorem. The construction above gives the most "classical" example of a standard Fano fibre space.

Example 1.2 (fibrations into Fano double covers). Let $S, \mathcal{E}, X, \pi$ be the same as in the previous example, $k = 1$. Consider a pair of integers $m \geq 2, l \geq 2$, satisfying the equality $m+l = M+1$. Assume that for some classes $A_Q, A_W \in \operatorname{Pic} S$ there are divisors

$$Q \sim mL_X + \pi^*A_Q \quad \text{and} \quad W_X \sim 2(lL_X + \pi^*A_W),$$

where $Q \subset X$ is a smooth subvariety, intersecting each fibre $\pi^{-1}(s)$, $s \in S$, by a hypersurface (that is, Q does not contain entire fibres of the projection π), and W_X cuts out on Q a smooth divisor that does not contain fibres of the projection $\pi_Q = \pi|_Q$. This collection of data determines the double cover $\sigma\colon V \to Q$, branched over the divisor $W = W_X \cap Q$. The fibres of the projection $\pi_Q \circ \sigma\colon V \to S$ are M-dimensional varieties and the general fibre is a smooth primitive Fano variety, which implies that

$$\operatorname{Pic} V = \mathbb{Z}K_V \oplus \sigma^*\pi_Q^* \operatorname{Pic} S,$$

that is, V/S is a standard Fano fibre space. A particular case of this construction (corresponding to the value $m = 1$) is a double cover $\sigma\colon V \to X$, branched over a smooth divisor $W_X \sim 2((M+1)L_X + \pi^*A_W)$ that does not contain fibres of the projection π (there is no divisor Q); this is a standard fibration into Fano double spaces of index one.

In each of the considered examples the fibres of the Fano fibre space $\pi\colon V \to S$ belong to a family $\mathcal{F}$ of M-dimensional schemes, the general element of which is a smooth Fano variety of index one. Each of these families admits a natural structure of a projective variety, so that to the fibre space V/S we can associate a map

$$S \to \mathcal{F},$$

sending a point $s \in S$ to the fibre $\pi^{-1}(s) \in \mathcal{F}$. In this connection, we get a geometric characterization of "twistedness" of the fibre space V/S as complexity of the map $S \to \mathcal{F}$. For instance, if $V \subset \mathbb{P}^1 \times \mathbb{P}^3$ is a smooth divisor of bidegree $(d, 3)$, so that the projection $\pi\colon V \to \mathbb{P}^1$ onto the first factor realizes V as a fibration into del Pezzo surfaces of degree 3, then the twistedness of this fibre space over the base $\mathbb{P}^1$ is easy to define as the value $d \geq 1$. The twistedness over the base is an intuitively clear degree of complexity of a fibre space, and it is natural to expect that the more twisted the fibre space is, the more rigid is its structure. Now let us give an example of a least twisted fibre space.

Example 1.3 (Fano direct products). Let $F_1, \dots, F_k$, $k \geq 2$, be primitive Fano varieties,

$$V = F_1 \times \cdots \times F_k$$

their direct product, $\pi_i\colon V \to S_i = \prod_{j \neq i} F_j$ the projection along $F_i, i \in \{1, \dots, k\}$. By assumption, $\pi_i\colon V \to S_i$ is a standard Fano fibre space. The corresponding map of twistedness $S_i \to \mathcal{F}_i$ is a map to the point.

Recall, that by a *structure of a rationally connected fibre space* (or, briefly, a rationally connected structure) on a rationally connected variety V we mean a birational map $V \overset{\chi}{\dashrightarrow} V^+$, where the variety V^+ is equipped with a morphism $\pi^+\colon V^+ \to S^+$, which is a rationally connected fibre space. A Fano fibre space is obviously a rationally connected fibre space. Now we can formulate the main problem of birational geometry of Fano fibre spaces as follows:

> *for a given standard Fano fibre space $\pi\colon V \to S$ describe all structures of a rationally connected fibre space on the variety V modulo the relation of fibrewise birational equivalence.*

Informally speaking, the more rationally connected structures there are on a given variety V, the more complicated is its birational geometry (in particular, the projective space $\mathbb{P}^n$ has the most complicated birational geometry). The group of birational self-maps $\operatorname{Bir} V$ acts on the set $RC(V)$ of rationally connected structures and of the greatest interest is the quotient set $\overline{RC}(V)$, introduced in Chapter 1. If V/S is a Fano fibre space over a base of positive dimension, then there is an important subgroup

$$\operatorname{Bir}(V/S) \subset \operatorname{Bir} V$$

of fibrewise (with respect to π) birational self-maps, equipped with the natural homomorphism

$$\operatorname{Bir}(V/S) \to \operatorname{Bir} S.$$

The kernel of the latter homomorphism is the group $\operatorname{Bir}(F_\eta)$ of the generic fibre (over the generic point of S with the residue field $\mathbb{C}(S)$).

As the first stage in the solution of the general problem that was formulated above, it is natural to identify those Fano fibre spaces V/S, that have exactly one structure of a rationally connected fibre space $\pi\colon V \to S$, or, at least, those, for which that structure is unique modulo the action of the group $\operatorname{Bir} V$. Such a uniqueness in most cases is a consequence of the fundamental property of *birational (super)rigidity*. As it was pointed out in Section 1.3, the informal general principle is that a sufficiently high twistedness over the base implies birational rigidity, the uniqueness of the structure V/S and reduces birational geometry of the variety V to birational geometry of the fibre F_η over the generic point. This principle will be realized below in various examples that are by now completely studied.

The method of maximal singularities, which forms a basis of the proof of almost all results of the current book, as a by-product solves other problems as well: describes the group $\operatorname{Bir} V$ and the group $\operatorname{Bir}(V/S)$. The modern version of the method makes it possible to start investigating varieties with several and even many *non-equivalent* structures of a rationally connected fibre space. Chapters 5, 6 are devoted (mainly) to varieties with a unique rationally connected structures, Chapters 7 and 8 are devoted to varieties with many structures.

In Chapters 1 and 2 we explained and illustrated the idea of *untwisting* of a birational map (or *untwisting* maximal singularities). Essentially, this is the universal idea of simplification of a complex object by means of subsequent elementary steps. For Fano varieties such steps are usually birational involutions (see Section 1.3). For Fano fibre spaces the most natural candidates for simplifying modifications are *fibrewise modifications*: here, apart from self-maps, operations of a different type emerge. The informal working principle is that a birational map $\chi\colon V \dashrightarrow V'$ of the total space of a Fano fibre space V/S onto the total space of the fibre space V'/S' should be simplified by fibrewise modifications until either it becomes, in a certain sense, "simple", or a new structure of a rationally connected fibre space appears on the modified space V, which is not compatible with the original structure V/S. In the latter case replacing the old structure by the new one makes the simplifying step.

Below we consider examples of simplifying modifications of both types.

1.2. Fibrewise birational maps. In the study of birational geometry of a Fano fibre space $\pi\colon V \to S$ of great importance are the fibrewise birational modifications, that is, commutative diagrams of maps

$$\begin{array}{ccccc} & V & \stackrel{\chi}{\dashrightarrow} & V^+ & \\ \pi & \downarrow & & \downarrow & \pi^+ \\ & S & \stackrel{\varphi}{\dashrightarrow} & S^+, & \end{array}$$

where χ and φ are birational maps. The most elementary type of such modifications is given by birational self-maps of the generic fibre, $S^+ = S$, $\varphi = \mathrm{id}$, $V^+ = V$. Following [**M72**] (see also [**Pukh98a**]), consider

Example 1.4 (fibrewise birational self-maps of varieties with a pencil of cubic surfaces). In this example the Fano fibre space is a particular case of the construction of Example 4.1. Let $\mathbb{P}(\mathcal{E}) \stackrel{\pi}{\to} \mathbb{P}^1$ be the projectivization of a locally free sheaf of rank 4 on $\mathbb{P}^1$, $V \subset \mathbb{P}(\mathcal{E})$ a smooth divisor, intersecting each fibre $\pi^{-1}(t)$ by a cubic surface $F_t \subset \mathbb{P}^3 = \pi^{-1}(t)$; assume that $V/\mathbb{P}^1$ is a standard Fano fibre space. Consider an arbitrary section $C \subset V$ of the projection $\pi\colon V \to \mathbb{P}^1$. For a general point $t \in \mathbb{P}^1$ a general line $L \subset \mathbb{P}^3 = G_t$, containing the point $C \cap F_t$, meets the cubic surface F_t at two more distinct points, say x, y. Set

$$\tau_C(x) = y.$$

Obviously, this defines a birational involution $\tau_C \in \operatorname{Bir} F_\eta \subset \operatorname{Bir} V$. Let $\alpha\colon V^* \to V$ be the blow up of the curve C, $E = \alpha^{-1}(C)$ the exceptional divisor, $\operatorname{Pic} V^* = \mathbb{Z}h \oplus \mathbb{Z}e \oplus \mathbb{Z}F$, where $h = -K_V$.

Lemma 1.1. *The birational involution τ_C extends to a biregular involution of an invariant open set $V^*\backslash Y$, $\operatorname{codim} Y \geq 2$, and its action on* $\operatorname{Pic} V^*/\mathbb{Z}F \cong \mathbb{Z}\bar{h} \oplus \mathbb{Z}\bar{e}$

is given by the relations

$$\tau_C^*\bar{h} = 3\bar{h} - 4\bar{e},$$
$$\tau_C^*\bar{e} = 2\bar{h} - 3\bar{e}.$$

Proof. See [**M72**]. Q.E.D.

Now let us consider an arbitrary bi-section $C \subset V$, that is, an irreducible curve which is a two-sheeted cover of the base $\mathbb{P}^1$. We define the involution τ_C by its action on the generic fibre F in the following way (see [**M72**] and [**Pukh98a**]). Let $\{a, b\} = C \cap F$, and $q = L_{ab} \cap F$ be the third point of intersection of the line in $\mathbb{P}^3$ that joins the points a and b, with the cubic surface F. The points q sweep out a curve $C^* \subset V$, a section of the morphism π, that is, $q = C^* \cap F$. The pencil of planes P in $\mathbb{P}^3$, containing the line L_{ab}, generates a pencil of elliptic curves $Q_P = P \cap F$ on the surface F. Set

$$\tau_C \,|_{Q_P}(x) = y,$$

where

$$x + y \sim 2q$$

on Q_P, that is, τ_C is the reflection on the elliptic curve Q_P from the point q. This defines the involution $\tau_C \in \operatorname{Bir} F_\eta \subset \operatorname{Bir} V$.

Let $\alpha\colon V^* \to V$ be the blow up of the curve C, $E = \alpha^{-1}(C)$ the exceptional divisor, $\operatorname{Pic} V^* = \mathbb{Z}h \oplus \mathbb{Z}e \oplus \mathbb{Z}F$, where again $h = -K_V$.

Lemma 1.2. *The birational involution τ_C extends to a biregular involution of an invariant open set $V^*\backslash Y$, $\operatorname{codim} Y \geq 2$, and its action on $\operatorname{Pic} V^*/\mathbb{Z}F \cong \mathbb{Z}\bar{h} \oplus \mathbb{Z}\bar{e}$ is given by the relations*

$$\tau_C^*\bar{h} = 5\bar{h} - 6\bar{e},$$
$$\tau_C^*\bar{e} = 4\bar{h} - 5\bar{e}.$$

Proof. Straightforward computations [**M72**]. Q.E.D.

The birational involutions constructed above are used for the study of geometry of the variety V in the following way. Let $\Sigma \subset |-nK_V + lF|$ be a mobile linear system on V, where $n \geq 1$, that is, Σ is not composed from the pencil of fibres. The curve $C \subset V$ is called a *maximal curve* of the system Σ, if the inequality

$$\operatorname{mult}_C \Sigma > n$$

holds. Assume that C is a *horizontal* maximal curve, that is, $\pi(C) = \mathbb{P}^1$ (in other words, C is not contained in a fibre F_t). It is easy to show that

$$(K_V^2 \cdot F) = 3 > \deg(\pi|_C) \in \{1, 2\},$$

that is, C is a section or bi-section of the projection π. By Lemmas 1.1 and 1.2 it is easy to see that the strict transform $\Sigma_1 = (\tau_C)_*\Sigma$ of the linear system Σ with respect to the involution τ_C satisfies the relation $\Sigma_1 \subset |-n_1K_V + l_1F|$, where $n_1 < n$. More precisely,

$$n_1 = 3n - 2\operatorname{mult}_C \Sigma,$$

if C is a section, and

$$n_1 = 5n - 4\operatorname{mult}_C \Sigma,$$

if C is a bi-section of the projection π. Since $n \in \mathbb{Z}_+$, in finitely many steps we come to a system without maximal curves. This is the first, easier step in the study of birational geometry of the variety V and in the proof of the main theorem on birational rigidity of the fibre space $V/\mathbb{P}^1$ (see Theorem 1.2.2 and Chapter 6). The second, harder, step (the exclusion of infinitely near maximal singularities) is discussed in Chapter 6.

Example 1.5 (fibrewise birational modifications of conic bundles). Let $\pi\colon V \to S$ be a standard conic bundle. Some examples of birational self-maps preserving the fibres were given in Chapter 1. However, what turns out to be productive to describe birational geometry of the variety V (see the Sarkisov theorem and its discussion that were given in Chapter 1) is not simplifying a linear system by birational self-maps as above but another approach, which we will now briefly describe. Consider the simplest situation: let the fibre $C = \pi^{-1}(p)$ over a point $p \in S$ be a non-singular conic, $C \cong \mathbb{P}^1$. Let us blow up simultaneously the point p on the surface S and the curve C on the variety V:

$$\begin{array}{ccccc} & V & \leftarrow & V^+ & \\ \pi & \downarrow & & \downarrow & \pi \\ & S & \leftarrow & S^+, & \end{array}$$

the projection $V^+ \to S^+$ is for simplicity denoted by the same symbol π. Let $E \subset V^+$ and $E_S \subset S^+$ be the exceptional divisors of these blow ups. Note that V^+/S^+ is again a standard conic bundle. Obviously, $E = E_S \times \mathbb{P}^1$. Let $\Sigma \subset |-nK_V + \pi^* A|$ be a mobile linear system, $\mu = \mathrm{mult}_C\,\Sigma$ and

$$\Sigma^+ \subset |-nK_{V^+} + \pi^*(A + (n-\mu)E_S)|$$

its strict transform on V^+. The fibre C is a *maximal curve* of the system Σ, if $\mu > n$, and this example shows how to remove all maximal curves of that type, modifying the conic bundle (if C is a component of a reducible fibre or the support of a non-reduced fibre, then the construction of the birational modification is much more complicated, see [**S80, S82**]). Let us now assume that some curve $\Gamma \subset E$, which is not a fibre of π, satisfies the condition

$$\nu + \mu > 2n,$$

where $\nu = \mathrm{mult}_\Gamma\,\Sigma^+ > n$, that is, it is an *infinitely near maximal curve* of the original system Σ. Since Σ^+ has no fixed components and for a general divisor $D^+ \in \Sigma^+$ we have

$$(D^+ \cdot \pi^{-1}(s)) = -n(K_{V^+} \cdot \pi^{-1}(s)) = 2n,$$

the curve Γ is a section of the ruled surface E/E_S, in particular, it is a smooth rational curve. Let $\varphi\colon V^\sharp \to V^+$ be the birational modification which is a composition of two operations: the blow up of the curve Γ and the subsequent fibrewise contraction of the strict transform of the ruled surface E. We obtain a new standard conic bundle $\pi\colon V^\sharp \to S^+$, for the strict transform of the linear system Σ on $V^\sharp$ we get

$$\Sigma^\sharp \subset |-nK_{V^\sharp} + \pi^*(A + (3n - 2\mu - \nu)E_S)|.$$

Note once again, that if a maximal singularity of the system Σ lies over a point of a singular or non-reduced fibre of the projection π, then the required modification is much more complicated than the construction described above. However, what has been said already makes it possible to explain the idea of the proof of the Sarkisov theorem [**S80, S82**] (Theorem 1.2.1). Let

$$\begin{array}{ccccc} & V & \stackrel{\chi}{\dashrightarrow} & V' & \\ \pi & \downarrow & & \downarrow & \pi' \\ & S & & S' & \end{array}$$

be a birational map between two conic bundles where V satisfies the condition $|4K_S+\Delta| \neq \emptyset$, and $\Delta \subset S$ is the discriminant divisor. Pulling back from the base S' a very ample linear system, we get a mobile system Σ' on V'. Let $\Sigma \subset |-nK_V+\pi^*A|$ be its strict transform on V. If χ is not fibrewise, then $n \geq 1$. So assume that $n \geq 1$. It is easy to check that the condition $|4K_S + \Delta| \neq \emptyset$ is invariant under fibrewise modifications (this is obvious in the example above). Therefore, applying fibrewise modifications, we may assume that the linear system Σ has no maximal singularities. Now it is easy to obtain from the condition of termination of canonical adjunction that in this case the class $A \in \operatorname{Pic} S$ is not effective (more precisely, not pseudo-effective). Let us compute the self-intersection

$$Z = (D_1 \circ D_2), \quad D_i \in \Sigma,$$

of the linear system Σ and push it down on S:

$$\pi_* Z \sim n^2 \pi_* K_V^2 + 4nA \sim -n^2(4K_S + \Delta) + 4nA.$$

From here by assumption it follows that

$$A \sim \frac{n}{4}(4K_S + \Delta) + \frac{1}{4n}\pi_* Z$$

is an effective class. This contradiction proves the Sarkisov theorem. For the details, see [**S82**].

1.3. Changing the structure of a Fano fibre space. In the class of Fano fibre spaces of dimension three and higher, fibrewise modifications appear much more seldom. The following example shows how much more rigid fibrewise birational geometry is in higher dimensions.

Example 1.6 (the absence of non-trivial fibrewise modifications in higher dimensions). Let C be a smooth affine curve with a marked point $p \in C$, and $C^* = C \backslash \{p\}$ the punctured curve. Let $\mathcal{V}(d), d \geq 2$, be the set of smooth divisors $V \subset X = C \times \mathbb{P}^M$, $M \geq 3$, each fibre of which

$$F_x = V \cap \{x\} \times \mathbb{P}^M,$$

$x \in C$, is a hypersurface of degree d. Set

$$X^* = C^* \times \mathbb{P}^M, \quad V^* = V \cap X^*,$$

so that V^* is obtained from V by removing the fibre F_p over the marked point. Assume that $d \geq 3$ and $V_1, V_2 \in \mathcal{V}(d)$. The following fact (proven in [**Pukh00a**]) is true.

Theorem 1.1. *Let $\chi^*\colon V_1^* \to V_2^*$ be a fibrewise isomorphism. Then χ^* extends to a fibrewise isomorphism $\chi\colon V_1 \to V_2$.*

Proof is given in Chapter 5.

The next example deals with varieties with two structures of Fano fibre spaces. The transition from one structure to another is the required simplifying modification. For other similar examples, see [**Pukh06a, Pukh09a**].

Example 1.7. Consider the following particular case of the construction of Example 1.1: let $\mathcal{E} = \mathcal{O}_{\mathbb{P}^1}^{\oplus(M-1)} \oplus \mathcal{O}_{\mathbb{P}^1}(1)^{\oplus 2}$ be a locally free sheaf of rank $(M+1)$ on $\mathbb{P}^1$, $X = \mathbb{P}(\mathcal{E})$ the corresponding $\mathbb{P}^M$-bundle over $\mathbb{P}^1$, $\operatorname{Pic} X = \mathbb{Z}L_X \oplus \mathbb{Z}R$, where R is the class of a fibre of the projection $\pi_X \colon X \to \mathbb{P}^1$, L_X the tautological class. Let $V \sim ML_X$ be a general divisor. It is a smooth variety, fibred by the projection $\pi = \pi_X|_V \colon V \to \mathbb{P}^1$ into Fano hypersurfaces of index one. Obviously, $V/\mathbb{P}^1$ is a standard Fano fibre space, $\operatorname{Pic} V = \mathbb{Z}K_V \oplus \mathbb{Z}F$, where F is the class of a fibre of the projection π and $K_V = -L$, $L = L_X|_V$. On the variety V, however, there is another structure of a rationally connected fibre space.

Consider the following locally free subsheaves:

$$\mathcal{E}_0 = \mathcal{O}_{\mathbb{P}^1}^{\oplus(M-1)} \hookrightarrow \mathcal{E} \quad \text{and} \quad \mathcal{E}_1 = \mathcal{O}_{\mathbb{P}^1}(1)^{\oplus 2} \hookrightarrow \mathcal{E}.$$

Obviously, $\mathcal{E} = \mathcal{E}_0 \oplus \mathcal{E}_1$. Let $\mathcal{L}_X$ be the tautological sheaf of Grothendieck of the bundle $\mathbb{P}(\mathcal{E})$ and $\Pi_0 \subset H^0(X, \mathcal{L}_X)$ the subspace, corresponding to the space of sections of the subsheaf $H^0(\mathbb{P}^1, \mathcal{E}_0) \hookrightarrow H^0(\mathbb{P}^1, \mathcal{E})$. Also set

$$\Pi_1 = H^0(X, \mathcal{L}_X \otimes \pi^* \mathcal{O}_{\mathbb{P}^1}(-1)) = H^0(\mathbb{P}^1, \mathcal{E}_1(-1)).$$

Let $x_0, \dots, x_{M-2}$ be some basis of the space Π_0, y_0, y_1 a basis of the space Π_1. Then the sections

$$x_0, \dots, x_{M-2}, y_0 t_0, y_0 t_1, y_1 t_0, y_1 t_1, \tag{85}$$

where t_0, t_1 is a system of homogeneous coordinates on $\mathbb{P}^1$, form a basis of the space $H^0(X, \mathcal{L}_X)$. It is easy to see that this complete linear system defines the morphism

$$\xi \colon X \to \overline{X} \subset \mathbb{P}^{M+2},$$

the image of which $\overline{X}$ is a quadric cone with the vertex space $\mathbb{P}^{M-2} = \xi(\Delta_X)$, the base of which is a non-singular quadric in $\mathbb{P}^3$, isomorphic to $\mathbb{P}^1 \times \mathbb{P}^1$, where $\Delta_X = \mathbb{P}(\mathcal{E}_0)$ is the base set of the pencil $|L_X - R|$, $\Delta_X = \mathbb{P}^{M-2} \times \mathbb{P}^1$, contracted by the map ξ onto the first factor. The morphism ξ is birational, more precisely,

$$\xi \colon X \setminus \Delta_X \to \overline{X} \setminus \xi(\Delta_X)$$

is an isomorphism and, moreover, ξ contracts $\Delta_X = \mathbb{P}^{M-2} \times \mathbb{P}^1$ onto the vertex space of the cone. Let

$$u_0, \dots, u_{M-2}, u_{00}, u_{01}, u_{10}, u_{11}$$

be the homogeneous coordinates on $\mathbb{P}^{M+2}$, corresponding to the ordered set of sections (85). The cone $\overline{X}$ is given by the equation

$$u_{00} u_{11} = u_{01} u_{10}.$$

On the cone $\overline{X}$ there are two pencils of M-planes, corresponding to the two pencils of lines on the smooth quadric in $\mathbb{P}^3$. Let $\tau \in \operatorname{Aut} \mathbb{P}^{M+2}$ be the automorphism, permuting the coordinates u_{01} and u_{10}, and not changing the other coordinates. Obviously, $\tau \in \operatorname{Aut} \overline{X}$ is an automorphism of the cone $\overline{X}$, permuting the pencils of M-planes. One of those pencils is the image of the pencil of fibres of the projection π, that is, the pencil $\xi(|R|)$. For the other pencil we have the obvious equality

$$\tau\xi(|R|) = \xi(|L_X - R|).$$

The automorphism τ induces an involutive birational self-map

$$\tau^+ \in \operatorname{Bir} X.$$

More precisely, τ^+ is a biregular automorphism outside a closed subset Δ_X of codimension two. Let $\varepsilon\colon \widetilde{X} \to X$ be the blow up of the smooth subvariety Δ_X. Obviously, the variety $\widetilde{X}$ is isomorphic to the blow up of the cone $\overline{X}$ at its vertex space $\xi(\Delta_X)$. It is easy to check that τ^+ extends to a biregular automorphism of the smooth variety $\widetilde{X}$. The linear systems $|kL_X|$, $k \in \mathbb{Z}_+$, are invariant under τ^+. In particular, for a general divisor $V \in |ML_X|$ its τ^+-image $V^+ = \tau^+(V)$ is a general divisor of the same linear system, in particular, V^+ is a smooth variety. Note that if $V \in |ML_X|$ is given by an equation

$$h(u_0, \dots, u_{M-2}, u_{00}, u_{01}, u_{10}, u_{11}),$$

then its image V^+ is given by the equation

$$h^+(u_*) = h(u_0, \dots, u_{M-2}, u_{00}, u_{10}, u_{01}, u_{11}),$$

where the coordinates u_{01} and u_{10} are permuted.

Therefore, we obtain two Fano fibre spaces, $V/\mathbb{P}^1$ and $V^+/\mathbb{P}^1$, related via the birational isomorphism $\tau^+\colon V \dashrightarrow V^+$, which is not fibrewise. That birational map is biregular in codimension one and acts on the Picard group in the following way:

$$(\tau^+)^* K_{V^+} = K_V, \quad (\tau^+)^* F^+ = -K_V - F,$$

where F^+ is the class of a fibre of the projection $V^+ \to \mathbb{P}^1$, so that $\operatorname{Pic} V^+ = \mathbb{Z}K_{V^+} \oplus \mathbb{Z}F^+$. By construction, the construction is involutive, that is, $(V^+)^+ = V$.

Note that the birational map $\tau^+\colon V \dashrightarrow V^+$ is the composition of the blow up of the subvariety $\Delta = \Delta_X \cap V$ of codimension two and the subsequent contraction of the exceptional divisor onto the subvariety $\Delta^+ = \Delta_X \cap V^+$.

The transition from the model V to the model V^+ by means of the birational map τ^+ is used as a simplifying modification in the following way. Let

$$\Sigma \subset |-nK_V + lF|$$

be a mobile linear system. If $l < 0$, then the linear system $\tau^+_*\Sigma \subset |-n_+K_{V^+} + l_+F^+|$ has parameters

$$n_+ = n + l, \quad l_+ = -l \geq 1.$$

Now the linear system is ready for applying the technique of excluding maximal singularities; see Chapter 5 for details.

2. The Sarkisov program

In this section we briefly discuss an alternative approach to birational classification of rationally connected fibre spaces, the Sarkisov program. We start by introducing the class of Mori fibre spaces, recalling some basic definitions and facts of the minimal model program. Then we formulate the problem of factorization of birational maps and give elementary examples of different types of links between Mori fibre spaces. After that we discuss the general program of factorization of birational maps into links and the level, to which it has been completely proved by now. We conclude by comparing the two approaches, the method of maximal singularities and the Sarkisov program.

2.1. Minimal models and Mori fibre spaces. The minimal model program (MMP) generalizes to higher (≥ 3) dimensions the classical theory of minimal models of algebraic surfaces [**AlgS**]. The purpose of MMP is to associate to every algebraic variety, by means of explicitly described birational modifications, a certain "model" with "good properties" with respect to the canonical class. While for a smooth projective surface it is sufficient to contract the exceptional lines (or (-1)-curves), to obtain either a minimal surface with a numerically effective canonical class or a ruled surface (that is, a $\mathbb{P}^1$-bundle over a smooth projective curve), in dimensions three and higher the situation is much more complicated:

— extremal contractions (the higher-dimensional analogues of the operation of contracting a (-1)-curve) inevitably produce singular varieties, even if the original variety was non-singular;

— a new type of birational modifications emerges, which are isomorphisms in codimension one, that is, outside a closed subset of codimension ≥ 2; it is these modifications that generate the worst technical complications;

— as it became clear starting from the mid-eighties, a technically more natural object is not an algebraic variety X, but a *pair* (or *log pair*) (X, Δ), where Δ is a *boundary*, which is an (effective, as a rule) Weil $\mathbb{Q}$-(or $\mathbb{R}$-)divisor on X, such that $K_X + \Delta$ is a $\mathbb{Q}$-(respectively, $\mathbb{R}$-)Cartier divisor. Working with pairs makes it possible to deal with all types of singularities and is well adjusted to constructing inductive procedures of MMP.

Let X be a normal projective variety, Δ an effective Weil $\mathbb{R}$-divisor, $\Delta = \sum d_i \Delta_i$, where $\Delta_i \subset X$ are distinct prime divisors. Let $f\colon Y \to X$ be a *log resolution* of the pair (X, Δ), that is, a sequence of blow ups such that Y is a smooth projective variety and

$$\bigcup \Delta_i^+ \bigcup E_j$$

is a normal crossings divisor, where $\bigcup E_j$ is the exceptional divisor of the morphism f, Δ_i^+ the strict transform of the divisor Δ_i on Y. Now writing

$$K_Y + \Delta^+ = f^*(K_X + \Delta) + \sum a_j E_j$$

(where $\Delta^+ = \sum d_i \Delta_i^+$; recall that $K_X + \Delta$ is a $\mathbb{R}$-Cartier divisor), we say, that the pair (X, Δ) is a klt-*pair* (Kawamata log terminal), if all $d_i < 1$ and all $a_j > -1$.

There are two most important types of klt-pairs:

1) the *minimal models* (or log terminal models), when $K_X + \Delta$ is numerically effective,

2) the *Mori fibre spaces*, when there is a morphism $\varphi\colon X \to S$ onto a normal projective variety S, where $\rho(X/S) = 1$, $\varphi_* \mathcal{O}_X = \mathcal{O}_S$ and $-(K_X + \Delta)$ is φ-ample.

The aim of MMP is to obtain, by means of birational modifications of a special type, from an arbitrary klt-pair (X, Δ) either a minimal model, or a Mori fibre space,

$$X_0 = X \overset{f_1}{\dashrightarrow} X_1 \overset{f_2}{\dashrightarrow} \cdots \overset{f_N}{\dashrightarrow} X_N, \tag{86}$$

$\Delta_0 = \Delta$, $\Delta_i = (f_i)_* \Delta_{i-1}$ and (X_N, Δ_N) is of type 1) or 2). Each birational map $f_{i+1}\colon X_i \dashrightarrow X_{i+1}$ is either a $(K_{X_i} + \Delta_i)$-extremal divisorial contraction, or a flip with respect to a small $(K_{X_i} + \Delta_i)$-extremal contraction $\varphi_i\colon X_i \to Y_i$, that is, the rational map $\varphi_i^+ = \varphi_i \circ f_{i+1}^{-1}\colon X_{i+1} \dashrightarrow Y_i$ is a small birational morphism and $(K_{X_{i+1}} + \Delta_{i+1})$ is relatively φ_i^+-ample. Existence of flips in the arbitrary dimension and the highest generality was proven in [**BCHM**]. The main difficulty in constructing the MMP was from the start the finiteness problem, that is, the problem of termination of a sequence of flips. Indeed, a sequence of divisorial contractions for obvious reasons cannot be infinite. Therefore the algorithm of MMP gives the desired result (a minimal model or a Mori fibre space) provided that a sequence of modifications of the flip type cannot be infinite. In [**BCHM**] the existence of minimal models was proven via a modification of the basic approach: instead of proving the finiteness of *any* sequence of flips, it is sufficient to construct (or prove the existence of) a *certain* sequence, which terminates after finitely many steps. That is what was done in [**BCHM**]. Now let us state the main result of that paper.

Theorem 2.1. *Let (X, Δ) be a* klt*-pair, where $K_X + \Delta$ is a $\mathbb{R}$-Cartier divisor and $\pi\colon X \to U$ a projective morphism of quasi-projective varieties. Assume that either Δ is π-big, and $K_X + \Delta$ is π-pseudoeffective, or $K_X + \Delta$ is π-big. Then:*

1) the pair (X, Δ) has a log terminal model over U,

2) if $K_X + \Delta$ is π-big, then (X, Δ) has a log canonical model over U,

3) if $K_X + \Delta$ is a $\mathbb{Q}$-Cartier divisor, then the $\mathcal{O}_U$-algebra

$$\bigoplus_{m\in\mathbb{N}} \pi_*\mathcal{O}_X(\llcorner m(K_X + \Delta)\lrcorner)$$

is finitely generated.

This rather technical result implies a number of important geometric facts. We will give three claims, all shown in [**BCHM**], of which the last one is most important for the present paper.

Corollary 2.1. *Let X be a smooth projective variety of general type. Then X has a minimal model, the canonical ring*

$$\bigoplus_{m\in\mathbb{N}} H^0(X, \mathcal{O}_X(mK_X))$$

is finitely generated and X has a model with a Kähler-Einstein metric.

Corollary 2.2. *Let (X, Δ) be a* klt*-pair and $\pi\colon X \to Z$ a small $(K_X + \Delta)$-extremal contraction. Then for π a flip exists.*

Corollary 2.3 [10]. *Let (X, Δ) be a* klt*-pair, whereas X is a $\mathbb{Q}$-factorial variety. Let $\pi\colon X \to U$ be a projective morphism of normal quasi-projective varieties where $K_X+\Delta$ is not π-pseudo-effective. Then some sequence of MMP modifications*

$$f = f_N \circ \cdots \circ f_1\colon X \dashrightarrow Y$$

gives a Mori fibre space $g\colon Y \to W$ over U.

The last claim immediately implies that any rationally connected variety X is birationally equivalent to a Mori fibre space

$$\begin{array}{ccc} X & \dashrightarrow & Y \\ & & \downarrow g \\ & & W, \end{array}$$

where Y is a variety with $\mathbb{Q}$-factorial terminal singularities, W is a projective normal variety and $(-K_Y)$ is g-ample. Similarly, for any rationally connected fibre space $\pi\colon X \to S$ there is a commutative diagram

$$\begin{array}{ccc} X & \dashrightarrow & Y \\ \pi\downarrow & & \downarrow g \\ S & \leftarrow & W, \end{array}$$

where the top arrow is a birational map and Y/W is a Mori fibre space. (Note that since termination of an *arbitrary* sequence of flips is still an open problem, Corollary 2.3 does *not* claim that *any* sequence of MMP-modifications gives a Mori fibre space.)

It follows from what was said, that all problems of birational geometry of rationally connected varieties could be set within the category of Mori fibre spaces: instead of finding out whether there is a birational equivalence $X_1 \dashrightarrow X_2$ between two rationally connected varieties, we can replace X_i by birationally equivalent Mori fibre space $\pi_i \colon Y_i \to S_i$ and consider the same problem for Y_1, Y_2. If X_1, X_2 are birationally equivalent, then to describe all birational maps $X_1 \dashrightarrow X_2$ is the same as to describe the group of birational self-maps $\operatorname{Bir} X_i = \operatorname{Bir} Y_i$. If $\pi \colon X \to R$ is a rationally connected fibre space, then the problem of description of the relation between the group $\operatorname{Bir} X$ and the group of fibrewise birational self-maps $\operatorname{Bir}(X/R)$ is also carried over to the corresponding Mori fibre space Y/S.

Restriction of the problems of birational geometry by the framework of the category of Mori fibre spaces has a number of clear advantages. For instance, it removes the asymmetry of the traditional approach, when birational maps $\chi \colon V \dashrightarrow V'$ are investigated, where the variety V belongs to the category of Fano fibre spaces and V' to the category of rationally connected varieties (or fibre spaces); whereas if we work with Mori fibre spaces (that is, with Fano fibre spaces with $\mathbb{Q}$-factorial terminal singularities and the relative Picard number one), then both varieties belong to the same category. Furthermore, the approach, motivated by MMP, makes it possible to set and solve the general problem of factorization of birational maps in a composition of elementary modifications (links), which we will consider in the next subsection.

Nevertheless, it is not clear, to what extent these advantages are essential and whether they justify replacing the traditional approach of the method of maximal singularities by the new one, to what extent the main definitions (for instance, that of the key concept of birational rigidity), corresponding to the ideology of the minimal model program, are "better" or "worse"; this issue will be discussed below.

2.2. The problem of factorization of birational maps. The Sarkisov program is the theory of factorization of birational maps between Mori fibre spaces in a composition of elementary modifications (links). Consider the diagram

$$\begin{array}{ccccc} & X & \overset{\chi}{\dashrightarrow} & X' & \\ \pi & \downarrow & & \downarrow & \pi' \\ & S & & S', & \end{array}$$

where χ is a birational map, X/S and X'/S' are Mori fibre spaces. It is required to construct a sequence of intermediate Mori fibre spaces $\pi_i \colon X_i \to S_i$, $i = 0, 1, \ldots, N$, that starts with $\pi_0 = \pi$ ($X_0 = X$ and $S_0 = S$) and ends with $\pi_N = \pi'$ ($X_N = X'$ and $S_N = S'$), and a sequence of elementary modifications (links) $\tau_i \colon X_{i-1} \dashrightarrow X_i$ such that

$$\chi = \tau_N \circ \cdots \circ \tau_1 \colon X_0 = X \dashrightarrow X_N = X'.$$

Recall that a *link* $\tau \colon X \dashrightarrow Y$ between Mori fibre spaces $\pi \colon X \to S$ and $\rho \colon Y \to T$ is a birational map of one of the following four types.

Type I (enlarging the base). There are, an extremal divisorial contraction $\varphi\colon Z \to X$, a birational map $\psi\colon Z \dashrightarrow Y$, which is a composition of flops (in particular, an isomorphism in codimension one) and an extremal contraction $\varepsilon\colon T \to S$ (in particular, $\rho(T/S)=1$) such that the following diagram of maps commutes:

(87)
$$\begin{array}{ccccc} & Z & \stackrel{\psi}{\dashrightarrow} & Y & \\ \varphi & \downarrow & & \downarrow & \rho \\ & X & & T & \\ \pi & \downarrow & & \downarrow & \varepsilon \\ & S & = & S, & \end{array}$$

where $\psi\circ\varphi^{-1}=\tau\colon X \dashrightarrow Y$. The simplest example of a link of that type was given above (Example 1.5); it corresponds to the blow up of a point $\varepsilon\colon T \to S$ on the base of the conic bundle and the blow up $\varphi\colon Z \to X$ of the fibre over that point. Let us give one more:

Example 2.1. Let $L\subset\mathbb{P}^3$ be a line, $\sigma\colon X\to\mathbb{P}^3$ its blow up, Π_L the pencil of planes in $\mathbb{P}^3$, containing the line L, Π_L^+ its strict transform on X. Obviously, Π_L^+ is the pencil of fibres of the morphism

$$\pi\colon X\to S=\mathbb{P}^1$$

with the fibre $\mathbb{P}^2$, that is, X/S is a Mori fibre space (and a Fano fibre space in the traditional sense). Now let $R\subset X$ be an arbitrary section of the projection π, $\varphi\colon Z\to X$ its blow up. The composite map

$$\pi\circ\varphi\colon Z\to\mathbb{P}^1$$

is a fibration into rational ruled surfaces of type $\mathbb{F}_1$. More precisely, let $E\subset Z$ be the exceptional divisor of the blow up φ, that is, $E=\varphi^{-1}(R)$. Obviously the projection $\mathbb{F}_1\to\mathbb{P}^1$ (the regularized projection of $\mathbb{P}^2$ from the point $R\cap\pi^{-1}(s)$, $s\in S$) generates the projection $p\colon Z\to E$, which is a $\mathbb{P}^1$-bundle, that is, Z/E is a Mori (Fano) fibre space. This gives the diagram (87) of a link of the first type with $T=E$, where $\varepsilon\colon T\to S$ is the projection $\varphi\colon E\to R$ with respect to the identification $\pi\colon R\to S$ and $Z=Y$, ψ is the identity map. In the described example the Mori fibre space Y/T is obtained from X/S by fibering the fibres $\pi^{-1}(s)$, $s\in S$. Accordingly, the fibres of the new fibre space Y/T are of a smaller dimension. This operation can also be interpreted as fibering the generic fibre F_η of the morphism $X\to S$ over the line $\mathbb{P}^1_\eta$ (which is the generic fibre of the morphism $T\to S$).

Type II (fibrewise modifications). In this case $S=T$. There are extremal divisorial contractions $\varphi\colon Z\to X$ and $\lambda\colon W\to Y$ and a birational isomorphism $\psi\colon Z\dashrightarrow W$, which is a composition of flops (an isomorphism in codimension one) such that the following diagram commutes:

$$\begin{array}{ccccc} & Z & \stackrel{\psi}{\dashrightarrow} & W & \\ \varphi & \downarrow & & \downarrow & \lambda \\ & X & \stackrel{\tau}{\dashrightarrow} & Y & \\ \pi & \downarrow & & \downarrow & \rho \\ & S & = & T. & \end{array}$$

The fibrewise self-maps of the pencils of cubic surfaces, considered above (Example 1.4) belong to this type.

Let us consider one more:

Example 2.2 (replacing the fibre). Let S be a smooth curve, $\pi\colon X \to S$ a $\mathbb{P}^n$-bundle, $x \in X$ an arbitrary point, and let $F = \pi^{-1}(\pi(x)) \cong \mathbb{P}^n$ be the fibre that contains that point. Let

$$\varphi\colon Z \to X$$

be the blow up of the point x, $E = \varphi^{-1}(x) \cong \mathbb{P}^n$ the exceptional divisor, $F^+ \subset Z$ the strict transform of the fibre F. Obviously, $R = F^+ \cap E$ is a hyperplane in $E = \mathbb{P}^n$ and the projection of the fibre $F = \mathbb{P}^n$ from the point x generates a $\mathbb{P}^1$-bundle

$$\lambda_F\colon F^+ \to R.$$

If $L \subset F$ is an arbitrary line, passing through the point x, then its strict transform $L^+ \subset F^+$ is a fibre of the projection λ_F. Since $K_Z = \varphi^* K_X + nE$, the following equalities hold:

$$(K_Z \cdot L^+) = -1, \quad (F^+ \cdot L^+) = -1$$

(taking into account that $(K_X \cdot L) = -(n+1)$), so that the numerical class of the curve L^+ generates the extremal ray $[L^+] \in \overline{NE}(Z)$. Let $\lambda\colon Z \to Y$ be the contraction of that ray. Obviously, $\lambda|_{F^+} = \lambda_F$ and λ contracts the divisor F^+. The image Y is again a $\mathbb{P}^n$-bundle over S, which is birationally isomorphic to the original one:

$$\tau = \lambda \circ \varphi^{-1}\colon X \dashrightarrow Y$$

is a link of the second type. Here $W = Z$ and ψ is an isomorphism. The fibre space Y/S had the same fibres over all points of the curve S, except for the point $\pi(x) \in S$. The fibre over that point is replaced by the exceptional divisor $E = \mathbb{P}^n$. For $n = 1$ the described procedure is the classical modification of a ruled surface.

Finally, the birational modification of a conic bundle, described in Example 1.5, also belongs to type II. (In fact, the links of that type were modelled by those modifications.) Note that by construction, a link of type II always induces a birational isomorphism of the generic fibres $F_\eta \dashrightarrow G_\eta$ of Mori fibre spaces X/S and Y/S.

Type III (shrinking the base). The links of this type are inverse to the links of type I, that is, in the diagram (87) the left-hand side and the right-hand side are swapped. More precisely, there are: an extremal divisorial contraction $\varphi\colon Z \to Y$, a birational map $\psi\colon X \dashrightarrow Z$, which is a composition of flops (an isomorphism in codimension one) and an extremal contraction $\varepsilon\colon S \to T$, $\rho(S/T) = 1$, such that the following diagram of maps commutes:

$$\begin{array}{ccccc} & X & \overset{\psi}{\dashrightarrow} & Z & \\ \pi & \downarrow & & \downarrow & \varphi \\ & S & & Y & \\ \varepsilon & \downarrow & & \downarrow & \rho \\ & T & = & T, & \end{array}$$

where $\varphi \circ \psi = \tau\colon X \dashrightarrow Y$. Inverting the construction of Example 1.5, we obtain a link of type III. From the geometric viewpoint, it is most interesting that a composition of two links which are of types I and III, gives, generally speaking, a birational modification which is not compatible with the structures of fibre spaces, that is, fibres are not mapped into fibres, as shown in the following

Example 2.3. In the notations of Example 2.1, let us choose the section $R \subset X$ in a special way: let $R = \sigma^{-1}(L^*)$, where $L^* \subset \mathbb{P}^3$ is a line that does not meet L, in particular, $\langle L, L^* \rangle = \mathbb{P}^3$. In this case the composition $\sigma \circ \varphi \colon Z \to \mathbb{P}^3$ is the blow up of a smooth reducible (non-connected) curve $L \cup L^*$ with the exceptional divisor $E_L \cup E$, where $E_L = \sigma^{-1}(L)$ is the exceptional divisor of the blow up σ. Contracting $E_L \subset Z$, we obtain a link of type III

$$\begin{array}{ccc} Y & \dashrightarrow & X^* \\ \downarrow & & \downarrow \\ T & \to & S^*, \end{array}$$

where X^* is $\mathbb{P}^3$ with the blown up line L^*, the curve $S^* = \mathbb{P}^1$ and $X^* \to S^*$ is the regularized projection from L^*. The composition of this link with the link of Example 2.1 gives a birational map

$$\begin{array}{ccc} X & \dashrightarrow & X^* \\ \downarrow & & \downarrow \\ S & & S^*, \end{array}$$

of $\mathbb{P}^2$-bundles over $S, S^* \cong \mathbb{P}^1$, which is not compatible with the structures of those fibre spaces. Note that in the special case under consideration, $T = S \times S^* = \mathbb{P}^1 \times \mathbb{P}^1$ and the projection $Y \to T$ is the regularization of the rational map

$$\mathbb{P}^3 \dashrightarrow L \times L^*,$$

that maps a point $x \in \mathbb{P}^3 \setminus (L \cup L^*)$ to the unique pair of points $(z, z^*) \in L \times L^*$ such that $x \in < z, z^* >$ (the lines L and L^* identify naturally with the curves S^* and S as sections of the Mori fibre spaces X^*/S^* and X/S, respectively). Note also that the diagram

$$\begin{array}{ccc} T & = & T \\ \downarrow & & \downarrow \\ S & & S^*, \end{array}$$

is an example of a link of type IV (in the dimension two), which we will now describe.

Type IV (changing the structure of a fibre space). This type is most interesting. There are: a birational map $\psi \colon X \dashrightarrow Y$, which is a composition of flops, and extremal contractions $s \colon S \to R$ and $t \colon T \to R$ such that the following diagram of maps commutes:

$$\begin{array}{ccccc} & X & \stackrel{\psi}{\dashrightarrow} & Y & \\ \pi & \downarrow & & \downarrow & \rho \\ & S & & T & \\ s & \downarrow & & \downarrow & t \\ & R & = & R. & \end{array}$$

The simplest example of a link of this type was given above (replacing one projection $\mathbb{P}^1 \times \mathbb{P}^1 \to \mathbb{P}^1$ by another). A non-trivial example of a link of type IV was described in Example 1.7. In that example the variety V has precisely two structures of a Fano fibre space and the structures are changed via a flop.

Today the strongest completely proved fact on factorization of birational maps into a composition of links is the following theorem proved in [**HM09**].

Theorem 2.2. *Let $\pi\colon X\to S$ and $\pi'\colon X'\to S'$ be Mori fibre spaces with $\mathbb{Q}$-factorial terminal singularities. The varieties X and X' are birationally equivalent if and only if when there exists a sequence of links*

$$\begin{array}{ccc} X_{i-1} & \stackrel{\tau_i}{\dashrightarrow} & X_i \\ \downarrow & & \downarrow \\ S_{i-1} & & S_i, \end{array}$$

$i=1,\dots,N$, connecting X/S and X'/S', that is, $X_0/S_0=X/S$ and $X_N/S_N=X'/S'$.

However, as we mentioned above, the main question is whether an *arbitrary* birational map $\chi\colon X\dashrightarrow X'$ can be decomposed into a composition of links. In the dimension three the answer if positive [**Co95**], in the dimension ≥ 4 the weaker claim, formulated above, is true. This comes from the different approaches to the problem of factorization of birational maps.

2.3. On the proof of the Sarkisov program. Let us briefly describe the original Sarkisov's approach [**S89, R91**], realized in [**Co95**]. Let $\chi\colon X\dashrightarrow X'$ be a birational map between the total spaces of Mori fibre space X/S and X'/S' of dimension three. One has to show that χ can be decomposed into a composition of elementary links, $\chi=\tau_N\dots\tau_1$. The proof, based on the original Sarkisov's ideas, is by producing an inductive algorithm, which associates with a birational χ, or, more precisely, a diagram

$$\begin{array}{cccc} & X & \stackrel{\chi}{\dashrightarrow} & X' & \\ \pi & \downarrow & & \downarrow & \pi' \\ & S & & S' & \end{array} \tag{88}$$

an *untwisting link* of one of the four types I–IV,

$$\begin{array}{ccccccc} & X & \stackrel{\tau}{\dashrightarrow} & X_1 & \stackrel{\chi\circ\tau^{-1}}{\dashrightarrow} & X' & \\ \pi & \downarrow & & \downarrow\ \pi_1 & & \downarrow & \pi' \\ & S & & S_1 & & S', & \end{array}$$

which decreases a certain invariant of the original birational map,

$$\delta(\chi\circ\tau^{-1})<\delta(\chi),$$

which cannot decrease infinitely.

Let us describe this invariant. The dimension $\dim X$ is now arbitrary ≥ 3. Since $\rho(X'/S')=1$, there is a positive integer m and a very ample divisor A' on the base S' such that

$$D'=-mK_X+\pi'^*A'$$

is a very ample divisor on X'. The very ample linear system $|D'|=\mathcal{H}'$ is fixed through the whole procedure of factorization of the map χ. Note that the map χ is fibrewise (that is, there exists a rational dominant map $\varepsilon\colon S\dashrightarrow S'$, making the diagram (88) a commutative one, $\varepsilon\pi=\pi'\chi$) if and only if the strict transform of the linear system $|\pi'^*A'|$ is pulled back from the base S, that is,

$$\chi_*^{-1}|\pi'^*A'|\subset|\pi^*A^+|$$

for some (mobile) divisor A^+ on the base S. Set

$$\mathcal{H}=\chi_*^{-1}|D'|$$

to be the strict transform of the system Σ' on X. Since X/S is a Mori fibre space, we get

$$\mathcal{H} \subset |-nK_X + \pi^* A|$$

for some $\mathbb{Q}$-Cartier divisor A on S. Since the denominators are bounded, we may assume that $n \in \mathbb{Z}_+$. Obviously, $n \geq 1$. The following claim holds.

Proposition 2.1. *The inequality $n \geq m$ holds. If $n = m$, then the map χ is fibrewise.*

Proof (see, for instance, [**Co95**]) is elementary. One has to repeat, almost word for word, the arguments that were used in the proof of Proposition 2.1.6. Let us outline it: let $\varphi\colon Z \to X$ be a resolution of singularities of the map χ, $\psi = \chi \circ \varphi\colon Z \to X'$ the composite map, a birational morphism,

$$\{E_i \,|\, i \in I\} \quad \text{and} \quad \{E'_j \,|\, j \in J\}$$

the sets of φ- and ψ-exceptional divisors, respectively. Arguing word for word as in Section 2.1, we get

$$\left(1 - \frac{n}{m}\right) \varphi^* K_X + \frac{1}{m} \varphi^* \pi^* A = \frac{1}{m} \psi^* \pi'^* A' + \sum_{j \in J} a'_j E'_j + \sum_{i \in I} \left(\frac{b_i}{m} - a_i\right) E_i,$$

where $a'_j > 0$, $a_i > 0$ (the singularities are terminal) and $b_i \geq 0$. Since E_i are φ-exceptional, the restriction of this equality onto the fibre of general position of the projection π shows that $n \geq m$, whereas, if $n = m$, then the strict transform of the linear system $|\pi'^* A'|$ with respect to χ is pulled back from the base S, and each ψ-exceptional divisor E'_j either is φ-exceptional, or its image $\varphi(E'_j)$ is pulled back from the base S. Q.E.D. for Proposition 2.1.

Recall [**Co95**], that the *canonical* (respectively, *log canonical*) *threshold* of the pair $(X, \mathcal{H})$ is the number

$$\operatorname{ct}(X, \mathcal{H}) = \sup\{\alpha \in \mathbb{Q}_+ \mid \text{the pair } (X, \alpha\mathcal{H}) \text{ is canonical}\}$$

(respectively, $\operatorname{lct}(X, \mathcal{H}) = \sup\{\alpha \in \mathbb{Q}_+ \mid \text{the pair } (X, \alpha\mathcal{H}) \text{ is log canonical}\}$). If $\varphi\colon Z \to X$ is a resolution of singularities of the pair $(X, \mathcal{H})$ with the set of exceptional divisors $\{E_i \,|\, i \in I\}$, then

$$\frac{1}{\operatorname{ct}(X, \mathcal{H})} = \max_{i \in I} \frac{\operatorname{ord}_{E_i} \varphi^* D}{a(E_i, X)}$$

for a general divisor $D \in \mathcal{H}$ and, similarly,

$$\frac{1}{\operatorname{lct}(X, \mathcal{H})} = \max_{i \in I} \frac{\operatorname{ord}_{E_i} \varphi^* D}{a(E_i, X) + 1}.$$

In particular, the inequality $\operatorname{lct}(X, \mathcal{H}) > \operatorname{ct}(X, \mathcal{H})$ takes place. If $\operatorname{ct}(X, \mathcal{H}) = \gamma$, then for each exceptional divisor E_i the inequality

$$\gamma \operatorname{ord}_{E_i} \varphi^* \mathcal{H}) \leq a(E_i, X),$$

holds, and, moreover, for at least one divisor E_i this inequality is an equality. Such divisors are called *crepant* divisors; the corresponding discerete valuations of the field of rational functions $\mathbb{C}(X)$ do not depend on the choice of resolution of singularities and are determined by the pair $(X, \mathcal{H})$. Let $e(X, \mathcal{H})$ be the number of crepant valuations ≥ 1.

Definition 2.1. The *degree* (or *Sarkisov degree*) of the pair $(X, \mathcal{H})$ is the triple of numbers

$$\delta(X, \mathcal{H}) = (n, \gamma = \mathrm{ct}(X, \mathcal{H}), e = e(X, \mathcal{H})).$$

The set of values of the degree δ is ordered in the following way (corresponding to the lexicographic order of the triples (n, γ^{-1}, e)):

$$\delta = (n, \gamma, e) > \delta_1 = (n_1, \gamma_1, e_1),$$

if either $n_1 < n$, or $n_1 = n$, but $\gamma_1 > \gamma$, or, finally, $n_1 = n$ and $\gamma_1 = \gamma$, but $e_1 < e$. Since (with the very ample linear system $\mathcal{H}'$ fixed) the map χ is uniquely determined by the system $\mathcal{H}$, we can write $\delta(\chi)$ instead of $\delta(X, \mathcal{H})$. At this stage in the Sarkisov program appears the key concept of a maximal singularity.

Definition 2.2. A *maximal singularity* of the linear system $\mathcal{H}$ (or of the birational map χ) in the sense of the minimal model program (briefly, an MMP-maximal singularity) is an exceptional divisor $E \subset Z$ of some resolution $\varphi\colon Z \to X$ of the pair $(X, \mathcal{H})$ (or the discrete valuation of the field of rational functions $\mathbb{C}(X)$, corresponding to that divisor), if the *Noether-Fano inequality* holds, then

$$\mathrm{ord}_E\, \varphi^*\mathcal{H} > na(E, X).$$

In [**Co95**] instead of the word combination "maximal singularity" the "base component of high multiplicity" is used. In the survey [**I01**] the "maximal singularity" is returned, however, one should remember, that in the classical theory the parameter n is the threshold of canonical adjunction. In the general situation, considered above, $c(\mathcal{H}) = n$, only if the divisor A is pseudo-effective on the base S. Thus between the concepts of a maximal singularity in the traditional approach and in the MMP-approach there are certain differences (ignored in [**Co95**]). To emphasize the point, we speak about MMP-maximal singularities (for the majority of problems, that are solved by now, these differences are inessential).

Obviously, an MMP-maximal singularity exists if and only if the inequality

$$\mathrm{ct}(X, \mathcal{H}) < \frac{1}{n}$$

holds.

Now ([**Co95**, Theorem 5.4]), if $\mathcal{H}$ has an MMP-maximal singularity, then applying $(K_Z + \alpha\mathcal{H}_Z)$-MMP to the extremal extraction $\varphi\colon Z \to X$ of one of the crepant discrete valuations, where the number α is specially selected, one gets a link $\tau\colon X/S \dashrightarrow X_1/S_1$ of type I or II, such that $(n_1, \gamma_1, e_1) < (n, \gamma, e)$, and moreover, if $n_1 = n$, then τ is fibrewise and induces birational isomorphisms of the bases $S \dashrightarrow S_1$ and generic fibres,

$$\begin{array}{ccccc} & X & \overset{\tau}{\dashrightarrow} & X_1 & \\ \pi & \downarrow & & \downarrow & \pi_1 \\ & S & \dashrightarrow & S_1, & \end{array}$$

that is, makes a "square" in the terminology of [**Co95**]. If, however, $\mathcal{H}$ has no MMP-maximal singularities (that is, the inequality $\mathrm{ct}(X, \mathcal{H}) \geq \frac{1}{n}$ holds), then a sequence of links of types III and IV is constructed

$$X/S \overset{\tau_1}{\dashrightarrow} X_1/S_1 \overset{\tau_2}{\dashrightarrow} \cdots \overset{\tau_k}{\dashrightarrow} X_k/S_k,$$

such that either $n = n_1 = \cdots = n_{k-1} > n_k$, or $n = n_1 = \cdots = n_k$ and the induced birational map is an isomorphism of the Mori fibre spaces. Finally, the last necessary fact, the finiteness of this procedure, that is, that the problem of factorization can be solved in finitely many steps, is proved via Alexeev's theorem that there are no accumulation points from below for the log canonical thresholds [**Al93**]. For a complete proof of the theorem on factorization see [**Co95**] or the survey [**I01**].

The algorithm, described above, in principle does not depend on dimension, however, today not all facts of MMP that are needed to prove it, are shown in dimension ≥ 4 (although the work is in progress and essential advances have been made; for instance, on the termination of thresholds see [**Bi07, ChPW07**]). All those facts are certain claims on finiteness (for instance, the termination of a sequence of flips), they are needed both for constructing links and for proving the finiteness of the factorization procedure. Using the approach that turned out to be so successful in [**BCHM**], Hacon and McKernan in [**HM09**] proved a weaker version of the Sarkisov program in the arbitrary dimension: instead of consructing an algorithm of factorization and proving its finiteness, they showed existence of *some* sequence of links, the composition of which gives a birational isomorphism $X \dashrightarrow X'$ of the total spaces of Mori fibre spaces X/S and X'/S' (about which it is assumed in advance, that they are birationally isomorphic).

The main result of [**HM09**] is as follows.

Theorem 2.2. *Assume that the Mori fibre spaces X/S and X'/S' are both products of the $(K_Z+\Phi)$-MMP for some* klt*-pair (Z, Φ). Then the induced birational map $X \dashrightarrow X'$ is a composition of links of types I–IV.*

It should be pointed out that if the total spaces of two Mori fibre spaces are birationally equivalent and have $\mathbb{Q}$-factorial terminal singularities, then they are products of the K_Z-MMP for some variety Z, so that the corresponding Mori fibre spaces satisfy the assumptions of Theorem 2.2.

The proof of Theorem 2.2 is based on the technique developed in [**BCHM**]; we refer an interested reader to the latter paper.

2.4. The method of maximal singularities and the factorization theory: a comparison. After what has been said it makes sense to compare the two approaches to investigating birational geometry of rationally connected fibre spaces: the classical method of maximal singularities and the program of factorization of birational maps, based on the theory of minimal models (the Sarkisov program). It should be noted that both approaches go back to the same ideas (developed in the works of M. Noether and his predecessors, and after that in the works of the Italian classics up to Fano, see Chapter 1), and for that reason in many respects are sufficiently close. Between these approaches there are, however, essential differences, on which we will now concentrate.

The main idea of the method of maximal singularities is to study the maximal singularities of a birational map $\chi\colon V \dashrightarrow V'$ or, what is equivalent, the maximal singularities of a mobile linear system Σ on V, which defines this birational map. The method works successfully, if it is possible to describe explicitly the potentially maximal singularities, which are then untwisted, as a rule, by birational self-maps $\tau \in \mathrm{Bir}\, V$ (in the relative case, by fibrewise birational self-maps $\tau \in \mathrm{Bir}(V/S)$). The point, why almost all potentially maximal singularities can be untwisted by means of self-maps, is not discussed, as it is an empirical fact. In those successfully studied cases, when the self-maps are not sufficient, each maximal singularity explicitly defines a transition to another structure of a Fano fibre space. The method of maximal singularities was modelled on the pioneer paper [**IM**]; in the past forty years the general approach did not change a lot, although the technical side has been transformed and made much stronger.

The MMP-approach was modelled on the proof of the Sarkisov theorem [**S80, S82**], taking into account the ideology of the Mori theory. That approach is about fifteen years "younger". Its main idea, described in the previous two subsections, is to simplify a birational map (or a mobile linear system) by means of elementary links. A link is constructed by applying MMP to a suitable log pair. A priori a link is a birational map between *distinct* Mori fibre spaces, that is, even in the case of a fibrewise link one comes to another model.

The method of maximal singularities is a (technically powerful and successfully working) scheme of arguments leading to very strong individual results, that is, results for particular explicitly given families of Fano varieties and fibre spaces. On the contrary, the Sarkisov program (in the form in which it is known since the paper [**Co95**] was published) is a general existence theorem, claiming the very fact that it is possible to factorize a birational map (or a somewhat weaker fact in dimension ≥ 4). Each of these two approaches has its advantages and disadvantages.

An obvious advantage of the Sarkisov program is its generality. From the description of the program and the proof of the main theorem follows a procedure of constructing factorizing links. Taking into account the rapid development of MMP since the mid-eighties, it is not surprising that the proof of the Sarkisov program in [**Co95**] generated a lot of optimism in respect of three-dimensional birational geometry. In 1993–95 it seemed to many people that solution of the problems of the classical birational geometry in dimension three, even such as the description of the Cremona group of rank three or the description of the group of birational self-maps of the three-dimensional cubic, or a proof of the rationality criterion for three-fold conic bundles, is a matter of the not so distant future. Something in that spirit was said (although in a rather cautious way) in [**Co95**]. The first thing to be done was to find new proofs of the known results that were earlier obtained by the method of maximal singularities. After that, it was meant to make further progress in the area where the classical methods did not work.

Today, after more than fifteen years, one can say that, so far, these hopes were not realized. The problems of describing birational maps between three-dimensional Mori fibre spaces turned out to be much more difficult than it was thought in the early nineties. One explanation of the optimism of those years is that the previous work of the Russian school of birational geometry was not known very well. Probably, for that reason the work of the Russian school was considered as belonging to the past: it was expected that the new techniques of the minimal model program would make it possible to easily overcome the difficulties that obstructed successful work of the method of maximal singularities. It should be acknowledged that some ideas that came from MMP turned out to be very fruitful indeed. It is, in the first place, the connectedness principle of Shokurov and Kollár and the inversion of adjunction that follows from that principle (see Section 3 below and Chapter

7), which made it possible to considerably simplify some parts of the classical techniques (compare, for instance, the exclusion of a maximal singularity over a quadratic point in [**Pukh89c**] and in [**Co00**]: the latter approach essentially does not need anything but a straightforward application of the inversion of adjunction).

However, as a whole, the attempts to realize the Sarkisov program for a wide class of three-dimensional Mori fibre spaces were so far not successful. Birational maps between particular three-dimensional varieties were in most cases studied using the well trodden approach of the classical method of maximal singularities, sometimes with certain technical improvements. This applies already to the very first, and best known, paper of that series [**CPR**], which was originally conceived as a first big-scale stage in the realization of the Sarkisov program. It turned out that the techniques described in the previous two subsections was useless for all 95 types of the weighted Fano hypersurfaces, because all potentially maximal singularities (on many varieties they simply do not exist) were untwisted by birational self-maps, which, in turn, were constructed by means of the known classical techniques. Nevertheless, [**CPR**] was sometimes interpreted as a result of application of the new MMP-techniques (see, for instance, the survey [**I01**]), which was, of course, an exaggeration. The paper [**CPR**] was written in the new language (and contains new proofs of certain known facts of the classic theory), but this seems to be the most that can be said. By its contents [**CPR**] is undoubtedly much closer to the series of papers that were previously done by the method of maximal singularities and directly followed the paper [**IM**] on the three-dimensional quartic; see Subsection 2.3.5.

Apart from [**CPR**], a few more papers were published, where the approach of the Sarkisov program was used for studying birational maps between three-dimensional Mori fibre spaces [**CoMe, Me04**]. In [**Co00**] some known results, that were earlier obtained by the method of maximal singularities, were presented in the language of the Sarkisov program. In [**CoMe**] that program really works: non-trivial links are constructed that connect the varieties under investigation with other Mori fibre spaces, however, the existence of such a link does give much information about the birational type of those varieties, although it does mean that the defining structure of a Mori fibre space is not unique, so that if the uniqueness of that structure is taken for the definition of birational rigidity, then it could be concluded that the variety is not birationally rigid. The paper [**BCZ**] which joins this series of papers contains certain non-trivial computations and constructions, also a number of interesting conjectures, but no progress towards their proof.

It makes sense to compare the results that were obtained in the course of attempts to apply the general theory [**Co95**] to the study of birational geometry of three-dimensional Mori fibre spaces with the results that were obtained during the same period of time in the framework of the classical method of maximal singularities in the dimension three [**Pukh98a, Grin00, Sob02, Grin03a, Grin04**], speaking not about the results in the arbitrary dimension. The reason for such contrast is, it seems, in the fact that to construct links one needs a precise description of all potentially maximal singularities, so that the work of the classical method of maximal singularities is inevitably included into the work of the Sarkisov program as the most important part. On the other hand, the above mentioned empirical fact takes place: for almost all (accessible for investigation) varieties the potentially realizable maximal singularities are untwisted by birational self-maps, which make the techniques of links in many cases unnecessary, as it is sufficient to define a birational map between two *given* varieties at the generic point, not taking care of how it could be decomposed into a sequence of elementary contractions/extractions and flips.

As an illustration, let us consider Example 1.4. Let $V \stackrel{\pi}{\to} \mathbb{P}^1$ be a standard fibration into cubic surfaces, $C \subset V$ a section of the projection π. Assume that C is a maximal curve of a mobile linear system $\Sigma \subset |-nK_V + lF|$, where $l \in \mathbb{Z}_+$. The extremal blow up, associated with the curve C, is simply its blow up in the usual sense, $\varphi\colon \widetilde{V} \to V$. Now the general theory guarantees that an application of the minimal model program to a suitable pair $(\widetilde{V}, \alpha\widetilde{\Sigma})$ leads to a link $V \dashrightarrow V^+/\mathbb{P}^1$, untwisting the maximal curve C. However, this general claim, which uses all the power of the program, is practically useless, since the construction of the link is obvious from the elementary geometric considerations and its proof does not require any efforts since $V = V^+$ and the link is well defined on the generic fibre. There are finitely many fibres $F_i = \pi^{-1}(t_i)$, $i = 1, \dots, k$, where through the point $C \cap F_i$ there is at least one line on F_i. The general theory guarantees that a sequence of flops, starting from $\widetilde{V}$, leads to a model $V^\sharp$, that admits an extremal contraction (so that the birational self-map of Example 1.4, associated with the section C, is a composition of two links, of type I and type III), however, the classical techniques gives at once the final result with considerably less effort. It is especially important here, that, using the classical approach, we stay on the same model, whereas the general theory provides an untwisting of the curve C, by means of, generally speaking, a transition to a new Mori fibre space. For that reason, whereas the classical method of maximal singularities requires only information on the biregular geometry of the *given* model of a Mori fibre space, for the Sarkisov program to work, one needs (as it is only natural to expect from a theory of such level of generality) information about *all* models of that dimension. Biregular classification of Mori fibre spaces even in dimension three is still very far from completion (in any reasonable sense). This is the reason why the general theory of the Sarkisov program has not found many applications so far.

In dimension three there are three classes of Mori fibre spaces: Fano varieties (with $\mathbb{Q}$-factorial terminal singularities), del Pezzo fibrations and conic bundles. By the mid-nineties, a lot was known on birational geometry of varieties of the first and third classes, however, del Pezzo fibrations remained a white spot. Attempts to study their geometry by means of the test class technique were made in the course of about ten years, but they turned out to be unsuccessful [**I95**] (we explain the reasons in the "Notes and references" section of Chapter 6). The situation had radically changed when the paper [**Pukh98a**] appeared, where the classical approach was essentially redesigned, in particular, the test class technique was replaced by the technique of counting multiplicities. However, as far as the author knows, up to a very recent time, attempts were made to obtain results on birational geometry of del Pezzo fibrations by means of methods similar to the proof of Sarkisov's theorem on the conic bundles (Example 1.5), that is, by means of fibrewise modifications. (Since it is precisely the maximal singularities, the center of which is contained in a fibre, present the biggest difficulty, see Chapter 6.) Those attempts were also unsuccessful. In contrast to the conic bundles, even the simplest fibrewise modifications of a pencil of cubic surfaces lead to varieties with complicated singularities and the study terminates at that point.

This is the situation today: introducing new ideas and new facts in the future can, of course, change it. A step in that direction has been made in [**ShCh11**]. This completes our short visit of the general theory of factorization of birational maps (the Sarkisov program); we now come back to the theory, techniques and results of the method of maximal singularities, the subject of the present book.

3. Birational rigidity of Fano fibre spaces

This section is an introduction to the method of maximal singularities for Fano fibre spaces combined with a brief overview of the main results. We introduce the K-condition and the K^2-condition for fibre spaces and show that they measure the "twistedness" of the fibre space over the base. After that we show how the information about the threshold of canonical adjunction implies the crucial geometric properties of the fibre space under consideration. This done, we proceed to an overview of the techniques and results: the quadratic method (which produced numerous theorems on birational rigidity of Fano fibre spaces over $\mathbb{P}^1$) and the linear method (which produced the theorem on Fano direct products and related results and made it possible to describe birational geometry of Fano double spaces).

3.1. The threshold of canonical adjunction. For an arbtrary rationally connected smooth projective variety X we denote by the symbol A^iX the Chow group of algebraic cycles of codimension $i \geq 1$ modulo numerical equivalence, $A^i_{\mathbb{R}}X = A^iX \otimes \mathbb{R}$. By the symbol A^i_+X we denote the closed cone in $A^i_{\mathbb{R}}X$, generated by the classes of effective cycles (the pseudo-effective cone). By the symbol $A^i_{\text{mob}}X$ we denote the closed cone in $A^i_{\mathbb{R}}X$, generated by the classes of mobile divisors (that is, such divisors D, that the complete linear system $|D|$ has no fixed components), the *mobile* cone.

Let us consider a standard Fano fibre space $\pi\colon V \to S$. Obviously, we get an inclusion $\pi^*A^i_{\text{mob}}S \subset A^i_{\text{mob}}V$. Furthermore,

$$A^1V = \mathbb{R}[K_V] \oplus \pi^*A^1S.$$

Definition 3.1. We say that the standard Fano fibre space $\pi\colon V \to S$ satisfies the *K-condition*, if

$$A^1_{\text{mob}}V \subset \mathbb{R}_+[-K_V] \oplus \pi^*A^1_+S.$$

In other words, V/S satisfies the K-condition if and only if for any mobile linear system $|-nK_V + \pi^*A|$ the class $A \in \text{Pic}\, S$ is pseudo-effective. If the pseudo-effective cone A^1_+S has a sufficiently simple structure, for instance, $A^1S = \mathbb{Z}H_S$, where H_S is the ample generator, so that $A^1_+S = \mathbb{R}_+[H_S]$ is the positive ray, or $S = S_1 \times \cdots \times S_k$, where $A^1S_i \cong \mathbb{Z}$, then it is easy to check that, in a certain sense, the "overwhelming majority" of standard Fano fibre spaces with the given fixed base S satisfies this condition. As an illustration let us consider the construction of Example 1.1, assuming that $A^1S = \mathbb{Z}H_S$.

In the notations of Example 1.1 we have $A_i \sim a_iH_S$ for some $a_i \in \mathbb{Z}$. Set $a = a_1 + \cdots + a_k$. Twisting the locally free sheaf $\mathcal{E}$, we may assume that it is generated by global sections, so that the tautological class L_X is numerically effective. Also set

$$bH_S \sim \det \mathcal{E} + K_S,$$

where the integral parameter b depends on X only. We get

$$K_V = -L + (a+b)H,$$

where for the brevity of notations $L = L_V$ and $H = \pi^* H_S$. Now if the linear system $|-nK_V + lH|$ is non-empty (the more so, mobile), then the inequality

$$\begin{array}{c}(-nK_V + lH) \cdot L^M \cdot H^{\dim S - 1} \geq 0 \\ \| \\ nL^{M+1} H^{\dim S - 1} + d(l - n(a+b)) H^{\dim S}\end{array}$$

holds, where $d = d_1 \cdot \dots \cdot d_k$, which immediately implies that for some $a_0 \in \mathbb{Z}$ for $a \geq a_0$ we have $l \in \mathbb{Z}_+$, that is, the K-condition holds for all standard Fano fibre space, satisfying the inequality $a_1 + \dots + a_k \geq a_0$.

This example shows that a "majority" of standard Fano fibre spaces of Example 1.1 satisfies a stronger condition:

$$A^1_+ V \subset \mathbb{R}_+[-K_V] \oplus \pi^* A^1_+ S.$$

The situation is similar for Fano double covers (Example 1.2). On the contrary, it is easy to see that Fano direct products (Example 1.3) do not satisfy the K-condition.

Now let us consider the most interesting case $S = \mathbb{P}^1$. To check the K-condition we use the fact that the *self-intersection* of a mobile class $z \in A^1_{\text{mob}} X$ is a pseudo-effective class of codimension two: $z^2 \in A^2_+ X$. Therefore, if $z^2 \notin A^2_+ X$, then $z \notin A^1_{\text{mob}} X$.

Definition 3.2. We say that a standard Fano fibre space $\pi \colon V \to \mathbb{P}^1$ satisfies the *K^2-condition*, if

$$K_V^2 \notin \operatorname{Int} A^2_+ V.$$

Let $\operatorname{Pic} V = \mathbb{Z} K_V \oplus \mathbb{Z} F$, where F is the class of a fibre of the projection π.

Proposition 3.1. *If a fibre space $\pi \colon V \to \mathbb{P}^1$ satisfies the K^2-condition, then it satisfies the K-condition, as well.*

Proof. The self-intersection of the class $-nK_V + lF$ is

$$n^2 K_V^2 + 2nl(-K_V \cdot F),$$

where in the brackets it is the anticanonical section of the fibre, that is, an effective cycle of codimension two. By the K^2-condition, the self-intersection cannot be pseudo-effective for $l < 0$, which is what was required. Q.E.D.

The importance of the K-condition can be seen from the following simple fact.

Proposition 3.2. *Assume that a standard Fano fibre space $\pi \colon V \to S$ satisfies the K-condition.*

(i) *For the threshold of canonical adjunction of a mobile linear system $\Sigma \subset |-nK_V + \pi^* A|$ we have the equality $c(\Sigma, V) = n$.*

(ii) *If the mobile linear system Σ satisfies the equality $c(\Sigma, V) = 0$, then Σ is a π-pullback of a mobile linear system Σ_S on the base S.*

(iii) *Assume in addition that the variety V is birationally superrigid: $c_{\rm virt}(\Sigma) = c(\Sigma, V)$ for any mobile linear system Σ. Then every structure of a rationally connected fibre space on V is compatible with $\pi: V \to S$, that is, for any birational map $\chi: V \dashrightarrow V'$, where $\pi': V' \to S'$ is a rationally connected fibre space, there is a rational dominant map $\varepsilon: S \dashrightarrow S'$ such that the following diagram commutes:*

$$\begin{array}{ccccc} & V & \stackrel{\chi}{\dashrightarrow} & V' & \\ \pi & \downarrow & & \downarrow & \pi' \\ & S & \stackrel{\varepsilon}{\dashrightarrow} & S'. & \end{array}$$

Proof. The claim (i) is obvious, because $\pi^* A \in A^1_+ V$ is a pseudo-effective class, whereas the class $\alpha K_V + \pi^* A$ for $\alpha > 0$ is negative on the fibres and cannot be pseudo-effective. The claim (ii) follows from (i). If the linear system Σ' on the variety V' is mobile and pulled back from the base S', then its strict transform $\Sigma = \chi_*^{-1}\Sigma'$ on V is a mobile linear system, satisfying the equality $c_{\rm virt}(\Sigma) = c(\Sigma', V') = 0$. By the asuumption on birational superrigidity and the claim (ii), it follows that Σ is a π-pullback of a mobile linear system on S. This proves the claim (iii) and completes the proof of the proposition. Q.E.D.

Now let us give an example of a non-trivial computation of the threshold of canonical adjunction on a variety that does not satisfy the K-condition.

Example 3.2. Consider the Fano fibre space $\pi: V \to \mathbb{P}^1$ of Example 1.7 (we use the notations of that example). If a mobile linear system Σ is a subsystem of the complete linear system $|-nK_V + lF|$ with $l \in \mathbb{Z}_+$, then, as above, we get $c(\Sigma) = n$. However, if $l < 0$, then the linear system $\Sigma^+ = \tau_*^+\Sigma$ is a subsystem of the complete linear system $|-n_+K_{V^+} + l_+F^+|$ on the standard Fano fibre space $V^+/\mathbb{P}^1$ from the same family. Now since $l_+ = -l \in \mathbb{Z}_+$, we have $c(\Sigma^+, V^+) = n_+$. However, $\tau^+: V \dashrightarrow V^+$ is a biregular isomorphism in codimension one (that is, an isomorphism of complements to closed sets $\Delta \subset V$, $\Delta^+ \subset V^+$ of codimension two). Therefore, $c(\Sigma, V) = c(\Sigma^+, V^+)$, that is, for $l < 0$ we have

$$c(\Sigma, V) = n_+ = n + l.$$

Another similar example is given by Fano fibre spaces $V/\mathbb{P}^1$ with a non-trivial (that is, non-compatible with the structure of the fibre space $\pi: V \to \mathbb{P}^1$) birational involution $\tau \in \operatorname{Bir} V$ [**Pukh04b**, Sec. 3.1] and [**Pukh06b**, Sec. 2.3]; in the latter case they are varieties, described in part (iii) of Theorem 3.1 below. Computations in those examples are completely similar to those considered above: if $\Sigma \subset |-nK_V + lF|$ ñ $l < 0$ is a mobile linear system, then applying the involution τ (which is biregular in codimension one), we transform Σ into a system $\Sigma^+ \subset |-n_+K_V + l_+F|$ with $l_+ \in \mathbb{Z}_+$, which makes it possible to compute the threshold $c(\Sigma, V) = c(\Sigma^+, V) = n_+$. For the details, see the papers mentioned above.

Today there are two main approaches to proving birational rigidity of Fano fibre spaces: the quadratic and the linear ones. The quadratic method is aimed at studying the self-intersection $Z = (D_1 \circ D_2)$ of a mobile linear system $\Sigma \ni D_i$; the method extends the techniques of proving birational rigidity of primitive Fano varieties, described in the previous chapters. Almost all results on birational geometry of standard Fano fibre spaces $V/\mathbb{P}^1$ are obtained by means of the quadratic method.

The linear method is aimed at studying singularities of an arbitrary divisor $D \in \Sigma$ or its restriction $D|_R$ onto a specially selected algebraic subvariety $R \subset V$. This method works in the proof of the theorem on Fano direct products [**Pukh05**], it is used in the proofs of the theorems on birational rigidity of fibre spaces $V/\mathbb{P}^1$ to simplify certain steps [**Pukh06a, Pukh08a**], also in the proof of the theorem on birational geometry of Fano double spaces of index two [**Pukh10**]. The quadratic technique is considered in full detail in Chapters 5 and 6, the linear technique in Chapters 7 and 8.

3.2. The quadratic method: the main results. Consider a standard Fano fibre space $\pi\colon V \to \mathbb{P}^1$, that is, $\operatorname{Pic} V = \mathbb{Z}K_V \oplus \mathbb{Z}F$, where F is the class of a fibre of the projection π. Assume in addition that the condition

$$A^2V = \mathbb{Z}K_V^2 \oplus \mathbb{Z}H_F, \tag{89}$$

holds, where $H_F = (-K_V \cdot F)$ is the class of an anticanonical section of the fibre. The K^2-condition can be weakened in the following way.

Definition 3.3. A standard Fano fibre space $V/\mathbb{P}^1$ satisfies the K^2-*condition of depth* $\varepsilon \geq 0$, if

$$K_V^2 - \varepsilon H_F \notin \operatorname{Int} A_+^2 V.$$

Sometimes for more clarity the K^2-condition of depth $\varepsilon = 0$ will be called the *strong* K^2-*condition.* If the depth is not specified, then it is assumed that it is equal to zero.

It is easy to see that the K-condition for the class of Fano fibre space under consideration takes the following form:

$$-K_V \notin \operatorname{Int} A_{\mathrm{mob}}^1 V.$$

Thus, it is obvious that if $\varepsilon_1 \leq \varepsilon_2$, then the K^2-condition of depth ε_1 implies the K^2-condition of depth ε_2.

Fibrations into Fano complete intersections. This class of varieties corresponds to the base $S = \mathbb{P}^1$ in Example 1.1. Let $a_* = \{0 = a_0 \leq a_1 \leq \cdots \leq a_{M+k}\}$ be a non-decreasing sequence of non-negative integers, $\mathcal{E} = \bigoplus_{i=0}^{M+k} \mathcal{O}_{\mathbb{P}^1}(a_i)$ a locally free sheaf on $\mathbb{P}^1$, $X = \mathbb{P}(\mathcal{E})$ the corresponding projective bundle in the sense of Grothendieck. Obviously, we get

$$\operatorname{Pic} X = \mathbb{Z}L_X \oplus \mathbb{Z}R, \quad K_X = -(M+k+1)L_X + (a_X - 2)R,$$

where L_X is the class of the tautological sheaf, R is the class of a fibre of the morphism $\pi\colon X \to \mathbb{P}^1$, $a_X = a_1 + \cdots + a_{M+k}$. Furthermore, we have $L_X^{M+k+1} = a_X$.

For a set of k integers $(b_1, \ldots, b_k) \in \mathbb{Z}_+^k$ let $G_i \in |\, d_i L_X + b_i R|$ be irreducible divisors such that the complete intersection

$$V = G_1 \cap \cdots \cap G_k \subset X$$

is a smooth subvariety. The projection $\pi\,|_V\colon V \to \mathbb{P}^1$ is denoted by the same symbol π, the fibre $\pi^{-1}(t) \subset V$ by the symbol F_t, the restriction $L_X\,|_V$ by the symbol L.

The fibre space $V/\mathbb{P}^1$ is a standard Fano fibre space, satisfying the condition (89). Obviously, $K_V = -L + (a_X + b_X - 2)F$, where $b_X = b_1 + \cdots + b_k$. It is easy to check the formulas

$$(L^M\cdot F)=(H_F\cdot L^{M-1})=d,\qquad L^{M+1}=d\left(a_X+\sum_{i=1}^{k}\frac{b_i}{d_i}\right),$$

where $d=d_1\dots d_k$ is the degree of the fibre. From here we get:

$$(-K_V\cdot L^M)=d\left(2-\sum_{i=1}^{k}\frac{d_i-1}{d_i}b_i\right)\ \text{and}\ (K_V^2\cdot L^{M-1})=d\left(4-a_X-\sum_{i=1}^{k}\frac{2d_i-1}{d_i}b_i\right).$$

Since the linear system $|L|$ is free, these formulas immediately imply

Proposition 3.3. (i) *If $a_X+\sum_{i=1}^k\frac{2d_i-1}{d_i}b_i\geq 4$, then the strong K^2-condition holds.*

(ii) *If $a_X+\sum_{i=1}^k\frac{2d_i-1}{d_i}b_i\geq 2$, then the K^2-condition of depth 2 holds.*

(iii) *If $\sum_{i=1}^k\frac{d_i-1}{d_i}b_i\geq 2$, then $-K_V\notin\operatorname{Int}A^1_+V$ and, furthermore, $-K_V\notin\operatorname{Int}A^1_{\mathrm{mob}}V$; if, in addition, the inequality above is strict, then $-K_V\notin A^1_+V$.*

Now let us formulate the main result.

Assume that the variety V is sufficiently general in its family.

Theorem 3.1. (i) *The variety V is birationally rigid, the projection $\pi\colon V\to\mathbb{P}^1$ is the only structure of a rationally connected fibre space on V and the groups of birational and biregular automorphisms coincide, $\operatorname{Bir}V=\operatorname{Aut}V$, if for the integral parameters of the variety V one of the following six cases takes place:*

- $a_X+b_V\geq 4$,
- $a_X=1, b_V=2$,
- $a_X=0, b_V=3$,
- $a_X=3, b_V=0$,
- $a_X=2, b_V=1$,
- $(a_*)=(0,\dots,0,2)$ *and* $b_V=0$.

(ii) *For $(a_*)=(0,\dots,0,1,1)$ and $b_V=0$ a general variety V is birationally superrigid. However, the K-condition does not hold: the linear system $|-K_V-F|$ is mobile and determines a rational map $\varphi\colon V\dashrightarrow\mathbb{P}^1$, the fibres of which are rationally connected. On the variety V there are precisely two non-trivial structures of a rationally connected fibre space: the morphism $\pi\colon V\to\mathbb{P}^1$ and the map φ. There exists a unique, up to a fibrewise isomorphism, Fano fibre space $\pi^+\colon V^+\to\mathbb{P}^1$ with the same parameters $(a_*)=(0,\dots,0,1,1)$ and $b_{V^+}=0$ and a birational isomorphism $\chi\colon V\dashrightarrow V^+$, biregular in codimension one, such that the following diagram commutes:*

$$\begin{array}{ccc} V & \stackrel{\chi}{\dashrightarrow} & V^+\\ \varphi\downarrow & & \downarrow\ \pi^+\\ \mathbb{P}^1 & = & \mathbb{P}^1.\end{array}$$

The correspondence $V\to V^+$ is an involution of the set of Fano fibre spaces of that type, that is, $(V^+)^+=V$.

(iii) *For $a_X=0$, $b_e=2$ for some $e\in\{1,\dots,k\}$ and $b_i=0$ for $i\neq e$ the variety V is birationally superrigid. However, the K-condition does not hold: the linear system $|-d_eK_V-F|$ is a pencil of rationally connected varieties. The group of birational self-maps $\operatorname{Bir}V$ is non-trivial and for a general variety V generated by the birational involution τ, which is biregular in codimension one, and moreover,*

$\tau_*|F| = |-d_e K_V - F|$. *On the variety* V *there are precisely two non-trivial structures of a rationally connected fibre space: the projection* $\pi\colon V \to \mathbb{P}^1$ *and the map* $\pi\tau\colon V \dashrightarrow \mathbb{P}^1$.

Fibrations into Fano cyclic covers. Let $a_* = \{0 = a_0 \le a_1 \le \cdots \le a_{M+1}\}$ be a non-decreasing sequence of non-negative integers, $\mathcal{E} = \bigoplus_{i=0}^{M+1} \mathcal{O}_{\mathbb{P}^1}(a_i)$ a locally free sheaf on $\mathbb{P}^1$, $X = \mathbb{P}(\mathcal{E})$ the corresponding projective bundle in the sense of Grothendieck. Obviously, we have

$$\operatorname{Pic} X = \mathbb{Z}L_X \oplus \mathbb{Z}R, \quad K_X = -(M+2)L_X + (a_X - 2)R,$$

where L_X is the class of the tautological sheaf, R is the class of a fibre of the morphism $\pi_X\colon X \to \mathbb{P}^1$, $a_X = a_1 + \cdots + a_{M+1}$, $L_X^{M+2} = a_X$. For some $a_Q, a_W \in \mathbb{Z}_+$ let

$$Q \sim mL_X + a_Q R \quad \text{and} \quad W_X \sim K(lL_X + a_W R)$$

be divisors on X, where $Q \subset X$ is a smooth subvariety, $W = W_X \cap Q$ a smooth divisor on Q. Let

$$\sigma\colon V \to Q$$

be the K-sheeted cyclic cover of the variety Q, branched over the divisor W. The projection $\pi_X\,|\,_Q$ will be denoted by the symbol π_Q, the projection $\pi_Q \circ \sigma\colon V \to \mathbb{P}^1$ by the symbol π. The fibre $\pi_Q^{-1}(t)$, $t \in \mathbb{P}^1$, will be denoted by the symbol G_t (or simply G, when it is clear, which point $t \in \mathbb{P}^1$ is meant), the fibre $\pi^{-1}(t) \subset V$ by the symbol F_t or F. Set $L_Q = L_X\,|\,_Q$ and $L = \sigma^* L_Q$, respectively. Obviously,

$$\operatorname{Pic} V = \mathbb{Z}L \oplus \mathbb{Z}F, \quad K_V = -L + (a_X + a_Q + (K-1)a_W - 2)F.$$

It is easy to check the formulas $(L^M \cdot F) = mK$, $L^{M+1} = K(ma_X + a_Q)$. From here we obtain $(-K_V \cdot L^M) = K((1-m)a_Q - m(K-1)a_W + 2m)$ and

$$(K_V^2 \cdot L^{M-1}) = K(-ma_X + (1-2m)a_Q - 2m(K-1)a_W + 4m).$$

For convenience we write the parameters of the cover V in the form

$$((a_1, \ldots, a_{M+1}),\ (a_Q, a_W))$$

and, moreover, among the numbers $a_1, \ldots, a_{M+1}$ we specify only non-zero values, if there are any, otherwise we write simply (0). These notations are convenient because only those covers V require an individual study which have almost all parameters equal to zero. Indeed, the explicit formulas above immediately imply the following:

Proposition 3.4. (i) *The variety* V *satisfies the strong* K^2*-condition, that is, the* K^2*-condition of depth* 0*, if one of the following cases takes place:*

- $a_W \ge 1$,
- $a_W = 0$, $a_Q \ge 3$,
- $a_W = 0$, $a_Q = 2$, $a_X \ge 1$,
- $a_W = 0$, $a_Q = 1$, $a_X \ge 3$,
- $a_W = a_Q = 0$, $a_X \ge 4$.

(ii) *If* $a_W = 0$, $a_Q = 2$, $a_X = 0$, *then the variety* V *satisfies the* K^2*-condition of depth* $\frac{2}{m}$.

(iii) *If* $a_W = 0$, $a_Q = 1$, *then the variety* V *satisfies the* K^2*-condition of depth* $\frac{1}{m}$ *for* $a_X = 2$ *and of depth* $(1 + \frac{1}{m})$ *for* $a_X = 1$.

(iv) *If $a_W = a_Q = 0$, then the variety V satisfies the K^2-condition of depth* 1 *for $a_X = 3$ and of depth* 2 *for $a_X = 2$.*

Proof. This follows immediately from the fact that for any irreducible subvariety Y the inequality $(Y \cdot L^{\dim Y}) \geq 0$ holds. Q.E.D.

Now let us formulate the main result.

We assume that the cyclic cover V is sufficiently general in the family constructed above.

Theorem 3.2. (i) *The variety V is birationally superrigid, the projection $\pi\colon V \to \mathbb{P}^1$ is the only structure of a rationally connected fibre space on V, and the groups of birational and biregular automorphisms of the variety V coincide, $\operatorname{Bir} V = \operatorname{Aut} V$, if the integral parameters of the variety V either satisfy any of the six conditions of Proposition 3.4(i), or are of one of the following six types:* $((2),(0,0))$, $((2),(1,0))$, $((1,1),(1,0))$, $((3),(0,0))$, $((1,2),(0,0))$, $((1,1,1),(0,0))$.

(ii) *The variety V of the type $((1,1),(0,0))$ is birationally superrigid. However, the K-condition does not hold: the linear system $|-K_V - F|$ is mobile and determines a birational map $\varphi\colon V \dashrightarrow \mathbb{P}^1$, the fibres of which are rationally connected. On the variety V there are precisely two structures of a rationally connected fibre space: the projection π and the map φ. There exists a unique (up to a fibrewise isomorphism) fibration into Fano cyclic covers $\pi^+\colon V^+ \to \mathbb{P}^1$ of the same type $((1,1),(0,0))$ and a birational isomorphism $\chi\colon V \dashrightarrow V^+$, biregular in codimension one, such that the following diagram of maps commutes:*

$$\begin{array}{ccccc} & V & \overset{\chi}{\dashrightarrow} & V^+ & \\ \varphi & \downarrow & & \downarrow & \pi^+ \\ & \mathbb{P}^1 & = & \mathbb{P}^1. & \end{array}$$

The correspondence $V \to V^+$ is an involution, that is, $(V^+)^+ = V$.

(iii) *The variety V of the type $((0),(2,0))$ is birationally superrigid. However, the K-condition does not hold: the linear system $|-mK_V - F|$ is mobile and determines a birational map, the fibres of which are rationally connected. The group of birational self-maps $\operatorname{Bir} V$ is strictly larger than the group of biregular automorphisms: it contains a non-trivial birational involution $\tau \in \operatorname{Bir} V \setminus \operatorname{Aut} V$ and, moreover, $\operatorname{Bir} V \cong (\mathbb{Z}/2\mathbb{Z}) \times (\mathbb{Z}/K\mathbb{Z})$, where $\mathbb{Z}/2\mathbb{Z} = \{\mathrm{id}, \tau\}$. On V there are precisely two structures of a rationally connected fibre space: the projection π and the rational map $\pi\tau\colon V \dashrightarrow \mathbb{P}^1$ and, moreover, $|-mK_V - F| = \tau_*|F|$.*

Note the obvious parallelism of Theorems 3.1 and 3.2 (their proofs, however, are essentially different, see Chapter 5). Birational (super)rigidity is also proved for some other families of Fano fibre spaces over $\mathbb{P}^1$; see [**Pukh98a, Pukh00b, Pukh00d, Sob01**] and Chapter 5. Theorems 3.1 and 3.2 describe the most "massive" families.

Varieties with a pencil of del Pezzo surfaces. That class of three-dimensional rationally connected varieties for a long time was out of reach. Birational rigidity of the overwhelming majority of these varieties was proved in [**Pukh98a**]. That paper was followed by [**Grin98, Grin00, Grin03a**], where birational geometry of almost all remaining types of varieties with a pencil of del Pezzo surfaces of degree 1 and 2 was completely described. That work was summarized in the survey [**Grin06**].

We will now state the results proven in [**Pukh98a**]. Let $V \stackrel{\pi}{\to} \mathbb{P}^1$ be a fibration into del Pezzo surfaces of degree $d \in \{1,2,3\}$, $\operatorname{Pic} V = \mathbb{Z}K_V \oplus \mathbb{Z}F$, where the variety V is smooth and for $d = 3$ sufficiently general.

Theorem 3.3. *Assume that the fibre space $V/\mathbb{P}^1$ satisfies the K^2-condition: $K_V^2 \notin \operatorname{Int} A_+^2 V$. Then the variety V is birationally rigid (superrigid for $d = 1$), the projection $\pi\colon V \to \mathbb{P}^1$ is the only structure of a rationally connected fibre space on the variety V, and the quotient group of the group of birational self-maps by the normal subgroup of birational self-maps of the generic fibre* $\operatorname{Bir} F_\eta$ *is finite, generically trivial.*

For $d = 1$ the group $\operatorname{Bir} F_\eta = \operatorname{Aut} F_\eta$ is finite, for $d = 2$ it is generated by the subgroup $\operatorname{Aut} F_\eta$ and the involutions, associated with sections of the fibre space π, for $d = 3$ it is generated by the subgroup $\operatorname{Aut} F_\eta$ and the involutions, associated with sections and bi-sections of the projection π, described in Example 1.4.

The proofs of Theorems 3.1 and 3.2 are given in Chapter 5, that of Theorem 3.3 is given in Chapter 6.

3.3. The linear method: the main results. The linear method is based on the theorem on "inversion of adjunction", proved by Shokurov in dimension three in [**Sh93**] and by Kollár in arbitrary dimension [**Kol93**]. Let us formulate a particular case of that result, which is used in the theory of birational rigidity. A discussion and the proof of inversion of adjunction are given in Chapter 7.

Theorem 3.4 (inversion of adjunction). *Let $x \in X$ be a germ of a $\mathbb{Q}$-factorial terminal variety, D an effective $\mathbb{Q}$-divisor, the support of which contains the point x. Let $R \subset X$ be an irreducible subvariety of codimension one, $R \not\subset$* $\operatorname{Supp} D$, *and, moreover, R is a Cartier divisor. Assume that the pair (X, D) is not canonical at the point x, but canonical outside that point, that is, the point x is an isolated centre of non-canonical singularities of that pair. Then the pair $(R, D_R = D|_R)$ is not log canonical at the point x.*

The inversion of adjunction (formulated above not in the most general way, but in the form in which it will be really needed) is used for excluding maximal singularities of mobile linear systems Σ on a rationally connected variety V under consideration in the following way. Let $R \subset V$ be an irreducible reduced Cartier divisor (as a rule, the variety V is smooth or has elementary singularities, so that the assumptions of Theorem 3.4 are satisfied automatically), $D \in \Sigma$ a general divisor. Since the linear system Σ is mobile, we get $R \not\subset \operatorname{Supp} D$, so that the restriction D_R is well defined. Now, if the pair $(R, \frac{1}{n}D_R)$ is log canonical at the point $x \in R$, then the pair $(V, \frac{1}{n}\Sigma)$ is canonical at the point x, that is, there are no maximal singularities of the system Σ, the centre of which is the point x. The procedure just described reduces studying birational geometry of the variety V to studying geometry of the divisor R, which sometimes essentially simplifies the work. Note that this procedure can be repeated (provided that the assumptions of Theorem 3.4 hold), reducing investigation of the singularities of the pair $(V, \frac{1}{n}\Sigma)$ to studying singularities of the pair $(R, \frac{1}{n}D_R)$, where $R \subset V$ is an irreducible subvariety (not necessarily a divisor). The most important case is restricting onto a fibre $R = \pi^{-1}(s)$ of a fibration $\pi\colon V \to S$.

There are three groups of results that make essential use of the linear method.

Fano direct products. Theorems on birational geometry of Fano direct products form the largest (up to now) group of results proven by the linear method [**Pukh05, Pukh06b, Pukh08a, Ch08, Pukh09b**].

Let F be a Fano variety of dimension ≥ 3 with $\mathbb{Q}$-factorial terminal singularities and the Picard number $\rho(F)=1$.

Definition 3.4. We say that the variety F satisfies the condition of *divisorial canonicity*, or the condition (C) (respectively, the condition of *divisorial log canonicity*, or the condition (L)), if for any effective divisor $D\in|-nK_F|$, $n\geq 1$, the pair

$$(F,\frac{1}{n}D) \tag{90}$$

has canonical (respectively, log canonical) singularities. If the pair (90) has canonical singularities for a general divisor $D\in\Sigma\subset|-nK_F|$ of any *mobile* linear system, then we say that F satisfies the condition of *mobile canonicity*, or the condition (M).

Explicitly, the condition (C) is formulated in the following way: for any birational morphism $\varphi\colon\widetilde{F}\to F$ and any exceptional divisor $E\subset\widetilde{F}$ the following inequality holds:

$$\nu_E(D)\leq na(E). \tag{91}$$

The inequality (91) is opposite to the Noether-Fano inequality. The condition (L) is weaker: it is required that the inequality

$$\nu_E(D)\leq n(a(E)+1) \tag{92}$$

holds. Recall that in (91) and (92) the number $a(E)$, that is, the discrepancy of an exceptional divisor $E\subset\widetilde{F}$ with respect to the model F. The inequality (92) is opposite to the *log Noether-Fano inequality*. The condition (M) means that (91) takes place for a general divisor D of any mobile system $\Sigma\subset|-nK_F|$ and any discrete valuation ν_E.

In another terminology, the condition (L) means that the *the global log canonical threshold* of the variety F is equal to one: $\mathrm{lct}(F)=1$. The condition (C) means that the *global canonical threshold* $\mathrm{ct}(F)=1$. The importance of these conditions is shown by the following fact [**Pukh05**].

Theorem 3.5. *Assume that the primitive Fano varieties $F_1,\dots,F_K$, $K\geq 2$, satisfy the conditions (L) and (M). Then their direct product*

$$V=F_1\times\cdots\times F_K$$

is a birationally superrigid variety, that is, for any mobile linear system Σ its virtual and actual thresholds of canonical adjunction coincide:

$$c(\Sigma)=c_{\mathrm{virt}}(\Sigma).$$

In particular:

(i) *All structures of a rationally connected fibre space on the variety V are projections onto direct factors. More precisely, let $\beta\colon V^\sharp\to S^\sharp$ be a rationally connected fibre space and $\chi\colon V\dashrightarrow V^\sharp$ a birational map. Then there exist a set of* *indices* $I = \{i_1, \ldots, i_k\} \subset \{1, \ldots, K\}$ *and a birational map*

$$\alpha\colon F_I = \prod_{i \in I} F_i \dashrightarrow S^\sharp$$

such that the following diagram commutes:

$$\begin{array}{ccccc} & V & \stackrel{\chi}{\dashrightarrow} & V^\sharp & \\ \pi_I & \downarrow & & \downarrow & \beta \\ & F_I & \stackrel{\alpha}{\dashrightarrow} & S^\sharp, & \end{array}$$

that is, $\beta \circ \chi = \alpha \circ \pi_I$*, where* $\pi_I\colon \prod_{i=1}^K F_i \to \prod_{i \in I} F_i$ *is the natural projection onto a direct factor. In particular, on the variety* V *there are no structures of a fibration into rationally connected varieties of dimension strictly smaller than* $\min\{\dim F_i\}$*. In particular,* V *has no structures of a conic bundle or a fibration into rational surfaces.*

(ii) *The groups of birational and biregular automorphisms of the variety* V *coincide:*

$$\operatorname{Bir} V = \operatorname{Aut} V.$$

(iii) *The variety* V *is non-rational.*

Theorem 3.5 is proven in [**Pukh05**] for smooth primitive Fano varieties, however, the proof is valid word for word in a more general case, described above. Obviously, the condition (C) is stronger than (L) and (M). To apply Theorem 3.5, it is sufficient to check that a Fano variety from a given family satisfies the condition (C) (or the both conditions (L) and (M)).

For generic Fano hypersurfaces $F_d \subset \mathbb{P}^d$, $d \geq 6$, and generic Fano double space of index one $F_2 \to \mathbb{P}^d$, $d \geq 3$, the condition (C) is shown in [**Pukh05**]. For generic Fano double hypersurfaces $F \to Q_m \subset \mathbb{P}^d$, $d \geq 7$, branched over $W = W_{2l}^* \cap Q_m$, where Q_m and W_{2l}^* are generic hypersurfaces of degrees m and $2l$, respectively, $m + l = d$, the condition (C) is shown in [**Pukh08a**]. For generic weighted Fano hypersurfaces of dimension three the condition (L) is checked in [**Ch08**].

The proof of Theorem 3.5 and of the divisorial canonicity of Fano hypersurfaces $F_d \subset \mathbb{P}^d$ is given in Chapter 7.

Fano fibrations over $\mathbb{P}^1$. Application of the linear method simplifies the proof of birational rigidity of fibre spaces $V/\mathbb{P}^1$, see [**Pukh08a, Pukh09b**]. Assume that $\Sigma \subset |-nK_V + lF|$ is a mobile linear system, where $l \in \mathbb{Z}_+$, and a fibre $F^* = \pi^{-1}(t_*)$ of the projection $\pi\colon V \to \mathbb{P}^1$ satisfies the condition (L). Then by the inversion of adjunction, the centre of any maximal singularity of the system Σ cannot be contained in the fibre F^*: it either covers the base $\mathbb{P}^1$, or is contained in another fibre. In particular, if all fibres satisfy the conditions (L) and (M), then the system Σ has no maximal singularities at all, which implies that

$$c_{\text{virt}}(\Sigma) = c(\Sigma, V) = n.$$

Arguments of that type can be used for exclusion of certain particular types of maximal singularities as well [**Pukh06a**].

Fano varieties of index two. This result formally comes out of the framework of the present book as it deals with *non-rigid* varieties (in fact, it is the first full fledged example of a complete description of birational geometry of non-rigid varieties of arbitrary dimension). However, it stands next to the previous results and makes essential use of the linear method. For that reason, we formulate it below and give a proof in Chapter 8.

Let $M \geq 5$ and $W = W_{2(M-1)} \subset \mathbb{P}^M$ be a smooth hypersurface of degree $2(M-1)$. Consider the double cover

$$\sigma\colon V \to \mathbb{P}^M,$$

branched over W. The variety V is a Fano variety of index two: $\operatorname{Pic} V = \mathbb{Z}H$, where H is the ample generator, $K_V = -2H$, the class H is the σ-pullback of the hyperplane in $\mathbb{P}$. On the variety V there are the following natural structures of a rationally connected fibre space: let $\alpha_P\colon \mathbb{P} \dashrightarrow \mathbb{P}^1$ be the linear projection from an arbitrary linear subspace P of codimension two, then the map

$$\pi_P = \alpha_P \circ \sigma\colon V \dashrightarrow \mathbb{P}^1$$

fibres V into $(M-1)$-dimensional Fano varieties of index 1. Assume that the variety V is sufficiently general.

Theorem 3.6. *Let $M \geq 5$ and $\chi\colon V \dashrightarrow Y$ be a birational map onto the total space of a rationally connected fibre space $\lambda\colon Y \to S$. Then $S = \mathbb{P}^1$ and for some isomorphism $\beta\colon \mathbb{P}^1 \to S$ and some subspace $P \subset \mathbb{P}$ of codimension two we have $\lambda \circ \chi = \beta \circ \pi_P$, that is, the following diagram commutes:*

$$\begin{array}{ccccc} & V & \stackrel{\chi}{\dashrightarrow} & Y & \\ \pi_P & \downarrow & & \downarrow & \lambda \\ & \mathbb{P}^1 & \stackrel{\beta}{\rightarrow} & S. & \end{array}$$

Corollary 3.1. (i) *On the variety V there are no structures of a rationally connected fibre space with the base of dimension ≥ 2. In particular, on V there are no structures of a conic bundle and del Pezzo fibration, and the variety V itself is non-rational.*

(ii) *Assume that there is a birational map $\chi\colon V \dashrightarrow Y$, where Y is a Fano variety of index $r \geq 2$ with factorial terminal singularities, such that* $\operatorname{Pic} Y = \mathbb{Z}H_Y$*, where $K_Y = -rH_Y$, where the linear system $|H_Y|$ is non-empty and free. Then $r = 2$ and the map χ is a biregular isomorphism.*

(iii) *The groups of birational and biregular automorphisms of the variety V coincide:* $\operatorname{Bir} V = \operatorname{Aut} V = \mathbb{Z}/2\mathbb{Z}$.

Proof of the corollary. The claim (i) and the equality $r = 2$ in (ii) are obvious (any linear subsystem of the projective dimension $\leq r-1$ in the complete linear system $|H_Y|$ defines a structure of a rationally connected fibre space on Y). Furthermore, the χ-pullback of a general divisor in the system $|H_Y|$ is a divisor in the linear system $|H|$ by Theorem 3.6, which completes the proof of the claim (ii). Part (iii) obviously follows from (ii). The proof is complete.

Notes and references

Given that this chapter is an introduction to the subsequently developed theory of birational rigidity of Fano fibre spaces over a non-trivial base, and that all references were given in the body of the chapter, there is very little we can add here.

We are not touching the complicated story of the first steps of the Sarkisov program; time does not seem to be ripe for that task as yet. Besides, with the recent progress of the minimal model program, some of those complications probably are not as acute as they were in the nineties. However, it is worth mentioning that we probably do not have an access to all work that was done in that area: some work may be unpublished or exist in the form of preprints not accessible to the community. One of the strongest motivations for the Sarkisov program was the task to find a proof for the hypothetical rationality criterion for three-dimensional conic bundles (see [**I87, I91b, I96a**]), which is widely believed to be true and which is one of the fundamental conjectures in three-dimensional birational geometry, similar to the classical Castelnuovo rationality criterion for surfaces. To a considerable extent the work [**MoPr08a, MoPr08b, MoPr09**] was motivated by Iskovskikh's attempts to find a proof of the rationality criterion.

The K^2-condition was first introduced in [**Pukh98a**]. The origins of the K-condition are difficult to trace, as it was implicitly used in earlier works on birational geometry of fibre spaces. It is clearly of decisive importance as it makes the starting point of the study, determining the value of the threshold of canonical adjunction.

CHAPTER 5

Fano Fibre Spaces Over $\mathbb{P}^1$

Summary

In this chapter we develop the techniques of the method of maximal singularities for the standard Fano fibre space over $\mathbb{P}^1$. As in the absolute case of a primitive Fano variety, in the relative case, in order to prove birational (super)rigidity, one needs to exclude (almost all) maximal singularities of a mobile linear system, and the infinitely near case presents, by far, the most difficult problem (more precisely, the case when the centre of the maximal singularity is a point). We use the same general approach as in Chapters 2 and 3: consider the self-intersection Z of the mobile system and apply the technique of counting multiplicities to show that Z is "too singular" to be an effective cycle of codimension two.

There are several issues that make the relative case considerably different from the absolute one, from the viewpoint of the techniques employed in the exclusion of maximal singularities. Namely,

- one should see the difference between *vertical* subvarieties (contained in the fibres) and *horizontal* ones (covering the base $\mathbb{P}^1$), as their singularities are treated in different ways,
- for the purpose of proving birational (super)rigidity of a fibre space the concept of a maximal singularity is not a sufficient one: it can be further refined in the sense that there is a special maximal singularity with stronger properties than just the usual Noether-Fano inequality (we call it a *supermaximal* singularity),
- when resolving a maximal singularity over a point o on V, one needs to use the information, whether the centres of the blow ups are contained in the strict transform of the fibre $F \ni o$ or not,
- treatment of the singular points of the fibres is considerably different (and more complicated) than that of non-singular points.

In Section 1 the first version of the techniques is developed, including the basic definitions and a sufficient condition of birational superrigidity, and some easier classes of Fano fibre spaces are proved to be birationally superrigid as an example.

Section 2 is devoted to fibrations into Fano hypersurfaces of index one. The fibres are familiar objects (studied in Chapter 3), but the new problems come from the singular points of the fibres and the additional regularity conditions. Section 3 does the same work for fibrations into Fano double hypersurfaces of index one: there the main new difficulty is in dealing with the singular points of the fibres coming from the singularities of the branch divisor, and a new technique is developed for them.

In Section 4 the second version of the technique of excluding the infinitely near case is developed, which is stronger and more efficient than the first one (for instance, it makes it possible to treat the fibre spaces with a considerable deviation from the K^2-condition), but requires more restrictive regularity conditions — that is why we present both versions of the techniques of excluding maximal singularities in this chapter.

Finally, in Section 5 we prove birational rigidity of fibrations into Fano complete intersections and cyclic covers (Theorems 4.3.1 and 4.3.2).

1. Sufficient conditions of birational rigidity. I

In this section, we develop the first version of the technique of the method of maximal singularities for Fano fibre spaces over $\mathbb{P}^1$. We start by introducing the important classes of horizontal and vertical subvarieties and state the sufficient conditions of birational superrigidity, Theorem 1.1 (Subsection 1.1). As a first application, we prove birational superrigidity of fibrations into Fano double spaces (Subsection 1.2) and three-dimensional quartics (Subsection 1.3), satisfying the K^2-condition. After that we introduce a refinement of the concept of a maximal singularity, defining a supermaximal singularity, and prove Theorem 1.1 (Subsections 1.4–1.6). Finally, we show that the K^2-condition can be slightly relaxed without any damage to birational superrigidity of the fibre space (Subsection 1.7).

1.1. Horizontal and vertical subvarieties. Let $\pi\colon V \to \mathbb{P}^1$ be a standard Fano fibre space, that is, V is a smooth projective variety,

$$\operatorname{Pic} V = \mathbb{Z}K_V \oplus \mathbb{Z}F,$$

where $F \in \operatorname{Pic} V$ is the class of a fibre and $(-K_V)$ is π-ample. We say that an irreducible subvariety $Y \subset V$ is *horizontal*, if $\pi(Y) = \mathbb{P}^1$, and *vertical*, if $\pi(Y) = t \in \mathbb{P}^1$ is a point. We define the *degree* of a horizontal subvariety $Y \subset V$ by the formula

$$\deg Y = (Y \cdot F \cdot (-K_V)^{\dim Y - 1}),$$

and the degree of a vertical subvariety $Y \subset \pi^{-1}(t)$ by the formula

$$\deg Y = (Y \cdot (-K_V)^{\dim Y}).$$

By this definition the degree of the variety V itself coincides with the degree of a fibre, $\deg V = \deg F$.

Assume in addition that the condition

$$A^2 V = \mathbb{Z}K_V^2 \oplus \mathbb{Z}H_F, \tag{93}$$

holds, where $H_F = (-K_V \cdot F)$ is the ample anticanonical section of the fibre, and that every fibre $F = F_t = \pi^{-1}(t)$, $t \in \mathbb{P}^1$, is a Fano variety with at most non-degenerate quadratic singularities and, moreover,

$$A^1 F = \operatorname{Pic} F = \mathbb{Z}K_F,$$

where $K_F = -H_F$ and H_F is considered as an element of the group $A^1 F$. The exception is made by fibrations into del Pezzo surfaces where the condition (93) should be modified. Fibrations into del Pezzo surfaces, as a special case, are considered in Chapter 6. In this chapter we assume that $\dim V \geq 4$.

When $\dim V \geq 6$, that is, when the fibres are at least five-dimensional, without loss of generality, we may assume that

$$A^2 F = \mathbb{Z} H_F^2,$$

where again H_F is considered as an element of $A^1 F$. All these conditions are satisfied for almost all families of standard Fano fibre spaces that are by now successfully studied. On the other hand, when the fibres are of dimension 3 and 4, the low dimension makes it possible to employ additional arguments.

Now let us formulate additional conditions, from which birational (super)rigidity of Fano fibre spaces is derived. These conditions should be understood as some conditions of general position for the fibre space $V/\mathbb{P}^1$; more precisely, generic (in their family) fibre spaces $V/\mathbb{P}^1$ satisfy these conditions.

Definition 1.1. The fibre space $V/\mathbb{P}^1$ satisfies

- the condition (v), if for every irreducible vertical subvariety Y of codimension two (that is, $Y \subset F_t$ is a prime divisor, $t = \pi(Y)$) and for every smooth point $o \in F_t$ the inequality

(94)
$$\frac{\operatorname{mult}_o Y}{\deg Y} \leq \frac{2}{\deg V}$$

  holds;
- the condition (vs), if for any vertical subvariety $Y \subset F_t$ of codimension 2 (with respect to V, that is, for a prime divisor on F_t), a singular point $o \in F_t$ and an infinitely near point $x \in \widetilde{F}_t$, where $\varphi \colon \widetilde{F}_t \to F_t$ is a blow up of the point o, $\varphi(x) = o$, $\widetilde{Y} \subset \widetilde{F}_t$ the strict transform of the subvariety Y on $\widetilde{F}_t$, the following estimates hold:
$$\frac{\operatorname{mult}_o Y}{\deg Y} \leq \frac{4}{\deg V}, \quad \frac{\operatorname{mult}_x \widetilde{Y}}{\deg Y} \leq \frac{2}{\deg V};$$
- the condition (h), if for any horizontal subvariety Y of codimension 2 and a point $o \in Y$ the following estimate holds:
$$\frac{\operatorname{mult}_o Y}{\deg Y} \leq \frac{4}{\deg V}.$$

Recall that the strong K^2-condition for the fibre space $V/\mathbb{P}^1$ means that $K_V^2 \notin \operatorname{Int} A_+^2 V \subset \mathbb{R}^2$, which is equivalent to the following property: for any $a \geq 1$ and $b \geq 1$ the class

$$aK_V^2 - bH_F$$

is not effective.

Now let us formulate the first sufficient condition of birational rigidity.

Theorem 1.1. *Assume that the smooth standard Fano fibre space $V/\mathbb{P}^1$ satisfies the strong K^2-condition and the conditions (v), (vs) and (h). Then $V/\mathbb{P}^1$ is a birationally superrigid Fano fibre space.*

1.2. Fibrations into double spaces. As a first and easy application of Theorem 1.1, let us consider fibrations into Fano double spaces of index one. Their construction is as follows. Consider the projective bundle $X = \mathbb{P}(\mathcal{E})$, where $\mathcal{E}$ is a locally free sheaf of rank $M+1$ on $\mathbb{P}^1$,

$$\mathcal{E} = \mathcal{O}_{\mathbb{P}^1} \oplus \bigoplus_{i=1}^{M} \mathcal{O}_{\mathbb{P}^1}(a_i),$$

where $0 \leq a_1 \leq \cdots \leq a_M$, $M \geq 3$, L_X is the class of the tautological sheaf and G is the class of a fibre $\cong \mathbb{P}^M$ in $\operatorname{Pic} X = \mathbb{Z}L_X \oplus \mathbb{Z}G$. Let

$$\sigma \colon V \to X$$

be the double cover branched over a smooth divisor $D \sim 2ML_X + 2mG$. Denoting the natural projection onto $\mathbb{P}^1$ by π, we obtain a presentation of the smooth projective variety V as a pencil $\pi \colon V \to \mathbb{P}^1$ of Fano varieties of degree 2. Obviously, $\operatorname{Pic} V = \mathbb{Z}K_V \oplus \mathbb{Z}F$, where $F = \sigma^*(G)$ is the class of a fibre and

$$K_V = \sigma^* \left(-L_X + (a_1 + \cdots + a_M - 2 + m)F\right),$$

so that

$$\pi_* \left(\mathcal{O}_V(-K_V)\right) \cong \mathcal{E} \otimes \mathcal{O}_{\mathbb{P}^1} \left(-\sum_{i=1}^{M} a_i + 2 - m \right).$$

The strong K^2-condition is satisfied, if

$$\left(K_V^2 \cdot \sigma^*(L)^{M-1}\right) \leq 0,$$

because the tautological linear system $|L_X|$ on X is base point free. Therefore, the strong K^2-condition is satisfied, if

$$m \geq 2 - \frac{1}{2} \sum_{i=1}^{k} a_i. \tag{95}$$

Assume that this inequality is true. Now the conditions (v), (vs) and (h) are satisfied in a trivial way, since $\deg V = \deg F = 2$.

Therefore, we have proved the following:

Theorem 1.2. *If the inequality* (95) *is satisfied, then the standard Fano fibre space $V/\mathbb{P}^1$ is birationally superrigid.*

1.3. Pencils of quartic three-folds. As another fairly elementary application of Theorem 1.1, let us consider fibrations into three-dimensional quartics. Their construction is as follows. Let

$$\mathcal{E} = \bigoplus_{i=0}^{4} \mathcal{O}_{\mathbb{P}^1}(a_i),$$

$0 = a_0 \leq a_1 \leq \cdots \leq a_4$, be a locally free sheaf of rank 5 on $\mathbb{P}^1$, $X = \mathbb{P}(\mathcal{E})$ its projectivization, $\pi \colon X \to \mathbb{P}^1$ the corresponding projective bundle (a fibration into $\mathbb{P}^4$). Let L_X be the class of the tautological sheaf, G the class of a fibre in $\operatorname{Pic} X = \mathbb{Z}L_X \oplus \mathbb{Z}G$.

Let $V \subset X$ be a smooth hypersurface in the linear system $|4L+mG|$, such that

$$\operatorname{Pic} V = \mathbb{Z}L \oplus \mathbb{Z}F,$$

where $L = L_X|_V$ and $F = G|_V$. This property is satisfied automatically if $4L+mG$ is an ample class. Assume that V satisfies the following conditions of general position:

(a) if $x \in F_t = \pi^{-1}(t)$ is a non-singular point of the fibre F_t, which is a three-dimensional quartic in $G_t \cong \mathbb{P}^4$, then the intersection $F_t \cap T_x F_t$ is a surface of degree 4 in $T_x F_t \cong \mathbb{P}^3$, the multiplicity of which at the point x is precisely 2;

(b) if $x \in F_t$ is a singular point of the fibre F_t, then it is a non-degenerate quadratic singularity, and F_t is a factorial three-dimensional variety, that is, every Weil divisor ia a Cartier divisor ($\operatorname{Pic} F_t = \mathbb{Z}H_F$, H_F is the class of a hyperplane section).

The conditions (a) and (b) are justified by the fact that in the space of quartics in $\mathbb{P}^4$ the subset of quartics W that have a non-singular point $x \in W$ such that

$$\operatorname{mult}_x(W \bigcap T_x W) \geq 3,$$

is of codimension 3, whereas a generic singular quartic satifies the condition (b).

Proposition 1.1. *Assuming the conditions* (a) *and* (b), *let* $Z \subset F_t$ *be an irreducible vertical surface,* $x \in Z$ *a point. Then*

(i) *if* $x \in F_t$ *is a non-singular point of the fibre, then*

$$\operatorname{mult}_x Z \leq \frac{1}{2} \deg Z;$$

(ii) *if* $x \in F_t$ *is a double point of the fibre, then*

$$\operatorname{mult}_x Z \leq \frac{2}{3} \deg Z,$$

provided that Z *is not the intersection* $F_t \cap T_x F_t$ *of the variety* F_t *with its tangent cone at the point* x. *Moreover, let* $\sigma\colon \widetilde{F}_t \to F_t$ *be the blow up of the point* x, $E = \sigma^{-1}(x) \cong \mathbb{P}^1 \times \mathbb{P}^1$ *the exceptional surface, and* $y \in E$ *an arbitrary point. Then, if* y *does not lie on the strict transform of a line on* V_t, *passing through the point* x, *then*

$$\operatorname{mult}_x Z + \operatorname{mult}_y \widetilde{Z} \leq \deg Z,$$

where $\widetilde{Z}$ *is the strict transform of the surface* Z *on* $\widetilde{V}_t$. *If the point* y *lies on the strict transform of a line on* V_t, *then the inequality*

$$3 \deg Z \geq 2 \operatorname{mult}_x Z + 4 \operatorname{mult}_y \widetilde{Z}$$

holds. In particular, in both cases,

$$\operatorname{mult}_y \widetilde{Z} \leq \frac{1}{2} \deg Z.$$

Proof. Let us show (i). Assume the converse: $2 \operatorname{mult}_x Z > \deg Z$. Consider the section of Z by a generic hyperplane in $\mathbb{P}^4$, containing the point x, a curve C. Obviously, $2 \operatorname{mult}_x C > \deg C$ (where the degree is understood in the sense of the projective space $\mathbb{P}^4$). Let $H = T_x F_t \cap F_t$ be the section of V_t by the tangent hyperplane. We get

$$\deg C = (C \cdot H) \geq (C \cdot H)_x \geq 2 \operatorname{mult}_x C,$$

unless C is contained in $T_x F_t$. Therefore, $Z \subset T_x F_t$ and for this reason $Z = T_x F_t \cap F_t$. Now we get a contradiction with condition (a).

To prove the first claim of part (ii), we use in a similar way the quadratic tangent cone at the singular point $x \in V_t$ instead of the tangent hyperplane as above.

Finally, let us prove the last claim. Let $R \subset \mathbb{P}^4$ be the (unique) line passing through the point x in the direction y. Consider first the case $R \not\subset F_t$. A general plane $P \supset R$ intersects the quartic F_t along a plane quartic C_P with the double point x, and its strict transform $\widetilde{C}_P$ contains the point y. If $dH_F - \nu E$ is the class of the surface $\widetilde{Z}$ in $\operatorname{Pic}\widetilde{F}_t$, then $\operatorname{mult}_x Z = 2\nu$, $\deg Z = 4d$ and

$$4d - 2\nu = \deg Z - \operatorname{mult}_x Z = (\widetilde{Z} \cdot \widetilde{C}_P)$$
$$\geq (\widetilde{Z} \cdot \widetilde{C}_P)_y \geq \operatorname{mult}_y \widetilde{Z}$$

which is what we need.

If $R \subset F_t$ is one of the 24 lines, consider the hyperplane $H \subset \mathbb{P}^4$, which is tangent to the quadratic cone $T_x F_t$ along the line R. If $P \subset H$, $P \supset R$ is a generic plane, then $P \cap F_t = R \cup C_P$, where C_P is the residual plane cubic, containing the point x. Moreover, its strict transform $\widetilde{C}_P$ contains the point y. If $C_P \not\subset Z$, then

$$3d - \nu = \frac{3}{4}\deg Z - \frac{1}{2}\operatorname{mult}_x Z = (\widetilde{Z} \cdot \widetilde{C}_P)$$
$$\geq (\widetilde{Z} \cdot \widetilde{C}_P)_y \geq \operatorname{mult}_y \widetilde{Z},$$

which is what we need. If $C_P \subset Z$, then $Z = H \cap F_t$, because the curves C_P sweep out $H \cap F_t$. However,

$$\operatorname{mult}_x(H \cap V_t) = 2 \leq \frac{1}{2}\deg(H \cap V_t).$$

This contradiction completes the proof of the proposition. Q.E.D.

Now, in order to get the strong K^2-condition, it is sufficient to require that

$$\left(K_V^2 \cdot L^2\right) \leq 0,$$

as the linear system $|L|$ is base point free. Therefore, the strong K^2-condition is satisfied, if the inequality

$$m \geq \frac{1}{7}\left(16 - 4\sum_{i=1}^{4} a_i\right). \tag{96}$$

holds. Assuming this, we note that the conditions (v), (vs) and (h) are satisfied either for the trivial reason that $\deg V = 4$ or due to Proposition 1.1 above. Therefore, the following theorem is proven.

Theorem 1.3. *In the assumptions of general position* (a), (b) *and of the inequality* (96), *the Fano fibre space $V/\mathbb{P}^1$ is birationally superrigid.*

The two examples considered above (fibrations into double spaces of index one and three-dimensional quartics) are so easy because the anticanonical degree of the fibre is low in both cases. For typical Fano fibre spaces over $\mathbb{P}^1$, it is much more difficult to establish the conditions (v), (vs) and (h). The main examples will be considered in Sections 2–5.

1.4. Start of the proof. In the remaining part of this section we prove Theorem 1.1. The key ingredient of the proof, apart from the standard technique of counting multiplicities, which was developed in Chapter 2, is the new concept of a *supermaximal singularity* of a mobile linear system on $V/\mathbb{P}^1$. We prove the existence of a supermaximal singularity, satisfying a stronger inequality than the standard Noether-Fano one, and obtain the required equality of the virtual and actual thresholds of canonical adjunction.

Let $\Sigma \subset |-nK_V + lF|$ be a mobile linear system. Since the strong K^2-condition holds, we get $l \in \mathbb{Z}_+$ and $c(\Sigma) = n$, whereas $n = 0$ if and only if the linear system Σ is composed from the pencil $|F|$, that is, it is pulled back from $\mathbb{P}^1$.

Assume that the inequality

$$c_{\mathrm{virt}}(\Sigma) < c(\Sigma) = n,$$

holds, that is, there is a model $V^\sharp$ of the variety V such that

$$c(\Sigma^\sharp, V^\sharp) < c(\Sigma, V),$$

where $\Sigma^\sharp$ is the strict transform of the system Σ on $V^\sharp$. By Proposition 2.1.2, the linear system Σ has a maximal singularity, that is, there exists a prime divisor $E \subset V^\sharp$, satisfying the Noether-Fano inequality

$$\nu_E(\Sigma) > n \cdot a(V, E). \tag{97}$$

Let $B = \mathrm{centre}_V(\nu_E) \subset V$ be the centre of the maximal singularity E on V. It is an irreducible subvariety of codimension at least two.

We use the standard notations for the resolution of the discrete valuation ν_E (see Chapter 2, formula (11)):

$$\begin{array}{rccc} \varphi_{i,i-1}\colon & V_i & \to & V_{i-1} \\ & \cup & & \cup \\ & E_i & \to & B_{i-1} \end{array} \tag{98}$$

is the sequence of blow ups with irreducible centres $B_{i-1} \subset V_{i-1}$, which is determined in a unique way by the following conditions:

(1) $V_0 = V$, $B_0 = B$, $i = 1, \ldots, K$;

(2) $B_j = \mathrm{centre}_{V_j}(\nu_E) \subset V_j$, $E_{j+1} = \varphi_{j+1,j}^{-1}(B_j)$;

(3) the valuation ν_{E_K} coincides with ν_E.

In other words, the birational map

$$V_K \dashrightarrow V^\sharp$$

is biregular at the generic point of the divisor E_K and transforms E_K into E. The symbol Σ^j below means the strict transform of the linear system Σ on V_j. As usual, we set

$$\nu_j = \mathrm{mult}_{B_{j-1}} \Sigma^{j-1}, \quad \delta_j = \mathrm{codim}\, B_{j-1} - 1$$

and define the structure of an oriented graph on the set of exceptional divisors $\{E_1, \ldots, E_K\}$, so that the Noether-Fano inequality takes the traditional form

$$\sum_{i=1}^{K} p_i \nu_i > n \sum_{i=1}^{K} p_i \delta_i.$$

Proposition 1.2. *The centre B of the maximal singularity ν_E on V is contained in some fibre $\pi^{-1}(t) = F_t$, $t \in \mathbb{P}^1$.*

Proof. Assume the converse, $\pi(B) = \mathbb{P}^1$. Let $F \subset V$ be a fibre of general position. It is easy to see that the restriction $\Sigma_F = \Sigma|_F$ of the linear system Σ onto F is a mobile linear system $\Sigma_F \subset |nH_F|$ with a maximal singularity $E_F = E|_F$. The simplest way to define this singularity is to restrict the sequence of blow ups (98) onto the fibre F and note that the discrepancy remains the same:

$$a(E|_F, F) = a(E, V);$$

precisely for the reason that B covers the base. The centre of the valuation E_F is $B_F = B \cap F$ (we assume both to be irreducible, otherwise an irreducible component of E_F and the corresponding component of B_F should be considered).

Now we have the following alternative: either $\operatorname{codim} B \geq 3$ or $\operatorname{codim} B = 2$. In the first case, $\operatorname{codim}_F B_F = \operatorname{codim} B \geq 3$, so that the $4n^2$-inequality for the generic fibre gives

$$\operatorname{mult}_B Z > 4n^2,$$

where $Z = (D_1 \circ D_2)$ is the self-intersection of the linear system Σ, $D_i \in \Sigma$ are generic divisors. Decompose Z into the vertical and horizontal parts:

$$Z = Z^v + Z^h,$$

collecting the vertical and horizontal components, respectively. Since B is a horizontal subvariety,

$$\operatorname{mult}_B Z^h > 4n^2,$$

however, $\deg Z^v = n^2 \deg V$ and thus we get a contradiction with the assumption (h).

Therefore, $\operatorname{codim}_F B_F = \operatorname{codim} B = 2$. Since $\operatorname{mult}_B \Sigma > n$, we conclude that B comes into the effective cycle Z^h with the multiplicity strictly higher than n^2. However,

$$Z^h = n^2 K_V^2 + \alpha H_F$$

with $\alpha \geq 0$ and $B = aK_V^2 + \beta H_F$ for some $a \geq 1$, $\beta \geq 0$, again since B is a horizontal subvariety. We conclude that the case $\operatorname{codim} B = 2$ is impossible, either. Q.E.D. for the proposition.

1.5. The supermaximal singularity. In the notations of the previous section, let $\mathcal{M} = \{T_1, \dots, T_k\}$ be the set of all prime divisors on $V^\sharp$, satisfying the Noether-Fano inequality. As we have proved, the centre $B_E = \operatorname{centre}(\nu_E)$ of each maximal singularity $E \in \mathcal{M}$ is contained in some fibre F_t. The set $\mathcal{M}$ is finite (since the model $V^\sharp$ is fixed), so that there is at most finite set of points $t \in \mathbb{P}^1$, the fibres F_t over which contain the centres of maximal singularities. Set $\mathcal{M}_t = \{E \in \mathcal{M} | B_E \subset F_t\}$,

$$e(E) = \nu_E(\Sigma) - na(E, V) > 0$$

for $E \in \mathcal{M}$. Recall that $\Sigma \subset |-nK_V + lF|$, $l \in \mathbb{Z}_+$.

Proposition 1.3. *The following inequality holds:*

$$\sum_{t \in \mathbb{P}^1} \max_{\{E \in \mathcal{M}_t\}} \frac{e(E)}{\nu_E(F_t)} > l. \tag{99}$$

Proof. Let $D^\sharp \in \Sigma^\sharp$ be a general divisor, that is, the strict transform on $V^\sharp$ of a divisor $D \in \Sigma$ of general position. By assumption, the linear system

$$|D^\sharp + nK_{V^\sharp}|$$

is empty. Therefore the linear system

$$|lF - \sum_{E \in \mathcal{M}} e(E)E|$$

is empty, too. On the other hand, by construction for $E \in \mathcal{M}_t$ the divisor

$$F_t - \nu_E(F_t)E$$

is effective, so that the divisor

$$\sum_{t\in\mathbb{P}_1}[\left(\max_{\{E\in\mathcal{M}_t\}}\frac{e(E)}{\nu_E(F_T)}\right)F_t-\sum_{E\in\mathcal{M}_t}e(E)E]$$

is also effective. This immediately implies the inequality (99). Q.E.D. for the proposition.

Now let $E\in\mathcal{M}_t\subset\mathcal{M}$ be a maximal singularity, and $B\subset F_t$ its centre.

Lemma 1.1. *The following estimate holds:*

$$\operatorname{codim}_{F_t}B\geq 2.$$

Proof. Assume the converse: $B\subset F=F_t$ is a prime divisor. Let $D\in\Sigma$ be a general divisor, D_F its restriction onto F. By the Noether-Fano inequality $\operatorname{mult}_B D>n$, so that

$$D_F=\alpha B+D^\sharp,$$

where $\alpha>n$ and $D^\sharp$ is an effective divisor on F. However,

$$D_F\sim nH_F,$$

which gives an immediate contradiction. Q.E.D. for the lemma.

Now let $D_i\in\Sigma$, $i=1,2$, be generic divisors, and consider the decomposition

$$Z=(D_1\circ D_2)=Z^v+Z^h$$

of the self-intersection of the system Σ into the vertical Z^v and horizontal Z^h parts. For the cycle Z^v we have a further decomposition

$$Z^v=\sum_{t\in\mathbb{P}^1}Z^v_t,\quad \operatorname{Supp}Z^v_t\subset F_t.$$

Recall that the fibre space $V/\mathbb{P}^1$ satisfies the K^2-condition: $K_V^2\notin\operatorname{Int}A^2_+V$. Then for some $\alpha\in\mathbb{Z}_+$ the following relation holds:

$$Z^h\sim n^2K_V^2+\alpha H_F,$$

so that

$$\deg Z^v=\sum_{t\in\mathbb{P}^1}\deg Z^v_t\leq(2n\deg V)l. \tag{100}$$

Proposition 1.4. *For some point $t\in\mathbb{P}^1$ there is a maximal singularity $E\in\mathcal{M}_t$, satisfying the estimate*

$$e(E)>\frac{\nu_E(F_t)}{2n\deg V}\deg Z^v_t. \tag{101}$$

Definition 1.2. A maximal singularity E, satisfying the inequality (101), is called a *supermaximal singularity.*

1.6. The technique of counting multiplicities. Fix a supermaximal singularity $E\in\mathcal{M}_t\subset\mathcal{M}$. As E and t are fixed throughout this subsection, until the end of the section we write F, Z^v, e, B instead of F_t, Z^v_t, $e(E)$, $B_E=\operatorname{centre}(E)$, respectively, omitting the reference to the point t. This simplifies our notations. A similar agreement will be used in Chapter 6.

Let us consider the sequence of blow ups (98) associated with the discrete valuation E. We use the notations of Subsections 1.1 and 1.4. The strict transforms

$(Z^h)^j$, $(Z^v)^j$ of the cycles Z^h, Z^v on V_j are well defined. By the symbol F^j we denote the strict transform of the fibre F on V_j. Set

$$N = \max\{i \mid B_{i-1} \subset F^{i-1}\}.$$

As $\varphi_{i,i-1}(B_i) = B_{i-1}$ for any $i = 1, \ldots, K-1$, the codimensions $\operatorname{codim} B_i$ do not increase. Set

$$L = \max\{i \mid \operatorname{codim} B_{i-1} \geq 3\} \leq K.$$

We also use the following notations: for $i \in \{1, \ldots, L\}$,

$$m_i^h = \operatorname{mult}_{B_{i-1}} (Z^h)^{i-1}, \quad m_i^v = \operatorname{mult}_{B_{i-1}} (Z^v)^{i-1},$$

where $m_i^h \leq m_{i-1}^h$ and $m_i^v \leq m_{i-1}^v$ for $i = 2, \ldots, L$. Set also

$$\mu_i = \operatorname{mult}_{B_{i-1}} F^{i-1}.$$

Note that

$$\mu_i = 0 \quad \text{and} \quad m_i^v = 0$$

for $i \geq N+1$ (if $N < L$), as for those values of i neither the strict transform F^{i-1} of the fibre nor (the more so) the effective cycle $(Z^v)^{i-1}$ contain the centre B_{i-1} of the blow up. Using the symbol p_i, as above, for the number of paths in the graph Γ of the resolution (98) from the vertex $E = E_K$ to E_i, we get

Proposition 1.5. *The following inequality holds:*

$$\sum_{i=1}^{L} p_i m_i^h + \sum_{i=1}^{\min\{N,L\}} p_i m_i^v \geq \sum_{i=1}^{K} p_i \nu_i^2 \geq \frac{(n \sum_{i=1}^{K} p_i \delta_i + e)^2}{\sum_{i=1}^{K} p_i}. \tag{102}$$

Proof. This inequality immediately follows from Proposition 2.2.4 by putting $a(i) = p_{Ki} = p_i$ for $i \leq L$, replacing the term

$$a(L) \sum_{i=L+1}^{K} \nu_i^2$$

by the term

$$\sum_{i=L+1}^{K} p_{Ki} \nu_i^2,$$

which is smaller than or equal to the previous one as the numbers $p_{Ki} = p_i$ are non-increasing, applying the Noether-Fano inequality in the form

$$\sum_{i=1}^{K} p_i \nu_i = n \sum_{i=1}^{K} p_i \delta_i + e, \tag{103}$$

$e > 0$, computing the minimum of the quadratic form

$$\sum_{i=1}^{K} p_i \nu_i^2$$

on the hyperplane (103) and, finally, separating in the left-hand side of the inequality the multiplicities of the horizontal and vertical parts. Q.E.D.

Let us continue the proof of Theorem 1.1. Set

$$\Sigma_l = \sum_{i=1}^{L} p_i, \quad \Sigma_u = \sum_{i=L+1}^{K} p_i, \quad \Sigma_f = \sum_{i=2}^{\min\{N,L\}} p_i.$$

Note that

$$\nu_E(F) = \sum_{i=1}^{N} p_i\mu_i \leq p_1\mu_1 + \mu_2\Sigma_f$$

by definition of the multiplicities μ_i. Obviously,

$$m_i^h \leq m_h = m_1^h = \operatorname{mult}_B Z^h.$$

Set also

$$d_h = \deg Z^h, \quad d_v = \deg Z_t^v$$

and introduce the coefficients

$$k_h = \frac{m_h}{d_h} \deg V, \quad k_v = \frac{\deg V}{\nu_E(F) d_v} \sum_{i=1}^{\min\{N,L\}} p_i m_i^v.$$

The next step is

Proposition 1.6. *The following estimate holds:*

$$(4 - k_h)\Sigma_l(\Sigma_l + \Sigma_u)n^2 + \Sigma_u^2 n + e^2 + 2(2 - k_v)\Sigma_l ne + 2(1 - k_v)\Sigma_u ne < 0.$$

Proof. In the left-hand side of the inequality (102) replace m_i^h by $m_h = k_h d_h / \deg V = k_h n^2$, which can only make the inequality sharper. In the right-hand side of (102), replace the numbers δ_i for $i \leq L$ by 2, which could also make the inequality only sharper. Finally, multiplying through by

$$\sum_{i=1}^{K} p_i = \Sigma_l + \Sigma_u,$$

and taking into account the definition of the coefficient k_v after easy computations we get

(104)
$$(4 - k_h)\Sigma_l(\Sigma_l + \Sigma_u)n^2 + (n\Sigma_u + e)^2 + 4\Sigma_l en$$
$$-k_v \frac{d_v \nu_E(F)}{\deg V}(\Sigma_l + \Sigma_u) \leq 0.$$

By Proposition 1.4, we can replace $d_v\nu_E(F)$ by $2ne \deg V$, making the inequality (104) a strict one.

We obtain exactly what we claimed.

Q.E.D. for the proposition.

Now we can complete the proof of Theorem 1.1.

Lemma 1.2. *The following estimates hold:* $k_h \leq 4$, $k_v \leq 2$.

Proof. The first inequality follows directly from the definition of the number k_h and the condition (h), which is by assumption satisfied for the fibre space $V/\mathbb{P}^1$.

Let us prove the second inequality. We get

$$\left(\frac{d_v}{\deg V}\right) k_v = \frac{\sum\limits_{i=1}^{\min\{N,L\}} p_i m_i^v}{\sum\limits_{i=1}^{N} p_i \mu_i} \leq \frac{p_1 m_1^v + \Sigma_f m_2^v}{p_1 \mu_1 + \Sigma_f}.$$

Now if $\dim B \geq 1$ or $B = o \in F$ is a non-singular point of the fibre, then obviously $\mu_1 = \cdots = \mu_N = 1$ and by the condition (v)

$$m_2^v \leq m_1^v \leq 2\frac{d_v}{\deg V},$$

which immediately implies the inequality $k_v \leq 2$. If $B = o \in F$ is a singular point of the fibre, then $\mu_1 \geq 2$ and by the condition (vs),

$$m_1^v \leq \frac{4d_v}{\deg V}, \quad m_2^v \leq \frac{2d_v}{\deg V},$$

whence we get again that $k_v \leq 2$. Q.E.D. for the lemma.

Combining Lemma 1.2 and Proposition 1.6, we obtain the inequality

$$\Sigma_u^2 n^2 - 2\Sigma_u ne + e^2 = (n\Sigma_u - e)^2 < 0.$$

Since this is impossible, we get a contradiction proving that the original assumption was false. Q.E.D. for Theorem 1.1.

1.7. Relaxing the K^2-condition. The method developed above works well for Fano fibre spaces $V/\mathbb{P}^1$ which do not satisfy the (strong) K^2-condition, if the deviation from the K^2-condition is not too great. Our technique is still effective and requires only a slight modification. Recall that a standard Fano fibre space $V/\mathbb{P}^1$ satisfies the *generalized K^2-condition of depth* $\varepsilon \geq 0$, if

$$K_V^2 - \varepsilon H_F \notin \operatorname{Int} A_+^2 V.$$

To compensate for the relaxed K^2-condition, we need to make the upper bound for the singularities of subvarieties on V sharper.

Definition 1.3. A standard Fano fibre space $V/\mathbb{P}^1$ satisfies the *generalized condition (h) of depth* $\delta \geq 0$, if for any horizontal subvariety $Y \subset V$ of codimension two and an arbitrary point $o \in Y$ the following inequality holds:

$$\frac{\operatorname{mult}_o}{\deg} Y \leq \frac{4 - \delta}{\deg V}.$$

Now let us formulate a sufficient condition of birational rigidity for fibre spaces $V/\mathbb{P}^1$ that do not satisfy the strong K^2-condition. Later (in Section 4) we will give a much stronger criterion, based on an improved technique of counting multiplicities. Here we present a straightforward generalization of Theorem 1.1 based on the idea of compensation for the deviation from the strong K^2-condition by tightening the condition (h).

Fix a mobile linear system $\Sigma \subset |-nK_V + lF|$ with $l \in \mathbb{Z}_+$.

Theorem 1.4. *Assume that the smooth standard Fano fibre space $V/\mathbb{P}^1$ satisfies the K^2-condition of depth $\varepsilon \geq 0$, the conditions (v), (vs) and the generalized* *condition (h) of depth $\delta \geq 2\varepsilon$. Then the equality*

$$c_{\text{virt}}(\Sigma) = c(\Sigma) = n$$

holds. In particular, if for any mobile linear system $\Sigma \subset |-nK_V + lF|$ we have $l \in \mathbb{Z}_+$, then the fibre space $V/\mathbb{P}^1$ is birationally superrigid.

Remark 1.1. Note that since the strong K^2-condition is, generally speaking, not satisfied, the K-condition is not automatic, that is, for the mobile linear system Σ we do not necessarily have $l \geq 0$.

Proof of Theorem 1.4. Assume that $c_{\text{virt}}(\Sigma) < n$. We repeat the proof of Theorem 1.1, only noting the required modifications. For the horizontal part of the self-intersection of the linear system Σ we get

$$Z^h \sim n^2 K_V^2 + \alpha H_F,$$

where the coefficient $\alpha \in \mathbb{Z}$ satisfies the inequality

$$\alpha \geq -\varepsilon n^2.$$

Therefore, for the vertical component we get

$$Z^v \sim (2nl - \alpha) H_F,$$

whence

$$\deg Z^v = \sum_{t \in \mathbb{P}^1} \deg Z^v_t \leq (2nl + \varepsilon n^2) \deg V. \tag{105}$$

Proposition 1.7. *For some point $t \in \mathbb{P}^1$ there exists a maximal singularity $E \in \mathcal{M}_t \neq \emptyset$, satisfying the estimate*

$$e(E) > \frac{\nu_E(F_t)}{2} \left(\frac{\deg Z^v_t}{n \deg V} - \varepsilon n \right) \tag{106}$$

Remark 1.2. For $\varepsilon = 0$ we get Proposition 1.4.

Proof of Proposition 1.7. Compare the inequalities (99) and (105). Replacing the number l in the right-hand side of the inequality (105) by the left-hand side of the inequality (99), we get

$$\sum_{t \in \mathbb{P}^1} \left[\deg Z^v_t - 2n \deg V \max_{\{E \in \mathcal{M}_t\}} \frac{e(E)}{\nu_E(F_t)} \right] < \varepsilon n^2 \deg V,$$

which immediately implies our claim. Q.E.D. for Proposition 1.7.

Remark 1.3. If there are a few maximal singularities the centres of which lie in the fibres over distinct points $t_1, \ldots, t_k$ then the claim of Proposition 1.7 can be improved: there is a maximal singularity $E \in \mathcal{M}_t$, $t \in \{t_1, \ldots, t_k\}$ satisfying the estimate

$$e(E) > \frac{\nu_E(F_t)}{2} \left(\frac{\deg Z^v_t}{n \deg V} - \frac{\varepsilon n}{k} \right).$$

However, this improvement is hardly useful, since to prove birational rigidity, that is, to realize the full scheme of the method of maximal singularities, the worst case should be considered.

Now we argue as in the proof of Theorem 1.1 with the only difference: the expression $d_v \nu_E(F)$ in the inequality (104) is replaced by the expression

$$(2ne + \varepsilon n^2 \nu_E(F)) \deg V,$$

which makes the inequality sharp. Taking into account that the conditions (h), (v) and (vs) still hold, so that the claim of Lemma 1.2 is true, we get the inequality

$$((4-k_h)\Sigma_l - k_v\varepsilon\nu_E(F))(\Sigma_l+\Sigma_u)n^2 + (n\Sigma_u - e)^2 < 0.$$

By the definition of the number k_v we get

$$k_v\nu_E(F) = \deg V \sum_{i=1}^{\min\{N,L\}} p_i \frac{m_i^v}{d_v} \leq 2\Sigma_l.$$

Therefore, the following inequality holds:

$$(4-k_h-2\varepsilon)\Sigma_l(\Sigma_l+\Sigma_u)n^2 + (n\Sigma_u - e)^2 < 0. \tag{107}$$

Now recall that by the generalized condition (h) of depth $\delta \geq 0$ the coefficient k_h satisfies the estimate

$$k_h \leq 4-\delta.$$

This completes our proof of Theorem 1.4. Q.E.D.

2. Pencils of Fano hypersurfaces

In this section we prove birational superrigidity of a wide class of Fano fibre spaces $V/\mathbb{P}^1$ with Fano hypersurfaces of index one as fibres. The proof is based on Theorem 1.1: all we need to do is to check that the assumptions of Theorem 1.1 are satisfied. To show this, we need the variety V to be sufficiently general in its family in the sense of the regularity conditions similar to those used in Section 3.1. We start with an explicit construction of fibrations into Fano hypersurfaces and the regularity conditions: now they should be satisfied by every fibre at every point, so the singular points of fibres also need to be included in the regularity conditions. After that, we formulate the main result (Theorem 2.1).

Subsections 2.2 and 2.3 check that for regular fibre spaces the conditions (v), (vs) and (h) are satisfied. Subsection 2.4 proves that a generic (in the sense of Zariski topology) fibre space $V/\mathbb{P}^1$ is regular. Finally, in Subsection 2.5 we prove Theorem 4.1.1, which makes our description of the birational type of fibrations into Fano hypersurfaces complete.

2.1. The explicit construction. Let $\mathcal{E}$ be a locally free sheaf of rank $M+1$ on the projective line $\mathbb{P}^1$, where $M \geq 5$,

$$\mathcal{E} = \bigoplus_{i=0}^{M} \mathcal{O}_{\mathbb{P}^1}(a_i),$$

$a_0 = 0 \leq a_1 \leq \cdots \leq a_M$, and let $X = \mathbb{P}(\mathcal{E})$ be the corresponding projective bundle, so that $\mathbb{P} = \mathbb{P}^M$ is the fibre of the canonical projection $\pi\colon X \to \mathbb{P}^1$. Let $L_X \in \operatorname{Pic} X$ be the class of the tautological sheaf and $G \in \operatorname{Pic} X$ be the class of a fibre $\cong \mathbb{P}$. For an arbitrary $m \in \mathbb{Z}$ define the family $\mathcal{V}(m)$ of codimension 1 subvarieties in X as the subset of the linear system $|ML+mG|$, consisting of smooth divisors. Here the parameter m measures "the degree of twistedness" of a Fano fibration $V \in \mathcal{V}(m)$ over the base $\mathbb{P}^1$. Projection of V onto $\mathbb{P}^1$ is denoted by the same symbol π. The generic fibre F_η of a Fano fibration $\pi\colon V \to \mathbb{P}^1$ is a Fano hypersurface of index 1 in $\mathbb{P}^M_{\mathbb{C}(t)}$, where t is a linear coordinate on the line. For $m \geq 1$ by the Lefschetz theorem $\operatorname{Pic} V = \operatorname{Pic} X|_V$, so that

$$\operatorname{Pic} V = \mathbb{Z}K_V \oplus \mathbb{Z}F,$$

where $F = G|_V$ is the class of a fibre. A particular fibre $\pi^{-1}(t)$ over a point $t \in \mathbb{P}^1$ is denoted by the symbol F_t.

Obviously,

$$K_V = -L_V + (a_1 + \cdots + a_M - 2 + m)F,$$

so that the higher m is, the "smaller" the anticanonical linear system $|-K_V|$ is. Since

$$0 \geq (K_V^2 \cdot L_V^{M-2}) = M(4 - a_1 - \cdots - a_M) - m(2M-1),$$

we conclude that if the inequality

$$m \geq \frac{M}{2M-1}\left(4 - \sum_{i=1}^{M} a_i\right)$$

holds, then the K^2-condition is satisfied.

Consider an arbitrary hypersurface $W \subset \mathbb{P}$ of degree M. Let $o \in W$ be any point, $z_* = (z_1, \ldots, z_M)$ a system of linear coordinates on $\mathbb{P}$ with the origin at the point o,

$$q = q_1 + q_2 + \cdots + q_M$$

the equation of the hypersurface W with respect to the coordinates z_*, decomposed into homogeneous polynomials q_i of degree i. Recall (Definition 3.1.3) that the point $o \in W$ is said to be a *regular point* of the hypersurface W, if the sequence $(q_1, \ldots, q_{M-1})$ is regular.

Definition 2.1. (i) The point p is said to be a *regular singular point* of the hypersurface W, if $q_1 \equiv 0$, the sequence

$$(q_2, \ldots, q_M)$$

is regular, the quadric $q_2 = 0$ in $\mathbb{P}^{M-1} = \mathbb{P}(T_p\mathbb{P})$ is smooth and, moreover, the system of equations

$$q_2 = \cdots = q_M = 0$$

defines in $\mathbb{P}^{M-1} = \mathbb{P}(T_p\mathbb{P})$ a finite set, consisting of exactly $M!$ distinct points, any M of which are linearly independent.

(ii) A Fano fibre space $\pi \colon V \to \mathbb{P}^1$, constructed above, is said to be *regular*, if any fibre F_t of the projection π is a $\mathbb{Q}$-factorial variety (so that the group of classes of Weil divisors, tensored by $\mathbb{Q}$, is isomorphic to $\mathbb{Q}$ and generated by the hyperplane section of $F_t \subset \mathbb{P}$), any point $p \in F_t$ of which is either a smooth regular point or a singular regular point.

In Chapter 3 it was shown that a general smooth hypersurface $W \subset \mathbb{P}$ of degree M is regular. Modifying the arguments of Chapter 3 a bit, we show below that the complement to the set of hypersurfaces W with at most double points, of which all the points are regular, is of codimension ≥ 2. A Fano fibre space $\pi \colon V \to \mathbb{P}^1$ can be looked at as a curve in the space $\mathbb{P}(H^0(\mathbb{P}, \mathcal{O}_{\mathbb{P}}(M)))$; thus, the general member of the system $|ML + mG|$ is a regular Fano fibre space.

The main result of this section is the following:

Theorem 2.1. *Let $\pi \colon V \to \mathbb{P}^1$ be a Fano fibre space, constructed above. Assume that it is regular and the (strong) K^2-condition is satisfied. Then V is birationally superrigid.*

2.2. Start of the proof. Let us consider the general fibre $F_t = \pi^{-1} \subset \mathbb{P}$, a smooth Fano hypersurface, of which each point is by assumption regular. Let $\Sigma \subset |-nK_V + lF|$ be a mobile linear system, $n \geq 1$. In Chapter 3 we proved (Section 3.1) that the restriction of Σ onto F_t cannot have a maximal singularity. Therefore, the centre of every maximal singularity of the system Σ is contained in some fibre F_t. In order to prove Theorem 2.1, we need to show that $V/\mathbb{P}^1$ satisfies the conditions (v), (vs) and (h).

Let us start with the condition (h). Let $Y \subset V$ be an irreducible horizontal subvariety of codimension two, $o \in Y$ a point, $F \ni o$ the corresponding fibre. By $Y_F = (Y \circ F)$ we denote the restriction of Y to F, an effective cycle of codimension two on F. Note that $\deg Y_F = \deg Y$.

If F is smooth at o, then by Proposition 3.1.1(i) — which applies to $Y \subset F$ because the fibre F satisfies the regularity condition at o — we get

$$\frac{\mathrm{mult}_o}{\deg} Y_F \leq \frac{4}{M},$$

which is what we need as $\mathrm{mult}_o Y \leq \mathrm{mult}_o Y_F$.

If F is singular at the point o, then

$$2 \,\mathrm{mult}_o Y \leq \mathrm{mult}_o Y_F.$$

On the other hand, repeating word for word the proof of Proposition 3.1.1(i) in the singular case, we get

$$\frac{\mathrm{mult}_o}{\deg} Y_F \leq \frac{4}{M},$$

which implies a much sharper estimate than we need:

$$\frac{\mathrm{mult}_o}{\deg} Y \leq \frac{2}{M}.$$

Thus the condition (h) is shown.

Now let us consider vertical subvarieties $Y \subset F$, that is, prime divisors in a fibre. First assume that $o \in Y$ is a smooth point of the fibre F.

Lemma 2.1. *Let $Y \subset F$ be a prime Weil divisor. Then the estimate*

$$\frac{\mathrm{mult}_o}{\deg} Y \leq \frac{2}{M}$$

holds, where the equality is attained at a unique cycle — that is, the intersection $T = F \cap T_oF$ with the tangent hyperplane. Moreover, for any prime divisor $Y \neq T$ the following inequality holds:

$$\frac{\mathrm{mult}_o}{\deg} Y \leq \frac{3}{2(M-1)}.$$

Proof. The first estimate is very easy to obtain. For $T = F \cap T_oF$ it is true by the regularity condition. Assume that $Y \neq T$. Then the effective cycle $(Y \circ T)$ has the multiplicity $\geq 2 \,\mathrm{mult}_o Y$ at the point o. Therefore,

$$\frac{2}{M} \geq \frac{1}{2} \frac{\mathrm{mult}_o}{\deg} (Y \circ T) \geq \frac{\mathrm{mult}_o}{\deg} Y,$$

which is exactly what we need.

Now let us consider the second claim of the lemma. By the regularity condition, for the hypertangent linear systems

$$\Lambda_k = |kH - (k+1)E|,$$

$k = 3, \ldots, M-2$, we have the estimate

$$\operatorname{codim}_T \operatorname{Bs} \Lambda_k|_T \geq k-1,$$

so that for general divisors $D_k \in \Lambda_k$ the algebraic cycle

$$Z = (T \circ D_3 \circ \cdots \circ D_{M-2})$$

is effective, and its components form an algebraic family of surfaces, sweeping out T. At the same time

$$\frac{\operatorname{mult}_o}{\deg} Z \geq \frac{2(M-1)}{3M} > \frac{1}{2},$$

so that (taking a hyperplane section, containing the point o), we conclude that T is swept out by an algebraic family of curves $\mathcal{C} = (C_s, s \in S)$, satisfying the estimate

$$\frac{\operatorname{mult}_o}{\deg} C_s \geq \frac{2(M-1)}{3M}.$$

Now if

$$\frac{\operatorname{mult}_o}{\deg} Y > \frac{3}{2(M-1)},$$

then $Y \supset C_s$ for any $s \in S$ and thus $Y = T$. Q.E.D.

The condition (v) has been shown.

2.3. Singular points of the fibres. Now we concentrate on proving the condition (vs). Let $o \in F$ be a (regular) singular point and $Y \subset F$ a prime Weil divisor.

Lemma 2.2. *The following estimate is true:*

$$\frac{\operatorname{mult}_o}{\deg} Y \leq \frac{3}{M}.$$

Moreover, the equality takes place for a unique cycle, that is, the intersection $Q = T_o F \cap F$, *where* $T_o F$ *is the projectivized tangent cone to the hypersurface* $F \subset \mathbb{P}$ *at the point* o. *If* $Y \neq Q$, *then*

$$\frac{\operatorname{mult}_o}{\deg} Y \leq \frac{8}{3M}.$$

Proof. Obviously, we get $\widetilde{Q} \in |2H - 3E|$, so that for Q we have the equality $(\operatorname{mult}_o / \deg) Q = 3/M$. Assume that $Y \neq Q$ is a prime Weil divisor. Then $(Y \circ Q)$ is an effective cycle of codimension 2 on F. As we mentioned above, the regularity condition makes it possible to repeat the proof of Proposition 3.1.1 word for word for the singular point o, so that we get the estimate

$$\frac{4}{M} \geq \frac{\operatorname{mult}_o}{\deg} (Y \circ Q) \geq \frac{3}{2} \frac{\operatorname{mult}_o}{\deg} Y,$$

thus $(\operatorname{mult}_p / \deg) Y \leq (8/3M)$, as required. Q.E.D.

Now in the notation of the condition (vs), let $\varphi \colon \widetilde{\mathbb{P}} \to \mathbb{P}$ be the blow up of the point o, $E = \varphi^{-1}(o) \cong \mathbb{P}^{M-1}$ the exceptional divisor, $\widetilde{F} \subset \widetilde{\mathbb{P}}$ the strict transform of F and $\widetilde{Y} \subset \widetilde{\mathbb{P}}$ the strict transform of Y.

Fix an arbitrary point $x \in E_F = E \cap \widetilde{F}$.

Proposition 2.1. *The following estimate holds:*

$$\frac{\mathrm{mult}_x \widetilde{Y}}{\deg Y} \leq \frac{2}{M}.$$

Note that Lemma 2.2 and Proposition 2.1 imply the condition (vs).

Proof. Assume the converse, that is, that the inequality

$$\frac{\mathrm{mult}_x \widetilde{Y}}{\deg Y} > \frac{2}{M}$$

holds. Since F is a $\mathbb{Q}$-factorial variety, Y is a $\mathbb{Q}$-Cartier divisor and so for certain $d, \nu \in \mathbb{Q}_+$ we have $\widetilde{Y} \in |dH_F - \nu E_F|$. Obviously, $\nu = \frac{1}{2}\,\mathrm{mult}_o Y$. Set $\mu = \mathrm{mult}_x \widetilde{Y}$. Recall that $E \subset \widetilde{E} = \mathbb{P}^{M-1}$ is a non-singular quadric hypersurface. Then $B = E \cap T_x E$ is a non-degenerate quadric cone with the vertex at the point x.

Lemma 2.3. *The following estimate holds:*

$$\mathrm{mult}_B \widetilde{Y} \geq \frac{1}{2}(\mu - \nu).$$

Proof is given below.

Now let us complete the proof of Proposition 2.1. Consider the hyperplane $H \subset \mathbb{P}$, containing the point o, such that $T_o H = T_x E \subset \widetilde{E}$. In accordance with Lemma 2.3 (and elementary intersection theory), we get

$$\mathrm{mult}_o (Y \circ H) \geq 2\nu + (\mu - \nu) > \frac{3}{M} \deg Y$$

(since $2\nu > \mu$). However, this is impossible by

Lemma 2.4. *For any hyperplane $H \subset \mathbb{P}$, containing the point o, and any effective $\mathbb{Q}$-Cartier divisor W on $F_H = F \cap H$ the following inequality holds:*

$$\frac{\mathrm{mult}_o}{\deg} W \leq \frac{3}{M},$$

whereas the equality is attained at the divisors mS_H, $m \in \mathbb{Z}_+$, only, where $S_H = F_H \cap T_o F = H \cap Q$.

Proof of the lemma is given below.

Therefore by Lemmas 2.3 and 2.4 we get a contradiction. We conclude that the initial assumption was wrong. Modulo the two lemmas the proposition (and thus Theorem 2.1) is proved. Q.E.D.

Proof of Lemma 2.3. Let $P \ni o$ be a general 4-plane in $\mathbb{P}$, the proper inverse image $\widetilde{P}$ of which on $\widetilde{\mathbb{P}}$ contains the point x. We add the lower index P to denote the restriction to P, e.g., set $Y_P = Y \cap P$, $F_P = F \cap P$, $\widetilde{E}_P = \widetilde{E} \cap \widetilde{P} \cong \mathbb{P}^3$, $E_P = E \cap \widetilde{P} \cong \mathbb{P}^1 \times \mathbb{P}^1$ etc. Obviously, $B_P = T_x E_P \cap E_P = L_1 + L_2$ is a pair of lines. Set $L = L_1$. It is sufficient to prove that

$$\mathrm{mult}_L \widetilde{Y}_P \geq \frac{1}{2}(\mu - \nu).$$

Let $\varphi_L \colon P_L \to \widetilde{P}$ be the blow up of the line L, F_L the proper inverse image of the (smooth) 3-fold hypersurface F_P on P_L, and $E_L \subset F_L$ the exceptional divisor. It is easy to see that

$$\mathcal{N}_{L/F_P} \cong \mathcal{O}_L \oplus \mathcal{O}_L(-1),$$

so that E_L is a ruled surface of the type $\mathbb{F}_1$. Let s and f be the classes of the exceptional section and the fibre in $\operatorname{Pic} E_L$, respectively. It is easy to see that $(E_L \cdot E_L) = -s - f$, since $(E_L \cdot s) = (E_L \cdot \varphi_L^* L) = 0$. Set $\gamma = \operatorname{mult}_L \widetilde{Y}_P$. The class of the proper inverse image of the divisor Y on F_L is obviously equal to $dH - \nu E_P - \gamma E_L$, whereas its restriction onto E_L is equal to $\gamma s + (\gamma + \nu) f$. This is the class of an effective curve on E_L, which contains the fibre over the point x with the multiplicity $\mu - \gamma$. Therefore,

$$\gamma + \nu \geq \mu - \gamma.$$

Thus, $2\gamma \geq \mu - \nu$, as required. Q.E.D.

Proof of Lemma 2.4. Note, first of all, that $\deg S_H = 2M$ and $\operatorname{mult}_p S_H = 6$, so that the divisor S_H does realize the limit case of our inequality.

First assume that $M \geq 9$, in which case the proof is easier. Suppose that a prime Weil divisor W on F_H satisfies the inequality

$$\frac{\operatorname{mult}_o}{\deg} W > \frac{3}{M}.$$

Then $W \neq S_H$ and the effective cycle $C = (W \circ S_H)$ of codimension 2 (on F_H) satisfies the estimate

$$\frac{\operatorname{mult}_p}{\deg} C > \frac{9}{2M}.$$

Now arguing as in Section 3.1, consider the hypertangent linear systems

$$\Lambda_k = |kL - (k+1)E|,$$

$k = 2, \dots, M-1$, where $L \in \operatorname{Pic} F_H$ is the class of a hyperplane section, and $E \subset \widetilde{F}_H \to F_H \ni o$ is the exceptional divisor of the blow up of the point o. Obviously, $\operatorname{codim}_{F_H} \operatorname{Bs} \Lambda_k \geq k - 1$ for $2 \leq k \leq M-2$ by the regularity condition. Choose general divisors $D_k \in \Lambda_k$ for $k = 4, \dots, M-2$ and look at the effective 1-cycle

$$Z^* = (C \circ D_4 \circ \cdots \circ D_{M-2}) \subset F_H.$$

This cycle satisfies the inequality

$$\frac{\operatorname{mult}_o}{\deg} Z^* > \frac{9}{2M} \cdot \frac{M-1}{4} \geq 1,$$

which is impossible. This contradiction proves the desired inequality for $M \geq 9$ (and also the claim on the extremality of the cycle S_H for $M \geq 10$).

Remark 2.1. F_H is a hypersurface of degree M in $\mathbb{P}^{M-1}$. It is exactly for this reason that, repeating the scheme of arguments of Section 3.1, we have to perform one step less, intersecting the cycle C with $D_4, \dots, D_{M-2}$, but not with D_{M-1}: the dimension does not allow us to do it. However, if M is sufficiently high, this makes no difference.

To obtain Lemma 2.4 in full, we need a few more delicate arguments. The proof given below works well for all the values $M \geq 5$. Note that $m\widetilde{W} \in |\alpha L - \beta E|$ for some $m, \alpha, \beta \in \mathbb{Z}_+ \setminus \{0\}$ with $\beta/\alpha > 3/2$. Thus any curve $C \subset F_H$, satisfying the inequality

$$\frac{\operatorname{mult}_o}{\deg} C > \frac{2}{3}, \tag{108}$$

is contained in W. Therefore, the lemma is implied by the following

Lemma 2.5. *The cycle S_H is swept out by a family of curves $\mathcal{C} = (C_t, t \in T)$, the general member of which is irreducible and satisfies the inequality* (108).

Proof of the lemma. As an example consider the most transparent case $M = 5$. The linear system of curves $\Lambda_4|_{S_H}$ on the surface S_H has by the regularity condition γ, $0 \leq \gamma \leq 4$, base lines of multiplicity one (since any 5 lines from the set of 5! solutions of the system of equations $q_2 = \cdots = q_5 = 0$ are linearly independent and therefore cannot lie in H). Throwing away these lines and taking an irreducible component of a general curve of the residual family (which has no fixed components), we get a family $\mathcal{C} = (C_t, t \in T)$, for which

$$\frac{\text{mult}_o}{\deg} C_t = \frac{30 - \gamma}{40 - \gamma} > \frac{2}{3},$$

as we need.

Now take an arbitrary $M \geq 6$. We argue in a similar way. Let γ, $0 \leq \gamma \leq M-1$ be the number of solutions of the system of equations $q_2 = \cdots = q_M = 0$, which lie in H. Set

$$\beta_k = \frac{3(k+4)}{4M}.$$

Let us construct by induction on $k = 1, \ldots, M-5$, a family of irreducible cycles $\mathcal{R}_k = (R_t^{(k)}, t \in T_k)$, which sweep out S_H, are of codimension k in S_H, and satisfy the inequality

$$\frac{\text{mult}_p}{\deg} R_t^{(k)} \geq \beta_k.$$

Assume that the family $\mathcal{R}_k$ has already been constructed and $k \leq M-6$. Let $R = R_t^{(k)}$ be a sufficiently general member of this family. Then by the regularity condition the linear system $\Lambda_{k+4}|_R$ has no fixed components, since

$$\operatorname{codim} \operatorname{Bs} \Lambda_{k+4}|_{S_H} \geq k + 2 = \operatorname{codim} R + 2.$$

A general divisor of the system $\Lambda_{k+4}|_R$ is irreducible, unless it is composed of a pencil; in any case the irreducible components of a general divisor, taken over all $R_t^{(k)}$, $t \in T_k$, form the desired family $\mathcal{R}_{k+1}$, since

$$\frac{\text{mult}_o}{\deg} R_t^{(k+1)} \geq \beta_k \frac{k+5}{k+4} = \beta_{k+1}.$$

Now take a general surface $R = R_t^{(M-5)}$ from the family $\mathcal{R}_{M-5}$. Restricting the linear system Λ_{M-1} onto R, we obtain a linear system of curves. It may have lines on F_H as fixed components. Throwing away these components (each of which is of multiplicity exactly 1), we get a family of curves $\mathcal{C} = (C_t, t \in T)$, sweeping out S_H. A general curve of this family is an irreducible component of a curve of the family $\mathcal{C}^* = (C_t^*, t \in T^*)$, which can be constructed directly as follows. Take the family of 1-cycles

$$(S_H \circ D_4 \circ \cdots \circ D_{M-1}),$$

where $D_i \in \Lambda_i$ are general divisors, and throw away γ lines of multiplicity one, lying in F_H, $0 \leq \gamma \leq M-1$. Consequently,

$$\frac{\text{mult}_o}{\deg} C_t = \frac{\text{mult}_o}{\deg} C_t^* \geq \frac{\frac{M!}{4} - \gamma}{\frac{M!}{3} - \gamma} > \frac{2}{3},$$

which is what we need. Q.E.D. for Lemma 2.4.

Proof of Theorem 2.1 is complete.

Remark 2.2. For M sufficiently large the ratio $(\mathrm{mult}_p / \deg)$ for the curves of the system $\mathcal{C}$ tends to be $3/4$. This means that, similarly to the smooth case (Lemma 2.1), there is a considerable gap in the set of possible values of the ratio $(\mathrm{mult}_p / \deg)$ for S_H and all the other prime Weil divisors $W \neq S_H$. For these cycles for $M \gg 0$ we get

$$\max \left\{ \frac{\mathrm{mult}_p}{\deg} W \;\middle|\; W \neq S_H \right\} \approx \frac{8}{3M}.$$

2.4. Regular hypersurfaces. Here we show that a generic fibre space $V/\mathbb{P}^1$, constructed in Subsection 2.1, is indeed regular.

Let $\mathcal{P}_{k,K}$ be the space of homogeneous polynomials of degree k in K variables, $\dim \mathcal{P}_{k,K} = \binom{k+K-1}{k}$. Consider the space of pairs

$$S = \{(x, F) \mid x \in \mathbb{P}, F \in H^0(\mathbb{P}, \mathcal{O}_{\mathbb{P}}(M)) \cong \mathcal{P}_{M,M+1}\}$$

with the natural projections $p\colon S \to \mathbb{P}$ and $q\colon S \to H^0(\mathbb{P}, \mathcal{O}_{\mathbb{P}}(M))$, the incidence hypersurface $I = \{(x, F) |\ F(x) = 0\} \subset S$ and set

$$Y^* = \{(x, F) \mid x \in F \quad \text{is a non-regular smooth point}\},$$

$$Y^{\mathrm{sing}} = \{(x, F) \mid x \in F \quad \text{is a non-regular singular point}\},$$

$$Y = Y^* \bigcup Y^{\mathrm{sing}}.$$

Proposition 2.2. *The set $q(Y)$ is a proper closed subset of codimension 2 in the space $H^0(\mathbb{P}, \mathcal{O}_{\mathbb{P}}(M))$.*

Proof. Set $I(x) = I \cap p^{-1}(x)$ and $Y(x) = Y \cap p^{-1}(x)$. It is sufficient to show that

$$\mathrm{codim}_{I(x)} Y(x) = M + 1.$$

Let us consider the smooth and singular points separately.

The case of singular points is easy: indeed, set

$$I^{\mathrm{sing}}(x) = \left\{ (x, F) \in I(x) \;\middle|\; \left.\frac{\partial F}{\partial z_*}\right|_x = 0 \right\}.$$

Since obviously $\mathrm{codim}_{I(x)} I^{\mathrm{sing}}(x) = M$, to show that $q(Y^{\mathrm{sing}})$ is of codimension two, it is sufficient to prove that $Y^{\mathrm{sing}}(x) \subset I^{\mathrm{sing}}(x)$ is a proper closed subset. But this is obvious, since the regularity condition is an open one, whereas for a general polynomial of the form

$$q_2 + \cdots + q_M$$

the origin is a regular double point.

So we concentrate on non-regular smooth points.

Set $Y^*(x) = Y^* \cap p^{-1}(x)$,

$$Y_k^*(x) = \left\{ (q_1, \ldots, q_k) \in \prod_{i=1}^{k} \mathcal{P}_{i,M} \;\middle|\; q_1 \neq 0, \mathrm{codim}\{q_1 = \cdots = q_k = 0\} < k \right\},$$

where we identify

$$p^{-1}(x) = \prod_{i=1}^{M} \mathcal{P}_{i,M} = \{(q_1, \ldots, q_M)\}$$

via a system of linear coordinates $(z_1, \dots, z_M)$ with the origin at x. Set also $R_1 = \{q_1 \neq 0\}$,

$$R_k = \prod_{i=1}^{k} \mathcal{P}_{i,M} \setminus Y_k^*, k \geq 2,$$

and (abusing the notations)

$$\pi\colon \prod_{i=1}^{k+1} \mathcal{P}_{i,M} = \prod_{i=1}^{k} \mathcal{P}_{i,M} \times \mathcal{P}_{k+1,M} \to \prod_{i=1}^{k} \mathcal{P}_{i,M}$$

to be the natural projection.

It was shown in Section 3.1 (see the proof of Lemma 3.1.3) that for $k = 2, \dots, M-2$ the estimate

$$\operatorname{codim} Y_k^* \bigcap \pi^{-1}(R_{k-1}) \geq \binom{M}{k} > M+1$$

holds, where the codimension is taken with respect to the space of polynomials $\prod_{i=1}^{k} \mathcal{P}_{i,M}$. Thus, the set of sequences $(q_1, \dots, q_M)$ with $q_1 \neq 0$, which fail to be regular at the k-th step, $k \leq M-2$, has a high codimension. Consider the set R_{M-2}. The system of equations

$$q_1 = \cdots = q_{M-2} = 0$$

defines a curve $C(q_*) \subset \mathbb{P}^{M-1} = \mathbb{P}(T_x\mathbb{P})$.

Lemma 2.6. *The set $\mathcal{L} \subset R_{M-2}$, for the points of which the curve $C(q_*)$ contains a line in $\mathbb{P}^{M-1}$ as a component, is of codimension $\frac{1}{2}(M-2)(M-3)$.*

Proof. Obvious computations.

For our purposes it is sufficient that $\mathcal{L} \subset R_{M-2}$ is a proper closed subset (i.e., is of codimension ≥ 1), since by the same proof of Lemma 3.1.3.

$$\operatorname{codim} Y_{M-1}^* \bigcap \pi^{-1}(\mathcal{L}) \geq M+1.$$

Therefore we may assume that $C(q_*)$ does not contain a line. To complete the proof (in the non-singular case) it is sufficient to show that

$$\operatorname{codim}_{\pi^{-1}(q_*)}(Y_{M-1}^* \bigcap \pi^{-1}(q_*)) \geq M+1.$$

Let $P \subset \mathbb{P}^{M-1}$ be a general $(M-4)$-plane, so that the restriction of the linear projection

$$\pi_P\colon \mathbb{P}^{M-1} \dashrightarrow \mathbb{P}^2$$

onto each component C of the curve $C(q_*)$ is a birational morphism and, moreover, $d = \deg \pi_P(C) = \deg C \geq 2$. Consequently, the kernel of the restriction map

$$\operatorname{res}_C\colon H^0(\mathbb{P}^2, \mathcal{O}_{\mathbb{P}^2}(M-1)) \to H^0(C, \mathcal{O}_C(M-1))$$

is of codimension

$$Md - \frac{d(d-1)}{2}$$

for $d \leq M-1$ (for $d \geq M$ the kernel is trivial). Thus there exists a subspace $W \subset \mathcal{P}_{3,M-1}$ of dimension $M+1$, intersecting $Y^* = Y_{M-1}^*$ by zero only. Therefore

$$\operatorname{codim}_{I(x)} Y^*(x) \geq M+1,$$

which is what we need.

Q.E.D. for Proposition 2.2.

2.5. Fibrewise birational modifications. Let us prove Theorem 4.1.1. Let $V \in \mathcal{V}(d)$, $F = V \cap \{p\} \times \mathbb{P}^M$ be the fibre over the marked point. Fix a local parameter t on the curve C in a neighborhood of the point p. The hypersurface $V \subset X$ in a neighborhood of the fibre F is given by the equation

$$f = f^{(0)} + tf^{(1)} + \cdots + t^j f^{(j)} + \ldots,$$

where $f^{(j)}$ are homogeneous polynomials of degree j in the homogeneous coordinates $(x_*) = (x_0 : \cdots : x_M)$ on $\mathbb{P}^M$. It is well known (see, for instance, [**IP, Pukh00a**]), that $\dim \mathrm{Sing}\, V \geq \dim \mathrm{Sing}\, F - 1$, so that the smoothness of the hypersurface V implies that, first, the hypersurface $F = \{f^{(0)} = 0\} \subset \mathbb{P}^M$ has at most zero-dimensional singularities and, second, for every point $x \in \mathrm{Sing}\, F$ we have $f^{(1)}(x) \neq 0$.

Take $V_1, V_2 \in \mathcal{V}(d)$ and let $\chi^* \colon V_1^* \to V_2^*$ be a fibrewise isomorphism outside the marked point $p \in C$. Since the fibres over points of general position $y \in C$ are smooth hypersurfaces of degree $d \geq 2$, over a point $y \in C^*$ the isomorphism χ_y^* is induced by an automorphism of the ambient projective space $\xi_y \in \mathrm{Aut}\, \mathbb{P}$. Therefore, $\chi^* = \xi^*|_{V_1}$, where $\xi_y^* = \xi_y$ is an algebraic curve $\xi^* \colon C^* \to \mathrm{Aut}\, \mathbb{P}$ of projective automorphisms. Let $\mathbb{P} = \mathbb{P}(L)$, where $L \cong \mathbb{C}^{M+1}$ is a linear space. The curve ξ^* lifts to a curve $\xi \colon C \to \mathrm{End}\, L$, $\xi(C^*) \subset \mathrm{Aut}\, L$. If $\xi(p) \in \mathrm{Aut}\, L$, then χ^* extends to a fibrewise (biregular) isomorphism $\chi = \xi|_{V_1}$, and the varieties V_1 and V_2 are fibrewise isomorphic. Assume that this is not the case: $\det \xi(p) = 0$.

Let $\sum_{i=0}^{\infty} t^i \xi^{(i)}$ be the Taylor series of the curve ξ. We may assume that $\xi^{(0)} \neq 0$. The next claim is a well-known fact of elementary linear algebra.

Lemma 2.7. *There exist curves of endomorphisms $\beta, \gamma \colon C \to \mathrm{End}\, L$ and a basis $(e_0, \ldots, e_M)$ of the space L such that $\beta(p), \gamma(p) \in \mathrm{Aut}\, L$ and in this basis the curve $\beta\xi\gamma^{-1} \colon C \to \mathrm{End}\, L$ is of the diagonal form:*

$$\beta\xi\gamma^{-1} \colon e_i \mapsto t^{w(e_i)} e_i, \tag{109}$$

where $w(e_i) \in \mathbb{Z}_+$.

Now replace V_1 by $\gamma(V_1)$ and V_2 by $\beta(V_2)$. We can simply assume that the fibrewise birational correspondence ξ is of the form (109) from the start. Let us show that if $m = \max\{w(e_i)\} \geq 1$, then this is impossible.

Let $\{a_0 = 0 < a_1 < \cdots < a_k\} = \{w(e_i), i = 0, \ldots, M\} \subset \mathbb{Z}_+$ be the set of weights of the diagonal transformation (109), $k \leq M$, $m = a_k$ the maximal weight. Consider the system of homogeneous coordinates $(x_0 : \cdots : x_M)$, dual to the basis (e_*). Define the weight of monomials in x_*, setting

$$w(x_0^{n_0} x_1^{n_1} \ldots x_M^{n_M}) = \sum_{i=0}^{M} n_i w(e_i).$$

Set $\mathcal{A}_i = \{x_j | w(e_j) = a_i\} \subset \mathcal{A} = \{x_0, \ldots, x_M\}$ to be the set of coordinates of weight a_i. We pay special attention to the sets $\mathcal{A}_* = \mathcal{A}_0$ and $\mathcal{A}^* = \mathcal{A}_k$ of coordinates of the minimal and maximal weights, respectively.

Now let $f = f^{(0)}(x) + tf^{(1)} + \ldots$ be the local over the base C equation of the hypersurface $V_2 \subset C \times \mathbb{P}$, $f^{(i)}$ are homogeneous polynomials of degree $d \geq 3$ in the coordinates x_*. The series

$$f_\xi = \sum_{l=0}^{\infty} t^l f_\xi^{(l)}(x) = \sum_{l=0}^{\infty} t^l f^{(l)}(t^{w(x_0)}x_0, \dots, t^{w(x_M)}x_M)$$

vanishes on V_1, and outside the marked fibre F_1, that is, for $t \neq 0$, gives an equation of V_1. Let $b \in \mathbb{Z}_+$ be the maximal power of the parameter t, dividing f_ξ. Then

$$t^{-b} f_\xi = g = \sum_{l=0}^{\infty} t^l g^{(l)}(x_0, \dots, x_M)$$

gives an equation of the hypersurface V_1 in the marked fibre X_p, as well.

Lemma 2.8. *For any $l \in \mathbb{Z}_+$ the polynomial $f^{(l)}$ is a linear combination of monomials of weight $\geq b - l$, and the polynomial $g^{(l)}$ is a linear combination of monomials of weight $\leq b + l$.*

Proof. Assume that the monomial x^I comes into the polynomial $f^{(l)}$ with a non-zero coefficient. Then it generates the component $t^{l+w(x^I)}x^I$ of the series f_ξ and, moreover, this component is generated *only* by this monomial in $f^{(l)}$. Therefore $l + w(x^I) \geq b$, as we claimed. Assume that the monomial x^I comes into $g^{(l)}$ with a non-zero coefficient. It comes from the monomial $t^{l+b}x^I$ of the series f_ξ, which, in turn, can come only from the monomial x^I in the polynomial f^α, where $\alpha + w(x^I) = l + b$. Q.E.D. for the lemma.

Let

$$\begin{aligned} P_* &= \{x_j = 0 \,|\, w(x_j) \geq 1\} = \mathbb{P}\langle e_j \,|\, w(x_j) = 0\rangle, \\ P^* &= \{x_j = 0 \,|\, w(x_j) \leq m - 1\} = \mathbb{P}\langle e_j \,|\, w(x_j) = m\rangle \end{aligned}$$

be the subspaces of the minimal and maximal weights, respectively.

Lemma 2.9. *If the inequality $b \geq m + 1$ holds, then $P_* \subset \operatorname{Sing} F_2$. If the inequality $m(d-1) \geq b + 1$ holds, then $P^* \subset \operatorname{Sing} F_1$.*

Proof. Assume that $b \geq m + 1$. The fibre $F_2 \subset \mathbb{P}$ over the marked point is given by the equation $f^{(0)} = 0$. By assumption, $f^{(0)}$ is a linear combination of the monomials of weight $\geq m + 1$. If the monomial x^I comes into $f^{(0)}$ with a non-zero coefficient, then x^I is divided by a quadratic monomial in the variables $\mathcal{A} \setminus \mathcal{A}_*$ (otherwise $w(x^I) \leq m$). Therefore all first partial derivatives of the polynomial $f^{(0)}$ vanish on P_*. Therefore $P \subset \operatorname{Sing} F_2$.

Similarly, if $b \leq m(d-1) - 1$, then any monomial x^I in $g^{(0)}$ are divided by a quadratic monomial in $\mathcal{A} \backslash \mathcal{A}^*$, otherwise we get $w(x^I) \geq m(d-1)$, which contradicts the assumption and Lemma 2.8. Q.E.D. for Lemma 2.9.

Note that for $d \geq 3$ the inequalities $b \leq m$ and $b \geq m(d-1)$ cannot hold simultaneously. Therefore, at least one of the two inequalities of Lemma 2.9 holds. Let $b \geq m + 1$. Since V_2 is non-singular, P_* is a point. Let $\mathcal{A}_* = \{x_0\}$, so that $P_* = (1, 0, \dots, 0)$. Again from the fact that V_2 is non-singular, we get $f^{(1)}(1, 0, \dots, 0) \neq 0$. Therefore, the monomial x_0^d comes into $f^{(1)}$ with a non-zero coefficient. By Lemma 2.8 we get $b \leq 1$. Therefore $m = 0$, a contradiction.

In the case $b \leq m(d-1) - 1$ the arguments are symmetric: V_1 is non-singular, P^* is a point $(0, \dots, 0, 1)$, $\mathcal{A}^* = \{x_M\}$ and $g^{(1)}(0, \dots, 0, 1) \neq 0$, so that $md \leq b + 1$, whence we get again $m = 0$, a contradiction.

Therefore, there are no non-trivial weights and ξ is a fibrewise biregular isomorphism. So $\chi = \xi|_{V_1}$ is a fibrewise isomorphism as well. Proof of Theorem 4.1.1 is complete.

3. Pencils of double hypersurfaces

In this section we prove birational superrigidity of another wide class of Fano fibre spaces $V/\mathbb{P}^1$ with Fano double hypersurfaces of index one as fibres. As in Section 2, we check that the assumptions of Theorem 1.1 are satisfied. A part of this work is similar to the case of Fano hypersurfaces considered in Section 2, but the crucial and most difficult part of the proof, when the singularity of a fibre comes from a singularity of the branch divisor, presents a new phenomenon that requires a new approach.

We start with an explicit construction of the fibre spaces (Subsection 3.1) and state the regularity conditions (Subsection 3.2), paying special attention to the regularity of the double points. This done, we formulate the main result of the section (birational superrigidity for the given class of Fano fibre spaces) and start the proof, constructing the hypertangent systems (Subsection 3.3). Non-singular points of fibres present no problem: their treatment in the absolute case (Section 3.2) is sufficient for the relative one. But checking the condition (vs) is really difficult (Subsections 3.4–3.9), especially when the singularity of the fibre comes from the singularity of the branch divisor, when a new technique needs to be developed (Subsections 3.6–3.9).

3.1. Explicit construction of the fibre space. The symbol $\mathbb{P}$ stands for the projective space $\mathbb{P}^{M+1}$ over the field of complex numbers $\mathbb{C}$, $M \geq 5$. Let

$$\mathcal{G} = \mathbb{P}(H^0(\mathbb{P}, \mathcal{O}_{\mathbb{P}}(m)))$$

be the space of all Fano hypersurfaces of degree m, $3 \leq m \leq M-1$, $\mathcal{W}$ the space of all hypersurfaces of degree $2l$ in $\mathbb{P}$, where $m + l = M + 1$. Let

$$\mathcal{F} = \{F \mid \sigma \colon F \stackrel{2:1}{\to} G\}$$

be the class of algebraic varieties realized as double covers of hypersurfaces $G \in \mathcal{G}$ branched over $W \cap G$, $W \in \mathcal{W}$. Set $\mathcal{F}_{sm} \subset \mathcal{F}$ to be the set of smooth double hypersurfaces corresponding to pairs $(G, W \cap G)$ of smooth varieties. Obviously, $F \in \mathcal{F}_{sm}$ is a smooth Fano variety of index 1 with the Picard group $\operatorname{Pic} F = \mathbb{Z}K_F$. Let $\mathcal{F}_{sm}^{reg} \subset \mathcal{F}_{sm}$ be the smooth subset consisting of varieties $F \in \mathcal{F}_{sm}$ satisfying the *regularity condition* (R1.1) and (R2.1) of Section 3.2. It was shown there (as a double cover is a particular case of a cyclic cover) that the set $\mathcal{F}_{sm}^{reg}$ is non-empty. Moreover, the following estimate holds:

$$\operatorname{codim}_{\mathcal{F}_{sm}}(\mathcal{F}_{sm} \setminus \mathcal{F}_{sm}^{reg}) \geq 2.$$

Set $\mathcal{F}_{sing} = \mathcal{F} \setminus \mathcal{F}_{sm}$, $\operatorname{codim}_{\mathcal{F}} \mathcal{F}_{sing} = 1$. Let $\mathcal{F}_{sing}^{reg}$ be the open subset in $\mathcal{F}_{sing}$, consisting of all singular double hypersurfaces satisfying the regularity conditions (R1.2) and (R2.2) for singular points formulated below at the end of this subsection. It is easy to see that the inequality

$$\operatorname{codim}_{\mathcal{F}}(\mathcal{F}_{\mathrm{sing}} \setminus \mathcal{F}_{\mathrm{sing}}^{\mathrm{reg}}) \geq 2$$

holds. Our aim is to construct and study Fano fibre spaces $V/\mathbb{P}^1$, each fibre $F_t = \pi^{-1}(t)$, $t \in \mathbb{P}^1$, of which is a variety from the family $\mathcal{F}$. Set

$$\mathcal{F}^{\mathrm{reg}} = \mathcal{F}^{\mathrm{reg}}_{sm} \cup \mathcal{F}^{\mathrm{reg}}_{\mathrm{sing}}.$$

By what was said above, $\operatorname{codim}_{\mathcal{F}}(\mathcal{F} \setminus \mathcal{F}^{\mathrm{reg}}) \geq 2$. Since the fibre space $V/\mathbb{P}^1$ can be looked at as a morphism $\mathbb{P}^1 \to \mathcal{F}$, that is, a curve in $\mathcal{F}$, for a general variety $V/\mathbb{P}^1$ we get:

$$F_t \in \mathcal{F}^{\mathrm{reg}}$$

for all points $t \in \mathbb{P}^1$. If this is the case, we say that the fibre space $V/\mathbb{P}^1$ is *regular.*

Let us describe an explicit construction of regular fibre spaces $V/\mathbb{P}^1$. For each fibre $F \in \mathcal{F}^{\mathrm{reg}}$ (singular or smooth) the anticanonical linear system $|-K_F|$ determines precisely the double cover $\sigma_F \colon F \to G \subset \mathbb{P}$. For this reason, $\pi_*\mathcal{O}(-K_V)$ is a locally free sheaf of rank $M+2$ on $\mathbb{P}^1$. It gives a locally trivial $\mathbb{P}$-fibration over $\mathbb{P}^1$. The variety V is realized as a double cover of a smooth divisor Q on $\mathbb{P}(\pi_*\mathcal{O}(-K_V))$. Namely, let

$$\mathcal{E} = \bigoplus_{i=0}^{M+1} \mathcal{O}_{\mathbb{P}^1}(a_i)$$

be a locally free sheaf, normalized by the condition that

$$a_0 = 0 \leq a_1 \leq \cdots \leq a_i \leq a_{i+1} \leq \cdots \leq a_{M+1}.$$

In particular, $\mathcal{E}$ is generated by global sections. Set $X = \mathbb{P}(\mathcal{E})$ to be its **Proj** in the sense of Grothendieck, $\pi_X \colon X \to \mathbb{P}^1$ the natural projection, $\mathcal{L}_X$ the tautological sheaf, $Q \subset X$ a smooth divisor on X, corresponding to a section

$$s_Q \in H^0\left(X, \mathcal{L}_X^{\otimes m} \otimes \pi_X^*\mathcal{O}_{\mathbb{P}^1}(a_Q)\right),$$

$a_Q \in \mathbb{Z}_+$. The symbol $\pi_Q \colon Q \to \mathbb{P}^1$ stands for the projection $\pi_X|_Q$. Obviously, $Q/\mathbb{P}^1$ is a smooth fibration into Fano hypersurfaces of degree m in $\mathbb{P}$. Let $W \subset X$ be an irreducible hypersurface, corresponding to a section

$$s_W \in H^0\left(X, \mathcal{L}_X^{\otimes 2l} \otimes \pi_X^*\mathcal{O}_{\mathbb{P}^1}(2a_W)\right),$$

$a_W \in \mathbb{Z}_+$, whereas $W_Q = W \cap Q$ is a smooth divisor on Q. We denote the fibre $\pi_Q^{-1}(t)$ over a point $t \in \mathbb{P}^1$ by the symbol G_t (or just G, when it is clear which point is meant or when it is inessential). Finally, set

$$\sigma \colon V \to Q$$

to be the double cover, branched over W_Q. The natural projection onto $\mathbb{P}^1$ will be denoted by π, the fibre $\pi^{-1}(t)$ by the symbol F_t (or just F). It is easy to see that

$$\operatorname{Pic} V = \mathbb{Z}K_V \oplus \mathbb{Z}F$$

and up to twisting by an invertible sheaf $\mathcal{O}_{\mathbb{P}^1}(k)$, $k \in \mathbb{Z}$, the sheaves $\mathcal{E}$ and $\pi_*\mathcal{O}(-K_V)$ on $\mathbb{P}^1$ coincide.

More precisely, let $L_X \in \operatorname{Pic} X$ be the class of the tautological sheaf $\mathcal{L}_X$, $L_Q = L_X|_Q$ its restriction to Q, so that $\operatorname{Pic} Q = \mathbb{Z}L_Q \oplus \mathbb{Z}G$. Set $L_V = \sigma^* L_Q$. It is easy to see that

$$K_V = -L_V + (a_1 + \cdots + a_M - 2 + a_Q + a_M)F.$$

By the Lefschetz theorem $A^2 V = \mathbb{Z}K_V^2 \oplus \mathbb{Z}H_F$, where $H_F = (-K_V \cdot F)$ is the class of a hyperplane section. The symbol H_F is used in the present paper in two different meanings: as a class of codimension two on V and as the hyperplane section of the fibre, that is, an element of A^1F. Every time it is clear which of the two concepts is meant.

It is easy to compute that

$$(K_V^2 \cdot L^{M-1}) = 2m(4 - a_1 - \cdots - a_{M+1} - a_Q - a_W) + 2a_Q.$$

Since $(H_F \cdot L^{M-1}) = 2m$ and the linear system $|L_V|$ is free, the inequality $(K_V^2 \cdot L^{M-1}) \leq 0$ implies that $K_V^2 \notin \operatorname{Int} A_+^2 V$, where $A_+^2 V \subset A^2 V \otimes \mathbb{R}$ is the closed cone of effective cycles of codimension two.

3.2. The regularity conditions. Now let us formulate the regularity conditions. For smooth points, they are just special cases of the regularity conditions (R1.1) and (R2.1) for Fano cyclic covers considered in Chapter 3. We recall the notations.

Let $\sigma\colon F \to G \subset \mathbb{P}$ be a Fano double hypersurface of index 1, $F \in \mathcal{F}$. The variety F is realized as a complete intersection of codimension two in the weighted projective space

$$\mathbb{P}(\underbrace{1, 1, \ldots, 1}_{M+2}, l),$$

F is of type $m \cdot 2l$ and given by the pair of equations

$$\tilde{f}(x_0, \ldots, x_{M+1}) = 0, \quad u^2 = \tilde{g}(x_0, \ldots, x_{M+1}),$$

where x_* are the coordinates of weight 1, u is the coordinate of weight l, $\tilde{f}$ is the equation of the hypersurface $G \subset \mathbb{P} = \mathbb{P}(1, \ldots, 1)$, $\tilde{g}$ is the equation of the hypersurface $W \cap \mathbb{P}$.

Let $o \in F$ be an arbitrary point. First, we draw the reader's attention to the following obvious fact:

$$o \neq (\underbrace{0, 0, \ldots, 0}_{M+2}, 1).$$

Thus we may assume that the point o lies in one of the standard affine charts $\mathbb{A}^{M+2}$ with the coordinates

$$z_i = x_i/x_0, \quad i = 1, \ldots, M+1, \quad y = u/x_0^l$$

and its z_*-coordinates are $(0, \ldots, 0)$. With respect to the coordinate system (z_*, y) the affine part of the variety F is given by the pair of equations

$$f = q_1 + \cdots + q_m = 0, \quad y^2 = g = w_0 + \cdots + w_{2l},$$

where q_i and w_j are homogeneous polynomials in z_* of degrees i and j, respectively. Set $p = \sigma(o) \in G$. The point p lies on the branch divisor W if and only if $w_0 = 0$. If $p \notin W$, then we normalize the second equation and assume that $w_0 = 1$.

Let us formulate first the regularity conditions outside the branch divisor. In this case the fibre F is given with respect to the affine coordinate system (z_*, y) with the origin of the z_*-system at $p = \sigma(o)$ by the equations

$$\begin{cases} f = q_a + \cdots + q_m = 0, \\ y^2 = g = 1 + w_1 + \cdots + w_{2l}, \end{cases}$$

where $a \geq 1$. Set

$$\sqrt{g} = (1 + w_1 + \cdots + w_{2l})^{1/2} = 1 + \sum_{i=1}^{\infty} \gamma_i (w_1 + \cdots + w_{2l})^i$$
$$= 1 + \sum_{i=1}^{\infty} \Phi_i(w_1, \ldots, w_{2l}),$$

where $\Phi_i(w_1(z_*), \ldots, w_{2l}(z_*))$ are homogeneous in z_* of degree $i \geq 1$,

$$\gamma_i = (-1)^{i-1} \frac{(2i-3)!!}{2^i i!} = (-1)^{i-1} \frac{(2i-3)!}{2^{2i-2} i!(i-2)!}$$

is the standard i-th coefficient of the Taylor expansion of the function $(1+s)^{1/2}$ at the point $s = 0$. Obviously,

$$\Phi_i(w_*) = w_i + A_i(w_1, \ldots, w_{i-1})$$

for $i \leq 2l$. For $i \geq 1$ set

$$[\sqrt{g}]_i = 1 + \sum_{j=1}^{i} \Phi_i(w_*), \quad g^{(i)} = g - [\sqrt{g}]_i^2.$$

It is easy to see that the first non-zero component of the polynomial $g^{(i)}$ is of degree $i+1$. More precisely, this component is equal to

$$g_{i+1} = 2\Phi_{i+1}(w_1(z_*), \ldots, w_{i+1}(z_*)).$$

Recall that the regularity condition (R1.1) at a smooth point $p \in G$ requires that if $m \leq 2l$, then the sequence

$$q_1, \ldots, q_m, g_{l+1}, \ldots, g_{2l-1}$$

is regular in $\mathcal{O}_{p,\mathbb{P}}$; if $m \geq 2l+1$, then the sequence

$$q_1, \ldots, q_{m-1}, g_{l+1}, \ldots, g_{2l}$$

should be regular in $\mathcal{O}_{p,\mathbb{P}}$. Here $a = 1$.

The regularity condition (R1.2) at a double point $p \in G$ outside the ramification divisor requires that if $2l \geq m+1$, then the system of $M-1$ homogeneous polynomials

$$q_2, \ldots, q_m, g_{l+1}, \ldots, g_{2l-1},$$

whereas if $2l \leq m$, then the system of homogeneous polynomials

$$q_2, \ldots, q_{m-1}, g_{l+1}, \ldots, g_{2l}$$

defines a curve in $\mathbb{P}^M = \mathbb{P}(T_p\mathbb{P})$, neither component of which is contained in a hyperplane. Furthermore, the system of M homogeneous equations

$$q_2 = \cdots = q_{m-1} = g_{l+1} = \cdots = g_{2l} = 0 \tag{110}$$

defines a non-zero subscheme Z_* in $\mathbb{P}^M$, such that for any hyperplane $P \subset \mathbb{P}^M$,

$$\deg(P \cap Z_*) < \lambda_{m,l} = \frac{m!(2l-1)!}{6(l-1)!}$$

for $m \geq 4$ and

$$\deg(P \cap Z_*) < \lambda_{3,l} = 12\frac{(2l-1)!}{(l+1)!}(l-2)$$

for $m = 3$. If the scheme Z_* is reduced, then this condition means simply that any set of $\lambda_{m,l}$ points is not contained in a hyperplane.

Remark 3.1. Since $w_0 = y(0) = 1$, in a neighborhood of the singular point $o \in F$ the equations

$$y - [\sqrt{g}]_i = 0 \quad \text{and} \quad \sigma^* g^{(i)} = 0$$

define the same divisor. Consider the system of equations (110) on the fibre F (and not on the projectivized tangent space $\mathbb{P}(T_p\mathbb{P})$). The system defines an effective 1-cycle C_* on F. By construction, its degree is equal to

$$\deg C_* = 2m!\frac{(2l-1)!}{(l-1)!},$$

whereas its multiplicity at the point $o \in F$ satisfies the estimate

$$\text{mult}_o C_* \geq m!\frac{(2l)!}{l!} = \deg C_*, \tag{111}$$

so that what we actually have in (111) is an equality and C_* is an algebraic sum of lines on F, that is, curves of the form $L \ni o$, the image $\sigma(L) \subset \mathbb{P}$ of which is a line and, moreover the morphism $\sigma\colon L \to \sigma(L)$ is an isomorphism. Considering the zero-dimensional scheme Z_* as an effective zero-dimensional cycle, we get by construction:

$$Z_* = \mathbb{P}(T_o C_*).$$

In particular, for any hyperplane $P \subset \mathbb{P}$ the one-dimensional part of the scheme

$$\{q_2 = \cdots = q_m = g_{l+1} = \cdots = g_{2l} = 0\} \cap \sigma^{-1}(P)$$

is of degree not higher than $\lambda_{m,l} - 1$. In other words, if all components of the cycle C_* are of multiplicity 1, then no more than $\lambda_{m,l} - 1$ of these lines are contained in $\sigma^{-1}(P)$.

Now let us consider the regularity conditions for points on the ramification divisor. In this case the variety F is given with respect to the affine coordinate system (z_*, y) by the system of equations

$$\begin{cases} f = q_1 + \cdots + q_m = 0, \\ y^2 = g = w_1 + \cdots + w_{2l}. \end{cases}$$

The regularity condition (R2.1) at a smooth point $o \in F$ on the ramification divisor requires that the sequence of homogeneous polynomials

$$q_1, \ldots, q_m$$

is regular in $\mathcal{O}_{p,\mathbb{P}}$ and the quadratic form q_2 does not vanish identically on the plane $\{q_1 = w_1 = 0\}$.

Note that since the point $o \in F$ is smooth, this plane is of codimension exactly two, that is, the linear forms q_1 and w_1 are linearly independent: the plane $\{q_1 = w_1 = 0\}$ is the tangent plane to the branch divisor $W \cap G$ of the morphism σ_F.

Finally, let us formulate the regularity condition (R2.2) at a double point $o \in F$. In this case we have the double cover $\sigma_F\colon F \to G$, branched over the divisor $W_G = W \cap G$. The first regularity condition is smoothness of the hypersurface G at the point $p = \sigma(o)$, that is, $q_1 \not\equiv 0$. Furthermore, the divisor W_G should have at the point p a non-degenerate quadratic singularity:

$$w_1 = \lambda q_1,$$

$\lambda \in \mathbb{C}$. For convenience of notations assume that $q_1 = z_{M+1}$. The quadratic polynomial

$$\bar{w}_2 = w_2|_{\{z_{M+1}=0\}}$$

is of the maximal rank. Let $E_G \cong \mathbb{P}^{M-1}$ be the exceptional divisor of the blow up $\varphi_G \colon \widetilde{G} \to G$ of the point p. Take $z_1, \dots, z_M$ for homogeneous coordinates on E_G and set

$$W_E = \{\bar{w}_2 = 0\}.$$

It is a non-singular quadratic hypersurface in E_G. Denote by the symbol $\bar{q}_i$ the restriction of the homogeneous polynomial q_i onto the hyperplane $q_{M+1} = 0$. Now the remaining part of the condition (R2.2) is as follows:

The system of homogeneous equations

$$\bar{q}_2 = \cdots = \bar{q}_m = 0$$

defines in $E_G \cong \mathbb{P}^{M-1}_{(z_1:\cdots:z_M)}$ an irreducible subvariety $Z_{2,\dots,m}$, which is an irreducible reduced complete intersection of codimension $(m-1)$. The quadric

$$\bar{q}_2 = 0$$

is smooth and distinct from W_E.

Definition 3.1. A Fano double hypersurface $F \in \mathcal{F}$ is *regular*, if each smooth point on it is regular in the sense of the corresponding condition (R1.1) or (R2.1) and each of its singular points is regular in the sense of the corresponding condition (R1.2) or (R2.2). Notation: $F \in \mathcal{F}^{\text{reg}}$.

The conditions (R1.1) and (R2.1) were introduced in Chapter 3, where it was shown that non-regular smooth cyclic covers form a closed subset of codimension at least two in the set of all smooth cyclic covers; this applies directly to the set of double hypersurfaces $\mathcal{F}_{sm}$. Moreover, it follows from the computations of Section 3.3 that the set of Fano double hypersurfaces F with at least one *smooth* non-regular point $o \in F$ is of codimension at least two in $\mathcal{F}$. Thus a general singular double hypersurface $F \in \mathcal{F}_{\text{sing}}$ has exactly one singular point whereas all its smooth points are regular. The singular point $o \in F$ is a non-degenerate double point. If $p = \sigma(o) \notin W_G$, then the fact that the condition (R1.2) is open implies that in a neighborhood of $F \in \mathcal{F}$ the following estimate holds:

$$\operatorname{codim}_{\mathcal{F}_{\text{sing}}}(\mathcal{F}_{\text{sing}} \setminus \mathcal{F}^{\text{reg}}_{\text{sing}}) \geq 1 \tag{112}$$

and thus

$$\operatorname{codim}_{\mathcal{F}}(\mathcal{F}_{\text{sing}} \setminus \mathcal{F}^{\text{reg}}_{\text{sing}}) \geq 2. \tag{113}$$

If $p = \sigma(o) \in W_G$, then in a similar way the fact that the condition (R2.2) is open implies the estimate (112) in a neighborhood of $F \in \mathcal{F}$. Thus the estimates (112) and (113) are global.

3.3. The main result. Start of the proof. Let us formulate the main result.

Theorem 3.1. *Assume that a regular fibre space $V/\mathbb{P}^1$ satisfies the K^2-condition. The the fibre space $V/\mathbb{P}^1$ is birationally superrigid.*

To prove the theorem, we need to show that regular Fano double covers satisfy the conditions (h), (v) and (vs). For the conditions (h) and (v) it is done in exactly the same way as for the Fano hypersurfaces in Section 2, taking into account the proof of Theorem 3.2.2 given in Chapter 3. The really new problem is how to prove the condition (vs) for a regular double cover.

We start with the case of a double point of a fibre outside the ramification divisor. This case is similar to that of a double point of a Fano hypersurface considered in Section 2, but in the case of a double cover some constructions need to be done in a different way, so we will consider this case in full detail.

Let $\varphi = \varphi_{F,o}\colon \widetilde{F} \to F$ be the blow up of the fibre at an arbitrary point o, $\varphi_G = \varphi_{G,p}\colon \widetilde{G} \to G$ the blow up of the fibre G at the point $p = \sigma(o)$, $E = E_F \subset \widetilde{F}$ and $E_G \subset \widetilde{G}$ the exceptional divisors. Let us consider the hypertangent linear systems

$$\Lambda_k = \Lambda_k^F = \varphi_*(|kH_F - (k+1)E|)$$

on F and

$$\Lambda_k^G = (\varphi_G)_*(|kH_G - (k+1)E_G|)$$

on G. The system Λ_k is the largest linear subsystem of the system $|kH_F|$, the strict transform of which satisfies the property

$$\widetilde{\Lambda}_k \subset |kH_F - (k+1)E|,$$

and similarly for G. In the general case one cannot assert that

$$\sigma^*\Lambda_k^G \subset \Lambda_k, \tag{114}$$

since if $p \in W_G$ is a smooth point of the branch divisor, then the double cover $\sigma\colon F \to G$ *does not* extend to a double cover $\widetilde{F} \to \widetilde{G}$ (there is a rational map of degree two between these varieties; this rational map has a fairly simple structure, however, it is not a finite morphism). But if $p \notin W_G$ or $p \in W_G$ is a double point of the branch divisor, then the inclusion (114) holds. The symbol Λ_k^E stands for the corresponding linear system on the exceptional divisor:

$$\Lambda_k^E = \widetilde{\Lambda}_k|_E \quad \text{or} \quad \Lambda_k^E = \widetilde{\Lambda}_k^G|_{E_G},$$

depending on the context. It is easy to see that

$$(\widetilde{\operatorname{Bs}\Lambda_k \circ E}) = \operatorname{Bs}\Lambda_k^E$$

in the scheme-theoretic sense, in particular, the corresponding effective algebraic cycles are equal, that is, the equality respects multiplicities.

Abusing out notations, we sometimes use the notion of a hypertangent system for a certain special subsystem of the hypertangent system, which permits an explicit description. Let $p \in G$ be a point, $z_1, \dots, z_{M+1}$, a system of linear coordinates with the origin at p, and assume that the hypersurface G is given by the equation

$$f = q_a + q_{a+1} + \dots + q_m = 0,$$

$a = 1$ or 2. Then

$$\Lambda_k^G \supset \left|\sum_{i=a}^{k} s_{k-i} f_i\right|, \tag{115}$$

where

$$f_i = q_a + \dots + q_i,$$

$k \geq a$ and s_j means an arbitrary homogeneous polynomial of degree j in the variables z_*. The inclusion (115) is obvious, since

$$f_i|_G = (-q_{i+1} - \cdots - q_m)|_G.$$

Now assume that $p \notin W_G$. The hypertangent system Λ_k is constructed in the same way as it was done for the hypertangent systems for cyclic covers (Section 3.2). Obviously, $\Lambda_k \supset \sigma^* \Lambda_k^G$, but in fact the system Λ_k is much larger. Since $p \notin W_G$, we may assume that the hypersurface $W_t = W \cap \mathbb{P}_t \subset \mathbb{P}$ is given by the equation $g(z_*) = 1 + w_1 + \cdots + w_{2l} = 0$, where $w_i(z_*)$ are homogeneous of degree i. Setting formally

$$\sqrt{g} = 1 + \sum_{i=1}^{\infty} \Phi_i(w_1, \ldots, w_{2l}), \tag{116}$$

where $\Phi_i(w_1(z_*), \ldots, w_{2l}(z_*))$ are homogeneous polynomials of degree i in z_*, write for $j \geq 1$,

$$[\sqrt{g}]_j = 1 + \sum_{i=1}^{j} \Phi_i(w_*(z_*)).$$

Now we get

$$\Lambda_k \supset \left| \sum_{i=a}^{k} s_{k-i} f_i + \sum_{i=l}^{\min\{k,2l-1\}} s^*_{k-i}(y - [\sqrt{g}]_i) \right|, \tag{117}$$

where s^*_{k-i} are homogeneous polynomials in z_* of degree $k-i$; if $k \leq l-1$, then the right-hand side is assumed to be equal to zero. The inclusion (117) follows from (115) and the following fact.

Lemma 3.1. *In the local coordinates z_* we get*

$$(y - [\sqrt{g}]_i)|_F = 2\Phi_{i+1}(w_*(z_*))|_F + \ldots,$$

where the dots stand for a formal series, the components of which are homogeneous polynomials of degree $i+2$ and higher in the variables z_*.

Proof. It is obvious, since $(y^2 - g)|_F \equiv 0$, $g(p) = 1$ and the formal decomposition (116) holds. Q.E.D.

Note that

$$\Phi_i(w_*) = \frac{1}{2} w_i + A_i(w_1, \ldots, w_{i-1}).$$

Now let us consider the case when $p = \sigma(o) \in W_G$. If the branch divisor is non-singular at the point p, then the local equation of the hypersurface W_t is of the form

$$g(z_*) = w_1 + \cdots + w_{2l} = 0,$$

where the linear forms q_1, w_1 are linearly independent. Since the inverse image of the divisor $\{w_1|_G = 0\}$ on F is obviously singular, we obtain:

$$\Lambda_k \supset \left| \sum_{i=1}^{k} s_{k-i} f_i + s_{k-1} w_1 \right|. \tag{118}$$

However, if $p = \sigma(o)$ is a singularity of the divisor W_G, then our methods of constructing hypertangent linear systems give at most the inclusion $\Lambda_k \supset \sigma^* \Lambda_k^G$, where Λ_k^G is given by the formula (115).

The regularity conditions make it possible to get a lower bound for the codimension of the base set of hypertangent systems. In the formulae below it is assumed that the segment $[a,b] \subset \mathbb{R}$ is an empty set when $b < a$. For an arbitrary point $o \in F$ set

$$\mathcal{M} = [a, m-1] \cap \mathbb{Z}_+ = \{a, \dots, m-1\},$$

where $a = \mathrm{mult}_p G \in \{1, 2\}$, $p = \sigma(o)$, and

$$\mathcal{L} = [l, 2l + a - 3] \cap \mathbb{Z}_+ = \{l, \dots, 2l + a - 3\}.$$

Thus the sets $\mathcal{M}$, $\mathcal{L}$ depend on the type of the point o. At each stage of the proof the point o is assumed to be fixed and the symbols $\mathcal{M}$, $\mathcal{L}$ mean the sets corresponding to this point.

For $e = \max\{m-1, 2l-1\}$ we denote the hypertangent linear system Λ_e by the symbol Λ_∞.

Proposition 3.1. *The following estimates hold:*

(i) *if* $p = \sigma(o) \notin W_G$ *is a smooth point of the hypersurface* G*, then*

$$\mathrm{codim}_o \,\mathrm{Bs}\,\Lambda_k \geq \mathrm{codim}_E \,\mathrm{Bs}\,\Lambda_k^E \geq \sharp[1,k] \cap \mathcal{M} + \sharp[1,k] \cap \mathcal{L},$$

in particular, $\dim_o \mathrm{Bs}\,\Lambda_\infty \leq 1$,

(ii) *if* $p = \sigma(o) \notin W_G$ *is a double point of the hypersurface* G*, then*

$$\mathrm{codim}\,\mathrm{Bs}\,\Lambda_k \geq \mathrm{codim}_E \,\mathrm{Bs}\,\Lambda_k^E \geq \sharp[2,k] \cap \mathcal{M} + \sharp[2,k] \cap \mathcal{L}$$

and, moreover, $\mathrm{Bs}\,\Lambda_\infty \subset C_*$. *Furthermore, let* $P \subset \mathbb{P}$, $P \ni p$, *be an arbitrary hyperplane,* $P_F = \sigma^{-1}(P \cap G)$ *the corresponding section of the fibre* F, $\Lambda_k^P = \Lambda_k|_{P_F}$ *the restriction of the linear system* Λ_k *onto* P_F. *Then for* $k \leq \max\{m, 2l\} - 2$,

$$\mathrm{codim}_{P_F} \mathrm{Bs}\,\Lambda_k^P \geq \sharp[2,k] \cap \mathcal{M} + \sharp[2,k] \cap \mathcal{L},$$

and $\dim \mathrm{Bs}\,\Lambda_\infty^P \leq 1$, *whereas if in the last inequality the equality holds, then the degree of the one-dimensional part of the basic subscheme* $\mathrm{Bs}\,\Lambda_\infty^P$ *does not exceed* $\lambda_{m,l}$.

(iii) *If* $p = \sigma(o) \in W_G$ *is a smooth point on the branch divisor* W_G*, then the following inequality holds:*

$$\mathrm{codim}_o \,\mathrm{Bs}\,\Lambda_k \geq \mathrm{codim}_E \,\mathrm{Bs}\,\Lambda_k^E \geq \sharp[1,k] \cap \mathcal{M} + 1,$$

(iv) *if* $p = \sigma(o) \in W_G$ *is a double point on the branch divisor* W_G*, then the following inequality holds:*

$$\mathrm{codim}_o \,\mathrm{Bs}\,\Lambda_k \geq \mathrm{codim}_E \,\mathrm{Bs}\,\Lambda_k^E \geq \sharp[1,k] \cap \mathcal{M}.$$

Proof. To obtain our claims, we replace the hypertangent linear systems Λ_k by their subsystems (115), (117) and (118), constructed above, and use the regularity conditions (Subsection 3.2). Q.E.D.

3.4. Scheme of the proof of the condition (vs)**.** Assume that there exists a prime divisor $Y \subset F = F_t$, satisfying the estimate

$$\frac{\mathrm{mult}_x \widetilde{Y}}{\deg Y} > \frac{1}{m}, \tag{119}$$

where $x \in E$ is an infinitely near point of the first order, that is, $E \subset \widetilde{F}$ is the exceptional divisor of the blow up of the point $o \in F$, $\varphi\colon \widetilde{F} \to F$. Here the singular point $o \in F$ is generated by a singularity of the hypersurface $G = G_t$, that is, $p = \sigma(o) \in G$ is a non-degenerate double point, $p \notin W$. Set $\sigma^{-1}(p) = \{o, o^+\}$ and let $\varphi_G \colon \widetilde{G} \to G$ be the blow up of the point p. The map σ extends in an obvious way to a morphism

$$\tilde{\sigma} \colon \widetilde{F} \setminus \{o^+\} \to \widetilde{G},$$

whereas on the exceptional divisor $E \subset \widetilde{F}$ the morphism $\tilde{\sigma}$ is an isomorphism, which makes it possible to identify E with the exceptional divisor of the blow up φ_G and thus consider E as embedded in $\mathbb{T} = \mathbb{P}(T_p\mathbb{P}) \cong \mathbb{P}^M$, that is, in the exceptional divisor of the blow up

$$\varphi_{\mathbb{P}} \colon \widetilde{\mathbb{P}} \to \mathbb{P}$$

of the point $p \in \mathbb{P}$. Depending on the context, one of the inclusions $E \subset \widetilde{F}$ or $E \subset \widetilde{G}$ will be meant.

Let us show that the assumption (119) leads to a contradictions. In order to do that, we will use the method developed in Section 2. The arguments break into a few steps. The **first step** is given by

Proposition 3.2. *There exists a hyperplane $P \subset \mathbb{P}$, $P \ni p$, such that $\sigma(Y) \not\subset P$ and the effective algebraic cycle $Y_P = (Y \circ_F P_F)$, where $P_F = \sigma^{-1}(P_G)$, $P_G = P \cap G$ is a hyperplane section, satisfies the estimate*

$$\frac{\mathrm{mult}_o}{\deg} Y_P > \frac{3}{2m}.$$

The symbol $\circ_F$ is used to emphasize that the cycle Y_P is constructed in the sense of the intersection theory on F, and not on V. The proof is word for word the same as the proof of Proposition 2.1 based on Lemma 2.3 above.

Step two. Consider the variety $P_F \subset F$. It is an irreducible variety of dimension $M-1$ with the double point $o \in P_F$. Let $\varphi_P \colon \widetilde{P} \to P_F$ be the blow up of the point o, $E_P \subset \widetilde{P}$ the exceptional divisor. Obviously, $\widetilde{P}$ embeds into $\widetilde{F}$, and E_P into E as a hyperplane section of the quadric E with respect to the embedding $E \hookrightarrow \mathbb{T}$. Since the variety $\widetilde{F}$ is factorial, the strict transform $\widetilde{Y}$ is a Cartier divisor. Therefore, the effective cycle $\widetilde{Y}_P = (\widetilde{Y} \circ \widetilde{P})$, that is, the strict transform of the cycle Y_P on $\widetilde{P}$, is a Cartier divisor,

$$\widetilde{Y}_P \sim aH_P - bE_P,$$

where H_P is the class of a hyperplane section. By Proposition 3.2, $b > \frac{3}{2}a$. By the regularity condition we get for the tangent divisor

$$T = \sigma^{-1}(T_pG \cap G)$$

that $\mathrm{mult}_o T = 6$, $\deg T = 4m$, so that for the class of its strict transform $\widetilde{T} \subset \widetilde{F}$ we get $\widetilde{T} \sim 2H - 3E$ and thus for its restriction $\widetilde{T}_P = \widetilde{T} \cap \widetilde{P}$ on $\widetilde{P}$ we get

$$\widetilde{T}_P \sim 2H_P - 3E_P.$$

Proposition 3.3. *Let $Z \sim \alpha H_P - \beta E_P$ be an effective Cartier divisor on $\widetilde{P}$. Assume that $\beta > \frac{3}{2}\alpha$. Then Z contains $\widetilde{T}_P$ as a component of positive multiplicity.*

Proof is given below.

Step three. Write the effective divisor $\widetilde{Y}_P$ in the following form:

$$\widetilde{Y}_P = c\widetilde{T}_P + Z,$$

where $c \in \mathbb{Z}_+$ and the effective divisor Z does not contain $\widetilde{T}_P$ as a component. Setting $Z \sim \alpha H_P - \beta E_P$, we obtain from the system of equations

$$a = 2c + \alpha, \quad b = 3c + \beta$$

and the condition $2b > 3a$, that

$$\beta > \frac{3}{2}\alpha.$$

By Proposition 3.3 this implies that $\widetilde{T}_P$ is a component of positive multiplicity of the divisor Z. A contradiction.

Thus we have proved that the estimate (119) is impossible which implies that the condition (vs) holds for the case of a singular point $o \in F$ outside the branch divisor. Q.E.D.

3.5. Mobile families of curves. Let us prove Proposition 3.3. We use the same method as we did in the proof of Lemma 2.4.

Lemma 3.2. *The divisor $T_P = T \cap P_F$ is swept out by a family of curves $\{C_\delta, \delta \in \Delta\}$, the general member of which is irreducible and satisfies the inequality*

$$\frac{\mathrm{mult}_o}{\deg} C_\delta > \frac{2}{3}. \tag{120}$$

First, let us obtain Proposition 3.3 from this fact. Let $\{\widetilde{C}_\delta, \delta \in \Delta\}$ be the strict transform of this family of curves on $\widetilde{P}$, $\widetilde{T}_P \subset \widetilde{P}$ the strict transform of the divisor T_P. Obviously,

$$(Z \cdot \widetilde{C}_\delta) = \alpha \deg C_\delta - \beta\, \mathrm{mult}_o\, C_\delta < 0,$$

since $\beta > \frac{3}{2}\alpha$. Therefore, $\widetilde{C}_\delta \subset Z$. However, the curves $\widetilde{C}_\delta$ sweep out $\widetilde{T}_P$, thus $Z \supset \widetilde{T}_P$. Q.E.D. for Proposition 3.3.

Proof of Lemma 3.2. The variety P_F is of dimension $m + l - 2$, the divisor $T_P \subset P_F$ is of dimension $m + l - 3$. We construct the required family of curves $(C_\delta, \delta \in \Delta)$, intersecting T_P with $m + l - 4$ hypertangent divisors. To order the construction procedure, let us introduce some new notations:

$$\mathcal{M} = \{2, \ldots, m-1\}, \quad \mathcal{L} = \{l, \ldots, 2l-1\},$$

$$c_e = \sharp[4, e] \cap \mathcal{M} + \sharp[3, e] \cap \mathcal{L}, \quad e \in \mathbb{Z}_+. \tag{121}$$

Here and below we assume silently that the segment $[a, b] \subset \mathbb{R}$ is the empty set when $b < a$. For $e \leq 2$ we get $c_e = 0$, for $e \geq \max\{m, 2l\} - 1$ we get that $c_e = m + l - 4$, provided that $l \geq 3$. Let us assume that this is the case and that $m \geq 4$. Note that for $m = 4$,

$$\sharp[4, e] \cap \mathcal{M} = 0,$$

since this set is empty. The cases $l = 2$ and $m = 3$ we will treat separately. Obviously, $c_{e+1} \geq c_e$. Define the *ordering function*

$$\chi\colon \{1, \ldots, m + l - 4\} \to \mathbb{Z}_+$$

by the formula

$$\chi([c_{e-1} + 1, c_e] \cap \mathbb{Z}_+) = e. \tag{122}$$

In accordance with our remark above, if $c_{e-1} = c_e$, then the formula (122) is meaningless, since the set $[c_{e-1} + 1, c_e]$ is empty. Note that $c_{e+1} - c_e \in \{0, 1, 2\}$

by the definition (121). It is easy to check that (122) gives a correct definition of an integer-valued function χ. Denote by the symbol Λ_i^P the restriction of the hypertangent system Λ_i onto P_F. Set

$$\Lambda^P = \prod_{i=1}^{m+l-4} \Lambda^P_{\chi(i)}.$$

Note that in this product the hypertangent system Λ_e can appear at most twice; see (121). Let

$$\mathbb{D} = \{D_i \in \Lambda^P_{\chi(i)},\ \ i = 1, \ldots, m+l-4\} \in \Lambda^P$$

be a general set of hypertangent divisors.

Definition 3.2. We say that a family of closed algebraic sets $(\Gamma_u, u \in U)$ of (co)dimension i on an algebraic variety Z is a dense mobile family if for a general $u \in U$ all irreducible components of the set Γ_u are of (co)dimension i and these components form a family of irreducible algebraic varieties sweeping out Z.

Lemma 3.3. *For $i = 1, \ldots, m+l-4$ the closed algebraic set*

$$R_i(\mathbb{D}) = \bigcap_{j=1}^{i} D_i \cap T_P$$

is for a general $\mathbb{D} \in \Lambda^P$ of codimension i in T_P. For $i = 1, \ldots, m+l-5$ the family of cycles

$$(R_i(\mathbb{D}), \mathbb{D} \in \Lambda^P)$$

is a dense mobile family of cycles of codimension i on T.

Proof. Set $R_0(\mathbb{D}) = T$ and argue by induction on $i = 1, \ldots, m+l-4$. Assume that the claim of the lemma is proved for $i \leq j \leq m+l-5$ (if $j = 0$, then there is nothing to prove). Set $\chi(j+1) = e$. By definition,

$$R_{j+1}(\mathbb{D}) = R_j(\mathbb{D}) \cap D_{j+1},$$

where $D_{j+1} \in \Lambda_e^P$ is a general divisor. By definition of the function χ we get

$$j + 1 \in [c_{e-1} + 1, c_e].$$

By Proposition 3.1, the following inequality holds:

$$\operatorname{codim}_{P_F} \operatorname{Bs} \Lambda_e^P \geq c_e + 1, \tag{123}$$

so that $\operatorname{codim}_{T_P} \operatorname{Bs} \Lambda_e^P|_{T_P} \geq c_e$, whereas $\operatorname{codim}_{T_P} R_j(\mathbb{D}) = j \leq c_e - 1$ by (121), (122). Therefore, neither of the irreducible components of the closed subset $R_j(\mathbb{D})$ is contained in the base set of the hypertangent system Λ_e^P. In particular,

$$R_j(\mathbb{D}) \not\subset D_{j+1}$$

and therefore $R_{j+1}(\mathbb{D})$ is a closed subset of pure codimension $j+1$ in T_P, which proves the first claim of the lemma. Now assume that $j \leq m+l-6$. Then either

$$e \leq \max\{m, 2l\} - 2,$$

so that by Proposition 3.1 we get the estimate

$$\operatorname{codim}_{P_F} \operatorname{Bs} \Lambda_e^P \geq c_e + 2,$$

which is stronger than the inequality (123), or $e = \max\{m, 2l\} - 1$, but in this case $c_e = c_{e-1} + 2$, since $j + 2 \in [c_{e-1} + 1, c_e]$, so that $\operatorname{codim}_{T_P} R_j(\mathbb{D}) = j = c_e - 2$. In any case for each irreducible component Z of the set $R_j(\mathbb{D})$ for $j \leq m+l-6$ we get $\operatorname{codim}_Z \operatorname{Bs}(\Lambda_e^P|_Z) \geq 2$, so that the linear system $\Lambda_e^P|_Z$ is mobile. This proves the second claim of Lemma 3.3. Q.E.D.

Let us complete the proof of Lemma 3.2. Consider the family of closed one-dimensional sets

$$(R(\mathbb{D}) = R_{m+l-4}(\mathbb{D}), \mathbb{D} \in \Lambda^P).$$

We can no longer assert that irreducible components of the set $R(\mathbb{D})$ form a mobile family of curves: at the last step, that is, when we make curves from surfaces, some fixed components can appear. However, in any case the following decomposition holds:

$$R(\mathbb{D}) = (T_P \circ D_1 \circ \cdots \circ D_{m+l-4}) = \sum_{\delta_i \in \Delta} C_{\delta_i} + \Phi, \tag{124}$$

where $(C_\delta, \delta \in \Delta)$ is a mobile family of curves, Φ an effective 1-cycle, that is, the fixed part of the family of curves $R(\mathbb{D})$, $\mathbb{D} \in \Lambda^P$. We get the equality of 1-cycles,

$$\Phi = \operatorname{Bs} \Lambda_\infty^P.$$

The family $(C_\delta, \delta \in \Delta)$ sweeps out T_F, *if it is non-empty*. However, by construction

$$\deg R(\mathbb{D}) = 4m \prod_{j=1}^{m+l-4} \chi(j) = 4m \left(\prod_{j=4}^{m-1} j \right) \left(\prod_{j=l}^{2l-1} j \right)$$
$$= \frac{2m!(2l-1)!}{3(l-1)!},$$

whereas by the regularity condition

$$\deg \Phi < \lambda_{m,l} = \frac{m!(2l-1)!}{6(l-1)!} < \deg R(\mathbb{D}).$$

Therefore, the family of irreducible curves $(C_\delta, \delta \in \Delta)$ is non-empty and sweeps out the divisor T.

Let us, finally, estimate the ratio $\operatorname{mult}_o / \deg$ for a general curve C_δ. As we mentioned above, $\operatorname{mult}_o \Phi = \deg \Phi$ (see Remark 3.1). Besides, for a general set $\mathbb{D} \in \Lambda^P$ the ratio

$$\frac{\operatorname{mult}_o}{\deg} C_{\delta_i}$$

(in the sense of the formula (124)) does not depend on i. Consequently,

$$\frac{\operatorname{mult}_o}{\deg} C_\delta = \frac{\operatorname{mult}_o R(\mathbb{D}) - \deg \Phi}{\deg R(\mathbb{D}) - \deg \Phi}.$$

However, by construction

$$\operatorname{mult}_o R(\mathbb{D}) \geq 6 \prod_{j=1}^{m+l-4} (\chi(j)+1) = 6 \left(\prod_{j=5}^{m} j \right) \left(\prod_{j=l+1}^{2l} j \right)$$
$$= \frac{m!}{4} \cdot \frac{(2l)!}{l!},$$

whence

$$\frac{\operatorname{mult}_o}{\deg} C_\delta \geq \frac{\frac{m!}{4} \cdot \frac{(2l)!}{l!} - \lambda_{m,l}}{\frac{2m!}{3} \cdot \frac{(2l-1)!}{(l-1)!} - \lambda_{m,l}} > \frac{2}{3}$$

in accordance with the choice of the number $\lambda_{m,l}$. This proves the lemma for $m \geq 5$, $l \geq 3$.

If $l = 2$, then the arguments presented above work with the only modification: instead of (121) one should use the formula

$$c_e = \sharp[3, e] \cap \mathcal{M} + \sharp[3, e] \cap \mathcal{L},$$

$e \in \mathbb{Z}_+$. In this case an independent hypertangent divisor adds into the linear system Λ_2^P and thus the codimension of its base set (and the codimension of the base set of all the subsequent hypertangent systems Λ_j^P, $j \geq 3$) exceeds by one the corresponding codimension in the case $l \geq 3$. It is this fact that makes it possible to change the definition of the number c_e and accordingly shift by one the function χ. The rest of the arguments are completely similar to the case $l \geq 3$ discussed above.

The case $m \geq 3$ is slightly harder. In order to obtain the needed codimension of the base set of a hypertangent system, one should use the following set of hypertangent divisors:

$$\mathbb{D} = \{D_i \in \Lambda_{l+i}^P \mid i = 1, \ldots, l - 1\}.$$

We draw the reader's attention to the fact that the first divisor in this set is taken from the linear system Λ_{l+1}^P, that is, in contrast to the case $m \geq 4$, which we considered above, we skip the system Λ_l^P. As a result, we obtain once again a mobile family of closed algebraic sets

$$R_k(\mathbb{D}) = \left(\bigcap_{j=1}^{k} D_j \right) \cap T$$

for $k \leq 2l - 2$, whereas irreducible components of the sets $R_k(\mathbb{D})$ form a family and sweep out T. Again we modify the family of curves $R_{2l-1}(\mathbb{D})$, deleting the fixed part Φ of degree $\deg \Phi < \lambda_{3,l} = 12(l-2)\frac{(2l-1)!}{(l+1)!}$ and obtain a family of irreducible curves $(C_\delta, \delta \in \Delta)$, sweeping out T and satisfying the estimate (120). Proof of Lemma 3.2 is now complete.

3.6. Singularity of a fibre on the branch divisor. We have the double cover

$$F = F_t \stackrel{\sigma}{\to} G = G_t \subset \mathbb{P} = \mathbb{P}^{M+1},$$

$G \subset \mathbb{P}$ is a smooth hypersurface of degree $m \leq M - 1$. At the point $p \in G$ the branch hypersurface $W_G = W \cap G$ has an isolated quadratic singularity, so that $o = \sigma^{-1}(p) \in F$ is an (isolated) non-degenerate double point of the fibre F. We get the following commutative diagram of maps

$$\begin{array}{ccccccc} E = E_F & \subset & \widetilde{F} & \stackrel{\tilde{\sigma}}{\to} & \widetilde{G} & \supset & E_G \\ & \varphi_F & \downarrow & & \downarrow & \varphi_G & \\ & & F & \stackrel{\sigma}{\to} & G, & & \end{array}$$

where φ_F and φ_G are the blow ups of the points $o \in F$ and $p \in G$, respectively, E_F and E_G are the exceptional divisors, $\tilde{\sigma}$ the double cover, branched over $\widetilde{W}_G \subset \widetilde{G}$, that is, over the strict transform of the divisor W_G. Thus,

$$\tilde{\sigma}_E = \tilde{\sigma}|_E \colon E \to E_G \cong \mathbb{P}^{M-1}$$

is the double cover, branched over the quadric $W_E = \widetilde{W_G} \cap E_G$. The symbol H_E stands for the hyperplane section of the quadric E with respect to the standard embedding $E \hookrightarrow \mathbb{P}^M$, $\operatorname{Pic} E = \mathbb{Z}H_E$. Let $W_t = W \cap \mathbb{P}_t$ be given by the equation

$$h = w_1 + w_2 + \cdots + w_{2l} = 0,$$

and G by the equation $f = q_1 + q_2 + \cdots + q_m = 0$ with respect to the affine coordinates $z_* = (z_1, \ldots, z_{M+1})$ with the origin at the point p. The divisor W_G has at p a non-degenerate quadratic singularity, so that $w_1 = \lambda q_1$, for simplicity of notation assume that $q_1 = z_{M+1}$. The quadratic polynomial $\bar{w}_2 = w_2|_{\{z_{M+1}=0\}}$ is of the maximal rank. Take $z_1, \ldots, z_M$ for homogeneous coordinates on E_G, then

$$\tilde{\sigma}_E \colon E \to E_G \cong \mathbb{P}^{M-1}$$

is branched over the non-singular quadric $W_E = \{\bar{w}_2 = 0\}$. For an arbitrary point $y \in E_G \setminus W_E$ let $C(y) \subset E_G$ be the cone consisting of all lines $L \subset E_G$ that contain y and touch W_E. More formally, let

$$\pi_y \colon E_G \setminus \{y\} \to \mathbb{P}^{M-2}$$

be the projection from the point y. Its restriction onto the quadric W_E,

$$\pi_y|_{W_E} \colon W_E \to \mathbb{P}^{M-2}$$

is a double cover, branched over a quadric $Q(y) \subset \mathbb{P}^{M-2}$. Now

$$C(y) = \overline{\pi_y^{-1}(Q(y))}.$$

Obviously, $C(y)$ is a quadric cone with the vertex at the point y. Since the quadric W_E is non-singular, the cone $C(y)$ has only one singularity, that is, the point y.

Denote the restriction of the polynomial q_i onto the hyperplane $q_{M+1} = 0$ by the symbol $\bar{q}_i$. By the regularity condition, the system of homogeneous equations

$$\bar{q}_2 = \cdots = \bar{q}_m = 0$$

defines in E_G an irreducible reduced complete intersection of codimension $(m-1)$, an irreducible subvariety $Z_{2,\ldots,m}$. Moreover, the quadric $\bar{q}_2 = 0$ is smooth and distinct from W_E.

Lemma 3.4. *Assume that the condition* (R2.2) *holds. Then the subvariety $Z_{2,\ldots,m}$ is not contained in a quadric cone $C(y)$, $y \in E_G \setminus W_E$, and in a tangent plane $T_y W_E$, $y \in W_E$.*

Proof. Set

$$Z_{2,\ldots,j} = \{z \in \mathbb{P}^{M-1} \mid \bar{q}_2 = \cdots = \bar{q}_j = 0\}.$$

It is easy to see that $Z_{2,\ldots,j}$ is an irreducible reduced complete intersection of codimension j. From the long exact cohomology sequence we obtain that

$$h^0(\mathcal{O}_{Z_2}(2)) = \cdots = h^0(\mathcal{O}_{Z_{2,\ldots,j}}(2)) = \cdots = h^0(\mathcal{O}_{Z_{2,\ldots,m}}(2)),$$

and moreover, the restriction map

$$H^0(\mathcal{O}_{\mathbb{P}^{M-1}}(2)) \to H^0(\mathcal{O}_{Z_{2,\ldots,m}}(2))$$

is surjective. This implies that $Z_{2,\ldots,m}$ is contained in one and only one quadric Z_2 and thus is not contained in any quadric cone $C(y)$, $y \in E_G \setminus W_E$. In a similar way, the restriction map

$$H^0(\mathcal{O}_{\mathbb{P}^{M-1}}(1)) \to H^0(\mathcal{O}_{Z_{2,\ldots,m}}(1))$$

is an isomorphism, so that $Z_{2,\dots,m}$ is not contained in a hyperplane, in particular, in a hyperplane of the form $T_y W_E$, $y \in W_E$. Q.E.D.

Now fix a prime divisor $R \subset F$, and let $\widetilde{R} \subset \widetilde{F}$ be its strict transform. Fix also an arbitrary point $x \in E$, lying outside the branch divisor of the cover $\tilde{\sigma}_E$, that is, $\tilde{\sigma}(x) \notin W_E$. (For a point $x \in E$ on the branch divisor the arguments given below work automatically with simplifications. The arguments of Subsection 3.5 above can also be used in this case, in contrast to the situation outside the branch divisor W_E.)

Proposition 3.4. *The following estimate holds:*

$$\mu = \operatorname{mult}_x \widetilde{R} \leq \frac{1}{m} \deg R. \tag{125}$$

Remark 3.2. For some $k \geq 1$ we have $R \sim H_F$, where $H_F = \sigma^* H_G$ is a hyperplane section. Since obviously $\deg R = 2mk$, the estimate (125) takes the form of the following inequality:

$$\mu \leq 2k.$$

3.7. Start of the proof of the condition (vs). Assume the converse: $\mu > 2k$. We have the presentation

$$\widetilde{R} \sim k\varphi_F^* H_F - \nu E,$$

whereas $\operatorname{mult}_o R = 2\nu$.

Lemma 3.5. *The following inequality holds: $\nu \leq 2k$.*

Proof. Assume the converse: $\nu > 2k$. Then

$$\frac{\operatorname{mult}_o}{\deg} R > \frac{2}{m}.$$

Set $\bar{R} = \sigma(R) \subset G$. It is a prime divisor on the smooth hypersurface $G \subset \mathbb{P}$. Since $\sigma\colon R \to \bar{R}$ is a finite morphism, we get the inequality

$$\frac{\operatorname{mult}_o}{\deg} \bar{R} > \frac{2}{m}.$$

However, this is impossible, since $p \in G$ is a regular point. Indeed, the tangent divisor $T_1^+ = T_p G \cap G$ satisfies the equality

$$\frac{\operatorname{mult}_p}{\deg} T_1^+ = \frac{3}{m},$$

so that $\bar{R} \neq T_1^+$ and $(\bar{R} \circ T_1^+)$ is an effective cycle of codimension two (T_1^+ is obviously irreducible).

Since $\operatorname{mult}_p T_1^+ = 2$, we get the inequality

$$\operatorname{mult}_p(\bar{R} \circ T_1^+) \geq 7 \operatorname{mult}_p \bar{R}.$$

Taking into account that $\deg(\bar{T} \circ T_1^+) = \deg \bar{R}$, we conclude that there exists an irreducible component Y_3 of the cycle $(\bar{R} \circ T_1^+)$, satisfying the estimate

$$\frac{\operatorname{mult}_p}{\deg} Y_2 \geq 2 \frac{\operatorname{mult}_p}{\deg} \bar{R}.$$

As usual, let $f = q_1 + q_2 + \cdots + q_m$ be the polynomial defining the hypersurface G with respect to the coordinate system z_* with the origin at the point p. Setting

$$f = q_1 + q_2 + \cdots + q_i,$$

let us construct the hypertangent systems

$$\Lambda_i^G = \left| \sum_{j=1}^{i} f_j s_{i-j}|_G = 0 \right|,$$

and consider the standard hypertangent divisors

$$T_i^+ = \{f_i|_G = 0\} \in \Lambda_i^G.$$

Set $T_i = \sigma^* T_i^+, \quad \Lambda_i = \sigma^* \Lambda_i^G$. These divisors and linear systems will be of crucial importance below. At the moment, note that by the regularity condition we get

$$\operatorname{codim}_G \operatorname{Bs} \Lambda_i^G = i$$

(in fact, $\operatorname{Bs} \Lambda_i^G = T_1^+ \cap \cdots \cap T_i^+$). Let

$$\mathbb{D} = (D_1, \ldots, D_{m-1}) \in \prod_{j=1}^{m-1} \Lambda_i^G$$

be a general set of divisors. Let us construct by induction a sequence of irreducible subvarieties Y_i, $i = 1, \ldots, m-1$, satisfying the following properties:

(i) $Y_1 = \bar{R}$, Y_0 was constructed above, $\operatorname{codim}_G Y_i = i$;

(ii) $Y_{i+1} \subset Y_i$, $R_i \not\subset D_{i+1}$, Y_{i+1} is an irreducible component of the closed set $Y_i \cap D_{i+1}$;

(iii) the following estimate holds:

$$\frac{\operatorname{mult}_p}{\deg} Y_{i+1} \geq \frac{i+1}{i+1} \cdot \frac{\operatorname{mult}_p}{\deg} Y_i.$$

It is possible to construct this sequence because

$$\operatorname{codim}_G \Lambda_{i+4}^G = i + 1 > \operatorname{codim}_G Y_i,$$

so that for a general divisor $D_{i+1} \subset \Lambda_{i+1}^D$ we have $Y_i \not\subset D_{i+1}$. One can ensure that the property (iii) holds since $\Lambda_j^G \subset |jH_G|$ and $\operatorname{mult}_p \Lambda_j^G = j + 1$.

Now for an irreducible subvariety $Y = Y_{m-1}$ we get the estimate

$$1 \geq \frac{\operatorname{mult}_p}{\deg} Y \geq \underbrace{\frac{m}{m-1} \cdot \frac{m-1}{m-2} \cdot \cdots \cdot \frac{4}{3} \cdot \frac{2}{1}}_{\displaystyle \stackrel{\|}{\frac{2m}{3}}} \cdot \frac{\operatorname{mult}_p}{\deg} \bar{R},$$

whence we get

$$\frac{\operatorname{mult}_p}{\deg} \bar{R} \leq \frac{3}{2m}.$$

Therefore the ratio $\operatorname{mult}_p / \deg$ attains its maximum at the tangent divisor $T_p G \cap G$ and this maximum is equal to 2. A contradiction. Q.E.D. for the lemma.

3.8. Hypertangent divisors and tangent cones. By what we have just proved, $\nu \leq 2k < \mu$. On the other hand,

$$\mu = \operatorname{mult}_x \widetilde{R} \leq \operatorname{mult}_x(\widetilde{R} \circ E) \leq \deg(\widetilde{R} \circ E) = 2\nu.$$

Set $B = T_x E \cap E$, where the quadric E is considered as embedded in $\mathbb{P}^M$ in the standard way. By Lemma 2.3,

$$\operatorname{mult}_B \widetilde{R} \geq \frac{1}{2}(\mu - \nu),$$

whereas for the effective cycle $R_E = (\widetilde{R} \circ E)$ we get

$$\operatorname{mult}_B R_E \geq \mu - \nu. \tag{126}$$

Set $\widetilde{T}_i, \widetilde{\Lambda}_i$ to be the strict transforms of the divisors T_i and linear systems Λ_i on the blow up $\widetilde{F}$ of the fibre F. It is easy to see that $\widetilde{T}_i \subset \widetilde{\Lambda}_i$, since by construction

$$\operatorname{mult}_o \Lambda_i = \operatorname{mult}_o T_i.$$

Set also

$$\mathbb{T}_i = (\widetilde{T}_i \circ E) = \widetilde{T}_i \cap E$$

to be the projectivized tangent cone to the divisor T_i at the point o. Recall that the quadric E is realized as the double cover $\tilde{\sigma}_E \colon E \to E_G \cong \mathbb{P}^{M-1}$, branched over the quadric W_E. For a system $(z_1, \ldots, z_{M+1})$ of affine coordinates on $\mathbb{P}$ with the origin at the point p we may assume that $q_1 \equiv v_{M+1}$ and therefore $(z_1, \ldots, z_M)$ can be taken for homogeneous coordinates on the projective space E_G. In terms of these coordinates the hypersurface $\mathbb{T}_i \subset E$ is given by the equation

$$(\tilde{\sigma}_E)^* q_{i+1}|_{E_G}.$$

Finally, set

$$\Lambda_i^E = \widetilde{\Lambda}_i|_E$$

to be the projectivized tangent system of the linear system Λ_i at the point o. Equations of divisors of this linear system are obtained by pulling back to E via $\tilde{\sigma}_E$ the equations

$$\sum_{j=1}^{i} \bar{q}_{j+1} \bar{s}_{i-j}, \tag{127}$$

where $\bar{\sharp}$ means the restriction of the polynomial $\sharp$ onto the hyperplane $z_{M+1} = 0$. Obviously,

$$\mathbb{T}_i \sim (i+1)H_E, \quad \Lambda_i^E \subset |(i+1)H_E|,$$

besides the equations (127) imply directly that

$$\operatorname{Bs} \Lambda_i = T_1 \cap \cdots \cap T_i, \quad \operatorname{Bs} \Lambda_i^E = \mathbb{T}_1 \cap \cdots \cap \mathbb{T}_i,$$

both equalities in the scheme-theoretic sense.

By the regularity condition the set $\mathbb{T}_1 \cap \cdots \cap \mathbb{T}_i$ is irreducible and not contained in the divisor B for all $i = 1, \ldots, m-1$. Let

$$\mathcal{L} = (L_2, \ldots, L_{m-1}) \in \Lambda_2 \times \cdots \times \Lambda_{m-1}$$

be a general set of hypertangent divisors. We denote the strict transform of the hypertanqent divisor L_j on $\widetilde{F}$ by the symbol $\widetilde{L}_j$ and its projectivized tangent cone at the point $o \in F$ by the symbol

$$\mathbb{L}_j = (\widetilde{L}_j \circ E).$$

For a general divisor $L_j \in \Lambda_j$ we get $\mathbb{L}_j = \widetilde{L}_j \cap E$.

Lemma 3.6. (i) *Let $Y \subset F$ be a fixed irreducible subvariety of codimension $l \leq m-2$. For a general divisor $L_{l+1} \in \Lambda_{l+1}$ we have $Y \not\subset L_{l+1}$.*

(ii) *Let $Y \subset E$ be a fixed irreducible subvariety of codimension $l \leq m-2$. For a general divisor $L_{l+1} \in \Lambda_{l+1}$ we have $Y \not\subset \mathbb{L}_{l+1}$.*

Proof. By the regularity condition

$$\operatorname{codim}_F \operatorname{Bs} \Lambda_{l+1} = l+1, \quad \operatorname{codim}_E \operatorname{Bs} \Lambda^E_{l+1} = l+1$$

and for a general divisor $L_j \in \Lambda_j$ we have $\mathbb{L}_j \in \Lambda^E_j$. Q.E.D. for the lemma.

Corollary 3.1. *For a general set $\mathcal{L}$ we have*

$$\operatorname{codim}_F(R \cap L_2 \cap \cdots \cap L_{m-1}) = m-1,$$
$$\operatorname{codim}_E(R_E \cap \mathbb{L}_2 \cap \cdots \cap \mathbb{L}_{m-1}) = m-1.$$

Proof. This follows immediately from the previous lemma. Q.E.D.

From this fact we obtain that the following effective algebraic cycles of codimension $m-1$ are well defined on F and E, respectively:

$$A^+ = (R \circ L_2 \circ \cdots \circ L_{m-1})$$

and

$$R^+_E = (R_E \circ \mathbb{L}_2 \circ \cdots \circ \mathbb{L}_{m-1}),$$

whereas (for a general set $\mathcal{L}$)

$$R^+_E = (\widetilde{R}^+ \circ E)$$

is the projectivized tangent cone to the cycle R^+ at the point o. Let us describe the structure of these effective cycles. First we get

$$\deg R^+ = 2km \cdot (m-5)! = 2km!,$$
$$\operatorname{mult}_o R^+ = \deg R^+_E = 2\nu \cdot 3 \cdot \cdots \cdot m = \nu m!.$$

Lemma 3.7. *Let Y be an irreducible component of the cycle R^+. If $Y \subset T_1$, then*

$$Y = T_1 \cap T_2 \cap \cdots \cap T_{m-1}.$$

Proof. By construction, the equation of the divisor L_i is of the form

$$f_1 s_{i-1} + f_2 s_{i-2} + \cdots + f_i s_0,$$

where s_j is a homogeneous polynomial of degree j in the coordinates z_*. Since the hypertangent divisors L_i are assumed to be general, we may assume that $s_0 \neq 0$ and thus normalize the equation by the condition that $s_0 = 1$. Assume that $Y \subset T_1$. Then the following polynomials vanish on Y:

$$\begin{array}{lllllll}
f_1, & & & & & & \\
f_1 s_{2,1} & + & f_2, & & & & \\
f_1 s_{3,2} & + & f_2 s_{3,1} & + & f_3, & & \\
 & & & & \cdots & & \\
f_1 s_{m-1,m-2} & + & \cdots & + & f_{m-2} s_{m-1,1} & + & f_{m-1},
\end{array}$$

where $s_{i,j}$ is a homogeneous polynomial of degree j. Thus

$$f_0|_Y \equiv f_2|_Y \equiv \cdots \equiv f_{m-1}|_Y \equiv 0,$$

so that $Y \subset T_1 \cap T_2 \cap \cdots \cap T_{m-1}$, but the latter set is irreducible and of the same dimension as Y. This proves Lemma 3.7.

3.9. Degrees and multiplicities. Set

$$T = T_1 \cap T_2 \cap \cdots \cap T_{m-1}, \quad \mathbb{T} = \mathbb{T}_1 \cap \mathbb{T}_2 \cap \cdots \cap \mathbb{T}_{m-1}.$$

Taking into consideration that $T = (T_1 \circ \cdots \circ T_{m-1})$ and $\mathbb{T} = (\mathbb{T}_1 \circ \cdots \circ \mathbb{T}_{m-1})_E$, it is easy to verify that

$$\deg T = \mathrm{mult}_o T = \deg \mathbb{T} = 2m!.$$

Now write

$$R^+ = aT + R^\sharp, \quad R^+_E = a\mathbb{T} + R^\sharp_E, \tag{128}$$

where $a \in \mathbb{Z}_+$, the effective cycle $R^\sharp$ is uniquely defined by the condition that it does not contain the subvariety T as a component, and

$$R^\sharp_E = (\widetilde{R}^\sharp \circ E)$$

is the projectivized tangent cone to $R^\sharp$ at the point o. Note that the irreducible subvariety $\mathbb{T}$, generally speaking, can come into the effective cycle $R^\sharp_E$ as a component.

Lemma 3.8. *The following estimate holds:*

$$2\,\mathrm{mult}_o R^\sharp \leq \deg R^\sharp.$$

Proof. Let Y be an irreducible component of the cycle $R^\sharp$. By construction, $Y \neq T$; therefore by Lemma 3.7 $Y \not\subset T_1$. Thus the closed subset

$$T_1 \cap \mathrm{Supp}\, R^\sharp$$

is of codimension m, so that the effective cycle $R^* = (R^\sharp \circ T_1)$ is well defined. Now we have a standard chain of estimates:

$$2\,\mathrm{mult}_o R^\sharp \leq \mathrm{mult}_o R^* \leq \deg R^* = \deg R^\sharp,$$

which is what we need. Q.E.D.

As in Corollary 3.1, Lemma 3.8 implies that the set

$$B \cap \mathbb{L}_2 \cap \cdots \cap \mathbb{L}_{m-1}$$

is of codimension $m-1$ in E. Denote by B^+ the part of the effective equidimensional cycle R^+_E, the support of which is contained in B:

$$R^+_E = \sum_{i \in I} r_i Y_i, \quad B^+ = \sum_{i \in I, Y_i \subset B} r_i Y_i.$$

Lemma 3.9. *The following estimate holds:*

$$\deg B^+ \geq (\mu - \nu) m!$$

Proof. Indeed, by (126) we get $R_E = (\mu - \nu)B + \Delta$, where Δ is an effective cycle. Furthermore,

$$\deg(B \circ \mathbb{L}_2 \circ \cdots \circ \mathbb{L}_{m-1}) = 2 \cdot 3 \cdot \cdots \cdot m = m!,$$

which proves the lemma. Q.E.D.

Lemma 3.10. *Let $Y \subset \mathbb{L}_2 \cap \cdots \cap \mathbb{L}_{m-1}$ be an irreducible subvariety of codimension $m-1$ in E. If $Y \subset \mathbb{T}_1$, then $Y = \mathbb{T}$.*

Proof. The equation of the divisor $\mathbb{L}_i$ with respect to the homogeneous coordinates z_* is of the form

$$q_2 s_{i-1} + \cdots + q_{i+1},$$

where s_j is a homogeneous polynomial of degree j. If $Y \subset \mathbb{T}_1$, then the following polynomials vanish on Y:

$$\begin{array}{llllllll} q_2, & & & & & & & \\ q_2 s_{2,1} & + & q_3, & & & & & \\ & & & \cdots & & & & \\ q_2 s_{m-1,m-2} & + & & \cdots & + & q_{m-1} s_{m-1,1} & + & q_m, \end{array}$$

where $\deg s_{i,j} = j$. Therefore,

$$q_2|_Y \equiv q_3|_Y \equiv \cdots \equiv q_m|_Y \equiv 0,$$

that is, $Y \subset \mathbb{T}$, and since the dimensions coincide, $Y = \mathbb{T}$. Q.E.D. for the lemma.

Corollary 3.2. *None of the components of the closed set*

$$B \cap \mathbb{L}_2 \cap \cdots \cap \mathbb{L}_{m-1}$$

are contained in $\mathbb{T}_1$.

Proof. Let Y be such component and $Y \subset \mathbb{T}_1$. By the previous lemma, $Y = \mathbb{T}$. Thus $\mathbb{T} \subset B$: a contradiction with the regularity condition. Q.E.D. for the corollary.

Finally, let us complete the proof of Proposition 3.4. From the presentations (128) we get

$$\deg R^+ = 2km! = 2am! + \deg R^\sharp,$$

$$\operatorname{mult}_o R^+ = \nu m! = 2am! + \operatorname{mult}_o R^\sharp.$$

By Corollary 3.2 the effective cycle B^+ lies entirely in $R^\sharp_E$. In particular,

$$\deg R^\sharp_E \geq \deg B^+ \geq (\mu - \nu)m!. \tag{129}$$

However, $\deg R^\sharp_E = \operatorname{mult}_o R^\sharp$. Applying Lemma 3.8, we obtain:

$$2(\nu m! - 2am!) \leq 2km! - 2am!.$$

Let us rewrite the inequality (129) in the form

$$\nu m! - 2am! \geq (\mu - \nu)m!.$$

Easy computations give us two inequalities,

$$k + a \geq \nu \quad \text{and} \quad 2\nu - 2a \geq \mu,$$

which imply the desired estimate (125) in an obvious way.

However, we assumed that $\mu > 2k$. The contradiction completes our proof of Proposition 3.4 and Theorem 3.1 as well. Q.E.D.

4. Sufficient conditions of birational rigidity. II

In this section we prove a modified sufficient condition of birational (super)rigidity for fibre spaces $V/\mathbb{P}^1$. The new condition (Theorem 4.1) is based on a somewhat stronger assumption than Theorem 1.1 (which is, however, not too restrictive) but gives a stronger claim that makes it possible to investigate families which do not satisfy the strong K^2-condition and, moreover, for which the deviation from the strong K^2-condition is considerable. To prove Theorem 4.1, we need to take into account the input into the multiplicities of the self-intersection of a mobile system Σ, made by different segments of the resolution of a maximal singularity, so that we have to go through the procedure of counting multiplicities again (Subsection 4.2).

4.1. Formulation of the sufficient conditions. Start of the proof. We go on working with a standard Fano fibre space $\pi\colon V \to \mathbb{P}^1$, that is, V is a smooth variety, $\operatorname{Pic} V = \mathbb{Z}K_V \oplus \mathbb{Z}F$, where F is the class of a fibre of the projection π. As in Section 1, we assume that the condition

$$A^2V = \mathbb{Z}K_V^2 \oplus \mathbb{Z}H_F, \tag{130}$$

holds, where $H_F = (-K_V \cdot F)$ is the ample anticanonical section of the fibre, and that every fibre $F = F_t = \pi^{-1}(t)$, $t \in \mathbb{P}^1$, is a Fano variety with at most non-degenerate quadratic singularities and, moreover,

$$A^1F = \operatorname{Pic} F = \mathbb{Z}K_F,$$

where $K_F = -H_F$ and H_F is considered as an element of the group A^1F. In addition, we assume that

$$A^2F = \mathbb{Z}H_F^2.$$

These conditions are satisfied for almost all families of standard Fano fibre spaces that are by now successfully studied. In Section 1 we derived birational (super)rigidity from the conditions (v), (vs) and (h); see Definition 1.1. Now we replace the condition (h) by the following condition for the fibres of the projection π.

Definition 4.1. The fibre space $V/\mathbb{P}^1$ satisfies

- the condition (f), if for every irreducible vertical subvariety Y of codimension three (that is, $\operatorname{codim}_F Y = 2$, $F = F_t \supset Y$) and every point of the fibre $o \in F$, the following inequality holds:

$$\frac{\operatorname{mult}_o Y}{\deg Y} \leq \frac{4}{\deg V}. \tag{131}$$

- the condition (fs) if for every singular point $o \in F$ of a fibre F and for any irreducible subvariety $Y \subset F$ of codimension 2 the following estimate holds:

$$\frac{\operatorname{mult}_o}{\deg} Y \leq \frac{6}{\deg V}, \tag{132}$$

  and, thus, for any infinitely near point of the first order $x \in E_F$, where $\varphi_o\colon \widetilde{F} \to F$ is the blow up of the point $o \in F$, $E_F \subset \widetilde{F}$ the exceptional divisor, the inequality

$$\frac{\operatorname{mult}_x \widetilde{Y}}{\deg Y} \leq \frac{3}{\deg V}$$

  holds, where $\widetilde{Y}$ is the strict transform of the subvariety Y on $\widetilde{F}$.

Remark 4.1. Let us draw the reader's attention to the fact that in the condition (f) we do not specify whether the point $o \in F$ is singular or smooth.

Recall that the degree $\deg Y$ of a vertical subvariety $Y \subset F_t$ is equal to

$$\deg Y = (Y \cdot (-K_V)^{\dim Y})$$

and the degree $\deg Y$ of a horizontal subvariety Y is the degree of the effective cycle $(Y \cdot F_t$ for a general $t \in \mathbb{P}^1$; finally, we denote the ratio of the multiplicity to the degree by the single symbol

$$\frac{\mathrm{mult}_o}{\deg} Y = \frac{\mathrm{mult}_o Y}{\deg Y}.$$

Now let us formulate the new sufficient condition of birational (super)rigidity.

Theorem 4.1. (i) *Assume that the standard Fano fibre space $V/\mathbb{P}^1$ satisfies the K^2-condition of depth 2, the conditions (v), (vs) and at least one of the conditions (f) or (fs) at any point $o \in V$. Then for any mobile linear system $\Sigma \subset |-nK_V + lF|$ with $l \in \mathbb{Z}_+$ its virtual and actual thresholds of canonical adjunction coincide:*

$$c_{\mathrm{virt}}(\Sigma) = c(\Sigma) = n.$$

In particular, if $V/\mathbb{P}^1$ satisfies the K-condition, then this fibre space is birationally superrigid.

(ii) *Assume that the standard Fano fibre space $V/\mathbb{P}^1$ satisfies the K^2-condition of depth 2 and conditions (v) and (f). If the centre of any maximal singularity of the mobile linear system $\Sigma \subset |-nK_V + lF|$ with $l \in \mathbb{Z}_+$ is not a singular point of a fibre, then the virtual and actual thresholds of canonical adjunction coincide:*

$$c_{\mathrm{virt}}(\Sigma) = c(\Sigma) = n.$$

In particular, if the latter assumption holds for any mobile linear system on V and the fibre space $V/\mathbb{P}^1$ satisfies the K-condition, then this fibre space is birationally superrigid.

Note that the claim (ii) is a simplified version of (i): the maximal singularities lying over the singular points of the fibres are not considered (it is assumed that they are excluded by some other method).

Proof of the claim (i). The first steps of the proof are identical to the beginning of the proof of Theorem 1.1 (Section 1). We fix a mobile linear system $\Sigma \subset |-nK_V + lF|$ with $l \in \mathbb{Z}_+$ and $n \in \mathbb{Z}_+$. Of course, $c(\Sigma) = n$, whereas $n = 0$ if and only if the linear system Σ is composed from the pencil $|F|$, so we assume that $n \geq 1$. We further assume that the inequality $c_{\mathrm{virt}}(\Sigma) < c(\Sigma) = n$ holds. This means that there is a model $V^\sharp$ of the variety V such that the strict transform $\Sigma^\sharp$ of the linear system Σ on $V^\sharp$ satisfies the inequality

$$c(\Sigma^\sharp, V^\sharp) < c(\Sigma, V),$$

so that the linear system Σ has a maximal singularity E on $V^\sharp$, satisfying the Noether-Fano inequality (97). We use the notations of Subsection 1.4 for the centre B of the maximal singularity on V and for the resolution (98) of the maximal singularity, and the subsequent notations for the oriented graph structure and the standard form of the Noether-Fano inequality.

The claim of Proposition 1.2 holds in the case under consideration because the condition (h) follows from (f) and (fs).

The claims of Proposition 1.3 and Lemma 1.1 are valid in our case (no modification of the proof is needed). In the sequel, we use these facts without special comments, also all the notations introduced in Subsections 1.4 and 1.5 up to Lemma 1.1. It is at that point that our arguments start to differ from those in Section 1.

4.2. The structure of the sequence of blow ups. Consider the self-intersection of the linear system Σ, that is, the effective algebraic cycle $Z = (D_1 \circ D_2)$, where $D_1, D_2 \in \Sigma$ are general divisors. Let $Z = Z^v + Z^h$ be the decomposition of the cycle Z into the vertical (Z^v) and horizontal (Z^h) parts. The cycle Z^v can be further decomposed as

$$Z^v = \sum_{t\in\mathbb{P}^1} Z^v_t, \quad \operatorname{Supp} Z^v_t \subset F_t.$$

Let $E \in \mathcal{M}_t$ be a maximal singularity over a point $t \in \mathbb{P}^1$. Let us study the structure of the sequence of blow ups, resolving the singularity E. This means, to break the sequence of blow ups into segments that determine multiplicities of the cycles Z^v_t, Z^h and $(Z^h \circ F_t)$ (the intersection of a horizontal cycle with a fibre is always well defined).

So long as we discuss a fixed singularity E, we omit the symbols t and E for simplicity of notation, for instance, we write F instead of F_t, Z^v instead of Z^v_t, e instead of $e(E)$ and so on.

So let us consider the sequence of blow ups (98), associated with E. We use the notations of Section 1:

$$N = \max\{i \,|\, B_{i-1} \subset F^{i-1}\} \quad \text{and} \quad L = \max\{i \,|\, \operatorname{codim} B_{i-1} \geq 3\} \leq N,$$

the numbers m^h_i, m^v_i, μ_i, etc., have the same meaning as in Subsection 1.6. The inequality (102) of Proposition 1.5 holds but unfortunately is not strong enough to prove Theorem 4.1. We need a more detailed and careful study of the resolution of the singularity E.

Set

$$\begin{aligned} J_s &= \{i \,|\, 1 \leq i \leq K, \quad \operatorname{codim} B_{i-1} \geq 4\}, \\ J_m &= \{i \,|\, 1 \leq i \leq K, \quad \operatorname{codim} B_{i-1} = 3\}, \\ J_u &= \{i \,|\, L+1 \leq i \leq K\}, \quad J_l = J_s \cup J_m. \end{aligned}$$

In turn, let us break the set J_m into two disjoint subsets, $J_m = J^+_m \coprod J^-_m$, where

$$J^+_m = \{i \in J_m \,|\, B_{i-1} \subset F^{i-1}\},$$

$J^-_m = J_m \setminus J^+_m = \{i \in J_m \,|\, B_{i-1} \not\subset F^{i-1}\}$. It might well turn out that the set J^+_m or J^-_m (or the whole set J_m) is empty. Furthermore, set

$$\Sigma_s = \sum_{i\in J_s} p_i, \quad \Sigma^{\pm}_m = \sum_{i\in J^{\pm}_m} p_i, \quad \Sigma_m = \Sigma^+_m + \Sigma^-_m,$$

whereas the symbol Σ_u retains its previous meaning. In the notation of Section 1 we get $\Sigma_l = \Sigma_s + \Sigma_m$. Now the inequality (102) can be rewritten as

$$\sum_{i\in J_l} p_i m^h_i + \sum_{i\in J_s\cup J^+_m} p_i m^v_i \geq \frac{((3\Sigma_s + 2\Sigma_m + \Sigma_u)n + e)^2}{\Sigma_s + \Sigma_m + \Sigma_u}. \tag{133}$$

Recall that $\mu_i = \operatorname{mult}_{B_{i-1}} F^{i-1}$, where $\mu_i = 1$ for $i \geq 2$ and for μ_1 there are two possible cases: $\mu_1 = 1$ or $\mu_1 = 2$.

Proposition 4.1. *The following estimate holds:*

$$\sum_{i\in J_s\cup J_m^+} \mu_i p_i m_i^h \leq p_1 \operatorname{mult}_B(Z^h\circ F) + (\Sigma_s - p_1)\operatorname{mult}_{B_1}(Z^h\circ F)^1. \tag{134}$$

In particular,

$$\sum_{i\in J_s\cup J_m^+} \mu_i p_i m_i^h \leq \Sigma_s \operatorname{mult}_B(Z^h\circ F). \tag{135}$$

Proof. The arguments below hold with obvious simplifications in the case when $J_m^+ = \emptyset$. So we assume that $J_m^+ \neq \emptyset$, so that, in particular, $J_s \subset \{1,\dots,N_f\}$.

Let us first consider the following general situation. Let $Y\subset V$ be an irreducible horizontal subvariety of codimension two, $Y^i\subset V_i$ its strict transform,

$$m_Y(i) = \operatorname{mult}_{B_{i-1}} Y^{i-1}$$

the corresponding multiplicity. Set $Y_F = (Y\circ F)$. This is an effective class of codimension two in the fibre F. Let $Y_F^i\subset V_i$ be its strict transform and

$$m_{Y,F}(i) = \operatorname{mult}_{B_{i-1}} Y^{i-1}.$$

Since the support of the cycle Y_F is contained in the fibre F, the numbers $m_{Y,F}(i)$ vanish for $i\in J_m^-$.

Lemma 4.1. *The following estimate holds:*

$$\sum_{i\in J_s\cup J_m^+} p_i m_Y(i)\mu_i \leq \sum_{i\in J_s} p_i m_{Y,F}(i). \tag{136}$$

Proof. We cannot just give reference to the technique of counting multiplicities developed in Chapter 2, as here we intersect a divisor and a subvariety whereas in Section 2.2 we considered the intersection of two divisors. However, the arguments are almost word for word the same as in Section 2.2. First, let us explain the intersection theory involved in this computation. Let X be an arbitrary smooth variety, $B\subset X$, $B\not\subset \operatorname{Sing} X$, an irreducible subvariety of codimension ≥ 2, $\sigma_B\colon X(B)\to X$ its blow up, $E(B)=\sigma_B^{-1}(B)$ the exceptional divisor. Let

$$Z=\sum m_i Z_i,\quad Z_i\subset E(B)$$

be a cycle of dimension k, $k\geq \dim B$. We define the *degree* of the cycle Z, setting

$$\deg Z = \sum_i m_i \deg\left(Z_i \bigcap \sigma_B^{-1}(b)\right),$$

where $b\in B$ is a point of general position, $\sigma_B^{-1}(b)\cong \mathbb{P}^{\operatorname{codim} B-1}$ and the degree in the right-hand side is the usual degree in the projective space.

Note that $\deg Z_i = 0$ if and only if $\sigma_B(Z_i)$ is a proper closed subset of the subvariety B.

Now let D be a prime Weil divisor on X, $Y\subset X$ an irreducible subvariety of dimension $l\leq \dim X-1$. Assume that $Y\not\subset D$ and that $\dim B\leq l-1$. The strict transforms of the divisor D and the subvariety Y on $X(B)$ are denoted by the symbols D^B and Y^B, respectively.

Lemma 4.2. (i) *Assume that* $\dim B\leq l-2$. *Then*

$$D^B\circ Y^B = (D\circ Y)^B + Z,$$

where $\circ$ *means the operation of taking the algebraic cycle of the scheme-theoretic intersection,* $\operatorname{Supp} Z \subset E(B)$ *and*

$$\operatorname{mult}_B(D \circ Y) = \operatorname{mult}_B D \cdot \operatorname{mult}_B Y + \deg Z.$$

(ii) *Assume that* $\dim B = l - 1$. *Then*

$$D^B \circ Y^B = Z + Z_1,$$

where $\operatorname{Supp} Z \subset E(B)$, $\operatorname{Supp} \sigma_B(Z_1)$ *does not contain* B *and*

$$D \circ Y = [(\operatorname{mult}_B D)(\operatorname{mult}_B Y) + \deg Z]\, B + (\sigma_B)_* Z_1.$$

Proof follows from the standard intersection theory [**Ful**]. Q.E.D.

Now let us prove Lemma 4.1. We construct a sequence of effective cycles of codimension three on the varieties V_i, setting

$$\begin{array}{rcl} Y \circ F & = & Z_0 \ (= Y_F), \\ Y^1 \circ F^1 & = & Z_0^1 + Z_1, \\ \vdots & & \\ Y^i \circ F^i & = & (Y^{i-1} \circ F^{i-1})^i + Z_i, \\ \vdots & & \end{array}$$

$i \in J_s$, where $\operatorname{Supp} Z_i \subset E_i$. Thus for any $i \in J_s$ we get:

$$Y^i \circ F^i = Y_F^i + Z_1^i + \cdots + Z_{i-1}^i + Z_i.$$

For any $j > i$, $j \in J_s$ set

$$m_{i,j} = \operatorname{mult}_{B_{j-1}}(Z_i^{j-1})$$

(the multiplicity of an irreducible subvariety along a smaller subvariety is understood in the usual sense; for an arbitrary cycle we extend the multiplicity by linearity).

Now set $d_i = \deg Z_i$. We get the following system of equalities:

$$\begin{array}{l} m_Y(1)\mu_1 + d_1 = m_{Y,F}(1), \\ m_Y(2)\mu_2 + d_2 = m_{Y,F}(2) + m_{1,2}, \\ \vdots \\ m_Y(i)\mu_i + d_i = m_{Y,F}(i) + m_{1,i} + \cdots + m_{i-1,i} \\ \vdots \end{array}$$

for all $i \in J_s$. Setting $S = \max\{i \in J_s\}$, look at the last equality in this sequence:

$$m_Y(S)\mu_S + d_S = m_{Y,F}(S) + m_{1,S} + \cdots + m_{S-1,S}.$$

If $J_m^+ \neq \emptyset$, then by the part (ii) of Lemma 4.2 we obtain

$$d_S \geq \sum_{i \in J_m^+} m_Y(i)\mu_i \deg(\varphi_{i-1,S})_* B_{i-1} \geq \sum_{i \in J_m^+} m_Y(i)\mu_i.$$

Recall (Section 2.2), that a function $a \colon J_s \to \mathbb{R}_+$ is said to be *compatible with the graph structure*, if

$$a(i) \geq \sum_{j \to i, j \in J_s} a(j)$$

for any $i \in J_s$. As usual, we will actually use only one function, compatible with the graph structure, namely $a(i) = p_i$.

Proposition 4.2. *Let $a(\cdot)$ be a function, compatible with the graph structure. Then the following inequality holds:*

$$\sum_{i \in J_s} a(i) m_{Y,F}(i) \geq \sum_{i \in J_s} a(i) m_Y(i) \mu_i + a(S) \sum_{i \in J_m^+} m_Y(i) \mu_i. \tag{137}$$

Proof is word for word the same as the proof of Proposition 2.2.4: multiply the i-th equality by $a(i)$ and put them all together. The arguments are identically the same as in Chapter 2. Q.E.D.

Putting in the inequality (137) $a(i) = p_i$ and recalling that for $j \geq S$ we have $p_j \leq p_S$, we complete the proof of Lemma 4.1. Q.E.D.

Now let us complete the proof of Proposition 4.1.

It is obvious that the inequality (136) remains valid if Y is an effective horizontal cycle of codimension 2 on V, that is, each component of the cycle Y is a horizontal subvariety, where the notations $m_Y(i)$ and $m_{Y,F}(i)$ naturally extend by linearity to the set of all effective horizontal cycles, and the left-hand and the right-hand sides of the inequality (136) are linear in $m_Y(\cdot), m_{Y,F}(\cdot)$, respectively.

Now set $Y = Z^h$ and take into account that

$$m_{Y,F}(i) \leq \operatorname{mult}_{B_1}(Z^h \circ F)^1$$

for $i \geq 2$. This implies the inequality (134). The second inequality of Proposition 4.1 follows from (134). Q.E.D. for the proposition.

Remark 4.2. The inequality (135) is more compact than (134), however, in some cases it is possible to get a stronger estimate for $\operatorname{mult}_{B_1}(Z^h \circ F)^1$ than for $\operatorname{mult}_B(Z^h \circ F)$.

4.3. Estimating multiplicities of the linear system. Let us complete the proof of Theorem 4.1. Recall that the fibre space $V/\mathbb{P}^1$ satisfies the conditions (v) and (vs), and at any point $o \in V$ satisfies at least one of the conditions (f) or (fs). Assume first that $B = \operatorname{centre}(\nu_E, V) \not\subset \operatorname{Sing} F$. In other words, either the fibre F is non-singular or B is not a singular point of this fibre. By the regularity condition, for any irreducible subvariety $Y \subset F$ of codimension two (with respect to F) the estimate (131) holds for a point $o \in B$ of general position. For all the multiplicities we have $\mu_i = 1$. The inequality (131) implies immediately the estimate

$$\sum_{i \in J_s \cup J_m^+} p_i m_i^h \leq 4n^2 \Sigma_s.$$

Since $m_i^h \leq m_1^h \leq 4n^2$, we get the inequality

$$\sum_{i \in J_l} p_i m_i^h \leq 4n^2 (\Sigma_s + \Sigma_m^-). \tag{138}$$

This is the required estimate of singularities of the horizontal component Z^h. Consider the vertical component Z^v. By the condition (v) the inequality

$$m_i^v \leq m_1^v \leq \frac{2}{\deg V} d_v, \tag{139}$$

holds, where $d_v = \deg Z_t^v$ is the degree of the vertical component of the self-intersection Z, contained in our fibre $F = F_t$. Now the K^2-condition of depth 2 gives the inequality

$$e(E) > \frac{\nu_E(F_t)}{2}\left(\frac{\deg Z_t^v}{n \deg V} - 2n\right); \tag{140}$$

see Proposition 1.7. Therefore,

$$\frac{d_v}{\deg V} < \frac{2en}{\nu_E(F)} + 2n^2. \tag{141}$$

Combining (139) and (141), we obtain the inequality

$$\sum_{i \in J_s \cup J_m^+} p_i m_i^v < 2n\left(\frac{2e}{\nu_E(F)} + 2n\right)(\Sigma_s + \Sigma_m^+).$$

Taking into account that by definition $\nu_E(F) = \sum_{i=1}^k p_i\mu_i \geq \Sigma_s + \Sigma_m^+$, we obtain finally

$$\sum_{i \in J_s \cup J_m^+} p_i m_i^v < 4ne + 4n^2(\Sigma_s + \Sigma_m^+). \tag{142}$$

Now the inequalities (102), (138) and (142) lead to the following estimate:

$$(4n^2(\Sigma_s + \Sigma_m^-) + 4ne + 4n^2(\Sigma_s + \Sigma_m^+))(\Sigma_s + \Sigma_m + \Sigma_u)$$

$$> ((3\Sigma_s + 2\Sigma_m + \Sigma_u)n + e)^2.$$

Taking into account that $\Sigma_m = \Sigma_m^+ + \Sigma_m^-$ (this is the crucial point), after some easy arithmetic we get the inequality

$$(n(\Sigma_s - \Sigma_u) + e)^2 < 0.$$

A contradiction.

The equality of the thresholds of canonical adjunction

$$c_{\text{virt}}(\Sigma) = c(\Sigma)$$

is proved for the non-singular case $B \not\subset \operatorname{Sing} F$.

Now let us consider the case when $B = o \in F$ is a singular point of the fibre. If the variety F satisfies the condition (f) at the point o, then the arguments of the previous section work well without any modifications. If this is not the case, then $\deg V = \deg F \geq 6$ (otherwise $\deg V \leq 4$ and the condition (f) holds automatically) and thus $\dim F \geq 4$, $\dim V \geq 5$. Therefore the discrepancy of the first exceptional divisor E_1 is at least 4 (we blow up a smooth point $o \in V$). By assumption, the condition (fs) holds, whence we get the inequality

$$m_i^h \leq m_1^h \leq 3n^2,$$

so that from the estimate (134) we obtain

$$\sum_{i \in J_l} p_i m_i^h \leq 3n^2(p_1 + \Sigma_s + \Sigma_m^-).$$

On the other hand, as we have mentioned above, $a(E_1, V) \geq 4$, so that one can replace in the inequality (102) the numerator by $(p_1 + 3\Sigma_s + 2\Sigma_m + \Sigma_u)$. Now arguing as in the non-singular case above we get the inequality

$$(n(\Sigma_s - \Sigma_u) + e)^2 + (\Sigma_s - 3p_1)(\Sigma_s + \Sigma_m + \Sigma_u)n^2 \\ + np_1(np_1 + 2(3\Sigma_s + 2\Sigma_m + \Sigma_u)n + 2e) < 0.$$

Obviously, $\Sigma_s - 3p_1 \geq -2p_1$ so that we again get a contradiction.

The proof of the equality of the thresholds of canonical adjunction $c_{\text{virt}}(\Sigma) = c(\Sigma)$ is complete. Proof of Theorem 4.1 is complete. Q.E.D.

5. Pencils of Fano complete intersections

The aim of this section is to explain the key steps of the proof of Theorems 4.3.1 and 4.3.2. The proof is based on the techniques of Theorem 4.1, part (ii), so that our work is reduced to checking the conditions (f), (v) and excluding the infinitely near maximal singularities lying over a singular point of a fibre. It is not hard to check the K^2-condition of depth 2 (see Propositions 4.3.3, 4.3.4) and we omit that step.

5.1. Fibrations into Fano complete intersections. Let us prove Theorem 4.3.1, part (i). Since the fibre space $V/\mathbb{P}^1$ is sufficiently general, we may assume that every fibre F at every point satisfies the regularity condition formulated in Definition 3.2.1, that is, the set of polynomials

(143) $$\{q_{i,j} \mid 1 \leq i \leq k, 1 \leq j \leq d_i, (i,j) \neq (k, d_k)\}$$

makes a regular sequence. Now Section 3.2 gives the condition (f). Let us prove the condition (v). Let $Y \subset F = F_t$ be a prime divisor, $o \in Y$ a point. Take a general hyperplane $H \subset \mathbb{P}$, tangent to F at the point o, that is, $H \supset T_oF$. Set $T = H \cap F$. By generality, $Y \neq T$, so that $Y_T = (Y \circ T)$ is a well-defined effective cycle of codimension two on F, and moreover,

$$\frac{\text{mult}_o}{\deg} Y_T \geq 2 \frac{\text{mult}_o}{\deg} Y.$$

Now the condition (f) implies the condition (v).

In order to prove part (i) of Theorem 4.3.1, we have to check that the centre of a maximal singularity of the system $\Sigma \subset |-nK_V + lF|$ cannot be a singular point of a fibre. There are finitely many such points on the variety V, so that we may assume that certain additional conditions of general position are satisfied. Let us formulate these conditions.

As we did in Section 3.2, take a system $(z_1, \dots, z_{M+k})$ of affine coordinates on $\mathbb{P} = \mathbb{P}^{M+k}$ with the origin at the point $o \in V$, which is a singular point of the fibre $F \ni o$, $F \subset \mathbb{P}$. By assumption, $o \in F$ is a non-degenerate quadratic singularity. Let us assume in addition, that the system of homogeneous equations $\{q_{i,j} = 0 \mid (i,j) \neq (k, d_k)\}$ defines a closed set of dimension two in $\mathbb{C}^{M+k}$ (respectively, a curve in $\mathbb{P}^{M+k-1}$), such that the linear span of each of its irreducible components is the linear space

$$T = \{q_{1,1} = q_{2,1} = \cdots = q_{k,1} = 0\}.$$

Note that if $x \in F$ is a singularity, then the linear forms $q_{i,1}$, $i = 1, \dots, k$, are linear dependent. The regularity of the point $x \in F$ means that, deleting from the set

(143) exactly one linear form, say $q_{1,e}$, we obtain a regular sequence, that is, the system of equations

(144) $$\{q_{i,j}=0\mid (i,j)\notin\{(1,e),(k,d_k)\}\}$$

defines a two-dimensional set in $\mathbb{C}^{M+k}$ (respectively, a curve in $\mathbb{P}^{M+k-1}$). In particular, $\operatorname{codim} T=k-1$ and the tangent cone $T_xF\subset T$ is a non-degenerate quadric. Moreover, it follows from the regularity condition that, replacing in the set (143) the linear form $q_{1,e}$ by an arbitrary linear form $l(z_1,\dots,z_{M+k-1})$, such that $l\,|_T\not\equiv 0$, we obtain a regular sequence, since neither component of the closed set (144) is contained in the hyperplane $l=0$.

Now assume that the singular point of the fibre $o\in F$ is the centre of a maximal singularity. Let $\lambda\colon F^+\to F$ be the blow up of the point o, $\lambda^{-1}(o)=E^+\subset F^+$ the exceptional divisor. The blow up λ can be looked at as the restriction of the blow up $\lambda_{\mathbb{P}}\colon\mathbb{P}^+\to\mathbb{P}$ of the point o on $\mathbb{P}$, so that $E^+\subset E$ is a non-singular quadric of dimension $M-1$, where $E=\lambda_{\mathbb{P}}^{-1}(o)\cong\mathbb{P}^{M+k-1}$ is the exceptional divisor.

Proposition 5.1. *There exists a hyperplane section B of the quadric $E^+\subset E$, satisfying the inequality*

$$\operatorname{mult}_B(\lambda^*\Sigma_F)>2n.$$

Proof is given below in Section 7.2, where we collect all facts related to the connectedness principle of Shokurov and Kollár. Q.E.D.

Let $D\in\Sigma_F=\Sigma\,|_F$ be an effective divisor on F, $D\in|nH_F|$. For the strict transform $D^+\subset F^+$ we have $\lambda^*D=D^++(\frac{1}{2}\operatorname{mult}_o D)E^+$. Taking into account that $\operatorname{mult}_B E^+=1$, we obtain from Proposition 5.1 the inequality

$$\operatorname{mult}_o D+2\operatorname{mult}_B D^+>4n.$$

Let $H\subset\mathbb{P}$ be a general hyperplane, containing the point o and cutting out B, that is, $H^+\cap E^+=(H^+\cap E)\cap E^+=B$, where $H^+\subset\mathbb{P}^+$ is the strict transform. Set $T=H\cap F$. The variety T is a complete intersection of type $(d_1,\dots,d_k)$ in $H=\mathbb{P}^{M+k-1}$ with an isolated quadratic singularity at the point o. The effective divisor $D_T=(D\circ T)$ on T satisfies the inequality

(145) $$\operatorname{mult}_o D_T>4n.$$

Obviously, $D_T\in|nH_T|$, where H_T is the hyperplane section of $T\subset\mathbb{P}^{M+k-1}$. By linearity, one may assume the divisor D_T to be prime, that is, an irreducible subvariety of codimension one.

Now we obtain a contradiction, repeating the arguments of Section 3.2 word for word: intersecting D_T with the hypertangent divisors, we construct a curve $C\subset T$, satisfying the inequality $\operatorname{mult}_o C>\deg C$, which is, of course, impossible. It remains to check that the technique of hypertangent divisors applies to our case. In Section 3.2 the following two facts were used:

1) the regularity condition for the complete intersection F at the point under consideration,

2) the irreducibility of the intersection $F\cap T_oF$ (it was derived from the relation $k<\frac{1}{2}\dim F$ by the Lefschetz theorem).

The arguments of Section 3.2 work in our case, if the conditions 1) and 2) hold.

As for the regularity condition, it is satisfied in our case due to the stronger regularity condition imposed on the singular point, formulated above. More precisely, the hyperplane section $T = H \cap F$ satisfies the usual regularity condition for any hyperplane $H \not\supset T_oF$.

Let us consider the condition 2). Recall that in this case $\operatorname{codim} T_oF = k-1$ (the point $o \in F$ is singular). Instead of the condition 2) we need the following fact: the intersection

$$T \cap T_oF = H \cap F \cap T_oF$$

is irreducible (and by the regularity condition has automatically the multiplicity exactly 2^k at the point o). This is true again by the Lefschetz theorem due to the inequality $k < \frac{1}{2} \dim F$, since $o \in F$ is a non-degenerate double point.

Now to complete the proof of the part (i) of Theorem 4.3.1, it remains to check that for the families under consideration the K-condition is satisfied.

The anticanonical class of the variety V is of the form $-K_V = L + \varepsilon F$, where $\varepsilon \leq 0$. This proves K-condition for all cases where $(a_*) = (0, \dots, 0, e)$, $e \in \{2, 3\}$ and $b_V = 0$. Indeed, the projection

$$\mathcal{O}_{\mathbb{P}^1}^{\oplus(M+k)} \oplus \mathcal{O}_{\mathbb{P}^1}(e) \to \mathcal{O}_{\mathbb{P}^1}^{\oplus(M+k)}$$

determines an embedding $\mathbb{P}^{M+k-1} \times \mathbb{P}^1 \hookrightarrow X$, the image of which is a smooth divisor $\Delta_X \cong \mathbb{P}^{M+k-1} \times \mathbb{P}^1$ and, moreover, the tautological class L is trivial on the lines $\{y\} \times \mathbb{P}^1$, $y \in \mathbb{P}^{M+k-1}$. Obviously, the divisor $\Delta = V \cap \Delta_X \subset V$ is of the form $\Delta_V \times \mathbb{P}^1$, where $\Delta_V \subset \mathbb{P}^M$ is a complete intersection of type $d_1 \cdot \dots \cdot d_k$ (because $b_1 = \dots = b_k = 0$ by assumption). Thus Δ is swept out by the lines $\{y\} \times \mathbb{P}^1$, $y \in \Delta_V$, on which the anticanonical class $-K_V$ is non-positive. Therefore, $-K_V \notin \operatorname{Int} A^1_{\text{mov}} V$.

If $(a_*) = (0, \dots, 0, 1, 2)$, then the class $L - 2F$ is effective. However,

$$((-K_V) \cdot (L - 2F) \cdot L^{M-1}) = ((L - F) \cdot (L - 2F) \cdot L^{M-1}) = 0,$$

which implies the K-condition.

If $(a_*) = (0, \dots, 0, 1, 1, 1)$, then the base set of the anticanonical system $\operatorname{Bs} |-K_V|$ is of codimension three. Therefore, for any mobile system $|-nK_V + lF|$ the class $((-K_V) \cdot (-nK_V + lF)^2)$ is effective. However,

$$((L - F)^3 \cdot L^{M-2}) = 0,$$

whence (taking into account that $-K_V = L - F$) the K-condition follows.

If $(a_*) = (0, \dots, 0, 2)$ and $b_V = 1$, that is, $b_e = 1$ for some $e \in \{1, \dots, k\}$ and $b_i = 0$ for $i \neq e$, we get $-K_V = L - F$ and, moreover, the class $L - 2F$ is effective. In this case the K-condition follows from the inequality

$$((L - F) \cdot (L - 2F) \cdot L^{M-1}) = d\left(\frac{1}{d_e} - 1\right) < 0.$$

Finally, consider the last case $(a_*) = (0, \dots, 0, 1, 1)$, $b_e = 1$, $b_i = 0$ for $i \neq e$. This case is somewhat harder than the previous ones. The projection

$$\mathcal{E} = \mathcal{O}_{\mathbb{P}^1}^{\oplus(M+k-1)} \oplus \mathcal{O}_{\mathbb{P}^1}(1)^{\oplus 2} \to \mathcal{O}_{\mathbb{P}^1}^{\oplus(M+k-1)}$$

defines an embedding $\mathbb{P}^{M+k-2}\times\mathbb{P}^1 \hookrightarrow X$, the image of which is a smooth subvariety $\Delta_X \cong \mathbb{P}^{M+k-2}\times\mathbb{P}^1$ of codimension two in X. Let

$$\alpha\colon \mathcal{O}_{\mathbb{P}^1}^{\oplus(M+k-1)} \to \mathcal{O}_{\mathbb{P}^1}^{\oplus(k+1)} \quad \text{and} \quad \beta\colon \mathcal{O}_{\mathbb{P}^1}(1)^{\oplus 2} \to \mathcal{O}_{\mathbb{P}^1}(1)$$

be generic surjective morphisms of sheaves. The morphism $\alpha\oplus\beta$ defines an embedding of projective bundles

$$i_{\alpha,\beta}\colon \mathbb{P}(\mathcal{O}_{\mathbb{P}^1}^{\oplus(k+1)}\oplus\mathcal{O}_{\mathbb{P}^1}(1)) \hookrightarrow X=\mathbb{P}(\mathcal{E}),$$

the image of which we denote by the symbol $S=S_{\alpha,\beta}$. The smooth subvariety $S\subset X$ is a $\mathbb{P}^{k+1}$-fibration over $\mathbb{P}^1$, the subvarieties $S_{\alpha,\beta}$ sweep out X and for generic projections α,β the intersection $V_S=V\cap S$ is a smooth surface on V. Set $\Delta_S=\Delta\cap V_S$. It is easy to see that Δ_S is a *curve* on V_S (although $\operatorname{codim}_X\Delta=2$), since the intersection $S\cap\Delta_X\cong\mathbb{P}^k\times\mathbb{P}^1$ is a *divisor* on S. Let us denote by the symbols L_S and F_S the restrictions of the classes L and F onto the surface V_S, respectively.

Lemma 5.1. (i) *For the curve Δ_S the linear equivalence $\Delta_S\sim L_S-F_S$ holds.*

(ii) *The self-intersection of the curve Δ_S satisfies the equality $\Delta_S^2=-d(1-\frac{1}{d_e})$.*

Proof. The equivalence (i) is obvious. Therefore,

$$\Delta_S^2=\left(S\cdot(L_X-R)^2\cdot\prod_{i\neq e}d_iL_X\cdot(d_eL_X+R)\right)=d\frac{1-d_e}{d_e},$$

which is what we claimed. Q.E.D. for the lemma.

Assume that the linear system $|-nK_V+lF|=|nL+(l-n)F|$ is mobile for some $l<0$. Then for generic projections α,β its restriction $|nL_S+(l-n)F_S|=|n\Delta_S+lF_S|$ is a non-empty system of curves on the surface V_S. Let C be a generic curve of this linear system. Now write

$$C=n_+\Delta_S+C^+,$$

where $n_+\le n$ and C^+ does not contain Δ_S as a component. We get

$$(C^+\cdot\Delta_S)=(n-n_+)\Delta_S^2+ld\ge 0,$$

which is impossible when $l<0$. This contradiction completes the proof of the K-condition for the family of fibre spaces under consideration.

Now the K^2-condition of depth 2 follows from part (ii) by a trivial check.

By Theorem 4.1, the proof of part (i) of Theorem 4.3.1 is complete.

5.2. Special cases. Let us prove part (ii) of Theorem 4.3.1. In this case $-K_V=L$, but the linear system $|L_X-R|$ is mobile, so that the system $|K_V-F|$ is mobile, either, and defines a rational map $\varphi\colon V\dashrightarrow\mathbb{P}^1$. It is easy to see that the complete linear system $|L_X|$ defines a birational morphism

$$\xi\colon X\to\bar{X}\subset\mathbb{P}^{M+k+2},$$

the image of which $\bar{X}$ is a quadric cone in $\mathbb{P}^{M+k+2}$ with the vertex space $\xi(\Delta_X)\cong\mathbb{P}^{M+k-2}$, where $\Delta_X\cong\mathbb{P}^{M+k-2}\times\mathbb{P}^1$ is a subvariety of codimension two, defined by the projection $\mathcal{E}\to\mathcal{O}_{\mathbb{P}^1}^{\oplus(M+k-1)}$, whereas $\xi\,|_{\Delta_X}\colon\Delta_X\to\mathbb{P}^{M+k-2}$ is the projection onto the direct factor. The base of the cone $\bar{X}$ is a non-singular quadric in $\mathbb{P}^3$, isomorphic to $\mathbb{P}^1\times\mathbb{P}^1$. The morphism ξ defines an isomorphism of open sets $X\setminus\Delta_X$ and $\bar{X}\setminus\xi(\Delta_X)$.

On the cone $\bar{X}$ there are exactly two pencils of $(M+k)$-planes, corresponding to the two pencils of lines on a smooth quadric in $\mathbb{P}^3$. It is easy to see that these pencils are $\xi_*|R|$ and $\xi_*|L_X - R|$. Since $\mathcal{O}_X(L_X) = \xi^*\mathcal{O}_{\mathbb{P}^{M+k+2}}(1)$, the equations of the divisors G_i are the pullbacks on X of generic polynomials of degrees $d_1, \ldots, d_k$:

$$G_i = \{\xi^* P_i = 0\}, \quad \text{where} \quad P_i \in H^0(\mathbb{P}^{M+k+2}, \mathcal{O}_{\mathbb{P}^{M+k+2}}(d_i)).$$

Let $\tau \in \operatorname{Aut} \mathbb{P}^{M+k+2}$ be an arbitrary biregular involution, an automorphism of order two, transforming $\bar{X}$ into itself, $\tau(\bar{X}) = \bar{X}$, fixing the vertex space $\xi(\Delta_X)$, and permuting the pencils of $(M+k)$-planes on $\bar{X}$. Set $X^+ = X$, $\xi^+ = \tau \circ \xi \colon X^+ \to \bar{X}$, $G_i^+ = \{(\xi^+)^* P_i = 0\}$ and

$$V^+ = G_1^+ \cap \cdots \cap G_k^+ \subset X^+.$$

Obviously, $V^+/\mathbb{P}^1$ is a Fano fibre space with the same parameters (a_*), (b_*) as $V/\mathbb{P}^1$, and for a sufficiently general V the fibre space $V^+/\mathbb{P}^1$ satisfies the regularity conditions (since those conditions actually mean just that the polynomials $P_1, \ldots, P_k$ are generic).

Now the identity map $\mathrm{id}_{\bar{X}}$ extends to a birational isomorphism $\chi \colon V \dashrightarrow V^+$, biregular outside $V \cap \Delta_X$, $V^+ \cap \Delta_{X^+}$, where $\Delta_{X^+} = \Delta_X$ with respect to the identification $X = X^+$. Thus χ is an isomorphism in codimension one. By construction, χ transforms the pencil $|-K_V - F|$ into the pencil $|F^+|$, whereas the pencil $|F|$ is transformed into the pencil $|-K_{V^+} - F^+|$. The action of χ on the Picard group is given by the formulas $\chi^* K_{V^+} = K_V$, $\chi^* F^+ = -K_V - F$. The self-duality of this construction is obvious.

Finally, let us prove the birational superrigidity. From part (ii) of Proposition 4.3.3 we get that V and V^+ satisfy the K^2-condition of depth 2. Let $\Sigma \subset |-nK_V + lF|$ be a mobile linear system. If $l \in \mathbb{Z}_+$, then we apply Theorem 4.1 and conclude that the thresholds coincide: $c_{\mathrm{virt}}(\Sigma) = c(\Sigma)$. Assume that $l < 0$. Consider the linear system $\Sigma^+ = \chi_*\Sigma$ on V^+. We have $\Sigma \subset |-nK_{V^+} + l_+F^+|$, where $n_+ = n + l$ and $l_+ = -l \geq 1$. Applying Theorem 4.1, we see again that the thresholds coincide: $c_{\mathrm{virt}}(\Sigma^+) = c(\Sigma^+)$. But χ is an isomorphism in codimension one, so that the thresholds of canonical adjunction on V and V^+ coincide, $c(\Sigma^+) = c(\Sigma)$. Since the virtual threshold of canonical adjunction is a birational invariant, we obtain

$$c_{\mathrm{virt}}(\Sigma) = c(\Sigma) = n + l.$$

This proves the birational superrigidity. Moreover, the argument above implies that if Σ is a mobile linear system and $c_{\mathrm{virt}}(\Sigma) = 0$, then either $\Sigma = |F|$ or $\Sigma = |-K_V - F|$. This implies that on V there are exactly two non-trivial structures of a fibre space into varieties of negative Kodaira dimension, namely, π and φ.

Q.E.D. for part (ii) of Theorem 4.3.1.

Remark 5.1. One can show that the structures π and φ are birationally distinct, that is, there is no birational self-map, transforming the fibres of π into the fibres of φ, whereas the groups of birational and biregular self-maps coincide: $\operatorname{Bir} V = \operatorname{Aut} V$.

Finally, let us prove the claim (iii) of Theorem 4.3.1. This case is quite similar to the one we have just studied. For $i \neq e$ we get $G_i = Q_i \times \mathbb{P}^1 \subset X = \mathbb{P} \times \mathbb{P}^1$, where $Q_i \subset \mathbb{P}$ is a hypersurface of degree d_i. Let

$$Q = \bigcap_{i \neq e} Q_i \subset \mathbb{P}$$

be a complete intersection of type $d_1 \cdot \dots \cdot d_{e-1} \cdot d_{e+1} \cdot \dots \cdot d_k$ in $\mathbb{P}$, a Fano variety of index $d_e + 1$. Obviously, V is a divisor on $Q \times \mathbb{P}^1$, given by an equation

$$A(x_*)u^2 + B(x_*)uv + C(x_*)v^2 = 0,$$

where $(u : v)$ are homogeneous coordinates on $\mathbb{P}^1$, (x_*) are homogeneous coordinates on $\mathbb{P}$, $A(\cdot)$, $B(\cdot)$ and $C(\cdot)$ are homogeneous polynomials of degree d_e restricted onto Q. Let $\sigma \colon V \to Q$ be the projection onto the direct factor; obviously, outside the subvariety $\Delta \subset V$ of codimension two, given by the system of equations $A = B = C = 0$, the morphism σ is a double cover. Let $\tau \in \operatorname{Bir} V$ be the corresponding Galois involution. The involution τ is well defined in codimension one, so that its action on the Picard group $\operatorname{Pic} V$ is well defined:

$$\tau^* K_V = K_V, \quad \tau^* F = -d_e K_V - F. \tag{146}$$

Indeed, let F_t, $t \in \mathbb{P}^1$, be a fibre of the projection π. Obviously,

$$\sigma^{-1}(\sigma(F_t)) = F_t \cup \tau(F_t),$$

however, $\sigma(F_t) \cong F_t \subset Q$ is cut out on Q by a hypersurface of degree d_e. Taking into account that $L = -K_V$, we obtain the second formula in (146).

Let us prove birational superrigidity of the variety V. According to Proposition 4.3.3, part (ii), the fibre space $V/\mathbb{P}^1$ satisfies the K^2-condition of depth 2 (in fact, of depth $\frac{2}{d_e} \leq 1$). Let $\Sigma \subset |-nK_V + lF|$ be a mobile linear system. If $l \in \mathbb{Z}_+$, then by Theorem 4.1 we have $c_{\text{virt}}(\Sigma) = c(\Sigma)$, which is what we need. If $l < 0$, then consider the system $\Sigma^+ = \tau_* \Sigma$. Since τ is an isomorphism in codimension one, we have $c(\Sigma^+) = c(\Sigma)$. Since the virtual threshold is a birational invariant, $c_{\text{virt}}(\Sigma^+) = c_{\text{virt}}(\Sigma)$. However, $\Sigma^+ \subset |-n_+ K_V + l_+ F|$, where by (11) $n_+ = n + l d_e$, $l_+ = -l \geq 1$. Applying to the system Σ^+ Theorem 4.1, we get $c_{\text{virt}}(\Sigma^+) = c(\Sigma^+)$, which implies birational rigidity by what has been said above.

The very same arguments (as in the case (ii) above) prove that there are exactly two non-trivial structures of a fibre space into varieties of negative Kodaira dimension on V, that is, π and $\pi\tau$.

Finally, if $\chi \in \operatorname{Bir} V$, then twisting by τ if necessary, one may assume that χ preserves the structure π, that is, transforms the fibres of F_t into the fibres $F_{\gamma(t)}$ for some isomorphism $\gamma \colon \mathbb{P}^1 \to \mathbb{P}^1$. However, for a generic variety V a general fibre F_t has the trivial group of birational (=biregular) self-maps and, moreover, a general fibre F_t is not isomorphic to any other fibre F_s, $s \neq t$, which implies that $\chi = \mathrm{id}_V$ is the identity map. Therefore, $\operatorname{Bir} V = \mathbb{Z}/2\mathbb{Z} = \{\mathrm{id}_V, \tau\}$. Q.E.D. for Theorem 4.3.1.

5.3. Fibrations into Fano cyclic covers. Let us prove Theorem 4.3.2. The conditions (f) and (v) for sufficiently general Fano cyclic covers are checked in the same way as for the complete intersections above, taking into account the additional hypertangent divisors, described in Section 3.2. Now it is sufficient to show that the centre of a maximal singularity cannot be a singular point of a fibre $o \in F$.

Let $\sigma\colon F \to G \subset \mathbb{P} = \mathbb{P}^{M+1}$ be a realization of the fibre as a K-sheeted cyclic cover. For a generic variety V the singular point $o \in F$ can belong to strictly one of the two types:

— when the hypersurface G has a non-degenerate quadratic singularity at the point $p = \sigma(o)$ (and the point p does not lie on the branch divisor),

— when the hypersurface G is non-singular at the point $p = \sigma(o)$, but the branch divisor $W \cap G$, where $W = W_{Kl} \subset \mathbb{P}$ is a hypersurface of degree Kl, has a quadratic singularity at the point p.

A singularity $o \in F$ of the first type is excluded as a possible centre of a maximal singularity in precisely the same way as in Section 2 (again, taking into account the additional hypertangent divisors for the K-sheeted cover). Referring to [**Pukh09a**] for the details, let us consider a singularity $o \in F$ of the second type. In this case, to exclude the maximal singularity, we need another method, similar to the method used in Section 3.

First, let us formulate the regularity condition for a singularity of the second type. Introducing a new coordinate u of weight l, we realize the fibre F as a complete intersection of the type $m \cdot Kl$ in the weighted projective space

$$\mathbb{P}(\underbrace{1,\dots,1}_{M+2},l).$$

Namely, F is given by the system of equations

$$\left\{\begin{array}{l} f(x_0,\dots,x_{M+1}) = 0, \\ u^K = g(x_0,\dots,x_{M+1}), \end{array}\right. \tag{147}$$

where $f(x_*)$ and $g(x_*)$ are homogeneous polynomials of degrees m and Kl, respectively. Recall that the integers m, l and K satisfy the relation $m+(K-1)l = M+1$.

Let $F \sim (f,g)$ be our variety, $o \in F$ an arbitrary point, $p = \sigma(o) \in G = \{f = 0\}$ its image on $\mathbb{P}$. Choose a system of affine coordinates $z_1,\dots,z_{M+1}$ with the origin at the point p. Without loss of generality, we may assume that $z_i = x_i/x_0$. Set $y = u/x_0^l$. Now the standard affine set $\mathbb{A}^{M+2}_{(z_1,\dots,z_{M+1},y)}$ is a chart for $\mathbb{P}(1,\dots,1,l)$. Abusing our notations, we use for the non-homogeneous polynomials, corresponding to f and g, the same symbols:

$$f = q_1 + \dots + q_m, \qquad g = w_0 + w_1 + \dots + w_{Kl},$$

where q_i, w_j are homogeneous components of degree i,j in the variables z_*, respectively, so that in the affine chart $\mathbb{A}^{M+2}_{(z_*,y)}$, introduced above, the variety F is given by the pair of equations $f = 0$, $y^K = g$ (replacing the system (147)). If the point $o \in F$ does not lie on the ramification divisor of the morphism σ, then we always assume that $w_0 = 1$. If the point $p \in G$ is non-singular, then, without loss of generality, we assume that $q_1 \equiv z_{M+1}$. In the latter case we set

$$\bar{q}_i = q_i\,|_{\{z_{M+1}=0\}} = q_i(z_1,\dots,z_M,0)$$

and $\bar{w}_j = w_j\,|_{z_{M+1}=0} = w_j(z_1,\dots,z_M,0)$ for $i,j \geq 2$.

Now let us formulate the regularity condition for the singular point $o \in F$. Here $w_0 = 0$, $q_1 = z_{M+1}$, $w_1 = \lambda z_{M+1}$, where $\lambda \in \mathbb{C}$ is a constant, that is, the point $p \in G$ is non-singular. We require the quadratic form $\bar{w}_2(z_1,\dots,z_M)$ to have the maximal rank M and the sequence

$$\bar{q}_2, \dots, \bar{q}_m, \bar{w}_2, \dots, \bar{w}_K$$

to be regular in $\mathcal{O}_{o,\mathbb{C}^M}$ and, moreover, the system of homogeneous equations

$$\bar{q}_2 = \dots = \bar{q}_m = \bar{w}_2 = \dots = \bar{w}_K = 0 \tag{148}$$

should define a closed algebraic set in $\mathbb{C}^M$, neither component of which is contained in a hyperplane.

Obviously, we may assume that $\lambda \in \{0,1\}$. Now, either $q_1 \equiv w_1 \equiv z_{M+1}$, or $q_1 \equiv z_{M+1}$ and $w_1 \equiv 0$. The germ of the variety $o \in F$ is analytically the germ of the hypersurface $y^K = \bar{w}_2(z_1,\dots,z_M) + \dots$, where $\bar{w}_i, \bar{q}_i$ are the restrictions of the polynomials w_i, q_i onto the hyperplane $z_{M+1} = 0$, in the space $\mathbb{C}^{M+1}_{(z_1,\dots,z_M,y)}$. Let $\varphi\colon F^+ \to F$ be the blow up of the point o, $E \subset F^+$ the exceptional divisor. It follows from what was said that E realizes naturally as a quadratic hypersurface, $E \subset \mathbb{P}^M_{(z_1:\dots:z_M:y)}$. Let $\varphi_G\colon G^+ \to G$ be the blow up of the point $p \in G$, $E_G = \varphi_G^{-1}(p) \subset G^+$ the exceptional divisor, $E_G \cong \mathbb{P}^{M-1}_{(z_1:\dots:z_M)}$. It is easy to see that the morphism σ extends to a rational map $\sigma^+\colon F^+ \dashrightarrow G^+$, whereas the restriction

$$\sigma_E = \sigma^+|_E\colon E \dashrightarrow E_G$$

is the projection of the quadratic cone $E \subset \mathbb{P}^M$, given in $\mathbb{P}^M_{(z_1:\dots:z_M:y)}$ by the equation $\bar{w}_2(z_1,\dots,z_M) = 0$, from its vertex $o^+ = (0:\dots:0:1)$, onto the smooth quadric $E_+ \subset E_G$, given in $E_G = \mathbb{P}^{M-1}$ by the very same equation $\bar{w}_2 = 0$. Therefore, σ^+ contracts generators of the cone E to points.

By the regularity condition, the system of homogeneous equations (148) cuts out on E_+ (and thus on E) a closed algebraic set, neither component of which is contained in a hyperplane.

Assume that there exist an effective divisor $R \in |nH_F|$ and a hyperplane section B of the quadric cone $E \subset \mathbb{P}^M$, satisfying the inequality

$$\nu + \mu > 2n, \tag{149}$$

where $R^+ \in |nH_F - \nu E|$, that is, $\nu = \frac{1}{2}\operatorname{mult}_o R$, and $\mu = \operatorname{mult}_B R^+$, $R^+ \subset F^+$ is the strict transform of the divisor R. By linearity of the inequality (149) one may assume the divisor R to be prime.

Lemma 5.2. *The following estimate holds:* $\nu \leq \frac{3}{2}n$.

Proof. Assume the converse: $\nu > \frac{3}{2}n$. Let

$$D_i = \sigma^*(f_i\,|\,_G), \quad f_i = q_1 + \dots + q_i, \quad i = 1,\dots,m-1$$

and

$$\Delta_i = \sigma^*(g_i\,|\,_G), \quad g_i = w_1 + \dots + w_i, \quad i = 2,\dots,K-1,$$

be hypertangent divisors on F, $\mathcal{D} = \{D_1,\dots,\Delta_{K-1}\}$, $\sharp\mathcal{D} = m + K - 3$. By the regularity condition we get

$$\operatorname{codim}_o\left(\bigcap_{D\in\mathcal{D}} D\right) = \sharp\mathcal{D}.$$

Lemma 5.3. *The prime divisors D_1 and R are distinct:* $D_1 \neq R$.

Proof. If $\nu \leq 2n$, then $\mu \geq 1$, that is, $R^+ \supset B$. By the regularity condition, $D_1^+ \not\supset B$. Therefore, $D_1^+ \neq R^+$.

If $\nu > 2n$, then, taking into account that $D_1^+ \in |H_F - 2E|$, we obtain again that $D_1 \neq R$. Q.E.D. for Lemma 5.3.

By the lemma the effective cycle $(R \circ D_1)$ of codimension two is well defined. The inequality

$$\frac{\mathrm{mult}_o}{\deg}(R \circ D_1) > \frac{6}{mK}$$

holds, so that there is an irreducible subvariety $Y \subset F$ of codimension two (an irreducible component of the cycle $(R \circ D_1)$), satisfying the inequality

$$\frac{\mathrm{mult}_o}{\deg} Y > \frac{6}{mK}.$$

Applying to the subvariety Y Proposition 3.2, we obtain the opposite inequality

$$\frac{\mathrm{mult}_o}{\deg} Y \leq \frac{2}{3} \cdot \ldots \cdot \frac{m-1}{m} \cdot \frac{3}{4} \cdot \ldots \cdot \frac{K-1}{K} = \frac{6}{mK}.$$

The contradiction just obtained shows, in addition, that the maximum of the ratio $\mathrm{mult}_o / \deg$ is attained at the divisor $\sigma^* D_1$ and equal to $4/mK$. Q.E.D. for Lemma 5.2.

Corollary 5.1. *The following inequality holds:* $\mu > \frac{1}{2}n$.

Thus the hyperplane section B is really present in the divisor $R^+ \subset F^+$. Now it is more convenient to use the technique of hypertangent linear systems (see Section 3.1). Set

$$\Lambda_i^G = \left| \left(\sum_{j=1}^{i} f_j s_{i-j} + \sum_{j=2}^{i} g_j r_{i-j} \right) \Big|_G = 0 \right|,$$

where s_k, r_k run through the set of all homogeneous polynomials in $z_1, \ldots, z_{M+1}$ of degree k. Furthermore, let

$$\Lambda_i = \sigma^* \Lambda_i^G \quad \text{and} \quad \Lambda_i^+$$

be the pullback of the system Λ_i^G on F and its strict transform on F^+, respectively. We get $\Lambda_i^+ \subset |iH_F - (i+1)E|$. Finally, let

$$\Lambda_i^E = \Lambda_i^+ \,|\, _E$$

be the projectivized tangent system of the linear system Λ_i at the point o.

Let H_E be the class of a hyperplane section of the cone E. We get $\Lambda_i^E \subset |(i+1)H_E|$. In the coordinate form

$$\Lambda_i^E = \sigma_E^* \left| \sum_{j=1}^{i} \bar{q}_{j+1} \bar{s}_{i-j} + \sum_{j=2}^{i} \bar{w}_{j+1} \bar{r}_{i-j} = 0 \right|,$$

where $\bar{s}_k = s_k \,|\, _{\{z_{M+1}=0\}}, \bar{r}_k = r_k \,|\, _{\{z_{M+1}=0\}}$. From this explicit presentation and the regularity condition we get at once that

$$\mathrm{codim}_E \operatorname{Bs} \Lambda_i^E = \min\{i, m-1\} + \min\{i, K-1\} - 1$$

and, moreover, neither component of the closed set $\operatorname{Bs} \Lambda_i^E$ is contained in a hyperplane. Note that for $i \geq \max\{m, K\} - 1$ we get $\operatorname{Bs} \Lambda_i = \operatorname{Bs} \Lambda_{i+1}$: it is precisely the closed set

$$\bar{q}_2 = \cdots = \bar{q}_m = \bar{w}_3 = \cdots = \bar{w}_K = 0.$$

Let

$$\mathcal{L} = (D_1, \dots, D_{m-1}, L_2, \dots, L_{K-1}) \in \prod_{i=1}^{m-1} \Lambda_i \times \prod_{i=2}^{K-1} \Lambda_i$$

be a general set of hypertangent divisors,

$$\mathcal{L}^+ = (D_1^+, \dots, L_{K-1}^+) \quad \text{and} \quad \mathcal{L}^E = (D_1^E, \dots, L_{K-1}^E) = \mathcal{L}^+|_E$$

its strict transform on F^+ and restriction onto the quadric cone E, respectively.

Set $R_E = R^+|_E$ to be the projectivized tangent cone of the divisor R at the point o. By the regularity condition the closed algebraic set

$$R_E \cap \left(\bigcap_{i=2}^{m-1} D_i^E \right) \cap \left(\bigcap_{i=2}^{K-1} L_i^E \right) \tag{150}$$

is of codimension precisely $m + K - 2$ with respect to E (note that in (150) the divisor D_1^E is omitted).

Therefore, the effective cycle

$$Y_E = (R_E \circ D_2^E \circ \cdots \circ D_{m-1}^E \circ L_2^E \circ \cdots \circ L_{K-1}^E)$$

of codimension $m + K - 2$ on E is well defined. Its H_E-degree is

$$\deg Y_E = \frac{1}{4} m! K! \deg R_E = \frac{1}{2} m! K! \nu.$$

Furthermore, by the regularity condition the closed set

$$R \cap \left(\bigcap_{i=2}^{m-1} D_i \right) \cap \left(\bigcap_{i=2}^{K-1} L_i \right)$$

is of codimension $m + K - 2$ with respect to F *in a neighborhood of the point* o. Thus we get the uniquely determined effective algebraic cycle Y of codimension $m + K - 2$ on F, each irreducible component of which contains the point o, and which coincides with the effective cycle

$$(R \circ D_2 \circ \cdots \circ D_{m-1} \circ L_2 \circ \cdots \circ L_{K-1})_U$$

on a suitable Zariski open set $U \subset F$, containing the point o. For a general set $\mathcal{L}$ we get: $Y_E = (Y^+ \circ E)$ is the projectivized tangent cone to Y at the point o. However, generally speaking, for the H_F-degree of the cycle Y we get only the inequality

$$\deg Y \leq (m-1)!(K-1)! \deg R = nm!K!,$$

since constructing the cycle Y, at the intermediate steps (the divisor R is successively intersected with D_2, L_2, D_3, L_3 and so on) we remove the components that do not contain the point o. Note that certain irreducible components of the cycles Y and Y_E can be contained in D_1 and D_1^E, respectively. Let us separate those components:

$$Y = Z + Y^\sharp, \quad Y_E = Z_E + Y_E^\sharp,$$

where $\operatorname{Supp} Z \subset D_1$ and Z is the maximal subcycle of the effective cycle Y with this property (that is, neither irreducible component of the cycle $Y^\sharp$ is contained in D_1), $Z_E = (Z^+ \circ E)$, $Y_E^\sharp = ((Y^\sharp)^+ \circ E) = Y_E - Z_E$. Obviously, $\operatorname{Supp} Z_E \subset D_1^E$, but irreducible components of the cycle $Y_E^\sharp$, generally speaking, *can* be contained in D_1^E. The following fact is of key importance.

Lemma 5.4. *The algebraic cycle*

$$Y_E^\sharp - \mu(B \circ D_2^E \circ \cdots \circ D_{m-1}^E \circ L_2^E \circ \cdots \circ L_{K-1}^E)$$

is effective. In particular, the following inequality holds:

$$\deg Y_E^\sharp \geq \frac{1}{2} m!K!\mu.$$

Proof. The first claim holds because by construction the algebraic cycle $R_E - \mu B$ is effective, so that the cycle $Y_E - \mu B_{\mathcal{L}}$ is effective, as well, where $B_{\mathcal{L}} = (B \circ D_2^E \circ \cdots \circ L_{K-1}^E)$. The support of the cycle $B_{\mathcal{L}}$ is a closed set $\operatorname{Supp} B_{\mathcal{L}}$ of pure codimension $m + K - 3$ with respect to E. For any irreducible component $\Delta \subset \operatorname{Supp} B_{\mathcal{L}}$ we get $\Delta \not\subset D_1^E$.

Indeed, assume the converse: $\Delta \subset D_1^E$. Then $\Delta \subset B$ and

$$\Delta \subset \left(\bigcap_{i=1}^{m-1} D_i^E \right) \cap \left(\bigcap_{i=2}^{K-1} L_i^E \right),$$

which contradicts the regularity condition.

Therefore, each irreducible component Δ of the cycle $B_{\mathcal{L}}$ cannot be a component of the cycle Z_E and so appears in the cycle $Y_E^\sharp$ only. This proves the first claim of the lemma. The second claim follows from the first one in a trivial way. Q.E.D. for the lemma.

Since the irreducble components of the cycle $Y^\sharp$ are not contained in the divisor D_1, the effective cycle $(Y^\sharp \circ D_1)$ is well defined and, moreover, the inequality

$$\frac{\operatorname{mult}_o}{\deg}(Y^\sharp \circ D_1) \geq 2 \frac{\operatorname{mult}_o}{\deg} Y^\sharp$$

holds. Moreover, since for any irreducible subvariety $\Delta \subset F$ we have the estimate $(\operatorname{mult}_o / \deg)\Delta \leq 1$, we get

$$\deg Y^\sharp \geq 2 \operatorname{mult}_o Y^\sharp = 2 \deg Y_E^\sharp \geq m!K!\mu.$$

On the other hand, we have the inequality $\deg Z \geq \operatorname{mult}_o Z = \deg Z_E$ (which is true for any effective cycle Z).

Combining these estimates, we get

$$\begin{aligned} m!K!n &\geq \deg Y = \deg Z + \deg Y^\sharp \\ &\geq \deg Z_E + 2 \deg Y_E^\sharp = \deg Y_E + \deg Y_E^\sharp \\ &\geq \frac{1}{2} m!K!\nu + \frac{1}{2} m!K!\mu = \frac{1}{2} m!K!(\nu + \mu). \end{aligned}$$

Therefore, $\nu + \mu \leq 2n$. Contradiction.

We have proved that a singular point of a fibre $o \in F$, lying on the ramification divisor of the morphism σ, cannot be the centre of a maximal singularity of a mobile linear system.

This completes the proof of Theorem 4.3.2. Q.E.D.

Notes and references

This chapter is based on the series of papers [**Pukh98a**] (Section 1), [**Pukh00d, Pukh00a**] (Section 2), [**Pukh04a**] (Section 3), [**Pukh04b**] (Section 4), [**Pukh06b, Pukh09a**] (Section 5); apart from those papers, birational rigidity of certain Fano fibre spaces over $\mathbb{P}^1$ was proved by the quadratic method in [**Pukh00b, Sob00, Sob01**].

CHAPTER 6

Del Pezzo Fibrations

Summary

The aim of this chapter is to review the known results on birational (super)rigidity of fibrations into del Pezzo surfaces of degrees 1, 2 and 3 over $\mathbb{P}^1$ and to prove Theorem 4.3.3. Fibrations into del Pezzo surfaces of higher degree $d \geq 4$ are not considered as for $d \geq 5$ they are rational and for $d = 4$ they are birational to conic bundles (see, for instance, [**Al87**]), so in any case not rigid and their birational geometry is outside the range of our technique. On the contrary, the cases of degree 1, 2 and 3 are by now very well studied. In Section 1 we reproduce the explicit constructions of del Pezzo fibrations and review the main known results on their birational geometry. Section 2 starts with a discussion of the technically more difficult cases when the strong K^2-condition does not hold. The remaining part of Section 2 and Section 3 deal with infinitely near maximal singularities for fibrations into cubic surfaces — the hardest and most interesting class of varieties of this type; there we give a proof of Theorem 4.3.3.

1. Explicit constructions and a summary of known results

In this section, we discuss explicit constructions of fibrations into del Pezzo surfaces of degrees 1, 2 and 3 and list the main facts known about their birational (super)rigidity. Subsection 1.1 treats the explicit constructions of these varieties: as double covers for $d = 1, 2$ and as divisors in $\mathbb{P}^3$-bundles over $\mathbb{P}^1$ for $d = 3$. In Subsection 1.2 we explicitly describe the fibrewise birational involutions of fibrations into del Pezzo surfaces of degree 2 and complete the untwisting procedure for the three degrees 1,2,3: every mobile linear system Σ can be untwisted to a mobile system Σ^* (on the same variety), without maximal curves. This is a simple but important step in the study of the birational type of del Pezzo fibrations. Finally, in Subsection 1.3 we summarize what is known about birational (super)rigidity of del Pezzo fibrations of degrees 1, 2 and 3.

1.1. Fibrations into del Pezzo surfaces. In this chapter we consider fibrations $\pi \colon V \to \mathbb{P}^1$, where V is a smooth projective variety, $\dim V = 3$, each fibre of π is an irreducible reduced del Pezzo surface and

$$\operatorname{Pic} V = \mathbb{Z}K_V \oplus \mathbb{Z}F,$$

where, as usual, F is the class of a fibre of π. In particular, the generic fibre (over the non-closed point) F_η is a non-singular del Pezzo surface over the non-closed field $\mathbb{C}(\mathbb{P}^1)$ with the Picard group $\operatorname{Pic} F_\eta = \mathbb{Z}K_{F_\eta}$. The integer $d = (K_V^2 \cdot F)$, the degree of the del Pezzo surface $\pi^{-1}(t) = F_t$ (or F_η) is assumed to be equal to 1, 2 or 3.

Let us consider the Chow group A^2V of curves (or 1-cycles) on V modulo numerical equivalence. It is well known that the projection π admits (plenty of) sections. Let $s \in A^2V$ be the class of a section. Now we can write the Chow group as

$$A^2V = \mathbb{Z}s \oplus \mathbb{Z}f,$$

where f is the class of a line in a fibre. The multiplication table for divisors and curves is given by the following obvious relations

$$(K_V \cdot f) = -1, \ (F \cdot s) = 1, \ (F \cdot f) = 0$$

(with $(K_V \cdot s)$ depending on the choice of the section).

Note that the difference between del Pezzo fibrations and the higher-dimensional fibre spaces considered in Chapter 5 (and the reason why the former need a special treatment) shows itself already on the level of the group A^2: the component $\mathbb{Z}(H \cdot F)$ (in higher dimensions) is to be replaced by $\mathbb{Z}f$. Speaking plainly, there are a lot more curves on a del Pezzo surface than just pluri-anticanonical divisors. In particular, treatment of singularities of codimension two cycles cannot be so uniform as in the higher-dimensional case.

Now let us describe briefly explicit geometric constructions of three-fold del Pezzo fibrations.

Fibrations into del Pezzo surfaces of degree 1. Here the relatively ample sheaf $\mathcal{O}_V(-2K_V)$ is generated by its sections on every fibre. Therefore, $\pi_*\mathcal{O}_V(-2K_V)$ is a locally free sheaf of rank 4. Twisting by $\mathcal{O}_V(m_0F)$, we can normalize that sheaf as

$$\mathcal{E} = \pi_*\mathcal{O}_V(-2K_V + m_0F) \cong \bigoplus_{i=0}^{3} \mathcal{O}_{\mathbb{P}^1}(a_i)$$

with $a_0 = 0$ and $a_i \in \mathbb{Z}_+$, $a_1 \leq a_2 \leq a_3$. Let $X = \mathbb{P}(\mathcal{E})$ be the corresponding projective bundle, $\pi_X \colon X \to \mathbb{P}^1$ the natural projection; X is a $\mathbb{P}^3$-bundle over $\mathbb{P}^1$ so that

$$\operatorname{Pic} X = \mathbb{Z}H_X \oplus \mathbb{Z}G, \tag{151}$$

where H_X is the class of the tautological sheaf and G is the class of a fibre. From the well-known properties of del Pezzo sufraces of degree 1 we deduce that the linear system $|-2K_V + m_0F|$ determines a finite morphism over $\mathbb{P}^1$ of degree 2

$$\varphi \colon V \to Q \subset X,$$

where $\pi_Q = \pi_X|_Q \colon Q \to \mathbb{P}^1$ is a fibration into non-degenerate quadric cones $Q_t = \pi_Q^{-1}(t) \subset \mathbb{P}^3$ (they are non-degenerate for every $t \in \mathbb{P}^1$). The morphism φ is branched over a smooth surface $R_Q = R \cap Q$, where $R \sim 3H_X + cG$ for some $c \in \mathbb{Z}$. The threefold $Q \subset X$ is a member of the linear system $|2H_X + aG|$ for some $a \in \mathbb{Z}$.

Let $T \subset Q$ be the curve of vertices of the cones Q_t and $\Gamma \subset V$ its set-theoretic preimage. For every $t \in \mathbb{P}^1$ the point $\Gamma \subset V$ is the base point of the anticanonical system of the fibre, $|-K_{F_t}|$; alternatively,

$$\Gamma = \operatorname{Bs}|-K_V + lF|$$

for $l \gg 0$. The curve T and the divisor R are disjoint: $T \cap R = \emptyset$. Let us denote by T_0 the uniquely determined section of the bundle $X/\mathbb{P}^1$, corresponding to the projection $\mathcal{E} \to \mathcal{O}_{\mathbb{P}^1}(a_0) = \mathcal{O}_{\mathbb{P}^1} \to 0$ (the "minimal" section of the bundle), and set $T \sim T_0 + el_X$, where l_X is the class of a line in $\mathbb{P}^3$. The following claim is true.

Proposition 1.1. *A non-empty family of varieties $V/\mathbb{P}^1$ exists if and only if one of the two cases takes place:*

(i) $e = 0, 2a_2 = a_1 + a_3$ *with* a_1, a_3 *even, so that*

$$Q \sim 2H_X - 2a_2G, \quad T \sim T_0, \quad R \sim 3H_X,$$

(ii) $e = a_1 > 0$ *is even,* $a_3 = 2a_2$ *and* $a_2 \geq 3a_1$, *here*

$$Q \sim 2H_X - 2a_2G, \quad T \sim T_0 + a_1l_X, \quad R \sim 3H_X - 3a_1G.$$

Proof is straightforward: note that the conditions for the Picard group $\operatorname{Pic} V$ and that $T \cap R = \emptyset$ are very restrictive in terms of the discrete parameters a_i, a, c, e above. See [**Grin00**] for the details.

Fibrations into del Pezzo surfaces of degree 2. Here the relatively ample sheaf $\mathcal{O}_V(-K_V)$ is generated by its sections on every fibre. For an appropriate $m_0 \in \mathbb{Z}$,

$$\mathcal{E} = \pi_*\mathcal{O}_V(-K_V + m_0F) \cong \mathcal{O}_{\mathbb{P}^1} \oplus \mathcal{O}_{\mathbb{P}^1}(a_1) \oplus \mathcal{O}_{\mathbb{P}^1}(a_2),$$

$0 \leq a_1 \leq a_2$. Set $X = \mathbb{P}(\mathcal{E})$ to be corresponding $\mathbb{P}^2$-bundle over $\mathbb{P}^1$ with the Picard group (151) (we use the same notation for all three cases $d = 1, 2, 3$ as there is no danger of misunderstanding). Set also $a = a_1 + a_2 \in \mathbb{Z}_+$, $H_X^3 = a$. The sheaf $\mathcal{O}_V(-K_V + m_0F)$ determines a double cover $\sigma \colon V \to X$, branched over a smooth divisor $W \sim 4H_X + 2mG$. The canonical class of V is

$$K_V = \sigma^*(-H_X + (m + a - 2)G).$$

Proposition 1.2. *A del Pezzo fibration $V/\mathbb{P}^1$ exists in one of the following three cases:*

(i) $2m + a \geq 3$,

(ii) $2m + a = 2$ *and the triple* (m, a_1, a_2) *belongs to the set*

$$\{(-1, 2, 2), (1, 0, 0), (0, 1, 1), (0, 0, 2), (-2, 2, 4), (-3, 2, 6)\},$$

(iii) $2m+a = 1$ *and the triple* (m, a_1, a_2) *belongs to the set* $\{(0, 0, 1), (-1, 1, 2)\}$.

Proof is straightforward. One needs to exclude all triples that do not satisfy one of the cases (i)–(iii). A simple example: the triple (0,0,0) does not correspond to a del Pezzo fibration because any variety V with these parameters is isomorphic to $\mathbb{P}^1 \times S$, where S is a del Pezzo surface of degree 2, hence $\operatorname{Pic} V \neq \mathbb{Z}K_V \oplus \mathbb{Z}F$. This type of argument works in several cases; in the other cases the variety V cannot be constructed (e.g., when $m + 2a_1 < 0$). See [**Grin00**] for the details.

Fibrations into del Pezzo surfaces of degree 3. Here the relatively ample sheaf $\mathcal{O}_V(-K_V)$ is generated by its sections on every fibre and thus determines an inclusion $V \hookrightarrow X$, where $X = \mathbb{P}(\mathcal{E})$ is a $\mathbb{P}^3$-bundle over $\mathbb{P}^1$,

$$\mathcal{E} = \mathcal{O}_{\mathbb{P}^1} \oplus \mathcal{O}_{\mathbb{P}^1}(a_1) \oplus \mathcal{O}_{\mathbb{P}^1}(a_2) \oplus \mathcal{O}_{\mathbb{P}^1}(a_3),$$

$0 \leq a_1 \leq a_2 \leq a_3$, $\operatorname{Pic} X = \mathbb{Z}H_X \oplus \mathbb{Z}G$. Since obviously $V \sim 3H_X + mG$, the family of del Pezzo fibrations is uniquely determined by the quadruple (m, a_1, a_2, a_3). Obviously,

$$K_V = (-H_X + (m + a - 2)G)|_V,$$

where $a = a_1 + a_2 + a_3 = H_X^4$. It is clear that for any $m \in \mathbb{Z}$ such that $m + a_1 > 0$ the quadruple (m, a_1, a_2, a_3) indeed determines a family of del Pezzo fibrations, whereas for any $m \in \mathbb{Z}$ such that $m + 3a_1 < 0$ the divisor of lines in the fibres is reducible and therefore $V/\mathbb{P}^1$ is not a del Pezzo fibration. The quadruple (0,0,0,0) is not possible as in this case the variety V is isomorphic to $\mathbb{P}^1 \times S$, where S is a cubic surface, so makes no del Pezzo fibration. For a discussion of possible parameters see [**BCZ**].

1.2. Fibrewise birational self-maps. It is well known (see [**M67**]) that del Pezzo surfaces of degree 1 over a non-closed field with the Picard group $\mathbb{Z}$ are birationally superrigid. In the case of del Pezzo fibrations over $\mathbb{P}^1$, this can be presented as follows. Let $\Sigma \subset |-nK_V + lF|$ be a mobile linear system on V, then there are no curves $C \subset V$, covering the base $\mathbb{P}^1$, such that $\operatorname{mult}_C \Sigma > n$. The latter claim is obvious: restricting Σ onto a fibre $F_t = \pi^{-1}(t)$ of general position, for a point $p \in C \cap F_t$ we have

$$n^2 = (D_1 \cdot D_2)_{F_t} \geq (\operatorname{mult}_C \Sigma)^2$$

for generic divisors $D_1, D_2 \in \Sigma$. For that reason $V/\mathbb{P}^1$ admits no non-trivial fibrewise birational self-maps.

Let us consider fibrations into del Pezzo surfaces of degree 2. The same argument as above shows that if $C \subset V$ is a horizontal curve satisfying the inequality $\operatorname{mult}_C \Sigma > n$ for a mobile system $\Sigma \subset |-nK_V + lF|$, then C is a section of the projection π. Moreover, the following claim is true:

Lemma 1.1. *C is not contained in the ramification divisor of the morphism σ.*

Proof. Assume the converse. Let $F = F_t$ be a fibre of general position. Take $p = C \cap F$, so that by assumption $\bar{p} = \sigma(p) \in R$. The map $\sigma\colon F \to \mathbb{P}^2$ is the double cover branched over the smooth quartic curve $R \cap \pi_X^{-1}(t) = R_t$. Let $\overline{L} = T_{\bar{p}} R_t \subset \mathbb{P}^2$ be the tangent line, $L = \sigma^{-1}(\overline{L})$. Note that the curve L may be reducible but none of its components is a base curve of Σ_F, since the latter linear system of curves has no fixed components. Now for a general divisor $D \in \Sigma$ we have

$$2n = (L \cdot D_F)_F \geq (L \cdot D_F)_p \geq 2 \operatorname{mult}_C D,$$

since $p \in \operatorname{Sing} L$. Q.E.D.

Now let us assume that C is a section of π, not contained in the ramification divisor of σ.

Put $\overline{C} = \sigma(C) \subset X$ and let $C^* \subset V$ be the residual component of the reducible curve $\sigma^{-1}(\overline{C})$. Denote by x^* the point $C^* \cap F$, by $\bar{p}\colon \mathbb{P}^2 \dashrightarrow \mathbb{P}^1$ the projection from the point $\bar{x}$, and by $p\colon F \dashrightarrow \mathbb{P}^1$ the composition $\bar{p} \circ \sigma|_F$. Obviously, the fibres of p form a pencil of elliptic curves Λ on F with two simple (that is, of multiplicity 1) base points x and x^*. Let $\Gamma \in \Lambda$ be a general curve. We define the birational automorphism $\tau_x \in \operatorname{Bir} F$ of the surface F, specifying (on Γ) that $\tau_x|_\Gamma$ is the reflection from the point x^*. In other words,

$$z + \tau_x(z) \sim 2x^*$$

on the smooth elliptic curve Γ. Finally, we define the involution $\tau_C \in \operatorname{Bir} V$, setting for a general fibre F,

$$\tau_C|_F = \tau_x\ ,$$

$x = C \cap F$. Thus the involution τ_C acts in the fibres of the pencil $|F|$.

Lemma 1.2. *Let* $\alpha\colon V^* \to V$ *be the blowing up of the curve* C, $E = \alpha^{-1}(C)$ *the exceptional divisor,* $\operatorname{Pic} V^* = \mathbb{Z}h \oplus \mathbb{Z}e \oplus \mathbb{Z}F$, $h = -K_V$. *The birational involution* τ_C *extends to a biregular involution of an invariant open subset* $V^* \backslash Y$, $\operatorname{codim} Y \geq 2$, *and its action on* $\operatorname{Pic} V^*/\mathbb{Z}F \cong \mathbb{Z}\bar{h} \oplus \mathbb{Z}\bar{e}$ *is given by the following relations:*

$$\tau_C^* \bar{h} = 3\bar{h} - 4\bar{e}, \quad \tau_C^* \bar{e} = 2\bar{h} - 3\bar{e}.$$

Proof. On a general fibre F^* of the morphism $\pi \circ \alpha\colon V^* \to \mathbb{P}^1$ (in other words, F^* is the blowing up of F at the point $x = C \cap F$) the map τ_x extends to a biregular involution $\tau_x \in \operatorname{Aut} F^*$, because x does not lie on a line in F. On the finite set of "bad" fibres, where the point $C \cap F$ lies on a line or is a singular point of the fibre itself, τ_C acts as a birational involution. The other assertions of the lemma are now obvious or can be checked by direct computations. Q.E.D.

Now assume that the curve C is a maximal curve of the mobile linear system $\Sigma \subset |-nK_V + lF|$, that is, the inequality $\nu = \operatorname{mult}_C \Sigma > n$ holds. Let us untwist the curve C: consider the strict transform $\Sigma_1 = (\tau_C)_* \Sigma \subset |-n_1 K_V + l_1 F|$. For convenience, set $\nu_1 = \operatorname{mult}_C \Sigma_1$.

Lemma 1.3. *The following relation holds:*

$$n_1 = 3n - 2\nu < n$$

and the curve C *is not maximal for the system* Σ_1.

Proof. In the notations of the previous lemma, the strict transform Σ_1^* of the system Σ_1 on V^* can be obtained by taking first the strict transform Σ^* of Σ on V^* and, second, by applying τ_C. Thus we get

$$n_1 \bar{h} - \nu_1 \bar{e} = \tau_C^*(n\bar{h} - \nu\bar{e}) = (3n - 2\nu)\bar{h} - (4n - 3\nu)\bar{e},$$

which implies both claims of the lemma. Q.E.D.

Now let us consider fibrations into del Pezzo surfaces of degree 3. Here the same argument shows that if $C \subset V$ is a horizontal curve satisfying the inequality $\operatorname{mult}_C \Sigma > n$, then C is either a section or bi-section of π; see Section 4.1 for the details and for the constructions of the involutions τ_C in both cases. Proofs of Lemmas 4.1.1 and 4.1.2 are similar to the proof of Lemma 1.2 above and make an elementary exercise. The untwisting procedure is also explained in Section 4.1.

Combining the discussions above, we come to the following conclusion. Let us denote by $B(V)$ the group of fibrewise birational self-maps of $V/\mathbb{P}^1$, namely

- if $d = 1$, set $B(V)$ to be the trivial group,
- if $d = 2$, set $B(V)$ to be the group, generated by the involutions τ_C for all sections C not lying in the ramification divisor of σ,
- if $d = 3$, set $B(V)$ to be the group, generated by the involutions τ_C for all sections and bi-sections C of the projection π.

Proposition 1.3. *For any mobile linear system* Σ *there exists a fibrewise birational self-map* $\chi^* \in B(V)$ *such that for the strict transform*

$$\Sigma^* = \chi^* \Sigma \subset |-n^* K_V + l^* F|$$

and any horizontal curve $C \subset V$ *the following inequality holds:*

$$\operatorname{mult}_C \Sigma^* \leq n^*.$$

Finally, we have the following easy fact.

Proposition 1.4. *For any mobile linear system $\Sigma \subset |-nK_V + lF|$ and any vertical curve $C \subset F = F_t$ the inequality $\nu_C = \mathrm{mult}_C \Sigma \leq n$ holds.*

Proof. Assume the converse: $\nu_C > n$ for some irreducible vertical curve $C \subset F$. Since

$$\emptyset \neq \Sigma|_F \subset |-nK_V + \nu_C C|,$$

we immediately get a contradiction in the case $d = 1$.

Let us assume that $d = 2$. Then C is a line on the del Pezzo surface F. Note first of all that C cannot pass through the singular points of the surface F. Indeed, otherwise any curve of the form $\sigma^{-1}(L)$, where $L \subset \mathbb{P}^2$ is a line ($\sigma\colon F \to \mathbb{P}^2$ is the double covering), passing through this singular point, will itself be singular at this point. Thus it will be a base curve for Σ, which is impossible.

Therefore, $C + C^* = \sigma^{-1}(L)$, where $L = \sigma(C) \subset \mathbb{P}^2$ is a line, and F is nonsingular along C, C^*. Set

$$\nu^* = \mathrm{mult}_{C^*} \Sigma|_F, \quad \nu = \mathrm{mult}_C \Sigma|_F \geq \nu_C,$$

so that the following inequalities hold:

$$((nh_F - \nu c - \nu^* c^*) \cdot c) \geq 0, \quad ((nh_F - \nu c - \nu^* c^*) \cdot c^*) \geq 0.$$

It follows immediately that $\nu, \nu^* \leq n$ and we get a contradiction again.

Finally, in the case $d = 3$ we prove the claim of Proposition 1.4 by restricting the system Σ onto the fibre F and applying the cone technique (Lemma 2.2.1). Q.E.D. for Proposition 1.4.

Therefore, we conclude that the inequality of Proposition 1.3 holds for the untwisted system Σ^* and any curve C, horizontal or vertical.

1.3. Summary of the known results. The main fact about birational geometry of del Pezzo fibrations is Theorem 4.3.3, proven in [**Pukh98a**]. It covers all but finitely many families of fibrations into del Pezzo surfaces of degree 1 and 2, and in some natural sense a "majority" of fibrations into cubic surfaces (for a fixed ambient projective bundle $\pi_X\colon X \to \mathbb{P}^1$, all but finitely many families of fibrations $V/\mathbb{P}^1$, realized as smooth divisors $V \subset X$ with $\pi = \pi_X|_V$ are covered by Theorem 4.3.3). As the K^2-condition implies birational rigidity (for $d = 1$, superrigidity), to complete birational classification of del Pezzo fibrations, one needs to study fibrations that do not satisfy the K^2-condition. Below we summarize what is known in the absence of that condition. The guiding principle is that if the deviation from the K^2-condition is not too significant, the rigidity still takes place. When that deviation overcomes a certain threshold, birational geometry gets out of reach of the method. We consider the cases $d = 1, 2$ and 3 separately.

Fibrations into del Pezzo surfaces of degree 1. Here we are in the situation described in Proposition 1.1. It is easy to check that the K^2-condition is satisfied in the case $e = 0$ for $a_2 \geq 4$ and in the case $e = a_1 > 0$ when $(a_1, a_2) \neq (2, 6)$. However, the technique of the proof of Theorem 4.3.3 still works in the remaining cases (see Section 2 below) and gives the following

Proposition 1.5. *The fibre space $V/\mathbb{P}^1$ is birationally superrigid, except for the case $(\varepsilon, a_1, a_2, a_3) = (0, 0, 1, 2)$.*

Let us consider the exceptional family. In the notation of Subsection 6.1, the ruled surface $S \subset X = \mathbb{P}(\mathcal{E})$, corresponding to the surjection $\mathcal{E} \to \mathcal{O}_{\mathbb{P}^1} \oplus \mathcal{O}_{\mathbb{P}^1}(a_1)$, is contained in the variety Q. It is easy to check that the linear system $|-2K_V - 2F|$ is base point free and determines a morphism contracting the surface $\sigma^{-1}(S)$, the image of which is a Fano three-fold known as the double Veronese cone. The explicit construction of the double cone is as follows.

Let $W_4 \subset \mathbb{P}^5$ be the Veronese surface and $\Delta \subset \mathbb{P}^6$ the cone over W_4 with the vertex $p \in \mathbb{P}^6 \setminus \mathbb{P}^5$. Then the double Veronese cone is given by the double covering $\lambda\colon Y \to \Delta$ branched over a non-singular surface $U_\Delta = U \cap \Delta$, where $U \subset \mathbb{P}^6$ is a cubic hypersurface, $U \not\ni p$. The double cone Y is a smooth Fano three-fold of index two. More precisely, $\operatorname{Pic} Y = \mathbb{Z}H_Y$, where $-K_Y = 2H_Y$ is the λ-pullback of the hyperplane section of the Veronese cone Δ.

Let $L \subset \Delta$ be any generator of the cone Δ, $L \ni p$. Take the curve $C = \lambda^{-1}(L)$: it is either a smooth elliptic curve or an irreducible rational curve of arithmetic genus 1. Identifying $W_4 = \mathbb{P}^2$, the line L corresponds to a point $l \in \mathbb{P}^2$. Obviously, the pullback of the pencil of lines on $\mathbb{P}^2$ through l (first to the cone Δ, then via λ to Y) is the linear system $|H_Y - C| \subset |H_Y|$ without fixed components, determining the rational map $\pi_C\colon Y \dashrightarrow \mathbb{P}^1$, the fibres of which are del Pezzo surfaces of degree 1. If $\varphi_C\colon Y_C \to Y$ is the blow up of C, then $\pi_C \circ \varphi_C\colon Y_C \to \mathbb{P}^1$ is a regular map, a del Pezzo fibration. It is easy to check that φ_C is the inverse map with respect to the contraction given by the linear system $|-2K_V - 2F|$ above (with $\sigma^{-1}(S)$ as the exceptional divisor).

Therefore, every point $l \in \mathbb{P}^2$ generates a structure of a rationally connected fibre space on V. For that reason, V cannot be birationally rigid. However, the following fact is true.

Theorem 1.1. *Let $\Sigma \subset |nH_Y|$ be a mobile linear system on Y. If the inequality $c_{\mathrm{virt}}(\Sigma) < c(\Sigma) = n/2$ holds, then there exists a unique generator L of the cone Δ such that*

$$\operatorname{mult}_C \Sigma > \frac{n}{2},$$

where $C = \lambda^{-1}(L)$, whereas for any other curve $\Gamma \neq C$ the inequality $2\operatorname{mult}_\Gamma \Sigma \leq n$ holds. For the strict transform Σ_C of the linear system Σ on Y_C the equality

$$c_{\mathrm{virt}}(\Sigma) = c_{\mathrm{virt}}(\Sigma_C) = c(\Sigma_C, Y_C)$$

holds. In particular, apart from the projections $\pi_C\colon Y \dashrightarrow \mathbb{P}^1$, there are no other structures of a rationally connected fibre space with a positive-dimensional base on Y. The groups of birational and biregular self-maps coincide: $\operatorname{Bir} Y = \operatorname{Aut} Y$; this group is finite, generically isomorphic to the Galois group of the double cover $\lambda\colon Y \to \Delta$.

The theorem gives a complete description of birational geometry of the Veronese double cone Y, and so of any fibration into del Pezzo surfaces of degree 1 with the parameters $(\varepsilon, a_1, a_2, a_3) = (0, 0, 1, 2)$. For the proof, see [**Grin03b, Grin04**]. Some comments on the technique are in Section 2 below.

Fibrations into del Pezzo surfaces of degree 2. Here we are in the situation described in Proposition 1.2. It is easy to check that the K^2-condition is satisfied if $2m + a \geq 4$, so all fibrations satisfying this condition are by Theorem 4.3.3 birationally rigid.

It easy to prove that if $2m + a = 3$, then $V/\mathbb{P}^1$ satisfies the K-condition: for any mobile linear system $|-nK_V + lF|$ we have $l \in \mathbb{Z}_+$. Now repeating the proof of Theorem 4.3.3 almost word for word (see Section 2 below), we get

Proposition 1.6. *Del Pezzo fibrations with* $2m + a = 3$ *are birationally rigid.*

Thus to complete the problem of birational rigidity for del Pezzo fibrations of degree 2, it remains to study the 8 cases of parts (ii) and (iii) of Proposition 1.2. In these cases excluding infinitely near maximal singularities gets especially hard and in some of them it is not completed. Here we list the known facts.

If $(m, a_1, a_2) \in \{(0, 0, 2), (-2, 2, 4), (-3, 2, 6)\}$, then it is shown that a sufficiently general fibration $V/\mathbb{P}^1$ is birationally rigid [**Grin03a**]. The conditions of general position, imposed on the branch divisor W, are needed to reduce the number of geometric configurations to be considered and the amount of computations. There is little doubt that any fibration in those families is birationally rigid.

If $(m, a_1, a_2) = (0, 1, 1)$, then the linear system $|-2K_V|$ determines a small contraction onto the double cone $V^+ \xrightarrow{\varphi} Q \subset \mathbb{P}^4$, where Q is a cone over a smooth quadric $\overline{Q} \subset \mathbb{P}^3$ and φ is branched over a divisor $R_Q = R \cap Q$, R is a quartic in $\mathbb{P}^4$. If the quartic R does not contain the vertex of Q, we get the variety, birational geometry of which was completely described in [**Grin98**]. Pulling back two pencils of lines on $\overline{Q}$ to V^+, we get two structures of fibrations into del Pezzo surfaces of degree 2 on V^+. Twisting one of these two structures by fibrewise birational self-maps of the other structure, one gets infinitely many structures of a del Pezzo fibration on V^+, and thus on V. Therefore, neither V nor any of its models can be biraionally rigid *as a fibre space over* $\mathbb{P}^1$. However, the variety V^+ (and thus V) is birationally rigid: for any mobile linear system Σ on V there exists a birational self-map $\chi^* \in \operatorname{Bir} V$ such that $c_{\mathrm{virt}}(\Sigma) = c((\chi^*)^{-1}\Sigma)$. In particular, any structure $V \dashrightarrow \mathbb{P}^1$ of a rationally connected fibre space can be transformed by an appropriate birational self-map $\chi^* \in \operatorname{Bir} V$ into exactly one of the two structures given by the two pencils of lines on $\overline{Q} \subset \mathbb{P}^3$. The variety V^+ has no structures of a conic bundle, in particular, it is not rational. Up to a finite group, the group of birational self-maps $\operatorname{Bir} V^+$ is isomorphic to the free product of $\operatorname{Bir} F_{\eta_1}$ and $\operatorname{Bir} F_{\eta_2}$, where F_{η_1}, F_{η_2} are generic fibres of the two default structures on V^+.

If the quartic R does contain the vertex of Q, the problem of the description of the birational type of V is open.

If $(m, a_1, a_2) = (-1, 2, 2)$, then there is an explicit construction of a birational isomorphism $V \dashrightarrow V^+$, which is a biregular isomorphism in codimension one, where V^+ is fibred over $\mathbb{P}^1$ into del Pezzo surfaces of degree 1. The variety V^+ has one terminal singularity, $V^+/\mathbb{P}^1$ is a del Pezzo fibration, the pencil of fibres on the original model is the system $|-K_V - F|$. The problem of birational rigidity in this case is open.

The situation is similar in the case $(m, a_1, a_2) = (-1, 1, 2)$, where the variety is birational to a singular Fano 3-fold: birational rigidity remains an open problem.

Note that the last two cases were declared non-rigid in [**Grin03a, Grin06**], but this is due to the difference in the definitions of birational rigidity, mentioned in Chapter 2. In both cases V has another structure of a Mori fibre space, that is why it was put on the non-rigid list in [**Grin03a, Grin06**]. However, existence of another structure by itself tells us very little about birational geometry of the variety under consideration, and in terms of Definition 2.1.3 both types can well be rigid.

If $(m, a_1, a_2) = (1, 0, 0)$, then the linear system $|-K_V - F|$ gives a projection $V \to \mathbb{P}^2$, which alternatively can be described as the composition of σ and the projection of $X = \mathbb{P}^1 \times \mathbb{P}^2$ onto $\mathbb{P}^2$. The fibres of this map $V \to \mathbb{P}^2$ are conics, so V cannot be birationally rigid.

Finally, if $(m, a_1, a_2) = (0, 0, 1)$, then the free linear system $|-2K_V - 2F|$ determines a birational morphism (a contraction of a ruled surface) onto a Fano double space V^+ of index two, that is, a Fano variety V^+ realized as a double cover of $\mathbb{P}^3$ (ramified over a non-singular quartic surface). Therefore, it is non-rigid, has infinitely many structures of rationally connected fibre space and an immence group of birational self-maps. It is known (shown by the transcendent methods, see [**Tikh83, Tikh86**]) that V^+ (and thus V) has no structures of a conic bundle (in particular, it is non-rational). Apart from that fact, there are no complete results on its birational geometry.

Fibrations into del Pezzo surfaces of degree 3. In the notation of Section 1.1 it is easy to see that for (a_1, a_2, a_3) fixed the K^2-condition is satisfied if $5m + 3a \geq 12$, since this is equivalent to the inequality

$$(K_V^2 \cdot (H_X)|_V) \leq 0,$$

which guarantees the K^2-condition. Therefore, Theorem 4.3.3 gives birational rigidity of a generic variety in the family if $m \geq \frac{3}{5}(4 - a)$. For (a_1, a_2, a_3) fixed there are finitely many values of m, corresponding to a non-empty family of divisors $V \sim 3H_X + mG$, that do not satisfy the K^2-condition.

Apart from this general result, the only known fact is Theorem 1.3.3. If $(a_1, a_2, a_3) = (0, 0, 0)$, then $X = \mathbb{P}^3 \times \mathbb{P}^1$ and $V \subset X$ is a divisor of bidegree $(3, m)$. If $m \geq 3$, then the K^2-condition holds and V is birationally rigid. If $m = 1$, then V is birational to $\mathbb{P}^3$ (more precisely, V is isomorphic to the blow up of $\mathbb{P}^3$ in the base curve of a given pencil of cubic surfaces), so V is not birationally rigid. The case $m = 2$ is done by Theorem 1.3.3: a generic variety V in this family is still birationally rigid. This completes our summary of the known facts.

2. Infinitely near maximal singularities

In this section, we discuss the proofs of the results collected in Section 1. Although the arguments given in the proofs of Theorems 5.1.1 and 5.1.4 work for del Pezzo fibrations, some modifications are needed due to the obvious differences between del Pezzo surfaces and Fano varieties. Apart from those modifications, in Subsection 2.1 we also discuss what arguments can be used in the proofs of birational rigidity when the deviation from the K^2-condition becomes considerable. While for the degrees $d = 1$ and 2 the claim of Theorem 4.3.3 can be proved by the arguments of the proof of Theorem 5.1.1, this is not the case for $d = 3$. Subsections 2.2 and 2.3 contain the first part of the proof of Theorem 4.3.3 for fibrations into cubic surfaces: in Subsection 2.2 we prove that there is a line on the cubic surface F, containing the centre $o \in F$ of the maximal singularity and in Subsection 2.3 we introduce the crucial construction of a staircase, associated with a line through the point o and with the mobile linear system Σ under investigation.

2.1. General remarks. Since the condition $A^2V = \mathbb{Z}s \oplus \mathbb{Z}f$ for the numerical Chow group is somewhat different from the corresponding condition (93) in the higher-dimensional case, we consider briefly the technique of excluding infinitely near maximal singularities: it would be somewhat confusing just to refer to the Theorems 5.1.1 and 5.1.4. The main difference of del Pezzo fibrations from their higher-dimensional analogs is that for a fibre F we have $\operatorname{Pic} F \neq \mathbb{Z}H_F$: numerical geometry of vertical curves inside the same fibre is much more rich than numerical geometry of codimension two cycles in higher dimensions. In some sense, this is an advantage, as we have more numerical invariants to use.

The generalized K^2-condition of depth ε is now formulated in terms of the class of a line $f \in A^2V$ rather than anticanonical section of the fibre:

$$K_V^2 - \varepsilon f \notin \operatorname{Int} A_+^2 V.$$

For $\varepsilon = 0$ this definition is the same as that of the strong K^2-condition in arbitrary dimension. Since the conditions (v), (vs) and (h) are satisfied in a trivial way for $d = 1, 2$, we obtain Theorem 4.3.3 for these two values of the degree of the fibre. Indeed, for any mobile linear system Σ on V by Propositions 1.3 and 1.4 we may assume that Σ has no maximal curves. Then the arguments identical to the proof of Theorem 5.1.1 (with significant simplifications as the dimension now is equal to 3) give the equality $c_{\text{virt}}(\Sigma) = c(\Sigma)$, which gives the claim of Theorem 4.3.3 for $d = 1, 2$. Note that in our exposition we reverse the historical course of events: the method was first developed for del Pezzo fibrations in [**Pukh98a**] and then generalized to Fano fibre spaces of arbitrary dimension in [**Pukh98a, Pukh00d, Pukh04a**] and other papers. Here we treat the del Pezzo case as a special case of the general class of problems.

The first observation is that it is always possible to find explicitly a set of "minimal" curves $\Lambda \subset V$, multi-sections of degree $q \geq 1$, which give the minimal effective class $\frac{1}{q}\Lambda = s_0 \in A^2_{\mathbb{Q}}V$ and yield the normalized presentation $A^2_{\mathbb{Q}}V = \mathbb{Q}s_0 \oplus \mathbb{Q}f$ where the class of any effective curve on V is $as_0 + bf$ with $a \in \mathbb{Z}_+$ and $b \in \mathbb{Q}_+$. It is important that the class s_0 corresponds to an explicitly described set of multi-sections. In particular, we can write down $K_V^2 = ds_0 + \beta f$ and $\beta \in \mathbb{Q}$ measures the deviation from the strong K^2-condition: the latter condition holds if and only if $\beta \leq 0$. Otherwise, the fibre space $V/\mathbb{P}^1$ satisfies the K^2-condition of depth β. Let us fix a mobile linear system $\Sigma \subset |-nK_V + lF|$.

The next (and easy) issue to deal with is the possible negativity of l. If the strong K^2-condition is satisfied, then $l \in \mathbb{Z}_+$ (Proposition 4.3.1). If $\beta \geq 1$, then in a number of cases it can be proved by direct computation that l is still non-negative, that is, the K-condition holds.

Example 2.1. The K-condition holds in the situation of Proposition 1.6, that is, for del Pezzo fibrations of degree $d = 2$ with $2m + a = 3$. Indeed, in this case s_0 is obtained from the minimal section Λ of the fibre space $X = \mathbb{P}(\mathcal{E})/\mathbb{P}^1$, corresponding to the surjection $\mathcal{E} \to \mathcal{O}_{\mathbb{P}^1} \to 0$, that is, $s_0 = \frac{1}{2}\sigma^*\Lambda$. Let us consider a general curve $C \subset V$, the class of which is $2s_0 + 2a_1 f$. Obviously, they sweep out surface $\sigma^{-1}(S)$ where $S \subset X$ is the ruled surface corresponding to the surjection $\mathcal{E} \to \mathcal{O}_{\mathbb{P}^1} \oplus \mathcal{O}_{\mathbb{P}^1}(a_1) \to 0$. Now the inequality

$$((-nK_V + lF) \cdot C) \geq 0$$

implies the desired fact $l \in \mathbb{Z}_+$.

In the cases when the K-condition is not satisfied and l can be negative, an application of a birational self-map or a birational map to another model resolves the problem.

Example 2.2. Let $V \subset \mathbb{P}^1 \times \mathbb{P}^3$ be a generic divisor of bidegree $(2, 3)$, that is, the variety, birational geometry of which is described by Theorem 1.3.3. Taking $(u_0 : u_1)$ and $(x_0 : x_1 : x_2 : x_3)$ for the homogeneous coordinates on the factors, the divisor V is given by one equation

$$A(x_*)u_0^2 + B(x_*)u_0u_1 + C(x_*)u_1^2 = 0,$$

A, B, C are polynomials of degree 3. For a generic V, the system of equations $\{A(x_*) = B(x_*) = C(x_*) = 0\}$ has 27 distinct solutions $p_i \in \mathbb{P}^3$ which determine 27 *horizontal lines* $R_i = \mathbb{P}^1 \times \{p_i\} \subset V$, $i = 1, \ldots, 27$. Obviously, the Galois involution τ of the morphism $V \overset{2:1}{\to} \mathbb{P}^3$ is well defined outside the horizontal lines, so that $\tau \in \operatorname{Bir} V$ acts on $\operatorname{Pic} V$ as follows:

$$\tau^* K_V = K_V, \quad \tau^* F = -3K_V - F.$$

Therefore, if $\Sigma \subset |-nK_V + lF|$ is a mobile linear system with $l < 0$, then $\tau^*\Sigma \subset |-(n+3l)K_V - lF|$, so that the coefficient at $(-K_V)$ went down ("untwisting") and the coefficient at F is now positive. It is essential here that τ is an automorphism outside a closed set of codimension 2.

Example 2.3. For the double cone $V^+ \overset{\varphi}{\to} Q \subset \mathbb{P}^3$, discussed in Section 1, there exist two distinct "default" structures of a del Pezzo fibration on V^+ given by the two pencils of planes on Q. It is easy to show that for any mobile linear system Σ^+ on V^+ its strict transform Σ_i on at least one of the two models V_i, $i = 1, 2$, of a del Pezzo fibrations over $\mathbb{P}^1$, satisfies the desired condition $l_i \in \mathbb{Z}_+$, where $\Sigma_i \subset |-n_iK_{V_i} + l_iF_i|$. If the coefficient at the fibre F_i at some step gets negative, then the untwisting is simply taking the strict transform of the system on the other model. See [**Grin98**] for the details.

Summing up, in all cases listed in Section 1 where the study has been completed, it is sufficient to consider a mobile linear system $\Sigma \subset |-nK_V + lF|$ with $l \in \mathbb{Z}_+$ and show that $c_{\text{virt}}(\Sigma) = c(\Sigma, V) = n$.

As we mentioned above, if $d = 1, 2$ and the strong K^2-condition is satisfied, then, repeating the proof of Theorem 5.1.1 word for word, we obtain the desired equality of the virtual and actual thresholds and complete the proof of Theorem 4.3.3. If $d = 3$ and the strong K^2-condition holds, we are in the case which is considered below (Subsections 2.2, 2.3 and the whole Section 3) in full detail. Here we will discuss the situation in the absence of the strong K^2-condition. The general idea of treating such cases is given by Theorem 5.1.4. Now we will explain how it works for del Pezzo fibrations.

Assuming the equality $K_V^2 = ds_0 + \beta f$, we write the self-intersection of the mobile system Σ:

$$Z = Z^h + Z^v = Z^h + \sum_{t \in \mathbb{P}^1} Z_t^v,$$

the cycle Z_t^v supported by curves in the fibre $\pi^{-1}(t) = F_t$. If $\beta \le 0$, then the strong K^2-condition is satisfied. Otherwise, there exists $q \in \mathbb{Q}_+$ such that

$$Z^h \sim dn^2 s_0 + qn^2 f, \quad Z^v \sim (2nld + (\beta - q)n^2)f.$$

The hardest cases correspond to the smallest possible value of q. Although there may exist several maximal singularities of the system Σ, lying over distinct fibres of π, we may assume that there is only one maximal singularity: this is the most difficult case (see Remark 5.1.3). Let $E \subset V^+$ be this maximal singularity, where $\varphi\colon V^+ \to V$ is a birational morphism of smooth projective varieties. Denote by $e = \nu_E(\Sigma) - na(E)$ the excess of the multiplicity of Σ along E, where $\nu_E = \operatorname{ord}_E \varphi^*$ is the discrete valuation, associated with E. As usual, for simplicity of notation we write F for the fibre of π, containing the point $\varphi(E) \in V$, Z^v for Z_t^v, etc.

In the usual way (see the proof of Theorem 5.1.4) we get the inequality

$$\deg Z^v < \frac{2nde}{\nu_E(F)} + (\beta - q)n^2.$$

Taking the resolution $\varphi_{i,i-1}\colon V_i \to V_{i-1}$, $i = 1, \ldots, K$, of the singularity E, where the centres $B_{i-1} \subset V_{i-1}$ of the blow ups $\varphi_{i,i-1}$ are points for $i = 1, \ldots, L < K$ and curves for $i = L+1, \ldots, K$, where the points B_{i-1} for $i = 1, \ldots, N \leq L$ lie on the strict transform F^{i-1} of the fibre F, we obtain the standard quadratic inequality

$$\sum_{i=1}^{L} p_i m_i^h + \sum_{i=1}^{N} p_i m_i^v \geq \frac{(2\Sigma_l n + \Sigma_u n + e)^2}{\Sigma_l + \Sigma_u}, \tag{152}$$

here $m_i^h = \operatorname{mult}_{B_{i-1}}(Z^h)^{i-1}$, $m_i^v = \operatorname{mult}_{B_{i-1}}(Z^v)^{i-1}$ and $\Sigma_l = \sum_{i=1}^{L} p_i$, $\Sigma_u = \sum_{i=L+1}^{K} p_i$, $p_i = p_{Ki}$. The deviation from the K^2-condition is potentially dangerous as it makes the "vertical" component on the left-hand side of the quadratic inequality larger. However, if this deviation is small enough, the standard technique works well.

Example 2.4. Let us consider the situation of Proposition 1.6. Here $d = 2$ and $\beta = 2$. Putting $\Sigma_f = \sum_{i=1}^{N} p_i$, we get the following obvious sequence of inequalities

$$\sum_{i=1}^{N} p_i m_i^v \leq m_1^v \Sigma_f \leq \Sigma_f \deg Z^v < \left(\frac{4ne}{\nu_E(F)} + \beta n^2\right) \Sigma_f$$

and the right-hand side is bounded from above by

$$4ne + 2\Sigma_l n^2$$

as $\nu_E(F) \geq \Sigma_f$ and $\Sigma_0 \geq \Sigma_f$ for obvious reasons. Since $m_i^h \leq m_1^h \leq dn^2$, we obtain the inequality

$$(\Sigma_l + \Sigma_u)(4n^2\Sigma_l + 4ne) > ((2\Sigma_l + \Sigma_u)n + e)^2$$

which simplifies to $-(n\Sigma_u - e)^2 < 0$ and gives a contradiction proving the desired equality $c_{\text{virt}}(\Sigma) = n$ and thus Proposition 1.6.

However, as β gets higher, obtaining a contradiction is not so straightforward and requires some extra work, which in some cases is computationally very difficult. Often it includes an extensive case-by-case study and rather tedious calculations. Here we only list the basic principles of this work.

(1) The horizontal component versus the vertical one. Since

$$\sum_{i=1}^{N} m_i^h \leq (Z^h \cdot F) = dn^2,$$

it is clear that if $N \geq 2$ (that is, more than one point B_{i-1} lies on the fibre F), then the input of the horizontal component into the quadratic inequality becomes insignificant. On the other hand, if $N = 1$ but $L \geq 2$, then the input of the vertical part Z^v becomes insignificant, in which case the deviation from the strong K^2-condition does not really matter (the vertical part is bigger than in the strong K^2-case but input of the vertical multiplicity m_i^h is not of essential importance). In some cases, the possibility $N = L = 1$ presents a real challenge. But typically, considering the worst situation, it is essential to do the case $N \geq 2$, where the multiplicities m_i^h, $i \geq 2$, can be ignored.

(2) The support of the horizontal component. Typically it is sufficient to consider the case when the horizontal component Z^h is supported on the curves of the class s_0. Otherwise, the effect of the deviation from the strong K^2-condition is essentially reduced.

Example 2.5. In the notations of Example 2.2, let $\Gamma \subset V$ be any irreducible curve, which is not one of the 27 horizontal lines $R_i \subset V$. Let (a, b) be the bidegree of $\Gamma \subset \mathbb{P}^1 \times \mathbb{P}^3$. Then for any point $p \in \Gamma$ the following obvious inequality holds:

$$\operatorname{mult}_p \Gamma \leq \min\{a, b\}.$$

Therefore, if $B_0 \notin \cup R_i$, then we obtain the estimate

$$m_1^h = \operatorname{mult}_{B_0} Z^h \leq n^2 \min\{d, q\},$$

which essentially reduces the problem to the same situation as when the strong K^2-condition holds. For the same reason, if $B_0 \in R_i$ for some $i \in \{1, \ldots, 27\}$, then we may assume that Z^h is (mostly) supported on R_i which again leaves as the only case essentially different from that of the strong K^2-condition the one with $m_i^h = 0$ for $i = 2, \ldots, L$.

Note that for some families of del Pezzo fibrations the curves of the numerical class as_0, $a \geq 1$, may form a positive dimensional family. For example, this is the case for fibrations into del Pezzo surfaces of degree 2 with the parameters $(m, a_1, a_2) = (0, 0, 2)$, where the minimal sections sweep out the surface which is the σ-preimage of the surface, corresponding to the surjection $\mathcal{E} \to \mathcal{O}_{\mathbb{P}^1} \oplus \mathcal{O}_{\mathbb{P}^1} \to 0$.

(3) The support of the vertical component. If $N \geq 2$ (especially if N is sufficiently high), then the most difficult case is when Z^v is concentrated on the (possibly reducible) curves in the system $|-K_F - B_0 - B_1|$. For other curves certain strong estimates of their multiplicities at the infinitely near points $B_1, \ldots, B_{N-1}$ make their input in the quadratic inequality inessential. However, a complete analysis of those special anticanonical curves can be a really hard task, including very tedious computations.

Using the observations (1)–(3), one can complete the study of infinitely near maximal singularities in all cases listed in Section 1.

Starting from this moment, we begin our proof of Theorem 4.3.3 for fibrations into cubic surfaces ($d = 3$), satisfying the K^2-condition.

2.2. Pencils of cubic surfaces: existence of a line. From now on, we work with a standard del Pezzo fibration $\pi\colon V \to \mathbb{P}^1$ with $(K_V^2 \cdot F) = 3$ and $K_V^2 \notin \operatorname{Int} A_+^2 V$, which is true if $5m + 3a \geq 12$ in the notation of Section 1. We also assume that the following condition of general position is satisfied:

> If the fibre $\pi^{-1}(t) \subset V$ over a point $t \in \mathbb{P}^1$ is a singular cubic surface, then it has exactly one singular point, which is a non-degenerate quadratic singularity. Moreover, exactly six lines on the cubic $\pi^{-1}(t) = F_t$ pass through this point.

This condition does not seem to be essential for the proof but it significantly reduces the number of cases to be considered.

Definition 2.1. A line $L \subset V$ (that is, a vertical curve of degree 1 or, in other words, a line $L \subset F \subset \mathbb{P}^3$ in the ordinary sense) is said to be *non-special*, if L does not contain singular points of the fibre $F \supset L$. Otherwise, L is said to be *special.*

When we fix a line $L \subset V$ (as we do below, excluding the infinitely near singularity), we refer to the cases when L does not contain or does contain the singular point of the fibre, as the non-special or special case, respectively.

Proposition 2.1. *Let $L \subset V$ be a special line, $F \supset L$ the corresponding fibre, a singular cubic surface. Let $\sigma : \widetilde{V} \to V$ be the blowing up of L, $\widetilde{F}$ the strict transform of the surface F on $\widetilde{V}$. Then $\widetilde{F}$ is a smooth surface, isomorphic to the blow up of F at the singular point.*

Proof. Straightforward local computations. Q.E.D.

For convenience of presenting our arguments, we use the following straightforward definition of the "intersection number" of two curves on a variety of arbitrary dimension. Let R be a non-singular projective curve on a non-singular algebraic variety Y, $C \subset Y$ any irreducible curve, $C \neq R$, so that $C \cap R$ is a finite set of points.

Definition 2.2. The *intersection number* $(C \cdot R)$ of the curves C and R is the integer given by one of the two equivalent constructions:

either we define it as

$$(C \cdot R) = \sum_{x \in C \cap R} \operatorname{mult}_x C,$$

where the sum is taken over all the points of intersection of C and R, *including infinitely near ones*, or we set

$(C \cdot R) = (\widetilde{C} \cdot E)$, where $\sigma\colon \widetilde{Y} \to Y$ is the blow up of the curve R with the exceptional divisor $E \subset \widetilde{Y}$, and $\widetilde{C} \subset \widetilde{Y}$ is the strict transform of the curve C.

We define the intersection number $(Z \cdot R)$ for any 1-cycle Z on Y, which does not contain R as a component, by linearity. If $S \subset Y$ is a smooth projective surface, containing R and the support of the cycle Z, then we get the intersection number $(Z \cdot R)$ on S in the usual sense. Of course, if $\dim Y \geq 3$, then the number $(C \cdot R)$ is by no means invariant under deformations of the curves C and R.

Now we fix a mobile linear system $\Sigma \subset |-nK_V + lF|$ with $l \in \mathbb{Z}_+$ and assume that Σ has no maximal curves on V but has a supermaximal singularity $E \subset V^+$, where $\varphi\colon V^+ \to V$ is a birational morphism, $\varphi(E) = x$ is a point on a fibre

$F = F_t \subset V$, satisfying the inequality

$$\deg Z_t^v < \frac{6ne}{\nu_E(F)}.$$

We omit the index t and write Z^v for the vertical component of the self-intersection Z, contained in the fibre F. We use the standard notations for the resolution of the supermaximal singularity E, described in Subsection 2.1 above.

Proposition 2.2. *There exists a line $L \subset F$, passing through the centre x of the supermaximal singularity. Moreover, if $Z^v = C + kL$, where the effective 1-cycle C does not contain the line L, then*

$$(C \cdot L) < \frac{4ne}{\nu_E(F)}.$$

Proof. Assume the converse: there is no line through the point x on the cubic surface F.

Lemma 2.4. *The following inequality holds:*

$$\text{mult}_x Z^v \geq \frac{4ne}{\nu_E(F)}.$$

Proof. Assume the converse. Repeating word for word the proof for del Pezzo fibrations of degree 2, we come to a contradiction via the quadratic inequality (152) as the only fact needed is the estimate

$$\text{mult}_x Z^v < \frac{4ne}{\nu_E(F)}.$$

Q.E.D. for the lemma.

Since, as we assumed, there is no line through x on F, this point is a non-singular point of the fibre $F \subset \mathbb{P}^3$. Moreover, the curve $R = T_x F \cap F$ is irreducible, its degree is equal to 3 and its multiplicity at the point x is equal to 2 precisely. If $C \subset F$ is any other curve, then

$$\deg C = (C \cdot R) \geq (C \cdot R)_x \geq 2\,\text{mult}_x C.$$

Thus for any curve $Q \subset F$ we get the inequality

$$\text{mult}_x Q \leq \frac{2}{3} \deg Q.$$

Therefore,

$$\text{mult}_x Z^v \leq \frac{2}{3} \deg Z^v < \frac{4ne}{\nu_E(F)}.$$

Contradiction with Lemma 2.1. Q.E.D.

Now assume that $x \in F$ is a non-singular point of the fibre. There exist k lines, lying on F, $1 \leq k \leq 3$, and passing through x. If $\text{mult}_x(F \cap T_x F) = 2$, then $k \leq 2$ and for any curve $C \subset F$, which is different from these k lines, we get the inequality

$$2\,\text{mult}_x C \leq \deg C.$$

If $\text{mult}_x(F \cap T_x F) = 3$, then a stronger inequality holds:

$$3\,\text{mult}_x C \leq \deg C.$$

We will treat each of the cases $k \in \{1, 2, 3\}$ separately.

The case of a single line. Assume that there is only one line on F, passing through x. In this case the point x is smooth on F and $T_xF \cap F = L + Q$, where $Q \subset F$ is a smooth conic. The arguments given above show that for any curve $C \subset F$, $C \neq L$, the following inequality takes place:

$$\mathrm{mult}_x C \leq \frac{1}{2} \deg C.$$

Write down $Z^v = C + kL$, where C is an effective 1-cycle, not containing L. Now

$$k + \frac{1}{2} \deg C \geq \frac{4ne}{\nu(F)} \quad \text{and} \quad k + \deg C < \frac{6n\epsilon}{\nu(F)}.$$

This obviously implies that

$$\deg C < \frac{4ne}{\nu(F)}.$$

Since $(C \cdot L) \leq \deg C$, Proposition 2.2 is proved in this case. Q.E.D.

The case of two lines. Assume that there are exactly two lines, L_1 and L_2 on F, passing through the point x. In this case x is a smooth point of the fibre F and $T_xF \cap F = L_1 + L_2 + L_3$, where L_3 is a line, different from L_1, L_2; note that $x \notin L_3$. Furthermore, note that one of the points $L_i \cap L_3$, $i = 1, 2$, can be singular on F.

Assume at first that this is not the case. Now write $Z^v = Q + k_1L_1 + k_2L_2 + k_3L_3$ and set $d = \deg Q$, $d_i = (Q \cdot L_i)$, $m = \mathrm{mult}_x Q$. We get the following two inequalities:

$$k_1 + k_2 + k_3 + d < \frac{6ne}{\nu_E(F)} \tag{153}$$

and

$$k_1 + k_2 + m \geq \frac{4ne}{\nu_E(F)}. \tag{154}$$

Thus, $2m \leq d$ and

$$d_1 + d_2 + d_3 = d.$$

It is easy to see that there is $i \in \{1, 2\}$ such that for $\{j\} = \{1, 2\} \backslash \{i\}$ the inequality

$$k_j + k_3 + d_i < \frac{4ne}{\nu_E(F)}$$

holds. Indeed, otherwise, for any $i \in \{1, 2\}$ we have the opposite inequality. Put them up together and add the inequality (154). We get

$$2(k_1 + k_2 + k_3) + m + d_1 + d_2 \geq \frac{12ne}{\nu_E(F)}.$$

This contradicts the inequality (153). So we may assume that $\nu(F)(k_2 + k_3 + d_1) < 4ne$. Setting $C = Q + k_2L_2 + k_3L_3$, we get exactly the claim of Proposition 2.2.

Finally, if the point $L_1 \cap L_3$ is singular on F, then, instead of the equality $d_1 + d_2 + d_3 = d$ one should use the inequality $d_1 + d_2 \leq d$, where $d_1 = (Q \cdot L_1)$ is understood in the sense of Definition 2.2 as the intersection number of the two curves on a smooth three-fold. Proof of Proposition 2.2 in the case of two lines through x is complete. Q.E.D.

The case of three lines. Assume that there are exactly three lines L_i, $i = 1, 2, 3$, on F, passing through x. Again x is smooth on F and $T_xF \cap F = L_1 + L_2 + L_3$.

Write down the 1-cycle $Z^v = Q + k_1L_1 + k_2L_2 + k_3L_3$ and set $d = \deg Q$, $d_i = (Q \cdot L_i)$, $m = \mathrm{mult}_x Q$. Again we get a set of inequalities,

$$3m \leq d,$$

$$k_1 + k_2 + k_3 + d < \frac{6ne}{\nu(F)},$$

$$d_1 + d_2 + d_3 = d,$$

$$k_1 + k_2 + k_3 + m \geq \frac{4ne}{\nu(F)}.$$

By means of the same elementary arithmetic as above we get that for some $i \in \{1, 2, 3\}$ and $\{j, l\} = \{1, 2, 3\} \backslash \{i\}$ the following inequality is true:

$$k_j + k_l + d_i < \frac{4ne}{\nu(F)}.$$

Setting $L = L_i$, $C = k_jL_j + k_lL_l + Q$, we get the claim of Proposition 2.2.

Our proof of Proposition 2.2 is now complete if $x \in F$ is a non-singular point. Assume that $x \in F$ is a non-degenerate quadratic singularity. Now we have to consider

The case of six lines. This case is harder and we need to retrace the work of the techniques of counting multiplicities in details. We use the notation for the resolution of the singularity E introduced above.

Since the six lines $L_1, \ldots, L_6$ have distinct directions at the point x, we can find among them a line (say L_1) such that

$$B_1 \notin L_i^1$$

for $i \neq 1$, if B_1 is a point. Now write down explicitly,

$$Z^v = kL_1 + R + Q,$$

where R consists of the multiple lines L_i, $i \neq 1$, whereas the 1-cycle Q does not contain any line, passing through x. We define the integers M and N, requiring that

$$B_{i-1} \in L_1^{i-1} \quad \text{for} \quad i = 1, \ldots, M$$

and

$$B_{i-1} \in F^{i-1} \quad \text{for} \quad i = 1, \ldots, N,$$

$M \leq N \leq L$, and set

$$q_i = \mathrm{mult}_{B_{i-1}} Q^{i-1}$$

for $i = 1, \ldots, N$. Obviously, $q_1 \geq q_2 \geq \ldots$ and

$$q = \deg Q \geq q_1 + q_2 + \cdots + q_N$$

(since Q does not contain the line L_1 as a component). Now, if the following inequality holds:

$$\left(\sum_{i=1}^{M} p_i\right) k + p_1 \deg R + \sum_{i=1}^{N} p_i q_i < 4ne,$$

then we immediately get a contradiction by means of the techniques of counting multiplicities, repeating the arguments for del Pezzo fibrations of degree 2 word for word. Therefore, the opposite inequality holds. Since $2q_2 \leq q$, the following inequality holds:

$$\left(\sum_{i=1}^{M} p_i\right) k + \frac{1}{2}\left(2p_1 + \sum_{i=2}^{N} p_i\right)(\deg R + \deg Q) \geq 4ne.$$

Comparing it with the inequality

$$\left(2p_1 + \sum_{i=2}^{N} p_i\right)(k + \deg R + \deg Q) < 6ne,$$

which is true by the definition of a supermaximal singularity, we get the following estimate:

$$k > \frac{2ne}{\nu_E(F)}.$$

Setting $C = R + Q$, we complete the proof of Proposition 2.2 in the case under consideration. Q.E.D.

Note that some other delicate arguments (which are based upon the properties of the integers p_j only) make it possible to obtain a stronger estimate for the integer $(C\cdot L)$ in the last case. However, to prove Theorem 4.3.3, we need only the inequality of Proposition 2.2.

2.3. The staircase, associated with a line. In order to exclude the supermaximal singularity, we need to develop the following basic construction.

Definition-construction 2.3. An infinite series of blow ups

$$\begin{array}{cccc} \sigma_i\colon & V^{(i)} & \to & V^{(i-1)} \\ & \cup & & \cup \\ & E^{(i)} & \to & L_{i-1}, \end{array}$$

$i \geq 1$, starting from $V^{(0)} = V$, where L_{i-1} is the centre of the i-th blow up, and $E^{(i)} = \sigma_i^{-1}(L_{i-1})$ is its exceptional divisor, $L_0 = L$, is said to be a *staircase, associated with the line* L, or, simply, an *L-staircase*, if the following conditions are satisfied:

> L_i is a curve for all $i \in \mathbb{Z}_+$, $E^{(i)}$ is a ruled surface of the type $\mathbb{F}_1$ over L_{i-1} and $L_i \subset E^{(i)}$ is the exceptional section (i.e., the (-1)-curve).

Obviously, by this definition the staircase is unique if it exists. Just below we show that it does. The segment of the staircase, consisting of the blow ups σ_i for $1 \leq i \leq M$, is called a (finite) staircase of length M.

It is convenient to prove the existence of the staircase together with some of its properties.

For convenience of notation denote by $E^{(0)}$ the fibre F of the morphism π, which contains L. The operation of taking the strict transform on the i-th step (i.e., on $V^{(i)}$) is denoted by adding the bracketed upper index i. For instance, the strict transform of the surface $E^{(i)}$ on $V^{(j)}$ for $j \geq i$ is written as $E^{(i,j)}$.

Set also s_i to be the class of L_i in $A^2(V^{(i)})$, $s_0 = f$, and $f_i \in A^2(V^{(i)})$ the class of the fibre of the ruled surface $E^{(i)}$ over a point of the curve L_{i-1}.

Abusing our notations, we sometimes treat s_i and f_i as numerical classes of curves on the ruled surface $E^{(i)}$:

$$A^1 E^{(i)} = \operatorname{Pic} E^{(i)} = \mathbb{Z}s_i \oplus \mathbb{Z}f_i,$$

so that, in particular, formulas like

$$(s_i \cdot s_i) = -1 \quad \text{and} \quad (s_i \cdot f_i) = 1$$

make sense.

In these notations we have the following

Proposition 2.3. (i) *For $i \geq 2$ the effective 1-cycle $(E^{(i-1,i)} \circ E^{(i)})$ is just the irreducible curve $E^{(i-1,i)} \cap E^{(i)}$. Its numerical class is equal to $(s_i + f_i)$. In particular, this curve does not intersect $L_i \sim s_i$. If the line L is non-special, then this statement is true for $i = 1$, too. If, on the contrary, L is special, then the 1-cycle $(F^{(1)} \circ E^{(1)})$ is a reducible curve. More precisely, it is the sum of the exceptional section L_1 and the fibre over the singular point of the surface F.*

(ii) *The following equalities hold:*

$$(E^{(i)})^3 = 1 \quad \textit{and} \quad (E^{(i)} \cdot L_i) = 0.$$

Taking into account the isomorphism $L_i \cong \mathbb{P}^1$, we can write down

$$\mathcal{N}_{L_i/V^{(i)}} \cong \mathcal{O}_{L_i} \oplus \mathcal{O}_{L_i}(-1).$$

In this presentation the first component is uniquely determined. It corresponds to the exceptional section $L_{i+1} \subset E^{(i+1)} = \mathbb{P}(\mathcal{N}_{L_i/V^{(i)}})$. For the second component we can take the one-dimensional subbundle, corresponding exactly to the curve $E^{(i,i+1)} \cap E^{(i+1)}$.

(iii) *The classes s_i and f_i satisfy the relations*

$$\sigma^* s_{i-1} = s_i \quad \textit{and} \quad \sigma_* f_i = 0$$

for $i \geq 1$.

Proof in the non-special case. Assuming that $L \cap \operatorname{Sing} F = \emptyset$, we prove simultaneously the existence of the staircase and Proposition 2.3. Let us consider the first step of the staircase, that is, the morphism

$$\sigma_1 \colon V^{(1)} \to V^{(0)} = V,$$

blowing up the line $L_0 = L \subset F$. We get the exact sequence

$$0 \to \mathcal{N}_{L/F} \to \mathcal{N}_{L/V} \to \mathcal{O}_V(F)\,|_L \to 0,$$

which can be rewritten in the following way:

$$0 \to \mathcal{O}_L(-1) \to \mathcal{N}_{L/V} \to \mathcal{O}_L \to 0.$$

We conclude that $E^{(1)}$ is a ruled surface of the type $\mathbb{F}_1$, $(E^{(1)})^3 = 1$, whence we get $(E^{(1)} \cdot E^{(1)}) \sim (-s_1 - f_1)$ and $(E^{(0,1)} \cdot L_1) = ((F - E^{(1)}) \cdot s_1) = 0$. Thus all the requirements of the definition of the staircase and parts (i)–(iii) of the proposition are satisfied for the first blow up.

We proceed by induction on $i \geq 1$. We have the exact sequence

$$0 \to \mathcal{N}_{L_i/E^{(i)}} \to \mathcal{N}_{L_i/V^{(i)}} \to \mathcal{O}_{V^{(i)}}(E^{(i)})\Big|_{L_i} \to 0.$$

Taking into account the facts which were already proved, this sequence can be rewritten as follows:

$$0 \to \mathcal{O}_{L_i}(-1) \to \mathcal{N}_{L_i/V^{(i)}} \to \mathcal{O}_{L_i} \to 0.$$

Again this implies that $E^{(i+1)} = \mathbb{P}(\mathcal{N}_{L_i/V^{(i)}})$ is a ruled surface of the type $\mathbb{F}_1$ and $(E^{(i+1)})^3 = 1$, so that

$$E^{(i+1)}|_{E^{(i+1)}} \sim (-s_{i+1} - f_{i+1}).$$

Thus $(E^{(i+1)} \cdot L_{i+1}) = 0$, parts (i) and (iii) of the proposition are satisfied in an obvious way. The proof is complete. Q.E.D.

Proof of Proposition 2.3 in the special case. Assume that the line L contains the double point $p \in F$ of the fibre. Again consider the first blow up:

$$\sigma_1 \colon V^{(1)} \to V^{(0)} = V.$$

As we have mentioned, the strict transform of the fibre $F^{(1)} = E^{(0,1)}$ is already a non-singular surface. It is easy to see that $(F^{(1)} \circ E^{(1)}) = L_1 + R$, where $R = \sigma_1^{-1}(p)$ is the fibre of σ_1 over the singular point, whereas L_1 is some section of the ruled surface $E^{(1)}$. Since $K_{F^{(1)}} = \sigma_1^* K_F$, we get

$$(L_1 \cdot L_1)_{F^{(1)}} = -1,$$

so that $(L_1 \cdot E^{(1)}) = 0$ and $(L_1 \cdot F^{(1)}) = 0$. Since

$$E^{(1)}\Big|_{E^{(1)}} = -F^{(1)}\Big|_{E^{(1)}} = -(L_1 + R),$$

we get

$$(L_1 \cdot L_1)_{E^{(1)}} = -1,$$

so that $E^{(1)}$ is a ruled surface of the type $\mathbb{F}_1$ and conditions (i)–(iii) of Proposition 2.3 are satisfied.

The rest of the arguments (for $i \geq 2$) just repeat word for word the non-special case.

The proof of the existence of the staircase and Proposition 2.3 is complete. Q.E.D.

Remark 2.1. (i) Since $E^{(i-1,i)}$ does not intersect L_i (for $i \geq 1$ in the non-special and for $i \geq 2$ in the special case), we get

$$E^{(i-1,i)} = E^{(i-1,i+1)} = \dots = E^{(i-1,j)} = \dots$$

for any $j \geq i$. In particular, if $C \subset E^{(i-1)}$ is a curve, which is not the exceptional section L_{i-1}, then its strict transforms on all the varieties $V^{(j)}$, $j \geq i$, are the same:

$$C^{(i)} = C^{(i+1)} = \dots = C^{(j)}.$$

(ii) Abusing our notation, we call an irreducible curve $C \subset E^{(i)}$, $i \geq 1$, a *horizontal* one, if $\sigma_i(C) = L_{i-1}$, and a *vertical* one, if $\sigma_i(C)$ is a point on L_{i-1}. We extend this definition to horizontal and vertical 1-cycles with the support in $E^{(i)}$. The *degree* of a horizontal curve C is equal to

$$\deg C = \deg \sigma_i|_C = (C \cdot f_i),$$

the *degree* of an (irreducible) vertical curve C is equal to $\deg C = (C \cdot L_i) = 1$. We define the degree of a horizontal (respectively, vertical) 1-cycle with the support in $E^{(i)}$ as its intersection with f_i (respectively, L_i). In particular, the degree of a vertical 1-cycle is just the number of lines (fibres) in it. Note that if an effective horizontal 1-cycle C does not contain the exceptional section L_i as a component, then its class in $A^1(E^{(i)})$ or $A^2(V^{(i)})$ is equal to $\alpha s_i + \beta f_i$, where $\alpha \geq 1$ and $\beta \geq \alpha$.

(iii) Obviously, the graph of the sequence of the blow ups σ_i is a chain. In particular,

$$K_{V^{(M)}} = \sigma_{M,0}^* K_V + \sum_{i=1}^{M} \sigma_{M,i}^* E^{(i)}$$

(where $\sigma_{i,j}$, as always, stands for the composition $\sigma_{j+1} \circ \cdots \circ \sigma_i$) and the discrepancy of the exceptional divisor $E^{(i)}$ is equal to i. In the non-special case

$$\sigma_{M,0}^* F = F^{(M)} + \sum_{i=1}^{M} E^{(i,M)},$$

whereas in the special case for $M \geq 2$,

$$\sigma_{M,0}^* F = F^{(M)} + E^{(1,M)} + 2 \sum_{i=2}^{M} E^{(i,M)}.$$

In the special case $F^{(1)} \cap E^{(1)}$ is equal to the reducible curve $L_1 + R$, $R = \sigma_1^{-1}(p)$ is the fibre over the double point of the surface F. There are five more lines on F apart from L (in accordance with the condition of general position), passing through the point p. Let Q be one of them, $Q^{(1)} \subset F^{(1)}$ be its strict transform on $V^{(1)}$. It is easy to see that the point

$$\left(Q^{(1)} \cap E^{(1)}\right) \in R$$

does not lie on the exceptional section L_1.

3. Completing the proof for the pencils of cubic surfaces

In this section we exclude the supermaximal singularity for a mobile linear system $\Sigma \subset |-nK_V + lF|$, $l \in \mathbb{Z}_+$, on a fibration $V/\mathbb{P}^1$ into del Pezzo surfaces of degree 3, thus completing the proof of Theorem 4.3.3.

3.1. The Noether-Fano inequality in terms of the staircase. In accordance with what was proved in Section 2, the centre x of the supermaximal singularity lies on a line $L \subset F$ and in the decomposition $Z^v = C + kL$, where the vertical 1-cycle C does not contain L, the following inequality is satisfied:

$$(C \cdot L) < \frac{4ne}{\nu_E(F)}.$$

Proposition 3.1. *There exists a finite L-staircase of length $M \geq 1$, satisfying the following conditions:*

(i) *for $i = 0, \ldots, M-1$ the centre of the discrete valuation ν_E on $V^{(i)}$ is a point $x_i \in L_i$, $x_0 = x$,*

(ii) *for* centre$(V^{(M)}, \nu_E)$ *there are the following three options:*

A) *it is a point x_M such that $x_M \notin L_M$ and $x_M \notin E^{(M-1,M)}$;*

B) *it is the line* $B = \sigma_M^{-1}(x_{M-1})$, *that is, the fibre of the ruled surface* $E^{(M)}$ *over* x_{M-1};

C) *it is the point* $x_M = E^{(M-1,M)} \cap \sigma_M^{-1}(x_{M-1})$.

Proof. If centre$(V^{(M)}, \nu_E) \subset E^{(i)}$, then $i = a(V, E^{(i)}) \leq a(V, E)$. Therefore, there exists an integer $M \geq 1$, such that for $i = 0, \dots, (M-1)$ the condition (i) above is satisfied, whereas for $i = M$ it is not satisfied. Now the subvariety centre$(V^{(M)}, \nu_E)$ is either a curve (and in this situation we get the case B)), or a point x_M, not lying on L_M (one of the cases A) and C)). Q.E.D. for the proposition.

Let us fix the staircase of the length M that we have just constructed.

Let us introduce a new parameter d_F, setting it to be equal to 1, if the case is non-special, and 2, if it is special. Denote the exceptional divisor $E^{(M)} \subset V^{(M)}$ by E_* in the cases A) and B). In the case C) denote $E^{(M)}$ by E_+ and $E^{(M-1,M)}$ by E_-.

The symbol $\Sigma^{(i)}$ stands for the strict transform of the system Σ on $V^{(i)}$; set

$$\lambda_i = \text{mult}_{L_{i-1}} \Sigma^{(i-1)},$$

so obviously $n \geq \lambda_1 \geq \lambda_2 \geq \dots$. Abusing our notation, we now set

$$\begin{array}{rccc} \varphi_{i,i-1}\colon & V_i & \to & V_{i-1} \\ & \cup & & \cup \\ & E_i & \to & B_{i-1}, \end{array}$$

$i = 1, \dots, K$, to be the resolution of the discrete valuation ν_E, *starting from* $V^{(M)}$: $V_0 = V^{(M)}$, and the blow ups $\varphi_{i,i-1}$ successively blow up the centres of ν_E on V_{i-1}. In other words, now we consider ν_E as a discrete valuation of the field $\mathbb{C}(V^{(M)})$ and take its resolution. As usual, we set

$$\nu_i = \text{mult}_{B_{i-1}} \Sigma^{i-1}$$

to be the multiplicity of the strict transform Σ^{i-1} of the linear system $\Sigma^{(M)}$ on V_{i-1} along the centre of the next blow up. We also use the following notations:

- $p_i = p(K, i)$ is the number of paths in the oriented graph of the resolution of the discrete valuation $\nu_E = \nu_{E_K}$, from the uppermost vertex K to the vertex i, $p_K = 1$;
- $N^* = \max\{i \,|\, 1 \leq i \leq K, B_{i-1} \subset E_*^{i-1}\}$ in the cases A) and B);
- $L = \max\{i \,|\, 1 \leq i \leq K, B_{i-1}$ is a point$\}$ (so that for $j \leq L$ the centre B_{j-1} is a point, whereas for $j \geq L+1$ the centre B_{j-1} is a curve) in the cases A) and C);
- $N = \min\{N^*, L\}$ in the case A) and $N = N^*$ in the case B);
- $N_\pm^* = \max\{i \,|\, 1 \leq i \leq K, B_{i-1} \subset E_\pm^{i-1}\}$ in the case C), where the signs $+$ or $-$ are chosen to be the same in the right-hand and in the left-hand sides;
- $N_\pm = \min\{N_\pm^*, L\}$ (in the right-hand side it is the minimum of two integers);
- $\Sigma_l = \sum_{i=1}^{L} p_i$ and $\Sigma_u = \sum_{i=L+1}^{K} p_i$ in the cases A) and C);
- $\Sigma_t = \sum_{i=1}^{K} p_i$, the total sum, in the case B);
- $\Sigma^* = \sum_{i=1}^{N^*} p_i$ and $\Sigma_* = \sum_{i=1}^{N} p_i$ in the cases A) and B);
- $\Sigma_\pm^* = \sum_{i=1}^{N_\pm^*} p_i$ and $\Sigma_\pm = \sum_{i=1}^{N_\pm} p_i$ in the case C).

In these notations we get the obvious equalities:

$$\nu_E(E_*) = \varepsilon = \Sigma^* \quad \text{and} \quad \nu(E_\pm) = \varepsilon_\pm = \Sigma^*_\pm.$$

Furthermore, in the non-special case A) or B) for $M \geq 2$ and in the special case A) or B) for $M \geq 3$ we have

$$\nu_E(F) = d_F \varepsilon,$$

whereas in the case C) (under the same restrictions on M),

$$\nu_E(F) = d_F(\varepsilon_+ + \varepsilon_-)$$

(see Proposition 3.1). In either of the cases A) and B) we get

$$\nu_E(\Sigma) = \nu_{E_*}(\Sigma)\nu_E(E_*) + \nu_E(\Sigma^{(M)})$$

and for the discrepancies we have the equality

$$a(V, E) = a(V, E_*)\nu_E(E_*) + a(V^{(M)}, E),$$

so that the Noether-Fano inequality takes the form

$$\sum_{i=1}^{K} p_i \nu_i = \varepsilon \sum_{i=1}^{M} (n - \lambda_i) + n \sum_{i=1}^{K} p_i \delta_i + e,$$

where $e > 0$ is the excess. In a similar way, in the case C) the Noether-Fano inequality takes the form

$$\sum_{i=1}^{K} p_i \nu_i = \varepsilon_+ \sum_{i=1}^{M} (n - \lambda_i) + \varepsilon_- \sum_{i=1}^{M-1} (n - \lambda_i) + n \sum_{i=1}^{K} p_i \delta_i + e.$$

In these notations, we can rewrite the quadratic inequality for the self-intersection of the linear system Σ as follows. Let $D_i^{(M)}$, $i = 1, 2$, be the strict transforms of generic divisors $D_1, D_2 \in \Sigma$ on $V^{(M)}$ and

$$Z^{(M)} = (D_1^{(M)} \circ D_2^{(M)})$$

the self-intersection of the system $\Sigma^{(M)}$. Set

$$m_i = \operatorname{mult}_{B_{i-1}}(Z^{(M)})^{i-1}$$

for $i \leq L$ in the cases A) and C). Applying the technique of counting multiplicities, we obtain the following estimate in the case A):

$$\sum_{i=1}^{L} p_i m_i \geq \frac{(2\Sigma_l n + \Sigma_u n + \varepsilon \sum_{i=1}^{M} (n - \lambda_i) + e)^2}{\Sigma_l + \Sigma_u}.$$

In the case C) we get the estimate

$$\sum_{i=1}^{L} p_i m_i \geq \frac{1}{\Sigma_l + \Sigma_u} \left(2\Sigma_l n + \Sigma_u n + \varepsilon_+ \sum_{i=1}^{M} (n - \lambda_i) + \varepsilon_- \sum_{i=1}^{M-1} (n - \lambda_i) + e \right)^2.$$

In the case B) the line B obviously comes into the 1-cycle $Z^{(M)}$ with the multiplicity at least

$$\sum_{i=1}^{K} \nu_i^2,$$

whereas the multiplicities ν_i satisfy the inequalities

$$\nu_i \geq \sum_{j \to i} \nu_j$$

(here the resolution of the maximal singularity ν_E is just a sequence of blowing ups of curves, covering each other). Computing the minimum of this quadratic form under the restrictions specified above and taking into account the Noether-Fano inequality, we obtain the estimate

$$\operatorname{mult}_B Z^{(M)} \geq \frac{\left(\Sigma_t n + \varepsilon \sum\limits_{i=1}^{M} (n - \lambda_i) + e\right)^2}{\sum\limits_{i=1}^{K} p_i^2}.$$

In order to proceed further, we need to look at the 1-cycle $Z^{(M)}$ more closely.

3.2. The self-intersection in terms of the staircase. Let us compute the cycle $Z^{(M)}$ explicitly. Set

$$z_i = (D_1^{(i)} \cdot D_2^{(i)}) \in A^2 V^{(i)}$$

to be the class of the effective 1-cycle $Z^{(i)} = (D_1^{(i)} \circ D_2^{(i)})$, that is, the self-intersection of the mobile system $\Sigma^{(i)}$. On the "zeroth" level of the staircase we have the decomposition $Z = Z^v + Z^h$. Let us trace down the changes that the 1-cycle $Z^{(k)}$ undergoes when k increases from $k = i - 1$ to $k = i$. Obviously, the components of the cycle $Z^{(i-1)}$, which are different from L_{i-1}, in the cycle $Z^{(i)}$ are replaced by their strict transforms (with the same multiplicities). However, the curve L_{i-1}, which comes into $Z^{(i-1)}$ with some multiplicity k_{i-1}, is replaced in the cycle $Z^{(i)}$ by an effective 1-cycle with the support in the exceptional divisor $E^{(i)}$. Let us break this sub-cycle into three parts.

Part 1): the cycle $C_h^{(i)}$ includes all the curves in $Z^{(i)}$, which are horizontal with respect to the morphism $\sigma_i \colon E^{(i)} \to L_{i-1}$ and which are different from the exceptional section L_i,

Part 2): the cycle $C_v^{(i)}$ includes all the vertical curves, that is, the fibres of σ_i over points of the curve L_{i-1},

Part 3): the exceptional section L_i with a certain multiplicity $k_i \in \mathbb{Z}_+$.

Therefore, we can write down

$$Z^{(i)} = (Z^{(i-1)})^{(i)} + C_h^{(i)} + C_v^{(i)} + k_i L_i.$$

In order to make our notations uniform, set $C_h^{(0)}$ to be the part of the cycle Z^v, which includes all the curves different from L. Set also

$$d_{h,v}^{(i)} = \deg C_{h,v}^{(i)}$$

(see Remark 2.1(ii) in the previous section). Now we get the following presentation of the cycles $Z^{(i)}$:

$$\begin{aligned} Z^{(0)} &= Z^h + Z^v = Z^h + C_h^{(0)} + k_0 L, \\ Z^{(1)} &= (Z^h)^{(1)} + C_h^{(0,1)} + C_h^{(1)} + C_v^{(1)} + k_1 L_1, \\ &\dots, \\ Z^{(i)} &= (Z^h)^{(i)} + C_h^{(0,a)} + C_h^{(1,2)} + C_v^{(1,2)} + \dots + \\ &+ C_h^{(i-1,i)} + C_v^{(i-1,i)} + C_h^{(i)} + C_v^{(i)} + k_i L_i. \end{aligned}$$

Here $a = 1$ in the non-special case and $a = 2$ in the special case. We write, for instance, $C_h^{(1,2)}$ instead of $C_h^{(1,i)}$, in accordance with Remark 2.1(i) of Subsection 2.3.

Now we can compute the classes z_i and, in particular, the most important class z_M. Obviously, the class of the cycle $C_v^{(i)}$ in $A^1 V^{(i)}$ is equal to $d_v^{(i)} f_i$, and the class of the cycle $C_h^{(i)}$ is equal to $d_h^{(i)} s_i + \beta_i f_i$, where the coefficients satisfy the important inequality

$$\beta_i \geq d_h^{(i)}$$

(see Remark 2.1, (ii) in Subsection 2.3). Furthermore, the class of the cycle $C_v^{(i,i+1)}$ is equal to

$$d_v^{(i)}(f_i - f_{i+1})$$

and the class of the cycle $C_h^{(i,i+1)}$ is equal to

$$d_h^{(i)} s_i + \beta_i f_i - (\beta_i - d_h^{(i)}) f_{i+1}.$$

Setting

$$\alpha_i = \left((Z^h)^{(i-1)} \cdot L_{i-1} \right)$$

in the sense of Definition 2.2 of the "intersection number", we can write

$$z_i^h = z_{i-1}^h - \alpha_i f_i,$$

where z_i^h is the numerical class of the horizontal cycle $(Z^h)^{(i)}$.

Lemma 3.1. *The following inequality is true:*

$$\alpha_i \leq \deg Z^h = 3n^2.$$

Proof. Since $L \subset F$, and $\deg Z^h$ is equal to $(Z^h \cdot F)$, this is obvious. Q.E.D.

Proposition 3.2. *The classes z_i satisfy the following system of relations:*

$$z_i = z_{i-1} - (2\lambda_i n + \lambda_i^2) f_i - \lambda_i^2 s_i.$$

Proof. We compute:

$$\begin{aligned} z_i &= (D^{(i)})^2 = (D^{(i-1)} - \lambda_i E^{(i)})^2 \\ &= z_{i-1} - 2\lambda_i (D^{(i-1)} \cdot L_{i-1}) f_i - \lambda_i^2 (s_i + f_i). \end{aligned}$$

It follows from what was proved in Section 2 that for any $j \in \mathbb{Z}_+$ we have the equality $(D^{(j)} \cdot L_j) = (D \cdot L) = n$. Q.E.D.

Proposition 3.3. *For $i \geq 2$ in the non-special and for $i \geq 3$ in the special case the integers k_i, α_i, β_i and $d^{(i)}_{h,v}$ satisfy the following system of relations:*

$$d_v^{(i)} + \beta_i = \alpha_i + d_v^{(i-1)} + (\beta_{i-1} - d_h^{(i-1)}) - 2\lambda_i n - \lambda_i^2.$$

For $i = 1$ both in the non-special and special cases we get

$$d_v^{(1)} + \beta_1 = \alpha_1 + (C_h^{(0)} \cdot L) - 2\lambda_1 n - \lambda_1^2,$$

whereas for $i = 2$ in the special case we get

$$d_v^{(2)} + \beta_2 = \alpha_2 + d_v^{(1)} + (\beta_1 - d_h^{(1)}) + (C_h^{(0,1)} \cdot L_1) - 2\lambda_2 n - \lambda_2^2.$$

Proof. Elementary computations. One needs to write down explicitly the class of the cycle $Z^{(i)}$ in terms of the parameters introduced above, and to use the previous proposition. The computations are straightforward and elementary. Q.E.D.

Proposition 3.4. *For any $i \geq 1$ in the non-special case we get the inequality*

$$d_v^{(i)} + \beta_i \leq (C_h^{(0)} \cdot L) + \sum_{j=1}^{i} (3n^2 - 2\lambda_j n - \lambda_j^2).$$

In the special case for $i \geq 2$ we get the inequality

$$d_v^{(i)} + \beta_i \leq (C_h^{(0)} \cdot L) + (C_h^{(0,1)} \cdot L_1) + \sum_{j=1}^{i} (3n^2 - 2\lambda_j n - \lambda_j^2),$$

and for $i = 1$ we get the estimate

$$d_v^{(1)} + \beta_1 \leq (C_h^{(0)} \cdot L) + 3n^2 - 2\lambda_1 n - \lambda_1^2.$$

Proof. One needs to apply i times the corresponding equality of Proposition 3.3 and to use Lemma 3.1. Q.E.D.

3.3. Computations in the case A). In this case it is clear that among all the curves, lying on the divisor

$$\bigcup_{i=0}^{M} E^{(i,M)},$$

only those can possibly contain the point x_M, which lie entirely in $E^{(M)}$ and are different from the exceptional section L_M. Therefore, we may write down

$$Z^{(M)} = (Z^h)^{(M)} + C_v^{(M)} + C_h^{(M)} + \dots,$$

where the dots stand for the sum of all the curves, which do not contain the point x_M. Set

$$W = C_v^{(M)} + C_h^{(M)},$$

and introduce the multiplicities

$$m_i^v = \text{mult}_{B_{i-1}} W^{i-1} \quad \text{and} \quad m_i^h = \text{mult}_{B_{i-1}} (Z^h)^{(M),i-1}$$

for $i \le L$, so that $m_i = m_i^v + m_i^h$. Obviously, the multiplicities m_i^v vanish for $N+1 \le i \le L$. Furthermore, we have $m_i^{h,v} \le m_1^{h,v}$, and similarly to Lemma 3.1 we get $m_1^h \le 3n^2$. Finally,

$$m_1^v \le d_v^{(M)} + d_h^{(M)} \le d_v^{(M)} + \beta_M,$$

so that, putting together all the previous estimates, we get the inequality

$$3n^2\Sigma_l + \Sigma_* \left((C_h^{(0)} \cdot L) + \sum_{i=1}^{M} (3n^2 - 2\lambda_i n - \lambda_i^2) \right)$$

$$\ge \sum_{i=1}^{L} p_i m_i^h + \sum_{i=1}^{N} p_i m_i^v \ge \frac{\left(2\Sigma_l n + \Sigma_u n + \varepsilon \sum_{i=1}^{M} (n - \lambda_i) + e \right)^2}{\Sigma_l + \Sigma_u},$$

if the case is the non-special one. In the special case for $M \ge 2$ one should add $(C_h^{(0,1)} \cdot L_1)$ to $(C_h^{(0)} \cdot L)$.

Let us first consider **the non-special case**. Replacing Σ_* by $\varepsilon = \Sigma^*$, we make our inequality stronger, and replacing $\varepsilon(C_h^{(0)} \cdot L)$ by $4ne$, we get a strict inequality. Subtract the left-hand side (the first line) of the quadratic inequality above from the right-hand side (in the second line) and look at the expression just obtained as a quadratic form in λ_i on the domain $0 \le \lambda_i \le n$. By symmetry, its minimum is attained somewhere on the diagonal line, that is, at $\lambda_i = \lambda$, $0 \le \lambda \le n$. Replace all the λ_i's by this value λ. Thus we get the strict inequality

$$\Phi < 0,$$

where the expression Φ by means of elementary arithmetic can be transformed as follows:

$$\begin{aligned} \Phi = (\Sigma_l^2 + \Sigma_l\Sigma_u + \Sigma_u^2)n^2 + M\varepsilon\Sigma_l(n-\lambda)^2 \\ -M\varepsilon\Sigma_u(n-\lambda)(n+\lambda) \\ +M^2\varepsilon^2(n-\lambda)^2 - 2e\Sigma_u n + 2M\varepsilon e(n-\lambda) + e^2. \end{aligned}$$

Since $\lambda \le n$, we can replace $(n+\lambda)$ by $2n$, preserving the strict inequality. However, it is easy to check that the last expression is the sum of the complete square

$$(\Sigma_1 n - M\varepsilon(n-\lambda) - e)^2$$

and a few non-negative components. Thus it cannot be negative. Our proof is complete (in the case under consideration). Q.E.D. for the non-special case A).

Now let us consider **the special case**. If $M \ge 2$, then we argue following the same scheme. One needs to take into account that here $\nu_E(F) = 2\varepsilon$, so that now we may replace the expression

$$\varepsilon\left((C_h^{(0)} \cdot L) + (C_h^{(0,1)} \cdot L_1) \right)$$

in the left-hand side of the big inequality above by $4ne$. The rest of the computations are the same as in the non-special case.

For $M = 1$ the computations are much easier.

Thus we obtained a contradiction showing that the case A) is impossible. Q.E.D.

3.4. Computations in the case B). As above, we consider in full detail only the non-special case. On one hand, we have the inequality

$$d_v^{(M)} \leq (C_h^{(0)} \cdot L) + \sum_{i=1}^{M} (3n^2 - 2\lambda_i n - \lambda_i^2).$$

On the other hand, the following estimate holds:

$$d_v^{(M)} \geq \operatorname{mult}_B Z^{(M)},$$

whereas for the multiplicity in the right-hand side a lower bound has been obtained above by means of the technique of counting multiplicities. Obviously,

$$\sum_{i=1}^{K} p_i^2 \leq p_1 \Sigma_t \leq \varepsilon \Sigma_t.$$

As in the case A), we may assume that all the λ_i's are equal to λ, $0 \leq \lambda \leq n$. Replacing $\varepsilon(C_h^{(0)} \cdot L)$ by $4ne$, we get the strict inequality

$$0 > \Sigma^2 n^2 - 2\Sigma ne + M^2 \varepsilon^2 (n - \lambda)^2 + e^2 + 2M\varepsilon(n - \lambda)e$$
$$-M\varepsilon(n - \lambda)\Sigma(n + \lambda),$$

which is still true if we replace $(n + \lambda)$ by $2n$. After this replacement, we obtain the inequality

$$0 > (\Sigma n - M\varepsilon(n - \lambda) - e)^2.$$

This contradiction completes the proof of our theorem in the case B), which is now shown to be impossible.

In the special case we proceed in the same manner. Here we must add $(C_h^{(0,1)} \cdot L_1)$ to $(C_h^{(0)} \cdot L)$. However, the multiplicity $\nu_E(F) = 2\varepsilon$ is twice as large now, so that eventually we come to the same strict inequality. This contradiction completes the proof in the case B). Q.E.D.

3.5. Computations in the case C). Let us assume first that one of the following three situations takes place:

- either we are in the non-special case,
- or we are in the special case and $M \geq 3$,
- or we are in the special case, $M = 2$ and the point $x = \operatorname{centre}(V, \nu_E)$ is not the singular point of the fibre F.

Under this assumption, it is clear that among the curves lying on the divisor

$$\bigcup_{i=0}^{M} E^{(i,M)},$$

only those can pass through the point x_M, which either lie entirely in $E^{(M)}$ and are different from the exceptional section L_M (exactly as it was in the case A)), or lie entirely in $E^{(M-1,M)}$ and are different from the exceptional section $E^{(M-1,M)} \cap E^{(M)}$ (which was already counted in the first group). Thus

$$Z^{(M)} = (Z^h)^{(M)} + W_- + W_+ + \dots,$$

where W_+ stands for the sum of all the curves in $E_+ = E^{(M)}$, which are different from L_M, W_- stands for the 1-cycle $C_v^{(M-1,M)} + C_h^{(M-1,M)}$, and the dots stand for the sum of all remaining curves, which do not pass through x_M. Let the symbol m_i^h mean the same as in the case A), and set

$$m_i^{\pm} = \mathrm{mult}_{B_{i-1}} W_{\pm}^{i-1},$$

$i = 1, \dots, L$, so that $m_i = m_i^+ + m_i^- + m_i^h$. Obviously, the multiplicities $m_i^{\pm}$ vanish for $N_{\pm} + 1 \leq i \leq L$. Similarly to the case A), we get $m_i^{h,\pm} \leq m_1^{h,\pm}$, $m_1^h \leq 3n^2$, $m_1^+ \leq d_v^{(M)} + \beta_M$, $m_1^- \leq d_v^{(M-1)} + \beta_{M-1}$, so that finally we come to the following inequality:

$$3n^2\Sigma_l + \Sigma_+ \left((C_h^{(0)} \cdot L) + \sum_{i=1}^{M} (3n^2 - 2\lambda_i n - \lambda_i^2) \right)$$

$$+\Sigma_- \left((C_h^{(0)} \cdot L) + \sum_{i=1}^{M-1} (3n^2 - 2\lambda_i n - \lambda_i^2) \right)$$

$$\geq \sum_{i=1}^{L} p_i m_i^h + \sum_{i=1}^{N_+} p_i m_i^+ + \sum_{i=1}^{N_-} p_i m_i^-$$

$$\geq \frac{1}{\Sigma_l + \Sigma_u} \left(2\Sigma_l n + \Sigma_u n + \varepsilon_- \sum_{i=1}^{M-1} (n - \lambda_i) + \varepsilon_+ \sum_{i=1}^{M} (n - \lambda_i) + e \right)^2,$$

provided that our case is the non-special one. In the special case for $M \geq 2$ one should add $(C_h^{(0,1)} \cdot L_1)$ to $(C_h^{(0)} \cdot L)$.

Let us consider **the non-special case**. Replacing $\Sigma_{\pm}$ by $\varepsilon_{\pm} = \Sigma_{\pm}^*$, we preserve the inequality, and replacing $(\varepsilon_+ + \varepsilon_-)(C_h^{(0)} \cdot L)$ by $4ne$, we make it a strict one. Subtract the left-hand side from the right-hand side and look at the resulting expression as a quadratic form in the two groups of variables, that is, λ_i^+ and λ_i^-, where we replace λ_i by $\lambda_i^{\pm}$ in accordance with the following rule: if a variable comes into the sum $\sum_{i=1}^{M-1}$, then we replace it by λ_i^-, and if it comes into $\sum_{i=1}^{M}$, then we replace it by λ_i^+. The new variables take their values in the domain $0 \leq \lambda_i^{\pm} \leq n$. By symmetry, the minimum of this quadratic form is attained at some point on the diagonal plane, that is, at $\lambda_i^{\pm} = \lambda_{\pm}$, $0 \leq \lambda_{\pm} \leq n$. Now replace all the $\lambda_i^{\pm}$'s by $\lambda_{\pm}$. The inequality is still strict. Thus we get

$$\Phi < 0,$$

where the expression Φ can be transformed by means of elementary arithmetic in the following way, where we set for convenience $M - 1 = M_-$, $M = M_+$:

$$\Phi = (\Sigma_l^2 + \Sigma_l\Sigma_u + \Sigma_u^2)n^2 + M_-\varepsilon_-\Sigma_l(n - \lambda_-)^2 + M_+\varepsilon_+\Sigma_l(n - \lambda_+)^2$$

$$-M_-\varepsilon_-\Sigma_u(n - \lambda_-)(n + \lambda_-) - M_+\varepsilon_+\Sigma_u(n - \lambda_+)(n + \lambda_+)$$

$$+(M_-\varepsilon_-(n - \lambda_-) + M_+\varepsilon_+(n - \lambda_+))^2$$

$$-2e\Sigma_u n$$

$$+2M_-\varepsilon_-(n - \lambda_-)e + 2M_+\varepsilon_+(n - \lambda_+)e + e^2.$$

Since $\lambda_{\pm} \leq n$, we can replace $(n + \lambda_{\pm})$ by $2n$, preserving the strict inequality. After this replacement, the expression for Φ turns out to be the sum of the complete square

$$(\Sigma_u n - M_-\varepsilon_-(n - \lambda_-) - M_+\varepsilon_+(n - \lambda_+) - e)^2$$

and a few non-negative components. Therefore, Φ cannot be negative. We come to a contradiction, completing our proof in the non-special case.

In the **special case** we use the same arguments, taking into account the equality $\nu_E(F) = 2(\varepsilon_- + \varepsilon_+)$.

Finally, if we are in the special case with $M = 1$ and $x = \text{centre}(V, \nu_E)$ is not the singular point of the fibre F, then the previous arguments work with simplifications.

Thus none of the assumptions, made at the beginning of this section, can hold and the only case to be considered is the special case with $M = 1$ or 2, when the point x is the singularity of the fibre. The case $M = 1$ is simpler. If $M = 2$, then the point x_M is the only common point of the following three divisors:

$$x_2 = F^{(2)} \cap E^{(1,2)} \cap E^{(2)}$$

(the intersection is transversal). Respectively,

$$Z^{(2)} = (Z^h)^{(2)} + C_h^{(0,2)} + C_h^{(1,2)} + C_v^{(1,2)} + C_h^{(2)} + C_v^{(2)} + k_2 L_2,$$

where all the 1-cycles but the last one may contain the point x_2.

This case is the only one, when our previous arguments formally do not work (because of the additional input, which is given by the 1-cycle $C_h^{(0,2)} \subset F^{(2)}$). Nevertheless, the general scheme of arguments, which was used in cases A)–C) above, works here, too. We just outline the principal steps of the proof.

Preserving the previous notations, set

$$N^* = \max\{i \mid 1 \leq i \leq K, B_{i-1} \subset E^{i-1}\} \quad \text{and} \quad N = \min\{N^*, L\},$$

also

$$\Sigma^* = \varepsilon = \sum_{i=1}^{N^*} p_i \quad \text{and} \quad \Sigma_* = \sum_{i=1}^{N} p_i \leq \Sigma^*.$$

Finally, we write $m_i^0 = \text{mult}_{B_{i-1}}(C_h^{(0,2)})^{i-1}$ for $1 \leq i \leq L$. Obviously, $m_i^0 = 0$ for $i \geq N+1$.

Now we get the following presentation:

$$\sum_{i=1}^{L} p_i m_i = \sum_{i=1}^{L} p_i m_i^h + \sum_{i=1}^{N_-} p_i m_i^- + \sum_{i=1}^{N_+} p_i m_i^+ + \sum_{i=1}^{N} p_i m_i^0.$$

For the four components in the right-hand side we get the following upper bounds:

$$\leq 3\Sigma_l n^2,$$

$$\leq \varepsilon_- \left((C_h^{(0)} \cdot L) + (3n^2 - 2n\lambda_1 - \lambda_1^2) \right),$$

$$\leq \varepsilon_+ \left((C_h^{(0)} \cdot L) + (C_h^{(0,1)} \cdot L_1) + \sum_{i=1}^{2} (3n^2 - 2n\lambda_i - \lambda_i^2) \right),$$

$$\leq \varepsilon m_1^0 \leq \varepsilon (C_h^{(0,1)} \cdot L_1),$$

respectively. It is because of the fourth component that this case is not formally covered by the previous arguments. However, here the multiplicity

$$\nu_E(F) = 2\varepsilon_+ + \varepsilon_- + \varepsilon$$

increases, too. Thus we are again able to replace all the components, into which $(C_h^{(0)} \cdot L)$ and $(C_h^{(0,1)} \cdot L_1)$ fall, by $4ne$. From now on we can just repeat the arguments, which were used in the "regular" case C). For $M = 1$ our computations work with considerable simplifications.

The proof of Theorem 4.3.3 is complete. Q.E.D.

Notes and references

For surfaces over a non-closed field rigidity-type results in the relative case, that is, for conic bundles (see [**I67, I70**]), immediately followed by the theorems on birational (super)rigidity of del Pezzo surfaces [**M67, M72**]. As soon as the first proof of birational (super)rigidity was completed for a Fano three-fold [**IM**], attempts were started to obtain similar results for three-fold fibre spaces, in the first place, for conic bundles over rational surfaces, such that it was rather clear what was to be expected. Ten years later Sarkisov proved his famous theorem on the uniqueness of the structure of the conic bundle when the discriminant divisor is sufficiently large [**S80**], and almost at once generalized it to arbitrary dimension [**S82**]. By that time, birational (super)rigidity was shown for several families of Fano three-folds [**I80**], and the only class of rationally connected three-folds that refused all attempts to be studied by the method of maximal singularities, was that of del Pezzo fibrations over $\mathbb{P}^1$. The situation was even more annoying that the fibrewise part of the investigation was completed more than ten years earlier, in the late 60s (see [**M72**]), so it was clear how the final results should look, but the infinitely near case turned out to be too difficult for the test class technique. Note that the approach that worked so well for conic bundles could not be employed for del Pezzo fibrations: for conic bundles, there was no need to exclude the infinitely near maximal singularities, as one can get rid of them via fibrewise birational modifications, remaining in the same nice class of standard conic bundles; see [**S80, S82**] for details. This approach simply did not work for del Pezzo fibrations (and at that time, there was no minimal model theory for three-folds, so it was not possible to use more complicated birational modifications than fibrewise birational self-maps). The only possibility to prove rigidity results was through exclusion of infinitely near maximal singularities by means of the test class technique, with the three-dimensional quartic as a model. This is what Iskovskikh and Khashin had been trying to do for about 5-7 years between 1985 and 1993. It seems that there is only one publication related to that work, [**I95**], but most of the calculations remained unpublished. As it is clear now, it was not possible to get anywhere close to the exclusion of an infinitely near maximal singularity in that way. What Iskovskikh and Khashin tried to do, was to construct a test class (see Section 2.3) with a negative intersection with the self-intersection of a mobile linear system with a maximal singularity. As we explained in Chapter 2, multiplication by a test class is essentially equivalent to comparing the singularities of the self-intersection Z (that is, the multiplicities of Z at a sequence of infinitely near points over the centre of the maximal singularity) with its degree $\deg Z$ in terms of some ample divisorial class. However, Z breaks into the horizontal Z^h and vertical Z^v parts, and their singularities should be estimated via two different "degrees": it is $(Z^h \cdot F) = \deg(Z^h \to \mathbb{P}^1)$ for the horizontal part and $(Z^v \cdot (-K_V))$ for the vertical one. Of course, one can find a class that works for both parts, say some very ample class on V, but the estimates obtained via such a class are by far too weak to provide a contradiction and exclude the maximal singularity.

As soon as the test class technique was replaced by that of counting multiplicities in [**Pukh00c**] (first published as a Warwick preprint in 1996), it became possible in the spring of 1996 to prove birational (super)rigidity of del Pezzo fibrations [**Pukh98a**]: the infinitely near case no longer resisted and Theorem 4.3.3

was at once proved in full). Further work on del Pezzo fibrations was done by Grinenko who completed the study of fibrations into surfaces of degree 1 [**Grin00, Grin03a, Grin03b, Grin04**] and nearly completed it for degree 2 [**Grin98, Grin00, Grin03a**] (except for the obviously rational fibrations, the most interesting case that remained uncompleted because it was too difficult was that of del Pezzo fibrations coming from the Fano double spaces of index two branched over a quartic surface), and Sobolev [**Sob02**]. Birational geometry of fibrations into cubic surfaces with a considerable deviation from the K^2-condition is still an open problem; see [**BCZ**]. Elliptic structures on del Pezzo fibrations were investigated in [**Ch05b**]. An attempt to develop a new approach towards the problem of birational rigidity of del Pezzo fibrations in the spirit of the minimal model program was recently made in [**ShCh11**]. Non-rationality of very general del Pezzo fibrations over $\mathbb{P}^1$ was earlier proved by means of the Clemens degeneration method in [**Bar**].

CHAPTER 7

Fano Direct Products

Summary

The aim of this chapter is to explain the linear method of proving birational superrigidity, mainly by the example of the theorem on Fano direct products (Theorem 4.3.5). Whereas the starting point of the quadratic method is to form the self-intersection $Z = (D_1 \circ D_2)$ of a mobile linear system (which obviously depends quadratically on the system), so that the main object of subsequent investigation is the set of singularities of that effective cycle of codimension two (including the infinitely near ones), the approach of the linear method is to study the singularities of a mobile linear system Σ by taking a generic divisor $D \in \Sigma$ and restricting it onto a suitable subvariety. The restriction of a mobile linear system onto a subvariety is not necessarily a mobile system, therefore, one has to study singularities of a possibly fixed and reducible divisor on a subvariety of the original variety. This is a clear disadvantage, however, in certain problems the linear method works and the quadratic one does not.

The main example of efficiency of the linear method is the proof of birational superrigidity of Fano direct products (Theorem 4.3.5), so far unaccessible for the quadratic method. It is given in Section 1. Predictably, to study singularities of a fixed divisor, without the assumption of mobility, one needs stronger and more restrictive regularity conditions for the variety under consideration, and their proof is harder than what was done in Chapter 3. However, as a by-product, the linear method radically simplifies (at the expense of sharpening the genericity conditions) the proof of birational superrigidity of Fano fibre spaces $V/\mathbb{P}^1$. This is explained in Section 3, which is devoted to proving the property of divisorial canonicity of singular Fano hypersurfaces of index one. The local facts (the inversion of adjunction and its immediate implications) are collected in Section 2.

1. Fano direct products

In this section we give a proof of Theorem 4.3.5 on birational superrigidity of Fano direct products and prove the divisorial canonicity of generic Fano hypersurfaces of index one. The proof is by induction on the number of direct factors. We represent $V = F_1 \times \cdots \times F_K$ as a standard Fano fibre space $V = F_i \times S_i$, splitting off one of the factors F_i and projecting V onto the product S_i of the remaining factors, and reduce the claim of Theorem 4.3.5 for V to the same claim for S_i. In order to be able to make this step of induction, we need the factors F_i to satisfy the conditions (L) and (M) of Subsection 4.3.3, which is a stronger assumption than just birational superrigidity of the factors. Here, for the first time, we work with a class of Fano fibre spaces over a base of arbitrary dimension. Theorem 4.3.5 is proved in Subsections 1.1–1.4. To make this result meaningful, we need to show that a generic Fano variety in the families under consideration indeed satisfies the conditions (L) and (M). We derive the stronger condition (C) of divisorial canonicity for Fano hypersurfaces and double spaces of index one in Subsections 1.5 and 1.6 from certain explicit regularity conditions, which are stronger than those introduced in Section 3.1. Therefore, we have to prove that generic Fano hypersurfaces and double spaces satisfy those regularity conditions, which is done in Subsection 1.7.

1.1. Maximal singularities of mobile linear systems. We prove Theorem 4.3.5 by induction on the number of factors K. When $K=1$, the theorem holds in a trivial way: the condition (M) means that mobile linear systems on the variety $F=F_1$ have no maximal singularities. This immediately implies birational superrigidity of the variety F.

Starting from this moment, $K\geq 2$.

Assume the converse: there is a moving linear system Σ on V such that the inequality $c_{\text{virt}}(\Sigma)<c(\Sigma)$ holds. By the definition of the virtual threshold of canonical adjunction it means that there exists a sequence of blow ups $\varphi\colon \widetilde{V}\to V$ such that the inequality

$$c(\widetilde{\Sigma})<c(\Sigma) \tag{155}$$

holds, where $\widetilde{\Sigma}$ is the strict transform of the linear system Σ on $\widetilde{V}$. To prove that the variety V is birationally superrigid, we must show that the inequality (155) is impossible, that is, to obtain a contradiction.

Let $H_i=-K_{F_i}$ be the positive generator of the group $\operatorname{Pic}F_i$. Set

$$S_i=\prod_{j\neq i}F_i,$$

so that $V\cong F_i\times S_i$. Let $\rho_i\colon V\to F_i$ and $\pi_i\colon V\to S_i$ be the projections onto the factors. Abusing our notations, we write H_i instead of $\rho_i^*H_i$, so that

$$\operatorname{Pic}V=\bigoplus_{i=1}^{K}\mathbb{Z}H_i \quad\text{and}\quad K_V=-H_1-\cdots-H_K.$$

We get $\Sigma\subset|n_1H_1+\cdots+n_KH_K|$, whereas $c(\Sigma)=\min\{n_1,\dots,n_K\}$. Without loss of generality we assume that $c(\Sigma)=n_1$. By the inequality (155) we get $n_1\geq 1$. Set $n=n_1$, $\pi=\pi_1$, $F=F_1$, $S=S_1$. We get

$$\Sigma\subset|-nK_V+\pi^*Y|,$$

where $Y=\sum_{i=2}^{K}(n_i-n)H_i$ is an effective class on the base S of the fibre space π.

Now we need to modify the birational morphism φ. For an arbitrary sequence of blow ups $\mu_S\colon S^+\to S$ we set $V^+=F\times S^+$ and obtain the following commutative diagram:

$$\begin{array}{ccccc} & V^+ & \stackrel{\mu}{\to} & V & \\ \pi_+ & \downarrow & & \downarrow & \pi \\ & S^+ & \stackrel{\mu_S}{\to} & S, & \end{array} \tag{156}$$

where π_+ is the projection and $\mu=(\mathrm{id}_F,\mu_S)$. Let $E_1,\dots,E_N\subset\widetilde{V}$ be all exceptional divisors of the morphism φ.

Proposition 1.1. *There exists a sequence of blow ups $\mu_S\colon S^+ \to S$ such that in the notations of the diagram (156) the centre of each discrete valuation E_i, $i = 1, \dots, N$, covers either S^+ or a divisor on S^+:*

$$\operatorname{codim}[\pi_+(\operatorname{centre}(E_i, V^+))] \leq 1.$$

Proof. Let $E \subset \widetilde{V}$ be the exceptional divisor of the birational morphism $\varphi\colon \widetilde{V} \to V$, $B = \varphi(E)$ the centre of the discrete valuation E on V. Assume that $\operatorname{codim}_S \pi(B) \geq 2$. Construct a sequence of commutative diagrams

$$\begin{array}{ccccc} & V_j & \stackrel{\varepsilon_j}{\to} & V_{j-1} & \\ \pi_j & \downarrow & & \downarrow & \pi_{j-1} \\ & S_j & \stackrel{\lambda_j}{\to} & S_{j-1}, & \end{array} \tag{157}$$

where $j = 1, \dots, l$, satisfying the following conditions:

1) $V_0 = V$, $S_0 = S$, $\pi_0 = \pi$;

2) $V_j = F \times S_j$, π_j is the projection onto the factor S_j, $\varepsilon_j = (\mathrm{id}_F, \lambda_j)$ for all $j \geq 1$;

3) λ_j is the blow up of the irreducible subvariety

$$B_{j-1} = \pi_{j-1}(\operatorname{centre}(E, V_{j-1})) \subset S_{j-1},$$

where $\operatorname{codim} B_{j-1} \geq 2$.

It is obvious that the properties 1)–3) determine the sequence of diagrams (157) in a unique way.

Lemma 1.1. *The following inequality holds: $l \leq a(E, V)$.*

Proof. Let $\Delta_j \subset V_j$ be the exceptional divisor of the morphism ε_j. By construction we get $\operatorname{centre}(E, V_j) \subset \Delta_j$, so that $\nu_E(\Delta_j) \geq 1$. Now we obtain

$$a(E, V) = a(E, V_l) + \sum_{j=1}^{l} \nu_E(\Delta_j) a(\Delta_j, V) \geq l.$$

Q.E.D. for the lemma.

Therefore the sequence of diagrams (157) terminates: we may assume that $\operatorname{centre}(E, V_l)$ covers a divisor on the base S_l. From this fact (by the Hironaka theorem on the resolution of singularities) Proposition 1.1 follows immediately. Q.E.D.

Let Σ^+ be the strict transform of the linear system Σ on V^+. Now the arguments break into two parts due to the following fact.

Proposition 1.2. *The following alternative holds:*

(i) *either the inequality $c(\Sigma^+) < c(\Sigma)$ is true,*

(ii) *or for a general divisor $D^+ \in \Sigma^+$ the pair $(V^+, \frac{1}{n}D^+)$ is not canonical and, moreover, for some $i = 1, \dots, N$ the discrete valuation E_i determines a non-canonical singularity of this pair.*

Remark 1.1. The alternative of Proposition 1.2 should be understood in the "and/or" sense: *at least one* of the two possibilities (i), (ii) takes place (or both).

Proof of Proposition 1.2. Consider the diagram of maps (156). Let $\tau\colon V^\sharp \to V^+$ be the resolution of singularities of the composite map

$$V^+ \stackrel{\mu}{\to} V \stackrel{\varphi^{-1}}{\to} \widetilde{V}.$$

Set $\psi = \varphi^{-1} \circ \mu \circ \tau\colon V^\sharp \to \widetilde{V}$. There exist an open set $U \subset V^\sharp$ and a closed set of codimension two $Y \subset \widetilde{V}$ such that

$$\psi_U = \psi|_U\colon U \to \widetilde{V} \setminus Y$$

is an isomorphism. Obviously, if $E \subset V^\sharp$ is an exceptional divisor of the morphism τ and $E \cap U \neq \emptyset$, then $E \cap U = \psi_U^{-1}(E_i)$ for some exceptional divisor E_i of the morphism φ.

Let Σ^+ and $\Sigma^\sharp$ be the strict transforms of the linear system Σ on V^+ and $V^\sharp$, respectively, $\Sigma_U = \Sigma^\sharp|_U$. If $D^\sharp \in \Sigma^\sharp$ is a general divisor, then

$$\widetilde{D} = \psi_U(D_U^\sharp) \in \widetilde{\Sigma}$$

is a general divisor of the linear system $\widetilde{\Sigma}$ (we make no difference between $\widetilde{\Sigma}$ and its restriction onto $\widetilde{V} \setminus Y$, since the set Y is of codimension two). We know that

$$\widetilde{D} + nK_{\widetilde{V}} \notin A^1_+\widetilde{V}$$

see (155). Therefore,

$$D_U^\sharp + nK_U \notin A^1_+U. \tag{158}$$

Let $\mathcal{E}$ be the set of exceptional divisors of the morphism τ with a non-empty intersection with U. By (158) we get

$$\tau^*(D^+ + nK^+)|_U - \sum_{E\in\mathcal{E}} (\nu_E(D^+) - na^+(E))E_U \notin A^1_+U,$$

where K^+ is the canonical class of V^+, $a^+(E) = a(E, V^+)$. Therefore, either

$$D^+ + nK^+ \notin A^1_+V^+,$$

and we are in the case (i) of Proposition 1.2, or there exists an exceptional divisor $E \in \mathcal{E}$, satisfying the Noether-Fano inequality $\nu_E(D^+) > n \cdot a^+(E)$, that is, the discrete valuation E realizes a non-canonical singularity of the pair $(V^+, \frac{1}{n}D^+)$. In the latter case we get part (ii) of the alternative described in Proposition 1.2, since $E \in \mathcal{E}$ and thus $E = E_i$ for some $i = 1, \ldots, N$ (as discrete valuations). Q.E.D. for Proposition 1.2.

1.2. Reduction to the base of the fibre space. Assume that the case (i) of the alternative of Proposition 1.2 takes place, that is, $D^+ + nK^+ \notin A^1_+V^+$. Let $z \in F$ be a point of general position. Set

$$S_z^+ = \{z\} \times S^+, \ S_z = \{z\} \times S.$$

It is clear that $K_z^+ = K^+|_{S_z^+}$ and $K_z = K_V|_{S_z}$ are the canonical classes $K_S^+ = K_{S^+}$ and K_S, respectively. Let

$$\Sigma_z = \Sigma|_{S_z} \quad \text{and} \quad \Sigma_z^+ = \Sigma^+|_{S_z^+}$$

be the restriction of the linear systems Σ, Σ^+ onto S_z and S_z^+. Take general divisors $D_z \in \Sigma_z$ and $D_z^+ \in \Sigma_z^+$. We get a mobile linear system Σ_z on the variety $S = F_2 \times \cdots \times F_K$. Moreover,

$$\Sigma_z \subset |n_2 H_2 + \cdots + n_K H_K|,$$

so that $c(\Sigma_z) = \min\{n_2, \ldots, n_K\} \geq n = c(\Sigma)$.

Lemma 1.2. *The following estimate holds:* $D^+ + nK^+ = \pi_+^*(D_z^+ + nK_z^+)$.

Proof. Set $\mathcal{E}_S$ to be the set of exceptional divisors of the morphism μ_S. The exceptional divisors of the morphism μ are $F \times E = \pi_+^* E$ for $E \in \mathcal{E}_S$. We get

$$K_S^+ = \mu_S^* K_S + \sum_{E \in \mathcal{E}_S} a_E E \quad \text{and} \quad K^+ = \mu^* K_V + \pi_+^*(\sum_{E \in \mathcal{E}_S} a_E E),$$

where $a_E = a(E)$ is the discrepancy of the divisor E. For some numbers $b_E \geq 0$ we get

$$D^+ = \mu^* D - \sum_{E \in \mathcal{E}_S} b_E \pi_+^* E,$$

whereas for a point $z \in F$ of general position

$$D_z^+ = \mu_S^* D_z - \sum_{E \in \mathcal{E}_S} b_E E.$$

Now taking into account that $D + nK_V = \pi^* Y$ and $D_z + nK_z = D_z + nK_S = Y$, we obtain the claim of the lemma. Q.E.D.

Corollary 1.1. $D_z^+ + nK_z^+ \notin A_+^1 S^+$.

Proof. Indeed, it is clear that

$$\pi_+^* A_+^1 S^+ \subset A_+^1 V^+.$$

Q.E.D. for the corollary.

Thus for the strict transform Σ_z^+ of the linear system Σ_z on S^+ we get the inequality $c(\Sigma_z^+) < c(\Sigma_z)$. The more so, $c_{\text{virt}}(\Sigma_z) < c(\Sigma_z)$. Therefore the variety S is not birationally superrigid. This contradicts the induction hypothesis.

1.3. Reduction to the fibre of the fibre space. End of the proof. By Proposition 1.2 and what was said above, for a general divisor $D^+ \in \Sigma^+$ the pair $(V^+, \frac{1}{n} D^+)$ is not canonical, that is, there exists a birational morphism $V^\sharp \to V^+$ and an exceptional divisor $E \subset V^\sharp$, satisfying the Noether-Fano inequality $\nu_E(\Sigma^+) > n \cdot a_E^+$, where $a_E^+ = a(E, V^+)$. Moreover, we can assume that the centre $B = \text{centre}(E, V)$ of the valuation E covers a divisor on the base or the whole base: $\text{codim}_{S^+} T \leq 1$, where $T = \pi_+(B)$.

Let $t \in T$ be a point of general position. The fibre $F_t = \pi_+^{-1}(t)$ cannot lie entirely in the base set $\text{Bs}\, \Sigma^+$ of the moving linear system Σ^+, since

$$\text{codim}_{V^+} \pi_+^{-1}(T) \leq 1.$$

Therefore, $\Sigma_t^+ = \Sigma^+|_{F_t}$ is a non-empty linear system on F, $\Sigma_t^+ \subset |nH| = |-nK_F|$ (if $T \subset S^+$ is a divisor, then Σ_t^+ can have fixed components). Let $D_t^+ \in \Sigma_t^+$ be a general divisor. By inversion of adjunction (see Theorem 1.8 and §2 of this chapter), the pair

$$(F, \frac{1}{n} D_t^+)$$

is not log canonical. We get a contradiction with the condition (L). This contradiction completes the proof of birational superrigidity of the variety V, the first and main claim of Theorem 4.3.5.

1.4. The structures of a rationally connected fibre space and birational self-maps. Let us prove the remaining claims of Theorem 4.3.5. Let $\beta\colon V^\sharp \to S^\sharp$ be a rationally connected fibre space, $\chi\colon V \dashrightarrow V^\sharp$ a birational map. Take a very ample linear system $\Sigma_S^\sharp$ on the base $S^\sharp$ and let $\Sigma^\sharp = \beta^*\Sigma_S^\sharp$ be a mobile linear system on $V^\sharp$, $c(\Sigma^\sharp) = 0$. Let Σ be the strict transform of the system $\Sigma^\sharp$ on V. By our remark, $c_{\text{virt}}(\Sigma) = 0$, so that by what we proved above we conclude that $c(\Sigma) = 0$. Therefore, in the presentation,

$$\Sigma \subset |-n_1H_1 - \cdots - n_KH_K|,$$

we can find a coefficient $n_e = 0$. We may assume that $e = 1$. Setting $S = F_2 \times \cdots \times F_K$ and $\pi\colon V \to S$ to be the projection, we get $\Sigma \subset |\pi^*Y|$ for a non-negative class Y on S. But this means that the birational map χ of the fibre space V/S onto the fibre space $V^\sharp/S^\sharp$ is fibrewise: there exists a rational dominant map $\gamma\colon S \dashrightarrow S^\sharp$ making the diagram

$$\begin{array}{ccccc} & V & \overset{\chi}{\dashrightarrow} & V^\sharp & \\ \pi & \downarrow & & \downarrow & \beta \\ & S & \overset{\gamma}{\dashrightarrow} & S^\sharp & \end{array}$$

commutative. For a point $z \in S^\sharp$ of general position let $F_z^\sharp = \beta^{-1}(z)$ be the corresponding fibre, $F_z^\chi \subset V$ its strict transform with respect to χ. By assumption, the variety F_z^χ is rationally connected. On the other hand,

$$F_z^\chi = \pi^{-1}(\gamma^{-1}(z)) = F \times \gamma^{-1}(z),$$

where $F = F_1$ is the fibre of π. Therefore, the fibre $\gamma^{-1}(z)$ is also rationally connected.

Thus we have reduced the problem of the description of rationally connected structures on V to the same problem for S. Now the claim (i) of Theorem 4.3.5 is easy to obtain by induction on the number of direct factors K. For $K = 1$ it is obvious that there are no non-trivial rationally connected structures. The second part of the claim (i) (about the structures of conic bundles and fibrations into rational surfaces) is obvious since $\dim F_i \geq 3$ for all $i = 1, \ldots, K$. Non-rationality of V is now obvious.

Let us prove the claim (ii) of Theorem 4.3.5. Set $RC(V)$ to be the set of all structures of a rationally connected fibre space with a non-trivial base on V. By the claim (i) we have

$$RC(V) = \{\pi_I\colon V \to F_I = \prod_{i\in I} F_i \,|\, \emptyset \neq I \subset \{1, \ldots, K\}\}.$$

The set $RC(V)$ has a natural structure of an ordered set: $\alpha \leq \beta$ if β factors through α. Obviously, $\pi_I \leq \pi_J$ if and only if $J \subset I$. For $I = \{1 \ldots, K\} \setminus \{e\}$ set $\pi_I = \pi_e$, $F_I = S_e$. It is obvious that $\pi_1 \ldots, \pi_K$ are the minimal elements of $RC(V)$.

Let $\chi \in \operatorname{Bir} V$ be a birational self-map. The map

$$\chi^*\colon RC(V) \to RC(V),$$

$$\chi^*\colon \alpha \longmapsto \alpha \circ \chi,$$

is a bijection preserving the relation $\leq$. From here it is easy to conclude that χ^* is of the form

$$\chi^*\colon \pi_I \longmapsto \pi_{I^\sigma},$$

where $\sigma \in S_K$ is a permutation of K elements and for $I = \{i_1, \ldots, i_k\}$ we define $I^\sigma = \{\sigma(i_1), \ldots, \sigma(i_k)\}$. Furthermore, for each $I \subset \{1, \ldots, K\}$ we get the diagram

$$\begin{array}{ccccc} & V & \stackrel{\chi}{\dashrightarrow} & V & \\ \pi_I & \downarrow & & \downarrow & \pi_{I^\sigma} \\ & F_I & \stackrel{\chi_I}{\dashrightarrow} & F_{I^\sigma}, & \end{array}$$

where χ_I is a birational map. In particular, χ induces birational isomorphisms $\chi_e\colon F_e \dashrightarrow F_{\sigma(e)}$, $e = 1, \ldots, K$. However, all the varieties F_e are birationally superrigid, so that all the maps χ_e are biregular isomorphisms. Thus

$$\chi = (\chi_1, \ldots, \chi_K) \in \operatorname{Bir} V$$

is a biregular isomorphism, too: $\chi \in \operatorname{Aut} V$. Q.E.D. for Theorem 4.3.5.

Remark 1.2. As we already mentioned in Chapter 4, the proof given above works word for word in a more general case, when all direct factors F_i are Fano varieties with $\mathbb{Q}$-factorial terminal singularities and Picard number $\rho(F_i) = 1$, satisfying the conditions (L) and (M). Unfortunately, there are very few cases when the condition (L) (or the stronger condition (C)) has been shown for a particular family of singular Fano varieties. Nevertheless, it should be kept in mind that the existence of $\mathbb{Q}$-factorial terminal singularities in itself creates no problem (normally it just makes proving the condition (L) more difficult).

1.5. Examples of divisorially canonical varieties. To apply Theorem 4.3.5, one needs to be able to show the divisorial (log) canonicity for particular families of Fano varieties. There are five main classes of varieties for which the divisorial (log) canonicity has been shown:

- generic Fano hypersurfaces of index 1 and dimension at least 5 [**Pukh05**];
- generic Fano double spaces of index 1 (and any dimension ≥ 3) [**Pukh05**];
- generic Fano complete intersections

$$F_{d_1\cdot\ldots\cdot d_k} \subset \mathbb{P}^{M+k}$$

  of index 1, that is, $d_1 + \cdots + d_k = M + k$, $2 \leq d_1 \leq \cdots \leq d_k$, where $M \geq 4k + 1$ and $d_k \geq 8$ [**Pukh06b**];
- generic 3-fold Fano hypersurfaces $V_d \subset \mathbb{P}(a_0, \ldots, a_4)$, $d = a_0 + \cdots + a_4$, from the standard list of 95 $\mathbb{Q}$-factorial terminal three-folds [**Ch08**], except for the double spaces already accounted for above and the quartics $V_4 \subset \mathbb{P}^4$, for which divisorial canonicity remains an open question; see [**Pukh05**] and [**ChPW07**];
- generic Fano double hypersurfaces of index 1, that is, varieties, realized as double covers $F \stackrel{\sigma}{\to} Q \subset \mathbb{P}^{M+1}$, where Q is a hypersurfaces of degree m and σ is branched over $W = W^* \cap Q$, $W^* \subset \mathbb{P}^{M+1}$ is a hypersurface of degree $2l$, where $m \geq 3$, $l \geq 2$ and $m + l = M + 1$ [**Pukh08a**].

In all the listed cases, except for that of weighted hypersurfaces, a generic variety was shown to be divisorially canonical. Now we will consider the first two classes (the higher-dimensional Fano hypersurfaces and Fano double spaces) as typical examples.

Fano hypersurfaces. Let $\mathbb{P} = \mathbb{P}^M$, $M \geq 3$, be the complex projective space. Set $\mathcal{F} = \mathbb{P}(H^0(\mathbb{P}, \mathcal{O}_{\mathbb{P}}(M)))$ to be the space of hypersurfaces of degree M. We will define a Zariski open subset $\mathcal{F}_{\text{reg}} \subset \mathcal{F}$ by explicit regularity conditions sharpening the conditions used in Chapter 3 for proving birational superrigidity of Fano hypersurfaces. Let $F = F_M \subset \mathbb{P} = \mathbb{P}^M$ be a smooth Fano hypersurface. For a point $x \in F$ fix a system of affine coordinates $z_1, \ldots, z_M$ with the origin at x and let

$$f = q_1 + q_2 + \cdots + q_M$$

be the equation of the hypersurface F, $q_i = q_i(z_*)$ are homogeneous polynomials of degree $\deg q_i = i$,

$$f_i = q_1 + \cdots + q_i$$

to be the left segments of the polynomial f, $i = 1, \ldots, M$.

The condition (R1.1): the sequence

$$q_1, q_2, \ldots, q_{M-1}$$

is regular in $\mathcal{O}_{x,\mathbb{P}}$, that is, the system of equations

$$q_1 = q_2 = \cdots = q_{M-1} = 0$$

defines a one-dimensional subset, a finite set of lines in $\mathbb{P}$, passing through the point x. This is the standard regularity condition, which was used in Chapter 3.

The condition (R1.2): the linear span of any irreducible component of the closed algebraic set

$$q_1 = q_2 = q_3 = 0$$

in $\mathbb{C}^M$ is the hyperplane $q_1 = 0$ (that is, the tangent hyperplane $T_x F$).

The condition (R1.3) The closed algebraic set

$$\overline{\{f_1 = f_2 = 0\} \cap F} = \overline{\{q_1 = q_2 = 0\} \cap F} \subset \mathbb{P} \tag{159}$$

(the bar means the closure in $\mathbb{P}$) is irreducible and any section of this set by a hyperplane $P \ni x$ is

- either also irreducible and reduced,
- or breaks into two irreducible components $B_1 + B_2$, where $B_i = F \cap S_i$ is the section of F by a plane $S_i \subset \mathbb{P}$ of codimension 3 and, moreover $\operatorname{mult}_x B_i = 3$,
- or is non-reduced and is of the form $2B$, where $B = F \cap S$ is the section of F by a plane S of codimension 3 and, moreover $\operatorname{mult}_x B = 3$.

Set $\mathcal{F}_{\text{reg}} \subset \mathcal{F}$ to be the set of Fano hypersurfaces, satisfying the conditions (R1.1–R1.3) at every point (in particular, every hypersurface $F \in \mathcal{F}_{\text{reg}}$ is smooth). It is clear that $\mathcal{F}_{\text{reg}}$ is a Zariski open subset of the projective space $\mathcal{F}$. The main result on Fano hypersurfaces is the following two-part

Theorem 1.1. (i) *Any hypersurface $F \in \mathcal{F}_{\text{reg}}$ is non-singular and satisfies the condition (C).*

(ii) *For $M \geq 6$ the set $\mathcal{F}_{\text{reg}}$ is non-empty.*

Proof of both claims is given below in Subsections 1.6 and 1.7.

Remark 1.3. For $M \geq 8$ in the condition (R1.3) we may require that the section of the set (159) by any hyperplane $P \ni x$ were irreducible and reduced, a general hypersurface satisfies this stronger condition. On the other hand, for

$M=4,5$ it is easy to show that for any hypersurface $F\in\mathcal{F}$ there is a point where the conditions (R1.2) and (R1.3) are not satisfied.

Now let us consider the second class of examples.

Fano double spaces. Once again, let $\mathbb{P}=\mathbb{P}^M$, $M\geq 3$, be the complex projective space. Set $\mathcal{W}=\mathbb{P}(H^0(\mathbb{P},\mathcal{O}_{\mathbb{P}}(2M)))$ to be the space of hypersurfaces of degree $2M$. We will define a Zariski open subset $\mathcal{W}_{\mathrm{reg}}\subset\mathcal{W}$ by explicit regularity conditions. Let $F\stackrel{\sigma}{\to}\mathbb{P}=\mathbb{P}^M$ be a Fano double space branched over a smooth hypersurface $W=W_{2M}\subset\mathbb{P}$ of degree $2M$, $M\geq 3$. For a point $x\in W$ fix a system of affine coordinates $z_1,\dots,z_M$ on $\mathbb{P}$ with the origin at x and set

$$w=q_1+q_2+\cdots+q_{2M}$$

to be the equation of the hypersurface W, $q_i=q_i(z_*)$ are homogeneous polynomials of degree $\deg q_i=i$. The regularity condition is formulated below in terms of the polynomials q_i. One should consider the three cases $M\geq 5$, $M=4$ and $M=3$ separately. For convenience of notation assume that $q_1\equiv z_1$. Set also

$$\bar{q}_i=\bar{q}_i(z_2,\dots,z_M)=q_i|_{\{z_1=0\}}=q_i(0,z_2,\dots,z_M).$$

Now let us formulate the regularity condition at the point x.

(R2) Let $M\geq 5$. We require that the rank of the quadratic form $\bar{q}_2$ is at least 2.

Assume that $M=3$ or 4. We require that the quadratic form $\bar{q}_2$ is non-zero and, moreover,

(i) either $\operatorname{rk}\bar{q}_2\geq 2$ (as above),

(ii) or $\operatorname{rk}\bar{q}_2=1$ and the following additional condition is satisfied. Without loss of generality, we assume in this case that

$$\bar{q}_2=z_2^2.$$

Now for $M=4$ we require that the following cubic polynomial in the variable t,

$$\bar{q}_3(0,1,t)=q_3(0,0,1,t)$$

has three distinct roots.

For $M=3$ we require that at least one of the following two polynomials in the variable z_3,

$$\bar{q}_3(0,z_3)\quad\text{or}\quad\bar{q}_4(0,z_3)$$

(they are of the form αz_3^m, $\alpha\in\mathbb{C}$, $m=3,4$) is non-zero.

Set $\mathcal{W}_{\mathrm{reg}}\subset\mathcal{W}$ to be the Zariski open set of hypersurfaces satisfying the condition (R2) at every point.

Similarly to the case of Fano hypersurfaces, we have the following:

Theorem 1.2. (i) *The Fano double space $\sigma\colon F\to\mathbb{P}$, branched over any hypersurface $W\in\mathcal{W}_{\mathrm{reg}}$, is smooth and satisfies the condition (C).*

(ii) *The Zariski open set $\mathcal{W}_{\mathrm{reg}}\subset\mathcal{W}$ is non-empty.*

Proof of both claims is given below in Subsections 1.6 and 1.7, respectively.

The arguments for Fano complete intersections and Fano double hypersurfaces are similar and we refer the reader to the papers [**Pukh06b, Pukh08a**]. The arguments for weighted Fano 3-fold hypersurfaces are based on the techniques of [**CPR**], we refer the reader to [**Ch08**]. On the whole, the examples listed above suggest that the following principle is true: if $\mathcal{F}$ is a family of birationally rigid Fano varieties with Picard number 1, then there exists a non-empty Zariski open subset $\mathcal{F}^\circ \subset \mathcal{F}$ such that any member $F \in calF^\circ$ is divisorially (log) canonical.

1.6. Proof of the divisorial canonicity. Let us prove the claims (i) of Theorems 1.1 and 1.2. First we consider

Fano hypersurfaces. We use the notations of Theorem 1.1 and the preceding regularity conditions. We must show that the condition (C) is satisfied for a regular Fano hypersurface $F \in \mathcal{F}_{\rm reg}$.

Let $\Delta \in |nH|$ be an effective divisor, $n \geq 1$, where $H \in \operatorname{Pic} F$ is the class of a hyperplane section, $K_F = -H$. We have to show that the pair $(F, \frac{1}{n}\Delta)$ has canonical singularities.

Assume the converse. Then for a certain sequence of blow ups $\varphi\colon F^+ \to F$ and an exceptional divisor $E^+ \subset F^+$ the Noether-Fano inequality

$$\nu_{E^+}(\Delta) > n \cdot a(E^+) \tag{160}$$

is satisfied. For a fixed E^+ the inequality (160) is linear in Δ, so that without loss of generality, we may assume that $\Delta \subset F$ is a prime divisor, that is, an irreducible subvariety of codimension 1. From (160) it follows easily that the centre $Y = \varphi(E^+)$ of the valuation E^+ on F satisfies the inequality $\operatorname{mult}_Y \Delta > n$. On the other hand, as we know (Lemma 2.2.1), for any irreducible curve $C \subset F$ the inequality $\operatorname{mult}_C \Delta \leq n$ holds. Thus $Y = x$ is a point. Let $\varepsilon\colon \widetilde{F} \to F$ be its blow up and $E \subset \widetilde{F}$ the exceptional divisor $E \cong \mathbb{P}^{M-2}$. By Proposition 2.3, which is formulated and proved below in Section 2, for some hyperplane $B \subset E$ the inequality

$$\operatorname{mult}_x \Delta + \operatorname{mult}_B \widetilde{\Delta} > 2n, \tag{161}$$

holds, where $\widetilde{\Delta} \subset \widetilde{F}$ is the strict transform of the divisor Δ.

Let $\mathbb{T} = \overline{T_x F} \subset \mathbb{P}$ be the tangent hyperplane at the point x. The divisor E can be naturally identified with the projectivization $\mathbb{P}(T_x\mathbb{T}) = \mathbb{P}(T_x F)$. There is a unique hyperplane $\mathbb{B} \subset \mathbb{T}$, $x \in \mathbb{B}$, such that $B = \mathbb{P}(T_x\mathbb{B})$ with respect to the above-mentioned identification. Let $\Lambda_{\mathbb{B}}$ be the pencil of hyperplanes in $\mathbb{P}$, containing $\mathbb{B}$, and $\Lambda_B = \Lambda_{\mathbb{B}}|_F \subset |H|$ its restriction onto F. Consider a general divisor $R \in \Lambda_B$. It is a hypersurface of degree M in $\mathbb{P}^{M-1}$, smooth at the point x. Let $\widetilde{R} \subset \widetilde{F}$ be the strict transform of the divisor R. Obviously,

$$\widetilde{R} \cap E = B.$$

Set $\Delta_R = \Delta|_R = \Delta \cap R$. It is an effective divisor on the hypersurface R.

Lemma 1.3. *The following estimate holds:*

$$\operatorname{mult}_x \Delta_R > 2n. \tag{162}$$

Proof. We have $(\widetilde{\Delta} \circ \widetilde{R}) = \widetilde{\Delta}_R + Z$, where Z is an effective divisor on E. According to the elementary rules of the intersection theory [**Ful**], $\operatorname{mult}_x \Delta_R = \operatorname{mult}_x \Delta + \deg Z$, since $\operatorname{mult}_x R = 1$. However, Z contains B with multiplicity at least $\operatorname{mult}_B \widetilde{\Delta}$. Therefore, the inequality (161) implies the estimate (162). Q.E.D. for the lemma.

Lemma 1.4. *The divisor $T_R = T_x R \cap R$ on the hypersurface R is irreducible and has multiplicity exactly 2 at the point x.*

Proof. The irreducibility is obvious (for instance, for $M \geq 6$ one can apply the Lefschetz theorem). By the condition (R1.2) the quadric $\{q_2|_E = 0\}$ does not contain a hyperplane in E as a component, in particular, it does not contain the hyperplane $B \subset E$. Thus the quadratic component of the equation of the divisor T_R, that is, the polynomial $q_2|_B$, is non-zero. Q.E.D. for the lemma.

Let us continue our proof of Theorem 1.1. By Lemmas 1.3 and 1.4 we can write

$$\Delta_R = aT_R + \Delta_R^\sharp,$$

where $a \in \mathbb{Z}_+$ and the effective divisor $\Delta_R^\sharp \in |n^\sharp H_R|$ on the hypersurface R satisfies the estimate $\operatorname{mult}_x \Delta_R^\sharp > 2n^\sharp$. Moreover, $\Delta_R^\sharp$ does not contain the divisor T_R as a component. Without loss of generality, we can assume the divisor $\Delta_R^\sharp$ to be irreducible and reduced.

Now consider the second hypertangent system (Chapter 3),

$$\Lambda_2^R = |s_0 f_2 + s_1 f_1|_R,$$

where s_i are homogeneous polynomials of degree i in the linear coordinates z_*. Its base set

$$S_R = \{q_1|_R = q_2|_R = 0\}$$

is by condition (R1.3) of codimension 2 in R and either irreducible and of multiplicity 6 at the point x, or breaks into two plane sections of R, each of multiplicity 3 at the point x. In any case, for a general divisor $D \in \Lambda_2^R$ we get $\Delta_R^\sharp \not\subset \operatorname{Supp} D$, so that the following effective cycle of codimension two on R,

$$\Delta_D = (D \circ \Delta_R^\sharp),$$

is well defined. Since $\operatorname{mult}_x D = 3$ and $\Lambda_2^R \subset |2H_R|$, the cycle Δ_D satisfies the estimate

$$\frac{\operatorname{mult}_x}{\deg}\Delta_D > \frac{3}{M}. \tag{163}$$

We can replace the cycle Δ_D by its suitable irreducible component and thus assume it to be an irreducible subvariety of codimension 2 in R. Comparing the estimate (163) with the description of the set S_R given above, we see that $\Delta_D \not\subset S_R$. This implies that $\Delta_D \not\subset T_R$. Indeed, if this were not true, we would have obtained

$$f_1|_{\Delta_D} \equiv q_1|_{\Delta_D} \equiv 0. \tag{164}$$

However, $\Delta_D \subset D$, so that for some $s_0 \neq 0$, $s_1 \neq 0$ (the divisor D is chosen to be general) we have

$$(s_0 f_2 + s_1 f_1)|_{\Delta_D} \equiv 0.$$

By (164) this implies that $f_2|_{\Delta_D} \equiv (q_1 + q_2)|_{\Delta_D} \equiv 0$ (since $s_0 \neq 0$ is just a constant), so that $\Delta_D \subset S_R$. A contradiction.

Thus $\Delta_D \not\subset T_R$. Therefore the effective cycle $\Delta^+ = (\Delta_D \circ T_R)$ is well defined. It satisfies the estimate

$$\frac{\operatorname{mult}_x}{\deg}\Delta^+ > \frac{6}{M}. \tag{165}$$

The effective cycle Δ^+ as a cycle on F is of codimension 4. Now recall the following fact (Section 3.1): if the Fano hypersurface F at the point x satisfies the regularity condition (R1.1), then for any effective cycle Y of pure codimension $l \leq M-2$ the inequality

$$\frac{\mathrm{mult}_x}{\deg} Y \leq \frac{l+2}{M}$$

holds. Therefore, the inequality (165) for an effective cycle of codimension 4 is impossible.

The proof of the claim (i) of Theorem 1.1 is complete. Q.E.D.

Now let us consider

Fano double spaces. We use the notation of Theorem 1.2 and the preceding regularity conditions. We must show that the condition (C) is satisfied for a Fano double space $\sigma\colon F \to \mathbb{P}$, branched over a hypersurface $W \in \mathcal{W}_{\mathrm{reg}}$.

Assume the converse: there is an irreducible divisor $D \in |nH|$, $H = -K_F = \sigma^* H_{\mathbb{P}}$, where $H_{\mathbb{P}}$ is the class of a hyperplane in $\mathbb{P}$, such that the pair $(F, \frac{1}{n}D)$ is not canonical. Recall the following fact [**Pukh00c**, Proposition 4.3]: for any irreducible curve $C \subset F$ the inequality

$$\mathrm{mult}_C D \leq n$$

holds. (Formally speaking, this inequality was proved in [**Pukh00c**] for a *mobile* divisor $D \in \Lambda \subset |nH|$, where Λ is a linear system without fixed components; however, the mobility was never used in the proof.) Now arguing as above, we conclude that the centre of a non-canonical singularity of the pair $(F, \frac{1}{n}D)$ is a point $x \in F$. Again we have the inequality

$$\mathrm{mult}_x D + \mathrm{mult}_B \widetilde{D} > 2n$$

for some hyperplane $B \subset E$ in the exceptional divisor $E \subset \widetilde{F}$ of the blow up $\widetilde{F} \to F$ of the point x.

Case 1. The point $z = \sigma(x) \notin W$. In this case let $P \subset \mathbb{P}$ be the hyperplane such that $P \ni z$ and

$$\widetilde{F}_P \cap E = B,$$

where $F_P = \sigma^{-1}(P)$. Such a hyperplane is unique. The divisor F_P is obviously irreducible and satisfies the equality

$$\mathrm{mult}_x F_P + \mathrm{mult}_B \widetilde{F}_P = 2.$$

Since $F_P \in |H|$, this implies that $F_P \neq D$. Therefore,

$$Z = (F_P \circ D)$$

is a well-defined effective cycle of codimension two on F. We get

$$\deg Z = 2n, \quad \mathrm{mult}_x Z \geq \mathrm{mult}_x D + \mathrm{mult}_B \widetilde{D} > 2n,$$

which is impossible. Thus the case $z \notin W$ is not realized.

Case 2. The point $z = \sigma(x) \in W$. In this case let

$$P = T_z W \subset \mathbb{P}$$

be the tangent hyperplane. Again, set $F_P = \sigma^{-1}(P)$.

We get $F_P \in |H|$, $\mathrm{mult}_x F_P \geq 2$. The divisor F_P is irreducible, so that if $F_P \neq D$, then for the cycle $Z = (F_P \circ D)$ we get again,

$$\deg Z = 2n, \quad \mathrm{mult}_x Z \geq 2\,\mathrm{mult}_x D > 2n, \tag{166}$$

since $\operatorname{mult}_x D \geq \operatorname{mult}_B \widetilde{D}$. Since there is no cycle Z, satisfying the condition (166), we conclude that

$$F_P = D.$$

Up to this moment we have never used the arguments of general position.

Lemma 1.5. *The pair (F, F_P) is canonical at the point x.*

Proof will be given below together with that of the claim (ii) of Theorem 1.2.

This contradiction completes the proof of the claim (i) of Theorem 1.2. Q.E.D.

1.7. Generic hypersurfaces and double spaces. Now let us prove the second claims of Theorems 1.1 and 1.2. The condition (C) of divisorial canonicity is considerably stronger than that of mobile canonicity (M), so it is only natural that the assumptions of genericity are stronger as well. We start with the harder case of

Fano hypersurfaces. We use the notation of Theorem 1.1 and the preceding regularity conditions. We must show that the closed set $\mathcal{F} \setminus \mathcal{F}_{\text{reg}}$ of Fano hypersurfaces, non-regular at at least one point, is of positive codimension in $\mathcal{F}$. Let $\mathcal{F}(x)$ be the set of hypersurfaces passing through a point $x \in \mathbb{P}$, $\mathcal{F}_{\text{reg}}(x) \subset \mathcal{F}(x)$ the set of hypersurfaces, regular at the point x. It suffices to show that

$$\operatorname{codim}_{\mathcal{F}(x)} [\mathcal{F}(x) \setminus \mathcal{F}_{\text{reg}}(x)] \geq M, \tag{167}$$

since $\mathcal{F}(x)$ is a divisor in $\mathcal{F}$. For the first condition (R1.1) it has already been proved in Section 3.1, so that we may assume that all hypersurfaces under consideration satisfy (R1.1). Fix a point $x \in \mathbb{P}$ and show that violation of the condition (R1.2) or (R1.3) imposes at least M conditions on the hypersurface $F \ni x$ (that is, on the polynomial f). As we shall see, the actual codimension of the set of non-regular hypersurfaces is much higher and, moreover, the estimate is sharper as M gets higher. Let $\varphi \colon \widetilde{F} \to F$ be the blow up of the point x, $E \subset \widetilde{F}$ the exceptional divisor, $E \cong \mathbb{P}^{M-2}$. We consider $z_2, \dots, z_M$ as homogeneous coordinates on E, assuming that $q_1 \equiv z_1$. Set

$$g_i = q_i|_E = q_i|_{\{z_1=0\}}.$$

Now the condition (R1.2) reads as follows: the linear span of any irreducible component of the set $\{g_2 = g_3 = 0\} \subset E$ is E.

It is easier to check the inequality (167) as M gets higher. We consider in full detail the hardest case $M = 6$. Here $E = \mathbb{P}^4$. It is easy to see that reducibility of the quadric

$$Q = \{g_2 = 0\} \subset E$$

imposes on the polynomial g_2 (and thus on q_2) 6 independent conditions. Thus, for a general hypersurface the quadric Q is irreducible at every point $x \in F$. Therefore, we get the following possibilities for Q:

- the quadric Q is smooth (this is true for a point $x \in F$ of general position),
- the quadric Q is a cone over a smooth quadric in $\mathbb{P}^3$ (this is the case for the points lying on a divisor $Z_1 \subset F$),
- the quadric Q is a cone with the vertex $\mathbb{P}^1$ over a smooth conic in $\mathbb{P}^2$ (this happens for the points $x \in Z_2$, where $Z_2 \subset F$ is a closed set of codimension 3).

Let us consider each of these three cases separately.

Assume that the quadric Q is smooth. We have

$$\operatorname{Pic} Q = \mathbb{Z} H_Q,$$

where H_Q is the hyperplane section. The condition (R1.2) is violated only in the case when the divisor

$$\{g_3|_Q = 0\} \subset Q$$

breaks into two components, one of which is a hyperplane section. It is easy to check that this imposes on the polynomial g_3 (and thus on q_3) 12 independent conditions, so that for a general hypersurface F the condition (R1.2) is not violated when the quadric Q is smooth.

Note that for an arbitrary $M \geq 6$ in the case of smooth quadric Q violation of the condition (R1.2) imposes

$$\frac{1}{6}(M^3 - 3M^2 - 10M + 24)$$

independent conditions on the polynomial q_3, so that the extra codimension of the set of non-regular hypersurfaces increases when M grows higher.

Now assume that the quadric Q is a cone with the vertex at a point $p \in E$ or a line $L \subset E$ and some component of the divisor $\{g_3|_Q = 0\}$ is contained in a hyperplane $P \subset E$. Consider the intersection $P \cap Q$. If this intersection is an irreducible quadric, then we argue as above in the smooth case. If $P \cap Q$ is reducible, then

$$P \cap Q = S_1 + S_2$$

for some planes $S_i \subset Q$. Vanishing of the polynomial g_3 on a plane $S \subset Q$ imposes $10 - 1 = 9$ independent conditions on g_3 and q_3, since the quadric Q contains a one-dimensional family of planes.

If $M \geq 7$, then the arguments are similar and the estimates stronger; see the example above. Thus we have proved that the condition (R1.2) holds at every point of a general hypersurface $F \subset \mathbb{P}$ for $M \geq 6$.

Let us consider the condition (R1.3). Let

$$\mathbb{T} = \overline{\{q_1 = 0\}} \subset \mathbb{P}$$

be the projective tangent hyperplane to the hypersurface F at the point x, $\mathbb{T} \cong \mathbb{P}^{M-1}$ and

$$Q = \{q_2|_{\mathbb{T}} = 0\} \subset \mathbb{T}$$

a quadratic cone with the vertex at the point x. To check the condition (R1.3) we must inspect a number of cases. The arguments are of the same type, for this reason we consider just a few main examples. Let us explain the scheme of arguments.

Fix the quadric Q and a hyperplane $P \ni x$, $P \subset \mathbb{T}$. Set $Q_P = Q \cap P$. We look at the intersection

$$Q_P \cap F$$

as a divisor on the quadric Q_P. Obviously, the quadric $Q_P \cap F$ is given by the equation

$$(q_3 + \cdots + q_M)|_{Q_P} = 0,$$

where the homogeneous polynomials $q_3, \ldots, q_M$ are arbitrary. Now to prove the proposition, it is necessary to estimate, how many conditions on the polynomials

$q_3, \dots, q_M$ imposes a violation of (R1.3) for each type of the quadric Q_P. Again we restrict ourselves by the hardest case $M = 6$.

Assume at first that Q is a cone over a non-singular three-dimensional quadric. In this case Q is a factorial variety. Let $P \ni x$ be an arbitrary hyperplane. For the quadric $Q_P = Q \cap P$ the two cases are possible:

(1) Q_P is a cone with the vertex x over a non-singular quadric $S \subset \mathbb{P}^3$;

(2) Q_P is a cone with the vertex at a line $L \ni x$ over a non-singular conic $C \subset \mathbb{P}^2$. This possibility realizes if and only if the hyperplane P is tangent to Q along the line L.

In turn, the case (1) breaks into two subcases.

(1A) The set $Q_P \cap F$ is reducible and at least one of its components, say A, does not contain the point x. For the divisor $A \subset Q_P$ we get

$$A \sim aH_P,$$

where H_P is the hyperplane section of the quadric Q_P and $1 \leq a \leq 4$. (A little bit later we explain why $a \leq 4$: formally $a \leq 5$.) Thus,

$$Q_P \cap F = A + B,$$

where $B \sim (6 - a)H_P$. The affine equation of the hypersurface $F \cap P \subset P$ takes the form

$$q_2^+ + \cdots + q_6^+ = 0,$$

where $q_i^+ = q_i|_P$: that is why $a \leq 4$. The equation $q_2^+ = 0$ defines the quadric Q_P. The set of sections of the quadric Q_P by hypersurfaces F of this form is of dimension 125 whereas, for $a \geq 2$,

$$h^0(Q_P, \mathcal{O}(a)) = \binom{a+4}{4} - \binom{a+2}{4}.$$

Now an easy calculation shows that with the hyperplane $P \subset \mathbb{T}$ fixed violation of the condition (R1.3) imposes on the set of polynomials $(q_3^+, \dots, q_6^+)$ at least 35 independent conditions. Since $P \ni x$ varies in a 4-dimensional family, we obtain finally that violation of the condition (R1.3) of the type (1A) at the point $x \in F$ imposes on f at least 31 conditions. Recall that we need just 6 conditions. Thus for a general hypersurface F the subcase (1A) is impossible.

(1B) Here the set $Q_P \cap F$ is reducible and each of its components contains the point x. In this case the following sets are reducible (or everywhere non-reduced): the projectivized tangent cone

$$C = \mathbb{P}T_x(Q_P \cap F)$$

and the intersection

$$C_\infty = Q_P \cap F \cap H_\infty$$

with the hyperplane at infinity $H_\infty \not\ni x$. Since Q_P is a cone over the smooth quadric $S \cong Q_P \cap H_\infty$, both curves C and C_∞ can be looked at as curves on the quadric $S \cong \mathbb{P}^1 \times \mathbb{P}^1$. The curve C is given by the polynomial q_3, the curve C_∞ by the polynomial q_6. Reducibility of the curve C (which is of bidegree $(3, 3)$) imposes $k \geq 3$ independent conditions on the polynomial q_3 and, moreover, the value $k = 3$ corresponds to the case of general position when $C = R + C^+$, where $R \subset S$ is a line and C^+ an irreducible curve of degree 5. Reducibility of the curve C_∞ (which is of bidegree $(6, 6)$) imposes $k_\infty \geq 6$ independent conditions on the polynomial

q_6 and, moreover, the value $k_\infty = 6$ corresponds to the case of general position when $C_\infty = R_\infty + C_\infty^+$, where $R_\infty \subset S$ is a line and C_∞^+ is an irreducible curve of degree 11. However, by the condition (R1.1) the hypersurface F (and thus $Q_P \cap F$) cannot contain a plane. Therefore a violation of the condition (R1.3) imposes on the polynomial f,

$$k + k_\infty \geq 3 + 6 + 1 = 10$$

independent conditions. (Taking into account the polynomials q_4 and q_5 gives a considerably higher codimension, but we do not need that.) Since the hyperplane P varies in a 4-dimensional family ($x \in P \subset \mathbb{T}$), we get finally at least 6 independent conditions on the polynomial f. Therefore, on a general hypersurface F a violation of the condition (R1.3) of the type (1) is impossible.

Let us consider the case (2). Here the hyperplane $P \subset \mathbb{T}$ varies in a 3-dimensional family. Assume that $Q_P \cap F$ is reducible. If $F \not\supset L$, where $L \cong \mathbb{P}^1$ is the vertex of the cone Q_P, then the arguments completely similar to the case (1A) above give the required estimate (167).

The condition $F \supset L$ means that

$$q_i|_L \equiv 0, \quad i = 3, 4, 5, 6,$$

which gives 4 additional independent conditions on f. Furthermore, it is easy to check that the case

$$\operatorname{mult}_L Q_P \cap F > 2$$

on a general hypersurface F is impossible. Let $S_1, S_2 \subset Q_P$, $S_i \supset L$ be a pair of general planes on Q_P. The restriction $F|_{S_i}$ is a plane curve of degree 6 with the line L as a component of multiplicity 1. Reducibility of the residual curves of degree 5 imposes on f at least 6 independent conditions. As a result, we get at least $7 = 6 + 4 - 3$ independent conditions on f at the point x. Recall that $Q_P \cap F$ cannot contain planes by the condition (R1.1).

The case when Q is a cone over a non-singular three-dimensional quadric is completed.

The other cases are inspected in a similar way. We do not give these arguments because they are of the same type and quite elementary. Let us consider in detail just one more case which is opposite to the case (1) above; it occurs for $M = 6$ only.

Let Q be a cone with the vertex at a plane $S \cong \mathbb{P}^2$ over a conic $C \subset \mathbb{P}^2$. Moreover, assume that the hyperplane P contains S, so that

$$Q_P = \Pi_1 + \Pi_2 \subset P \quad \text{or} \quad Q_P = 2\Pi$$

is a pair of 3-planes or a double 3-plane. The set of points $x \in F$ where this situation is possible is two-dimensional. The hyperplane P varies in a two-dimensional family: $S \subset P \subset \mathbb{T}$. The hypersurface F cannot contain planes because of the condition (R1.1).

If $\Pi_1 \neq \Pi_2$, then $\Pi_1 \cap \Pi_2 = S$ and

$$F \cap Q_P = F \cap \Pi_1 + F \cap \Pi_2.$$

Reducibility of the surface $F \cap \Pi_i$ imposes on F at least 25 conditions so that we may assume that both surfaces $F \cap \Pi_i$ are irreducible. Obviously,

$$\operatorname{mult}_x F \cap \Pi_i \geq 3.$$

However, by the condition (R1.2) we have

$$\operatorname{mult}_x F \cap Q_P = \operatorname{mult}_x F \cap \Pi_1 + \operatorname{mult}_x F \cap \Pi_2 = 6.$$

Therefore the condition (R1.3) holds in this case, as well.

If $\Pi_1 = \Pi_2 = \Pi$, the arguments are similar.

This completes our proof for $M = 6$. For $M = 7$ the arguments are simpler and for $M \geq 8$ one can argue as in the case (1) above.

Q.E.D. for part (ii) of Theorem 1.1.

Now let us consider the case of

Fano double spaces. We use the notation of Theorem 1.2 and the preceding regularity conditions. We must show that the closed set $\mathcal{W} \setminus \mathcal{W}_{\mathrm{reg}}$ of branch hypersurfaces, non-regular at at least one point, is of positive codimension in $\mathcal{W}$. The set of quadratic forms of rank ≤ 1 in the variables $z_2, \ldots, z_M$ is of codimension

$$c(M) = \frac{(M-1)(M-2)}{2}.$$

When $M \geq 5$ we have $c(M) \geq M$, so that (R2) holds at every point $x \in W$ for a sufficiently general polynomial w.

Assume that $M = 4$. Here $c(4) = 3 = \dim W$, so that for a general hypersurface W the condition (R2), (i) is violated at a finite set of points $x \in W$. For a general cubic polynomial q_3 the polynomial $\bar{q}_3(0, 1, t)$ has three distinct roots: a multiple root gives a condition of codimension 1 for $\bar{q}_3$. Therefore (R2) holds for $M = 4$.

Assume that $M = 3$. Here $c(3) = 1$, $\dim W = 2$, so that there is a one-dimensional set of points $x \in W$ where the conditon (i) is violated. It is easy to see that the conditions

$$\bar{q}_3(0, z_3) \equiv 0 \quad \text{and} \quad \bar{q}_4(0, z_3) \equiv 0$$

are independent.

This completes the proof of the claim (ii) of Theorem 1.2. Q.E.D.

It remains to prove Lemma 1.5. Since the hypersurface W is non-singular, the point x is an isolated singularity of the hypersurface $W_P = W \cap P$, where $P = T_x W$. The irreducible divisor F_P is the double cover of the hyperplane P branched at W_P. Set $p = \sigma^{-1}(x) \in F$, $\varphi\colon \widetilde{F} \to F$ the blow up of the point p, $E \subset \widetilde{F}$ the exceptional divisor. Assume that the pair (F, F_P) is not canonical. By Proposition 2.3, for some hyperplane $B \subset E$ the inequality

$$\operatorname{mult}_p F_P + \operatorname{mult}_B \widetilde{F}_P \geq 3 \tag{168}$$

holds. However, $\operatorname{mult}_p F_P = 2$ so that the projectivized tangent cone

$$\mathbb{P}(T_p F_P) = (\widetilde{F}_P \circ E) \tag{169}$$

must contain the hyperplane B. On the other hand, the tangent cone $T_p F_P$ is given by the equation

$$y^2 = \bar{q}_2(z_2, \ldots, z_M)$$

(in the notation of Subsection 1.5). For $M \geq 5$ by the condition (R2) we get $\operatorname{rk} \bar{q}_2 \geq 2$, so that the quadric (169) is irreducible and cannot contain B. This proves the lemma for $M \geq 5$.

When $M = 3, 4$ the inequality (168) holds only for those points p where $\operatorname{rk} \bar{q}_2 = 1$ and only for the two hyperplanes

$$B = \{y \pm z_2 = 0\}.$$

Assume that $M = 4$. Local computations show that $\widetilde{F}_P$ has exactly three singular points: they lie on the line

$$\{y = z_2 = 0\}$$

and correspond to the roots of the polynomial $\bar{q}_3(0, 1, t)$. Moreover, these three points are non-degenerate quadratic singularities. Thus the three-fold F_P has terminal singularities, so that the pair (F, F_P) is terminal, and also, canonical, contrary to our assumption.

Assume that $M = 3$. Here $E = \mathbb{P}^2$ and

$$\widetilde{F}_P \cap E = L_+ + L_- = \{y = \pm z_2\}$$

is a pair of lines. The only singularity of the surface $\widetilde{F}_P$ over the point p is the point $p^* = L_+ \cap L_-$ of the intersection of these lines. If $\bar{q}_3(0, z_3) \neq 0$, then p^* is a non-degenerate quadratic singularity, so that its blow up

$$\varphi^* \colon F_P^* \to \widetilde{F}_P$$

gives a non-singular surface F_P^*. If $\bar{q}_3(0, z_3) \equiv 0$, then the exceptional divisor E^* of the blow up φ^* is a pair of lines, $E^* = L_+^* + L_-^*$, so that the only singularity of the surface F_P^* over the point p^* is the point $p^\sharp = L_+^* \cap L_-^*$.

It is easy to check that by the condition $\bar{q}_4(0, z_3) \neq 0$ the point $p^\sharp$ is a non-degenerate quadratic singularity.

Thus in any case the surface F_P is canonical.

Q.E.D. for Lemma 1.5 and Theorem 1.2.

2. Inversion of adjunction

In this section, we formulate the connectedness principle of Shokurov and Kollár and consider its geometric applications. First, we reproduce the well-known proof of the inversion of adjunction which follows from the connectedness principle (Subsection 2.1). After that, we prove a geometric fact which can be looked at as a divisorial analog of the $8n^2$-inequality for the self-intersection of a mobile system (Proposition 2.3 in Subsection 2.2); the proof of divisorial canonicity in Section 1 was based on that fact. Finally, we use the connectedness principle and the inversion of adjunction to study isolated non-canonical singularities of pairs, centred at isolated hypersurface singularities (Subsections 2.3–2.5), that is, maximal singularities, the centre of which is an isolated hypersurface singularity, and obtain an important inequality for the multiplicity of the centre (Proposition 2.7).

2.1. The connectedness principle. Inversion of adjunction. Let X, Z be normal varieties or analytic spaces and $h \colon X \to Z$ a proper morphism with connected fibres and $D = \sum d_i D_i$ a $\mathbb{Q}$-divisor on X.

Theorem 2.1 (the connectedness principle). *Assume that D is effective ($d_i \geq 0$) and the class $-(K_X + D)$ is h-numerically effective and h-big. Let*

$$f \colon Y \overset{h}{\to} X \overset{h}{\to} Z$$

be a resolution of singularities of the pair (X, D). *Set*

$$K_Y = g^*(K_X + D) + \sum e_i E_i.$$

The support of the $\mathbb{Q}$*-divisor* $\sum_{e_i \le -1} e_i E_i$, *that is, the closed algebraic set*

$$\bigcup_{e_i \le -1} E_i,$$

is connected in a neighborhood of any fibre of the morphism f.

Proof. See [**Kol93**, Chapter 17]. It has been also reproduced in the survey [**I01**] and in [**Pukh02b**] for a particular case (in which the arguments follow the same scheme as in [**Kol93**]). Q.E.D.

The connectedness principle has numerous applications, which we will now consider. The first application is Theorem 4.3.4 (inversion of adjunction).

Proof of Theorem 4.3.4. We use the notation of Theorem 4.3.4. Let $D = \sum_{i\in I} d_i D_i$ be an effective $\mathbb{Q}$-divisor, $d_i \in \mathbb{Q}_+$ for all $i \in I$. Since the pair (X, D) is canonical outside the point x, we get the inequality $d_i \le 1$ for all $i \in I$. Replacing D by $\frac{1}{1+\varepsilon} D$ for a small $\varepsilon \in \mathbb{Q}_+$, we may assume that $d_i < 1$ for all $i \in I$.

Let $\varphi\colon \widetilde{X} \to X$ be a resolution of singularities of the pair $(X, D + R)$. Now write

$$K_{\widetilde{X}} = \varphi^*(K_X + D + R) + \sum_{j\in J} e_j E_j - \sum_{i\in I} d_i \widetilde{D}_i - \widetilde{R}, \tag{170}$$

where E_j, $j \in J$, are all exceptional divisors of the morphism φ, $\widetilde{D}_i$ and $\widetilde{R}$ are the strict transforms of the divisors D_i, R on $\widetilde{X}$, respectively. Set

$$b_j = \mathrm{ord}_{E_j} \varphi^* D, \quad a_j = a(E_j, X),$$

$j \in J$. In these notations for $j \in J$ we get $e_j = a_j - b_j - r_j$, where $r_j = \mathrm{ord}_{E_j} \varphi^* R$. Obviously,

$$\varphi^{-1}(x) = \bigcup_{j\in J^+} E_j$$

for some subset $J^+ \subset J$. Recall that by assumption R is a Cartier divisor, containing the point x, which implies (it is a key point) that for $j \in J^+$,

$$r_j = \mathrm{ord}_{E_j} \varphi^* R \ge 1.$$

Furthermore, by assumption, the pair (X, D) is not canonical, but canonical outside the point x. Therefore, among the indices $j \in J^+$ there is an index k such that $a_k < b_k$. For this index we have $e_k < -1$.

Now by the connectedness principle we get: there is an index $l \in J$, such that $e_l < -1$ and

$$E_l \cap \widetilde{R} \neq \emptyset.$$

Now from (170) by the adjunction formula we get

$$K_{\widetilde{R}} = (K_{\widetilde{X}} + \widetilde{R})|_{\widetilde{R}} = \varphi_R^*(K_R + D_R) + (\sum_{j\in J} e_j E_j|_{\widetilde{R}} - \sum_{i\in I} d_i \widetilde{D}_i|_{\widetilde{R}}),$$

where $\varphi_R = \varphi|_{\widetilde{R}}\colon \widetilde{R} \to R$ is the restriction of the sequence of blow ups φ onto R. By what was said, in the last bracket there is at least one prime divisor of the form

$E_l|_{\widetilde{R}}$, where $l \in J^+$, the coefficient at which is strictly less than -1. Q.E.D. for Theorem 4.3.4.

The following version of the inversion of adjunction is useful.

Proposition 2.1. *Let $x \in X$ be a germ of a smooth variety, D an effective $\mathbb{Q}$-divisor, the pair (X, D) is not canonical, but canonical outside the point x, that is, the point x is an isolated centre of non-canonical singularities of the pair. Let $R \ni x$ be a non-singular divisor where $T_x R$ is a hyperplane of general position in $T_x X$. Then the pair (R, D_R) is not log canonical, but canonical outside the point x.*

Proof. In the notation of the proof of Theorem 4.3.4, the index k, which realizes a non-log canonical singularity of the pair (R, D_R), lies in J^+: by the assumption of general position, the divisor R does not contain any centres of singularities of the pair (X, D) outside the point x. Q.E.D. for the proposition.

Here is one more application of the connectedness principle.

Proposition 2.2. *Let $x \in X$ be a germ of a smooth variety, D an effective $\mathbb{Q}$-divisor, the pair (X, D) is not canonical at the point x, but canonical outside that point. Let $\lambda\colon X^+ \to X$ be the blow up of the point x, $E = \lambda^{-1}(x) \subset X^+$ the exceptional divisor, D^+ and R^+ the strict transforms of the divisors D and R, respectively. Furthermore, let $\mu\colon \widetilde{X} \to X^+$ be a resolution of singularities of the pair $(X^+, D^+ + R^+)$,*

$$\varphi = \lambda \circ \mu\colon \widetilde{X} \to X$$

the composite map. Now write

$$K_{\widetilde{X}} = \varphi^*(K_X + D + R) + \sum_{j \in J} e_j E_j - \sum_{i \in I} d_i \widetilde{D}_i - \widetilde{R}, \tag{171}$$

where E_j, $j \in J$, are all exceptional divisors of the morphism φ, $\widetilde{D}_i$ and $\widetilde{R}$ are the strict transforms of the divisors D_i and R on $\widetilde{X}$, respectively.

Then the following alternative takes place:

(1) *either* $\operatorname{mult}_x D > \dim X$,

(2) *or the set*

$$\mu\Big(\bigcup_{b_j > a_j + 1} E_j\Big) \subset E$$

is connected.

Proof. By the assumptions the claim follows immediately from the connectedness principle. Q.E.D.

2.2. Further applications of the connectedness principle. First, let us show the following useful fact.

Proposition 2.3. *Assume that the pair (X, D) is the same as in Proposition 2.2, $\lambda\colon X^+ \to X$ is the blow up of the point x, $E = \lambda^{-1}(x) \subset X^+$ is the exceptional divisor, D^+ the strict transform of the divisor D. Then the following alternative takes place:*

1) *either* $\operatorname{mult}_x D > 2$,

2) *or there is a hyperplane $B \subset E$, which is uniquely determined by the pair (X, D) such that the following inequality holds:*

$$\operatorname{mult}_x D + \operatorname{mult}_B D^+ > 2. \tag{172}$$

Proof. Canonicity is stronger than log canonicity. Therefore one can apply inversion of adjunction (in the form of Proposition 2.1) several times, subsequently restricting the pair (X, D) onto smooth subvarieties

$$R_1 \supset R_2 \supset \cdots \supset R_k,$$

where $R_1 \subset X$ is a smooth divisor, $R_{i+1} \subset R_i$ is a smooth divisor, $x \in R_k$ and $R_k \not\subset \operatorname{Supp} D$. All the pairs

$$(R_i, D|_{R_i})$$

are not log canonical at the point x. Thus Proposition 2.1 holds for a generic smooth germ $R \ni x$ of codimension $k \le \dim X - 1$. In particular, it holds for a general surface $S \ni x$. Thus the pair

$$(S, D_S = D|_S)$$

has at the point x an isolated (for a general S) non-log canonical singularity. Let us consider the two-dimensional case more closely. Let $x \in S$ be a germ of a smooth surface, $C \subset S$ a germ of an effective (possibly reducible) curve, $x \in C$. Consider a sequence of blow ups

$$\varphi_{i,i-1}\colon S_i \to S_{i-1},$$

$S_0 = S$, $i = 1, \ldots, N$, $\varphi_{i,i-1}$ blows up a point $x_{i-1} \in S_{i-1}$, $E_i = \varphi_{i,i-1}^{-1}(x_{i-1}) \subset S_i$ is the exceptional line. For $i > j$ set

$$\varphi_{i,j} = \varphi_{j+1,j} \circ \cdots \circ \varphi_{i,i-1}\colon S_i \to S_j,$$

$\varphi = \varphi_{N,0}$, $\widetilde{S} = S_N$. We assume that the points x_i lie one over another, that is, $x_i \in E_i$, and that $x_0 = x$, so that all the points x_i, $i \ge 1$, lie over x:

$$\varphi_{i,0}(x_i) = x \in S.$$

Let Γ be the graph with the vertices $1, \ldots, N$ and oriented edges (arrows) $i \to j$, that connect the vertices i and j if and only if $i > j$ and

$$x_{i-1} \in E_j^{i-1},$$

where for a curve $Y \subset S_j$ its strict transform on S_a, $a \ge j$, is denoted by the symbol Y^a. Assume that the point x is the centre of an isolated non-log canonical singularity of the pair $(S, \frac{1}{n}C)$ for some $n \ge 1$. This means that for some exceptional divisor $E \subset \widetilde{S}$ the log Noether-Fano inequality

$$\nu_E(C) = \operatorname{ord}_E \varphi^* C > n(a_E + 1), \tag{173}$$

holds, where a_E is the discrepancy of E. Without loss of generality, we may assume that $E = E_N$ is the last exceptional divisor.

As usual, for $i > j$ let the symbol p_{ij} denote the number of paths in the graph Γ from the vertex i to the vertex j, for $i < j$ set $p_{ij} = 0$, as always $p_{ii} = 1$. In terms of the numbers p_{ij} the log Noether-Fano inequality (173) takes the traditional form

$$\sum_{i=1}^{N} p_{Ni}\mu_i > n\Big(\sum_{i=1}^{N} p_{Ni} + 1\Big), \tag{174}$$

where $\mu_i = \operatorname{mult}_{x_{i-1}} C^{i-1}$.

Proposition 2.4. *Either $\mu_1 > 2n$ (that is, the first exceptional divisor $E_1 \subset S_1$ already gives a non-log canonical singularity of the pair $(S, (1/n)C)$), or $N \geq 2$ and the following inequality holds:*

$$\mu_1 + \mu_2 > 2n.$$

Proof. If $N = 1$, then $\mu_1 > 2n$ by means of log Noether-Fano inequality. Assume that $\mu_1 \leq 2n$, then $N \geq 2$. Obviously, $\mu_1 > n$. If $\mu_2 \geq n$, then $\mu_1 + \mu_2 > 2n$, as we claim. So assume that $\mu_2 < n$. Then for each $i \in \{2, \dots, N\}$ we have $\mu_i \leq \mu_2 < n$ (since the point x_{i-1} lies over x_1). Therefore from the inequality (174) we get

$$p_{N1}(\mu_1 - n) + \sum_{i=2}^{N} p_{Ni}(\mu_2 - n) > n.$$

However,

$$p_{N1} = \sum_{j \to 1} p_{Nj} \leq \sum_{i=2}^{N} p_{Ni},$$

such that

$$\sum_{i=2}^{N} p_{Ni}(\mu_1 + \mu_2 - 2n) > n.$$

Therefore $\mu_1 + \mu_2 > 2n$. Q.E.D. for the proposition.

Let us now complete the proof of Proposition 2.3. Consider a general surface $S \ni x$. The pair (S, D_S) is not log canonical, but log canonical outside the point x. By Proposition 2.4, either $\operatorname{mult}_x D_S > 2$, but in this case $\operatorname{mult}_x D > 2$, so that the first of the two cases of Proposition 2.3 takes place, or the pair (S^+, D_S^+) (that is, the strict transform of the pair (S, D_S) on X^+) is not log canonical, but log canonical outside some proper closed connected subset

$$Z_S \subset E_S = E \cap S^+ \cong \mathbb{P}^1.$$

Obviously, Z_S is a point $y_S \in E_S$. Since the surface S is general, there is a hyperplane $B \subset E$ such that

$$y_S = B \cap S^+.$$

By Proposition 2.4, the inequality

$$\operatorname{mult}_x D_S + \operatorname{mult}_{y_S} D_S^+ > 2$$

holds. This immediately implies the inequality (172) and Proposition 2.3. Q.E.D.

2.3. Isolated hypersurface singularities. As one more application of the connectedness principle, consider a germ $x \in X$ of an isolated terminal singularity with the following properties. Let

$$\varphi\colon X^+ \to X$$

be the blow up of the point x, $E = \varphi^{-1}(x)$ the exceptional divisor, which is irreducible and reduced. The varieties X, X^+ and E have $\mathbb{Q}$-factorial terminal singularities. Let $\delta = a(E, X)$ be the discrepancy of E, D an effective $\mathbb{Q}$-divisor on X, D^+ its strict transform on X^+. Define the number $\nu_E(D)$ by the formula

$$\varphi^* D = D^+ + \nu_E(D) E.$$

Proposition 2.5. *Assume that the pair (X, D) is not canonical at the point x, which is an isolated centre of a non-canonical singularity of this pair. Assume also that for some integer $k \geq 1$ the inequality*

$$\nu_E(D) + k \leq \delta \tag{175}$$

holds. Then the pair (X^+, D^+) is not log canonical and there is a non-log canonical singularity $\widetilde{E} \subset \widetilde{X}$ of that pair (where $\widetilde{X} \to X^+$ is some model), of which

$$\text{centre}(\widetilde{E}, X^+) \subset E$$

is of dimension $\geq k$.

Proof. Assuming $X \subset \mathbb{P}^N$ to be projectively embedded, consider a *generic* linear subspace $P \subset \mathbb{P}^N$ of codimension k, containing the point x. Let Λ_P be the linear system of hyperplanes, containing P, and Λ the corresponding linear system of sections of the variety X. Let $\varepsilon > 0$ be a sufficiently small rational number of the form $\frac{1}{K}$ and

$$\{H_I \,|\, i \in I\} \subset \Lambda$$

a set of $\sharp I = Kk$ generic divisors. Set

$$R = D + \sum_{i \in I} \varepsilon H_i,$$

and let R^+ be the strict transform of R on X^+.

Obviously, the pair (X^+, D^+) is not log canonical. The centre of any of its non-log canonical singularities is contained in E. Furthermore, being non-log canonical is an open property, so that, slightly decreasing the coefficients in D, we may assume that the strict version of the inequality (175) holds, that is, $\nu_E(D) + k < \delta$ (whereas other assumptions still hold).

Now consider the pair (X^+, R^+) (we still assume that $X \ni x$ is a germ, so that all constructions are local in a neighborhood of the point x). It is non-log canonical, and all its non-log canonical singularities are non-log canonical singularities of the pair (X^+, D^+), with the exception of one additional singularity, the germ $(P \cap X)^+$ of the section of X by the plane P, that is, the base set of the system Λ. By the strict version of the inequality (175), the class $-(K_{X^+} + R^+)$ is obviously φ-nef and φ-big, so that, applying the connectedness principle (to $X = X^+$, $Z = X$, $h = \varphi$, $D = R^+$), we conclude that the union of the centres of non-log canonical singularities of the pair (X^+, R^+) on X^+ is connected. Since P is generic, this is only possible if $(P \cap X)^+$ intersects some centre of a non-log canonical singularity of the pair (X^+, D^+), which should be of dimension at least k. Q.E.D. for the proposition.

The fact which we have just proven will be applied to our case of a hypersurface singularity $x \in X$ with a smooth exceptional divisor.

2.4. Singularities of pairs on a smooth hypersurface. Let $X \subset \mathbb{P}^N$ be a smooth hypersurface of degree $m \in \{2, \ldots, N-1\}$, $D \in |\,lH_X\,|$ an effective divisor, which is cut out on X by a hypersurface of degree $l \geq 1$. (So that H_X is the class of a hyperplane section of X.) The following fact and its proof are well known.

Proposition 2.6. *For any $n \geq l$ the pair $(X, \frac{1}{n}D)$ is log canonical.*

Proof. We may consider the case $n = l$. Assume the converse: the pair $(X, \frac{1}{n}D)$ is not log canonical. Since for any curve $C \subset X$ the inequality $\text{mult}_C D \leq$

n holds, the centre of a non-log canonical singularity of the pair $(X, \frac{1}{n}D)$ can only be a point. Let $x \in X$ be such a point. Consider now a general projection $\pi\colon \mathbb{P}^N \dashrightarrow \mathbb{P}^{N-1}$. Its restriction onto X is a finite morphism $\pi_X\colon X \to \mathbb{P}^{N-1}$ of degree m, which is an analytic isomorphism at the point x, and one may assume that

$$\pi_X^{-1}(\pi_X(x)) \cap \operatorname{Supp} D = \{x\}.$$

This implies that the germ of the pair $(X, \frac{1}{n}D)$ at the point x and the germ of the pair $(\mathbb{P}^{N-1}, \frac{1}{n}\pi(D))$ at the point $\pi(x)$ are analytically isomorphic. In particular, the point $\pi(x)$ is an isolated centre of a non-log canonical singularity of the pair $(\mathbb{P}^{N-1}, \frac{1}{n}\pi(D))$. However, this is impossible.

Being non-log canonical is an open property, so that for a rational number $s < n^{-1}$, sufficiently close to n^{-1}, the pair

$$(\mathbb{P}^{N-1}, s\pi(D))$$

still has the point $\pi(x)$ as an isolated centre of a non-log canonical singularity. Let $P \subset \mathbb{P}^{N-1}$ be a hyperplane, not containing the point $\pi(x)$. By the inequality $smn+1 < N$ the $\mathbb{Q}$-divisor $-(K_{\mathbb{P}^{N-1}} + s\pi(D) + P)$ is ample, so that one may apply to the pair

$$(\mathbb{P}^{N-1}, s\pi(D) + P)$$

the connectedness principle of Shokurov and Kollár (in the notation of Theorem 2.1, $X = \mathbb{P}^{N-1}$, Z is a point, for the $\mathbb{Q}$-divisor D we take $s\pi(D) + P$, the conditions of Theorem 2.1 are satisfied in a trivial way by what was said above) and obtain a contradiction: the point $\pi(x)$ is an isolated centre of a non-log canonical singularity and the divisor P comes into the $\mathbb{Q}$-divisor $s\pi(D) + P$ with the coefficient one, however, $\pi(x) \notin P$, so that the connectedness is violated. A contradiction. Q.E.D. for Proposition 2.6.

2.5. The weak local inequality for an isolated hypersurface singularity. Let $x \in X$ be a germ of isolated hypersurface terminal singularity. More precisely, if $\varphi\colon X^+ \to X$ is the blow up of the point x, $\varphi^{-1}(x) = E \subset X^+$ is the exceptional divisor, we assume that X^+ and E are smooth, whereas E is isomorphic to a smooth hypersurface of degree $\mu = \operatorname{mult}_o V$ in $\mathbb{P}^M$.

Furthermore, let $D \ni x$ be a germ of a prime divisor, $D^+ \subset X^+$ its strict transform, $D^+ \sim -\nu E$ for $\nu \in \mathbb{Z}_+$, so that the following equality holds:

$$\operatorname{mult}_o D = \mu\nu.$$

Proposition 2.7. *Assume that the pair $(X, \frac{1}{n}D)$ is not canonical at the point x, which is an isolated centre of a non-canonical singularity of that pair. Then the following inequality holds*:

$$\nu > n. \tag{176}$$

Proof. Assume the converse: $\nu \leq n$. Then the pair $(X^+, \frac{1}{n}D^+)$ is not canonical and, moreover, the centre of any non-canonical singularity of this pair (that is, of any maximal singularity of the divisor D^+) is contained in the exceptional divisor E. By the inversion of adjunction the pair $(E, \frac{1}{n}D_E^+)$, where $D_E^+ = D^+\,|\,_E$, is not log canonical. Let $H_E = -E\,|\,_E$ be the generator of the Picard group $\operatorname{Pic} E$, that is, the hyperplane section of E with respect to the embedding $E \subset \mathbb{P}^M$. We get

$$D_E^+ \sim -\nu E\,|\,_E = \nu H_E.$$

Since $\nu \le n$, the non-log canonicity of the pair $(E, \frac{1}{n} D_E^+)$ contradicts Proposition 2.6. Q.E.D.

As one more application of the connectedness principle, consider the following local situation. Let $x \in X$ be a germ of a quadratic singularity, $\dim X \ge 3$. Let us blow up the point x:

$$\lambda \colon X^+ \to X,$$

and denote by the symbol E the exceptional divisor $\lambda^{-1}(x)$, which we consider as a quadric hypersurface $E \subset \mathbb{P}^{\dim X}$. Furthermore, let D be an effective $\mathbb{Q}$-Cartier divisor on the variety X and D^+ its strict transform on X^+. Assuming the exceptional quadric E to be irreducible, define the number $\beta \in \mathbb{Q}_+$ by the relation

$$D^+ \sim \lambda^* D - \beta E.$$

Proposition 2.8. *Assume that the rank of the quadric hypersurface E is at least 4 and the pair (X, D) has the point x as an isolated centre of a non-canonical singularity, that is, it is non-canonical, but canonical outside the point x. Then the inequality $\beta > 1$ holds.*

Proof. If $\dim X = 3$, then by assumption the point $x \in X$ is a non-degenerate quadratic singularity, and this fact is well known. (If $\beta \le 1$, then the pair (X^+, D^+) is non-canonical, so that by inversion of adjunction the pair (E, D_E^+) is not log canonical, but $E \cong \mathbb{P}^1 \times \mathbb{P}^1$ and D_E^+ is an effective curve of bidegree (β, β), which is impossible.) If $\dim X \ge 4$, then, restricting D onto a generic hyperplane section $Y \ni x$ of the variety X with respect to some embedding $X \hookrightarrow \mathbb{P}^N$, and repeating this procedure $\dim X - 3$ times, we reduce the problem (by inversion of adjunction) to the already considered case $\dim X = 3$. Proof of the proposition is complete. Q.E.D.

3. Fano varieties with elementary singularities

In this section we show that generic Fano hypersurfaces and double spaces remain divisorially canonical when they acquire elementary singularities. This is especially important for applications to birational rigidity of Fano fibre spaces over $\mathbb{P}^1$ as every fibre space contains singular fibres. In the case of double spaces, it is sufficient to assume that the singularities are non-degenerate quadratic points; in the case of hypersurfaces, stronger assumptions about the singularities are needed, including the regularity conditions.

3.1. Singular Fano hypersurfaces and double spaces. It is natural by itself to try to extend Theorems 1.1 and 1.2 to singular varieties. However, there is a more compelling reason for this task: let us consider a standard Fano fibre space $\pi \colon V \to \mathbb{P}^1$, the fibres F_t, $t \in \mathbb{P}^1$, of which belong to a given family $\mathcal{F}$ of Fano varieties. The following claim is obvious.

Proposition 3.1. *Assume that every fibre F_t, $t \in \mathbb{P}^1$, is a factorial terminal Fano variety, satisfying the condition $\operatorname{Pic} F_t = \mathbb{Z} K_{F_t}$ and the conditions (L) and*

(M). Then for every mobile linear system $\Sigma \subset |-nK_V + lF|$ with $l \in \mathbb{Z}_+$ its virtual and actual thresholds of canonical adjunction coincide:

$$c_{\rm virt}(\Sigma) = c(\Sigma) = n.$$

In particular, if the fibre space $V/\mathbb{P}^1$ satisfies the K-condition, that is, $-K_V \notin \operatorname{Int} A^1_{\rm mov} V$, then $V/\mathbb{P}^1$ is birationally superrigid.

In order to apply Proposition 3.1, we need a Fano fibre space, *every fibre* of which is divisorially (log) canonical. Since there always are (finitely many) singular fibres, we need divisorial (log) canonicity for (generic) singular Fano varieties. This motivates the main results of this section, Theorems 3.1 and 3.2.

The first one is easy.

Theorem 3.1. *The double space $\sigma\colon V \to \mathbb{P}^M$, $M \geq 3$, branched over a hypersurface $W \subset \mathbb{P}^M$ of degree $2M$ with isolated non-degenerate quadratic singularities, satisfying the regularity conditions* (R2) *(that were used in Theorem 1.2) at every smooth point, satisfies the condition (C).*

Proof is not hard: the theorem follows immediately from Proposition 2.7 (and from the fact that the multiplicity of an irreducible subvariety Y at every point does not exceed its anticanonical degree $(Y \cdot (-K_V)^{\rm dim Y})$.

Let us give a formal proof. Assume that the pair $(V, \frac{1}{n}D)$ is not canonical. By what was proven in Section 1, only a singular point o can be the centre of a non-canonical singularity. The divisor $D \sim -nK_V$ can be assumed to be irreducible. By Proposition 2.7, we get the inequality

$$\operatorname{mult}_o D > 2n,$$

however, the anticanonical degree of the divisor D is

$$\deg D = (D \cdot (-K_V)^{\rm dim V - 1}) = 2n.$$

Therefore, $\operatorname{mult}_o D > \deg D$, which is impossible. This contradiction proves the theorem. Q.E.D.

Note that, using the arguments of Subsection 1.7, it is easy to check that a generic hypersurface $W \in \mathcal{W}$ with a fixed singular point $o \in \mathbb{P}^M$ is regular at every smooth point, so that varieties described in the theorem do exist and form a non-empty Zariski subset in the space of singular Fano double spaces.

Now let us consider Fano hypersurfaces:

$$V = V_{M+1} \subset \mathbb{P} = \mathbb{P}^{M+1}.$$

(Note that our notations have slightly changed: M is replaced by $M+1$. This is to simplify some formulas in the computations below.) For the dimension, we assume that $M = \dim V \geq 8$. This restriction is connected with the techniques of the proof of Theorem 3.2 below: for the smaller values of M the arguments do not work. Let us formulate the regularity condition for a non-degenerate double point $o \in V$.

Let $z_1, \dots, z_{M+1}$ be a system of affine coordinates on the space $\mathbb{P}$ with the origin at the point o,

$$f = q_2 + q_3 + \dots + q_{M+1} = 0$$

the equation of the hypersurface V, decomposed into homogeneous components, where $q_2(z_*)$ is a non-degenerate quadratic form. We say that V is *regular* at the point o, if the following conditions hold:

(R2.1) the sequence $q_2, \dots, q_{M+1}$ is regular in $\mathcal{O}_{o,\mathbb{P}}$, that is, the system of equations

$$q_2 = \dots = q_{M+1} = 0$$

defines a finite set of points in $\mathbb{P}^M$ (corresponding to the lines in $\mathbb{P}$, passing through the point o and lying on V),

(R2.2) for any $k \in \{2, 3, 4, 5\}$ and any linear subspace $P \subset \mathbb{P}$ of codimension two, containing the point o, the closed algebraic set

$$V \cap P \cap \{q_2 = 0\} \cap \dots \cap \{q_k = 0\} \tag{177}$$

is irreducible and has multiplicity precisely $(k+1)!$ at the point o.

Remark 3.1. The condition that the multiplicity of the set (177) at the point o is $(k+1)!$, in terms of the polynomials q_i means that the intersection of the closed set

$$\{q_2 = 0\} \cap \dots \cap \{q_{k+1} = 0\} \subset \mathbb{P}^M$$

with any linear subspace of codimension two is of codimension $k+2$ in $\mathbb{P}^M$ (and of degree $(k+1)!$).

The methods of Section 1 combined with the methods of [**Pukh02a**] make it possible to show that a generic hypersurface $V \subset \mathbb{P}$ with a fixed double point $o \in \mathbb{P}$ satisfies the conditions (R1.1–R1.3) at every point $x \neq o$, $x \in V$. The main fact, that is, that the condition (R1.1) holds, was shown in Chapter 3. The additional conditions (R1.2–R1.3) are checked directly: the presence of a fixed singularity does not affect the arguments of Section 1. Finally, the fact that a generic hypersurface $V \ni o$, singular at the point o, satisfies at that point the conditions (R2.1) and (R2.2), is obvious. Therefore, a generic hypersurface $V \in \mathcal{F}$ with a fixed double point $o \in \mathbb{P}$ is regular at every point, in the sense of the conditions (R1.1–R1.3) or (R2.1–R2.2), respectively.

The main result of this section is

Theorem 3.2. *Assume that the Fano hypersurface $V \subset \mathbb{P}$ of degree $M+1 \geq 9$ is regular at every point, smooth or regular. Then the variety V satisfies the condition (C).*

Proof of Theorem 3.2 breaks into two main parts: the local and the global ones. As usual, the converse is assumed so that we need to obtain a contradiction. In the local part we study a germ of a quadratic singularity. In the global part we show that the local information obtained at the previous step is incompatible with the global properties of the Fano hypersurface under consideration, which gives the desired contradiction and completes the proof.

3.2. Effective divisors on quadrics. Let $Q \subset \mathbb{P}^M$ be a non-degenerate quadric, $H_Q \in \operatorname{Pic} Q$ the class of a hyperplane section, and $B \subset Q$ an irreducible subvariety.

Definition 3.1. We say that an effective divisor D on Q satisfies *the condition $H(n)$ (with respect to B)*, where $n \geq 1$ is a fixed integer, if for any point of general position $p \in B$ there exists a hyperplane $F(p) \subset E_p$ in the exceptional divisor $E_p = \varphi_p^{-1}(p)$ of the blow up $\varphi_p \colon Q_p \to Q$ of the point p, for which the inequality

$$\operatorname{mult}_p D + \operatorname{mult}_{F(p)} \widetilde{D} > 2n, \tag{178}$$

holds, where $\widetilde{D} \subset Q_p$ is the strict transform of the divisor D.

Note that the divisor D is not assumed to be irreducible and the number n does not depend on the point p. The hyperplane $F(p)$ depends algebraically on the point p; this is assumed everywhere in the sequel without special comments. Let $l \geq 1$ be the degree of a hypersurface in $\mathbb{P}^M$ that cuts out D on Q, that is,

$$D \sim lH_Q.$$

Proposition 3.2. *Assume that* $\dim B \geq 3$. *Assume furthermore that the effective divisor* D *satisfies the condition* $H(n)$ *with respect to* B. *Then the following alternative takes place: either*

1) *the inequality* $l > 2n$ *holds (this case will be referred to as the simple one), or*

2) *there exists a hyperplane section* $Z \subset Q$, *containing entirely the subvariety* B, *such that for a point of general position* $p \in B$ *in the notations above,*

$$F(p) = \widetilde{Z} \cap E_p,$$

where $\widetilde{Z} \subset Q_p$ *is the strict transform, whereas* Z *comes into the divisor* D *with the multiplicity*

$$a > 2n - l$$

(that is, the presentation $D = aZ + D^*$ *takes place, where* D^* *does not contain* Z *as a component; this case we will call the hard one).*

Proof. Let us assume at once that the simple case is not realized, that is, $l \leq 2n$. In the notation above for a point of general position $p \in B$ let $\Lambda_p \subset |H_Q|$ be the pencil of hyperplane sections, the strict transform of which $\widetilde{\Lambda}_p$ on Q_p cuts out on E_p the hyperplane $F(p)$, that is,

$$\widetilde{\Lambda}_p \cap E_p = F(p).$$

The exceptional divisor E_p is the projectivization of the tangent space $T_pQ \cong \mathbb{C}^{M-1}$. Let

$$[F(p)] \subset T_pQ$$

be the hyperplane, the projectivization of which is $F(p)$. Consider now T_pQ as an *embedded* tangent space (in some affine chart $\mathbb{C}^M \subset \mathbb{P}^M$) and let $\overline{T_pQ} \subset \mathbb{P}^M$ be its closure (that is, the hyperplane in $\mathbb{P}^M$, tangent to Q at the point p). Respectively, let

$$\overline{F(p)} = \overline{[F(p)]} \subset \overline{T_pQ}$$

be the closure of the subspace $[F(p)]$. This is a linear subspace in $\mathbb{P}^M$ of codimension two. It is easy to see that the base set (and subscheme) of the pencil Λ_p is

$$\operatorname{Bs} \Lambda_p = \overline{F(p)} \cap Q.$$

Denote this subset by the symbol $\Theta(p)$. Set

$$Z = \overline{\bigcup_{p \in B} \Theta(p)}$$

(where the union is taken over the points of some open subset of B, whereas the overline means the closure).

Note that $\Theta(p)$ is a quadric in $\overline{F(p)}$ with at least one singular point p and at most a line of double points (containing p). Since $\dim B \geq 3$, for a pair of distinct points of general position $p_1 \neq p_2$ we get $\Theta(p_1) \neq \Theta(p_2)$, which implies that either $Z = Q$ or Z is a prime divisor on Q.

Let $R \in \Lambda_p$ be a general element of the pencil, $D_R = D\,|\,_R$ the restriction of the divisor D. By the inequality (178) we get

$$\operatorname{mult}_p D_R > 2n.$$

The variety R is smooth at the point p. We get the presentation

$$D_R = a\Theta(p) + D_R^*,$$

where D_R^* does not contain $\Theta(p)$ as a component, $a \in \mathbb{Z}_+$ is some non-negative integer. Since Λ_p cuts out $\Theta(p)$ with the multiplicity one, we get

$$a = \operatorname{mult}_{\Theta(p)} D.$$

In particular, if $a \geq 1$, then Z is a divisor on Q.

Lemma 3.1. *The inequality $a \geq 1$ holds.*

Proof. Assume the converse: $a = 0$. It is easy to see that

$$\Theta(p) = R \cap T_p R.$$

The intersection $\Theta(p) \cap D_R$ is of codimension two on R, so that the effective cycle $(\Theta(p) \circ D_R)_R$ is well defined. Therefore,

$$2l = \deg(\Theta(p) \circ D_R) \geq \operatorname{mult}_p(\Theta(p) \circ D_R) > 4n,$$

so that, contrary to the assumption above, the inequality $l > 2n$ holds. The contradiction proves the lemma. Q.E.D.

Now we get

$$D = aZ + D^*,$$

where the divisor D^* does not contain Z as a component. Set

$$Z \sim l_Z H_Q, \quad D^* \sim l^* H_Q,$$

so that the following equality holds:

$$l = al_Z + l^*.$$

Lemma 3.2. *$Z \subset Q$ is a hyperplane section: $l_Z = 1$.*

Proof. Since Z is a prime divisor, the set

$$\Delta = \bigcup_{p \in B} \overline{F(p)}$$

(where the union is taken over an open subset of B) cannot be dense in $\mathbb{P}^M$. Since $\dim B \geq 3$, in this union an, at least, two-dimensional family of linear subspaces of codimension two is present. By the following elementary lemma, the closure of Δ in $\mathbb{P}^M$ is a hyperplane. Q.E.D. for the lemma.

Lemma 3.3. *If the surface $S \subset \mathbb{P}^3$ contains a two-dimensional family of lines, then S is a plane.*

Proof. There is a one-dimensional family of lines through a generic point of the surface S (actually, just one point is sufficient). The case of a cone is obvious. Q.E.D. for the lemma.

By Lemma 3.2, for a general point $p \in B$ we get

$$\Theta(p) = Z \cap R$$

(the left-hand side is contained in the right-hand side, however, in the right-hand side we have a section of Q by a linear subspace of codimension two). Therefore, $D^*_R = D^* \mid_R$. Obviously,

$$\deg D^+_R = 2(l - a) = 2l^*.$$

Arguing as in the proof of Lemma 3.1, that is, intersecting D^*_R with Z (or with $\Theta(p)$ inside R), we obtain the estimate

$$2 \operatorname{mult}_p D^*_R \leq \deg D^*_R,$$

which in terms of our integral parameters gives the estimate

$$2a + l^* = a + l > 2n.$$

The divisor Z contains B and has at most one singular point. Therefore, for a general point $p \in B$ the divisor Z is non-singular at p and thus

$$F(p) = \widetilde{Z} \cap E_p,$$

as we claimed. Q.E.D. for Proposition 3.2.

Remark 3.2. If $\dim B \geq 4$, then the same arguments, word for word, give the claim of Proposition 3.2 for the case when Q is a cone over a non-degenerate quadric.

3.3. The hyperplane section of high multiplicity. Let $o \in V$ be a germ of a non-degenerate quadratic singularity, $\varphi \colon V^+ \to V$ the blow up of the point o, and $E = \varphi^{-1}(o) \subset V^+$ the exceptional divisor. Let $D \ni o$ be a germ of an effective divisor, where the pair $(V, \frac{1}{n}D)$ has the point o as an isolated centre of a non-canonical singularity. Let $D^+ \subset V^+$ be the strict transform of D on V^+, $D^+ \sim -lE$. Assume that $l \leq 2n$, so that the pair $(V^+, \frac{1}{n}D^+)$ is not log canonical. Let $S \subset E$ be a centre of a non-log canonical singularity, which has the maximal dimension (in particular, S is an isolated centre of a non-log canonical singularity), $\dim S \geq 3$. In particular, the inequality

$$\operatorname{mult}_S D^+ > n, \tag{179}$$

holds. The following fact is true.

Proposition 3.3. *One of the following two cases takes place:*

1) *S is a hyperplane section of the quadric E,*

2) *there is a hyperplane section $Z \supset S$ of the quadric E, satisfying the inequality*

$$\operatorname{mult}_Z D^+ > \frac{2n - l}{3}. \tag{180}$$

Proof. If $S \subset E$ is a prime divisor, then by the inequality (179) we get $l > nl_S$, where $S \sim l_S H_E$ and $H_E = -E \mid_E$ is the hyperplane section of the quadric E. Since by assumption $l \leq 2n$, this implies that $l_S = 1$, that is, S is a hyperplane section (case 1)).

Assume that $\operatorname{codim}_E S \geq 2$, that is, the case 1) is not realized. Since the pair $(V^+, \frac{1}{n}D)^+$ is not log canonical at S, for a generic point $p \in S$ there is a hyperplane $\Pi(p) \subset E_p^+$ in the exceptional divisor $E_p^+ = \varphi_p^{-1}(p)$ of the blow up

$$\varphi_p \colon V_p^+ \to V^+$$

of the point p, satisfying the inequality

$$\operatorname{mult}_p D^+ + \operatorname{mult}_{\Pi(p)} D_p^+ > 2n, \tag{181}$$

where $D_p^+ \subset V_p^+$ is the strict transform with respect to the blow up of the point p. Let $E(p) \subset V_p^+$ be the strict transform of the exceptional quadric.

Lemma 3.4. *For a point p of general position,*

$$\Pi(p) \neq E(p) \cap E_p^+.$$

Proof. Assume the converse. Then for the restriction $D_E^+ = (D^+ \circ E)$ we get the inequality

$$\operatorname{mult}_p D_E^+ > 2n,$$

which is true for almost all points $p \in S$. Since the point p runs through a set of positive dimension, for a general point the divisor D_E^+ does not contain the hyperplane section $E \cap T_p E$ as a component. This, as in the proof of Lemma 3.1, immediately implies that $l > 2n$. This is a contradiction. Q.E.D. for the lemma.

For a point of general position $p \in S$ set

$$F(p) = \Pi(p) \cap E(p).$$

The following inequalities hold:

$$\operatorname{mult}_p D^+ + \operatorname{mult}_{F(p)} D_p^+ > 2n, \tag{182}$$

which is weaker than (181), and

$$\operatorname{mult}_p D_E^+ + \operatorname{mult}_{F(p)} \widetilde{D}_E^+ > 2n, \tag{183}$$

which follows from (182), where $\widetilde{D}_E^+ \subset E(p)$ is the strict transform. All geometric objects, participating in the inequality (183), are defined in terms of the quadric E, that is, they do not require any involvement of V^+ and V_p^+.

Let us apply Proposition 3.2 to the divisor D_E^+ on the quadric E. By our assumptions, the simple case does not realize, so that there exists a hyperplane section $Z \supset S$, cutting out the hyperplane $F(p)$ on $E_p = E(p) \cap E_p^+$, that is, $F(p) = \widetilde{Z} \cap E_p$. The section Z comes into the divisor D_E^+ with a multiplicity strictly higher than $2n - l$. In particular, $\operatorname{mult}_Z D^+ > 0$.

Now let us prove the inequality (180). Both the assumptions and all claims of Proposition 3.3 are local at the point o. A non-degenerate quadratic singularity is analytically equivalent to the germ of the quadric

$$\{z_1^2 + \cdots + z_{M+1}^2 = 0\} \subset \mathbb{C}^{M+1},$$

so that one can assume that the divisor D is given by an equation

$$f = q_l(z_*) + q_{l+1}(z_*) + \ldots,$$

where the quadric $q(z_*)=\sum_{i=1}^{M+1}z_i^2$ divides none of the polynomials q_i (if $q_i\neq 0$). The affine coordinates z_* can be looked at as homogeneous coordinates on the exceptional divisor of the blow up of the origin

$$o=(0,\dots,0)\in\mathbb{C}^{M+1},$$

and then $\{q=0\}\subset\mathbb{P}^M$ is precisely the exceptional quadric E. The divisor D_E^+ is given by the equation

$$q_l\,|\,_E=0.$$

Let $\lambda(z)=0$ be the equation of the hyperplane section $Z\subset E$. In terms of the coordinates z_* Proposition 3.2 asserts that for some $a_l>2n-l$,

$$q_l=\lambda^{a_l}g+qw,$$

where $g(z_*)$ and $w(z_*)$ are homogeneous polynomials of the corresponding degrees. Replacing q_l by q_l-qw, we may assume that λ^{a_l} divides the polynomial q_l. This implies that the strict transform D_l^+ of the divisor

$$D_l=\{q_l\,|\,_V=0\}$$

is of multiplicity $\geq a_l>2n-l$ along Z. However, the divisor D_l is the intersection of the cone V with the cone $\{q_l=0\}$ (the polynomial q_l is homogeneous), so that the inequality (183) for $D_E^+=(D_l^+)\,|\,_E$ implies the inequality (182) for D_l^+, where $D_l^+\subset V^+$ is the strict transform of D_l on V^+. Therefore, the following claim is true:

Both divisors D^+ and D_l^+ satisfy the inequality (182) *for a point of general position $p\in S$.*

By linearity of the inequality (182) (and the obvious fact that

$$\operatorname{mult}_p D_l^+=\operatorname{mult}_p(D_l^+)\,|\,_E=\operatorname{mult}_p D_E^+\geq\operatorname{mult}_p D^+,$$

and similarly for $F(p)$), the divisor

(184) $$\varphi^*(f_{l+1}\,|\,_V=0)-lE,$$

where

$$f_{l+1}=q_{l+1}(z_*)+q_{l+2}(z_*)+\cdots=f-q_l(z_*),$$

again satisfies that inequality. Let $k\geq 1$ be the first index, for which $q_{l+k}(z_*)\,|\,_E\not\equiv 0$,

$$D_{\geq l+k}=\{f_{l+k}\,|\,_V=0\},$$

$f_{l+k}=\sum_{i=l+k}^{\infty}q_i(z_*)$. For the strict transform $D_{\geq l+k}^+$ we get $D_{\geq l+k}^+\sim-(l+k)E$. Then, as we mentioned above, the divisor $D_{\geq l+k}^++kE$ satisfies the inequality (182) at the point of general position $p\in S$. Since

$$\operatorname{mult}_p E=\operatorname{mult}_{F(p)}E(p)=1,$$

the divisor $D_{\geq l+k}^+$ by itself satisfies the inequality

$$\operatorname{mult}_p D_{\geq l+k}^++\operatorname{mult}_{F(p)}(D_{\geq l+k}^+)_p>2(n-k)$$

(where the lower index p for a divisor means, as usual, the strict transform on the variety V_p^+).

This makes it possible to prove the inequality (180) by decreasing induction on $l\leq 2n$. The base of induction is the case $l=2n$: in that case Z comes into D_E^+

with a positive multiplicity, that is, $\mathrm{mult}_Z D^+ > 0$, which is what we claim in 2) for $l = 2n$.

If, in the notation above, the inequality

$$l + k \leq 2(n - k)$$

holds, then by the induction hypothesis we get

$$\mathrm{mult}_Z D^+_{\geq l+k} > \frac{2(n-k) - (l+k)}{3} = \frac{2n-l}{3} - k,$$

so that the divisor (184), obtained by subtraction from the equation f the equation of the divisor D^+_E, contains Z with a multiplicity strictly higher than

$$k + \mathrm{mult}_Z D^+_{\geq l+k} > \frac{2n-l}{3}$$

(since the divisor (184) contains with the multiplicity k the exceptional quadric E), whence, taking into account the inequality $\mathrm{mult}_Z D^+_l > 2n - l$, we obtain the required inequality (180).

If the inequality

$$l + k > 2(n - k)$$

holds, then we cannot apply the induction hypothesis, but in that case the estimate

$$k > \frac{2n-l}{3}$$

holds, so that, arguing as above, we obtain the inequality (180) all the same, simply because the divisor (184) contains the exceptional quadric E with the multiplicity k. Q.E.D. for Proposition 3.3.

Remark 3.3. The claim of Proposition 3.3 remains true for a germ of an elementary degenerate singularity, when E is a cone over a non-degenerate quadric, if we assume that $\dim S \geq 4$. No changes in the proof are needed. One can consider a general hyperplane section of the germ $o \in V$ (containing the point o) and apply Proposition 3.3 to that section.

The local part of the proof of Theorem 3.2 is now complete and we come over to the global part.

3.4. Estimating the multiplicity at the double point. Let $V \subset \mathbb{P} = \mathbb{P}^{M+1}$ be a hypersurface of degree $M + 1$, where $M \geq 8$, $o \in V$ is an isolated quadratic singularity, satisfying the regularity conditions (R2.1–R2.2). Set

$$\varphi \colon V^+ \to V$$

to be the blow up of the point o, $E = \varphi^{-1}(o)$ the exceptional quadric. By the symbol H we denote the class of a hyperplane section of V. We use the notation of Subsection 3.1, in particular, $q_2 = 0$ is the equation of the tangent cone at the point o.

Let $D \sim nH$ be an effective divisor, D^+ its strict transform, $D^+ \sim nH - \nu E$ for some $\nu \geq 1$.

Proposition 3.4. *The inequality $\nu \leq \frac{3}{2}n$ holds.*

Proof. This inequality is linear in the divisor D, so that, without loss of generality, we may assume that D is a prime divisor. Assume the converse: $\nu > \frac{3}{2}n$.

For the first hypertangent divisor $D_2 = \{q_2 \mid_V = 0\}$ we get

$$D_2^+ \sim 2H - 3E,$$

so that D_2 and D are distinct prime divisors. Therefore, the set-theoretic intersection $D \cap D_2$ is of codimension two and the effective cycle $Y = (D \circ D_2)$ is well defined. It satisfies the inequality

$$\frac{\mathrm{mult}_o}{\deg} Y \geq \frac{3}{2} \cdot \frac{2\nu}{n(M+1)} > \frac{9}{2(M+1)}.$$

Now let us consider the standard hypertangent systems

$$\Lambda_k = \left| \sum_{i=o}^{k-2} s_i f_{k-i} = 0 \right|,$$

where $f_j = q_2 + \cdots + q_j$ is a left segment of the equation of V at the point o, s_i runs through the set of all homogeneous polynomials in the coordinates z_* of degree i. By the regularity condition (R2.1), for $r = 2, \ldots, M$ we get

$$\mathrm{codim}\, \mathrm{Bs}\, \Lambda_k = k - 1,$$

so that in the usual way we construct a sequence of irreducible subvarieties

$$Y_2 = Y, Y_3, \ldots, Y_{M-1}$$

of codimension $\mathrm{codim}\, Y_i = i$, where $Y_{i+1} \subset Y_i$ is an irreducible component of the effective cycle $(Y_i \circ D_{i+2})$, $D_j \in \Lambda_j$ is a hypertangent divisor of general position, and for Y_{i+1} we take a component with the maximal ratio $\mathrm{mult}_o / \deg$. The effective cycle $(Y_i \circ D_{i+2})$ is well defined, because $\mathrm{codim}\, \mathrm{Bs}\, \Lambda_{i+2} = i + 1$, so that a general divisor D_{i+2} does not contain Y_i. We obtain the estimate

$$\frac{\mathrm{mult}_o}{\deg} Y_{i+1} \geq \frac{i+3}{i+2} \cdot \frac{\mathrm{mult}_o}{\deg} Y_i,$$

so that for the last subvariety, that is, the curve Y_{M-1}, we obtain the inequality

$$\frac{\mathrm{mult}_o}{\deg} Y_{M-1} \geq \frac{\mathrm{mult}_o}{\deg} Y \cdot \frac{5}{4} \cdot \ldots \cdot \frac{M+1}{M} > \frac{9}{8},$$

which is, of course, impossible. The contradiction proves the proposition. Q.E.D.

3.5. Reduction to a hyperplane section. Let us go back to the **proof of Theorem 3.2.** Assume that the pair $(V, \frac{1}{n}D)$ has the point o as an isolated centre of a non-canonical singularity. By linearity of the Noether-Fano inequality we may assume that D is a prime divisor. According to Proposition 3.4, which we have just proven, $\nu \leq \frac{3}{2}n$, so that we are in the situation of Subsection 3.3. The pair $(V^+, \frac{1}{n}D^+)$ is not log canonical, some subvariety $S \subset E$ is the centre of a non-log canonical singularity of that pair. We assume that S has the maximal dimension among all centres of such singularities, so that $\dim S \leq 4$.

Proposition 3.5. *S is of codimension at least two in the exceptional quadric E.*

Proof. If $S \subset E$ is a prime divisor, then, in accordance with Proposition 3.3, S is a hyperplane section of the quadric E. Let $P \ni o$ be the unique hyperplane in $\mathbb{P}$, cutting out S on E, that is,

$$V_P^+ \cap E = S,$$

where $V_P = V \cap P$, $V_P^+ \subset V^+$ is the strict transform. The pair

$$(V, V_P)$$

is canonical, so that $D \neq V_P$ and the set-theoretic intersection $D \cap V_P$ is of codimension two. For the effective cycle $D_P = (D \circ V_P)$ of codimension two we get

$$\operatorname{mult}_o D_P = \operatorname{mult}_o D + 2 \operatorname{mult}_S D^+ > 4n,$$

so that, arguing as in the proof of Proposition 3.4, we construct a sequence of irreducible subvarieties

$$Y_2, Y_3, \ldots, Y_{M-1},$$

where $\operatorname{codim} Y_i = i$, the subvariety Y_2 is an irreducible component of the cycle D_P, satisfying the inequality

$$\frac{\operatorname{mult}_o}{\deg} Y_2 \geq \frac{\operatorname{mult}_o}{\deg} D_P > \frac{4}{M+1},$$

and obtain a contradiction:

$$\frac{\operatorname{mult}_o}{\deg} Y_{M-1} > \frac{4}{M+1} \cdot \frac{5}{4} \cdot \ldots \cdot \frac{M+1}{M} = 1,$$

which is impossible. Q.E.D. for the proposition.

We conclude that the second case of Proposition 3.3 takes place: there is a hyperplane section $Z \supset S$ of the exceptional quadric E, satisfying the inequality (180). Let $P \subset \mathbb{P}$ be the unique hyperplane, cutting out Z on E (in the same sense as in the proof of Proposition 3.5) and $V_P = V \cap P \neq D$. For the effective cycle $D_P = (D \circ V_P)$ of codimension two we get

$$\operatorname{mult}_o D_P \geq \operatorname{mult}_o D + 2 \operatorname{mult}_Z D^+ > \frac{4}{3}(l+n) > \frac{8}{3} n.$$

Unfortunately, this estimate is insufficient for excluding the maximal singularity in the same way as the one that we used in the proof of Propositions 3.4 and 3.5. (The lower bound for $(\operatorname{mult}_o / \deg) Y_{M-1}$ turns out to be smaller than one, which does not allow one to get a contradiction.) However, we can consider the pair

$$(V_P, \frac{1}{n} D_P).$$

By inversion of adjunction its strict transform

$$(V_P^+, \frac{1}{n} D_P^+)$$

with respect to the blow up of the point o is not log canonical, whereas the subvariety $S \subset E_P = Z$ is a centre of a non-log canonical singularity of that pair. We may assume that

$$\operatorname{mult}_o D_P^+ \leq 4n,$$

otherwise we could obtain a contradiction using word for word the same argument as in the proof of Proposition 3.5. Without loss of generality, we may assume that S is a maximal centre of a non-log canonical singularity of the pair $(V_P^+, \frac{1}{n} D_P^+)$. By Remark 3.3 (and the inequality $\dim S \geq 4$), we may apply Proposition 3.3 to the latter pair and obtain the alternative: either

1) S is a hyperplane section of the quadric E_P, or

2) there is a hyperplane section $Z^+ \supset S$ of the quadric E_P, satisfying the inequality

$$\text{mult}_{Z^*} D_P^+ > \frac{2n - l^*}{3}, \tag{185}$$

where $D_P^+ \sim nH_P - l^* E_P$. Recall that the integer l^* satisfies the inequality

$$l^* > \frac{2}{3}(l+n) > \frac{4}{3}n.$$

Let $R \subset P = \mathbb{P}^M$ be the unique hyperplane, cutting out on E_P in the case 1) the subvariety S, in the case 2) the subvariety Z^*.

Assume that the case 1) takes place. By linearity of the inequalities

$$\text{mult}_o D_P > \frac{8}{3}n \quad \text{and} \quad \text{mult}_S D_P^+ > n \tag{186}$$

in the divisor D_P and by the fact that the divisor $V_R = V_P \cap R$ does not satisfy them, we may assume that the divisor D_P (which is, possibly, reducible) does not contain V_R as a component (otherwise, we delete this component, which could only make both inequalities (186) sharper). For this reason, the intersection $D_P \cap V_R$ is of codimension two on V_P and the effective algebraic cycle

$$D_R = (D_P \circ V_R)$$

is well defined and satisfies the inequality

$$\text{mult}_o D_R > \text{mult}_o D_P + 2\,\text{mult}_S D_P^+ > \frac{14}{3}n. \tag{187}$$

By linearity of the last inequality we may assume that $D_R = Y$ is an irreducible variety, that is, a prime divisor on V_R. However, V_R is a section of the hypersurface V by the linear subspace $R \subset \mathbb{P}$ of codimension two. Let $D_2|_R = \{q_2|_{V_R} = 0\}$ be the first hypertangent divisor of the variety V_R. According to the regularity condition (R2.2), $D_2|_R$ is irreducible and does not satisfy the inequality (187), that is,

$$Y \neq D_2\,|\,_R,$$

and for that reason $Y \not\subset D_2 = \{q_2|_V = 0\}$. Let Y_4 be an irreducible component of the effective algebraic cycle $(Y \circ D_2)$ with the maximal ratio $\text{mult}_o / \deg$. We have the inequality

$$\frac{\text{mult}_o}{\deg} Y_4 > \frac{7}{M+1}.$$

Now we argue as above: construct a sequence of irreducible subvarieties

$$Y_4, Y_5, \dots, Y_{M-1},$$

$\text{codim}_V Y_i = i$, Y_{i+1} is an irreducible component of the effective cycle $(Y_i \circ D_{i+2})$, where $D_{i+2} \in \Lambda_{i+2}$ is a generic hypertangent divisor. For Y_{M-1} we get the estimate

$$\frac{\text{mult}_o}{\deg} Y_{M+1} > \frac{7}{M+1} \cdot \frac{M+1}{6} = \frac{7}{6},$$

which is impossible. The contradiction excludes the case 1).

Finally, consider the hardest case 2). Again we use the linearity of the conditions that are satisfied by the divisor D_P on V_P: the inequality

$$\text{mult}_o D_P > \frac{8}{3}n$$

and the existence of a non-log canonical singularity of the pair $(V_P^+, \frac{1}{n}D_P^+)$ with the centre S. As above, set $V_R = V_P \cap R$. Since $\mathrm{mult}_o V_R = 2 < \frac{8}{3}$ and the pair (V_P^+, V_R) is log canonical, we may assume that the divisor D_P does not contain the hyperplane section V_R as a component. For this reason, the effective cycle of codimension two (with respect to V_P) $D_R = (D_P \circ V_R)$ is well defined. This cycle satisfies the inequality

$$\mathrm{mult}_o D_R \geq \mathrm{mult}_o D_P + \mathrm{mult}_{Z^*} D_P^+ > \frac{28}{9}n. \tag{188}$$

As in the case 1), the inequality (188) is sufficient to conclude (by the regularity condition (R2.2)), that the component $Y = Y_3$ of the cycle D_R with the maximal ratio $\mathrm{mult}_o / \deg$ is not contained in the divisor $D_2 = \{q_2|_V = 0\}$, so that the effective cycle

$$(Y_3 \circ D_2)$$

is well defined. There is an irreducible component Y_4 of this cycle, satisfying the inequality

$$\frac{\mathrm{mult}_o}{\deg} Y_4 > \frac{14}{3(M+1)}. \tag{189}$$

By the regularity conditions (R2.2) this procedure can be repeated three times more. Let us consider in more detail the first step. The subvariety

$$W_{2\cdot 3} = \{q_2 = 0\} \cap \{q_3 = 0\} \cap V_R$$

is irreducible, of degree $\deg W_{2\cdot 3} = 6(M+1)$ and multiplicity

$$\mathrm{mult}_o W_{2\cdot 3} = 24,$$

so that $W_{2\cdot 3} \neq Y_4$. Since by construction,

$$Y_4 \subset \{q_2 = 0\} \cap V_R,$$

this implies that $Y_4 \not\subset \{q_3 = 0\}$, so that

$$Y_4 \not\subset D_3 = \{(q_2 + q_3)|_V = 0\}$$

and the effective cycle $(Y_4 \circ D_3)$ of codimension 5 is well defined. Some irreducible component Y_5 of this cycle satisfies the inequality

$$\frac{\mathrm{mult}_o}{\deg} Y_5 > \frac{56}{9(M+1)}.$$

In the same way, consider the irreducible subvariety

$$W_{2\cdot 3\cdot 4} = W_{2\cdot 3} \cap \{q_4 = 0\}$$

and construct an irreducible subvariety $Y_6 \subset V$ of codimension 6, satisfying the inequality

$$\frac{\mathrm{mult}_o}{\deg} Y_6 > \frac{70}{9(M+1)}.$$

Finally, consider the subvariety

$$W_{2\cdot 3\cdot 4\cdot 5} = W_{2\cdot 3\cdot 4} \cap \{q_5 = 0\}$$

and construct an irreducible subvariety $Y_7 \subset V$ of codimension 7, satisfying the inequality

$$\frac{\mathrm{mult}_o}{\deg} Y_7 > \frac{28}{3(M+1)}.$$

If $M=8$, we get a contradiction. If $M\geq 9$, we apply the technique of hypertangent divisors in the traditional version (as in the proof of Proposition 3.4), intersecting Y_7 with the generic hypertangent divisors $D_9,\dots$. For the irreducible curve Y_{M-1} we obtain the estimate

$$\frac{\mathrm{mult}_o}{\deg}Y_{M-1} > \frac{28}{3(M+1)}\cdot\frac{M+1}{9}=\frac{28}{27}.$$

This contradiction completes the case 2), which is now excluded.

Q.E.D. for Theorem 3.2.

Notes and references

Theorem 4.3.5 was formulated and proved in [**Pukh05**]; in Section 1 we follow that paper. The property of divisorial canonicity was later shown for several other classes of primitive Fano varieties in [**Pukh06b, Pukh08a**]; in [**Pukh07a**] some explicit examples (of double spaces of index one) were constructed. In [**Ch08**] it was shown that generic weighted three-fold hypersurfaces (with the exception of three-dimensional quartics) are also covered by Theorem 4.3.5 in the following sense: they satisfy the condition (L) of divisorial log canonicity. As it was shown in [**CPR**] that every mobile linear system on a generic weighted hypersurface can be untwisted to a system without maximal singularities (see Subsection 2.3.5), repeating the proof of Theorem 4.3.5 word for word, one obtains birational rigidity of direct products of the weighted hypersurfaces and a complete description of their groups of birational self-maps. For many other families of primitive Fano varieties, for which birational (super)rigidity is known, divisorial (log) canonicity of generic members is still an open problem.

The connectedness principle and inversion of adjunction were first used in the theory of birational rigidity by Corti in [**Co00**]. Gradually more and more applications of those important facts were discovered, including the $8n^2$-inequality and its divisorial analog, Proposition 2.3.

The work on divisorial canonicity of singular primitive Fano varieties, presented in Section 3, was motivated by the desire to make the linear method of proving birational rigidity for Fano fibre spaces $V/\mathbb{P}^1$ working uniformly for non-singular and singular fibres.

CHAPTER 8

Double Spaces of Index Two

Summary

In this chapter, we study birational geometry of generic Fano double spaces V of index two and dimension $M \geq 6$ and prove Theorem 4.3.6 for these values of M. These varieties admit $2(M-1)$-dimensional of rationally connected structures over $\mathbb{P}^1$, corresponding to the pencils of hyperplanes in $\mathbb{P}^M$ (or linear subspaces of codimension two in $\mathbb{P}^M$), so they are not birationally rigid. However, the method of maximal singularities provides a complete description of their birational geometry: to obtain the equality of the virtual and actual thresholds of canonical adjunction for a given mobile linear system Σ, we take its strict transform on the Fano fibre space, corresponding to the uniquely determined (by this system Σ) rationally connected structure, which, in turn, corresponds to a maximal subvariety of Σ. Therefore, the proof breaks into two major parts: first, to prove the existence of this maximal subvariety (which is necessarily the preimage of a linear subspace of codimension two in $\mathbb{P}^M$), second, to prove the coincidence of the virtual and actual thresholds of canonical adjunction of the strict transform of Σ on the corresponding (possibly singular) Fano fibre space. In more detail, the strategy of the proof of Theorem 4.3.6 is described in Subsection 1.1.

Assuming the existence of a maximal subvariety of the required type, we give the second part of the proof in Section 1. Here we use the linear technique in the form of Sections 7.2 and, to some extent, 7.3. The The proof of Theorem 4.3.6 is then completed modulo the most difficult part on the existence of a maximal subvariety, which is shown in the subsequent Sections 2–4.

In Section 2 we show that if a mobile linear system Σ has a maximal subvariety B of codimension 2, then B is the inverse image of a linear subspace of codimension two in $\mathbb{P}^M$. After that we exclude maximal singularities, the centre of which is a subvariety of codimension 3. In Section 3 we further improve the technique of counting multiplicities. Using this improvement, we exclude the infinitely near case in full and complete the first part of the proof of Theorem 4.3.6 in Section 4. Our technique makes use of certain conditions of general position for the branch divisor of the double cover $V \to \mathbb{P}^M$, listed in Subsection 1.1, so that we need to show that a generic branch hypersurface satisfies these conditions; this is done in Section 5.

1. Half-anticanonical pencils

We start the section with an outline of the proof of Theorem 4.3.6, listing the main steps and genericity assumptions (Subsection 1.1). After that, assuming the key technical fact, the existence of a very special maximal subvariety of a mobile linear system Σ, claimed by Proposition 1.1, we complete the proof of Theorem 4.3.6

(in the form of Theorem 1.1, which immediately implies the desired Theorem 4.3.6). The proof makes use of the linear method, which is applied to the strict transform of Σ on the Fano fibre space, corresponding to the unique half-anticanonical pencil, determined by the maximal subvariety of Σ.

1.1. The main result and the strategy of the proof. Throughout the chapter, the integer $M \geq 6$ is fixed. The symbol $\mathbb{P}$ stands for the projective space $\mathbb{P}^M$. Let $W = W_{2(M-1)} \subset \mathbb{P}$ be a smooth hypersurface of degree $2(M-1)$. There exists a uniquely determined double cover

$$\sigma\colon V \to \mathbb{P},$$

branched over W. It can be explicitly defined as the hypersurface, given by the equation

$$x_{M+1}^2 = f(x_0, \ldots, x_M)$$

in the weighted projective space $\mathbb{P}^{M+1}(1, \ldots, 1, M-1)$, where $f(x_*)$ is a homogeneous polynomial of degree $2(M-1)$, the equation of the hypersurface W. Recall (Subsection 4.3.3) that the variety V is a Fano variety of index two:

$$\operatorname{Pic} V = \mathbb{Z}H,$$

where H is the ample generator, $K_V = -2H$, the class H is the pullback via σ of a hyperplane in $\mathbb{P}$. On the variety V there are the following natural structures of a rationally connected fibre space: let $\alpha_P\colon \mathbb{P} \dashrightarrow \mathbb{P}^1$ be the linear projection from an arbitrary linear subspace P of codimension two, then the map

$$\pi_P = \alpha_P \circ \sigma\colon V \dashrightarrow \mathbb{P}^1$$

fibres V into $(M-1)$-dimensional Fano varieties of index one. Theorem 4.3.6 claims that these are the only structures of a rationally connected fibre space on V.

The claim of Theorem 4.3.6 will be derived from another fact, a much more technical and less visual Theorem 1.1 on the thresholds of canonical adjunction of mobile linear systems on the variety V. Let us describe the strategy of the proof. For an arbitrary linear subspace $P \subset \mathbb{P}$ of codimension two let V_P be the blow up of the subvariety $\sigma^{-1}(P) \subset V$ (it is irreducible by the conditions of general position, see Proposition 1.3 below). For a mobile linear system Σ on V the symbol Σ_P stands for its strict transform on V_P.

The following technical fact is crucial.

Theorem 1.1. *Assume that $M \geq 5$ and for a mobile linear system Σ the inequality*

$$c_{\text{virt}}(\Sigma) < c(\Sigma, V) \tag{190}$$

holds. Then there exists a uniquely determined linear subspace $P \subset \mathbb{P}$ of codimension two, satisfying the inequality

$$\operatorname{mult}_{\sigma^{-1}(P)} \Sigma > c(\Sigma, V),$$

whereas for the strict transform Σ_P the equality

$$c_{\text{virt}}(\Sigma) = c_{\text{virt}}(\Sigma_P) = c(\Sigma_P, V_P)$$

holds.

Theorem 4.3.6 is derived from Theorem 1.1 in a few lines, see Subsection 1.2. Almost all of this chapter is devoted to proving Theorem 1.1 in the assumption that $M \geq 6$, as we need only those values. Let us fix a mobile linear system Σ, satisfying the inequality (190). Taking, if necessary, a symmetric power of Σ, we may assume that

$$\Sigma \subset |2nH| = |-nK_V|,$$

where $n \geq 1$ is a positive integer. The system Σ (and the integer n) are fixed throughout the chapter, with the exception of the technical Section 3, where the notations are independent; we usually point this out but it is always clear from the context. Obviously,

$$c(\Sigma, V) = n.$$

Now let us state the Noether-Fano inequality for the system Σ. Applying Proposition 2.1.2, we obtain a birational morphism $\varphi\colon \widetilde{V} \to V$ and an irreducible exceptional divisor $E \subset \widetilde{V}$, satisfying the estimate

$$\operatorname{ord}_E \varphi^*\Sigma > na(E, V). \tag{191}$$

As usual, the divisor E (or the corresponding discrete valuation of the field of rational functions of the variety V) is called a *maximal singularity* of the linear system Σ. If φ is the blow up of an irreducible subvariety $B \subset V$ (and in that case $E = \varphi^{-1}(B)$), then the latter is called a *maximal subvariety* of the system Σ. In that latter case (191) is equivalent to the inequality

$$\operatorname{mult}_B \Sigma > n(\operatorname{codim} B - 1).$$

An equivalent formulation of the Noether-Fano inequality is that the pair

$$\left(V, \frac{1}{n}\Sigma\right) \tag{192}$$

is not canonical and the prime divisor $E \subset \widetilde{V}$ is a non-canonical singularity of this pair. The key difference from the cases considered in the previous chapters of this book (and the main source of trouble) is that the system Σ is a subsystem of the system $|2nH|$, not $|nH|$, which makes the Noether-Fano inequality "twice weaker". The bigger part of the chapter is devoted to proving the following fact.

Proposition 1.1. *There exists a unique linear subspace $P \subset \mathbb{P}$ of codimension two, such that the subvariety $\sigma^{-1}(P)$ is a maximal subvariety of the system Σ.*

Proposition 1.1 is proved in Sections 2–4 in the "negative" version: assuming that the system Σ has no maximal subvarieties of the form $\sigma^{-1}(P)$, where $P \subset \mathbb{P}$ is a linear subspace of codimension two, we exclude one by one all possibilities for a maximal singularity of the system Σ, thus coming to a contradiction with Proposition 2.1.2. The arguments of Section 2 exclude also the possibility that the system Σ has two maximal subvarieties, $\sigma^{-1}(P_1)$ and $\sigma^{-1}(P_2)$, where $P_1 \neq P_2$ are distinct linear subspaces of codimension two in $\mathbb{P}$.

Now we blow up the maximal subvariety $\sigma^{-1}(P)$ and on the new (generally speaking, singular) model complete the proof of Theorem 1.1, which directly implies Theorem 4.3.6. This part of our work, although it is the concluding one in the sense of the proof as a whole, is based on Proposition 1.1 only and is independent of the contents of Sections 2–4. So it is done in this section (in the assumption that Proposition 1.1 holds).

Section 5 is devoted to proving the conditions of general position, which the double space V is supposed to satisfy. There is no doubt that these conditions are unnecessary, that is, that Theorems 1.1 and 4.3.6 are true for any smooth double space of index two. However, the conditions of general position are essentially used in the proof. Some of those conditions could have been at least relaxed, however, this would have made our proof, hard and long as it is, even more complicated. On the other hand, we do not use the conditions of general position at all or use them not to the full extent in those cases where it does not make the proof too complicated. Some types of maximal singularities are excluded under the only assumption that the variety V is smooth.

As we mentioned above, we assume that $M \geq 6$; when $M = 5$, some complicated geometric situations cannot be just ruled out by dimension count (there are not enough moduli of V) and for that reason computations sometimes become very difficult. It is for that reason that we excluded the case $M = 5$ from our considerations, although our proof works in that case as well; see [**Pukh10**].

Now let us discuss the conditions of general position for the double space V, that is, for the branch divisor $W \subset \mathbb{P}$, a hypersurface of degree $2(M-1)$. We shall use several conditions of general position, which are formulated below and proved in Section 5 (we show that a generic hypersurface W satisfies these conditions indeed).

The first main condition deals with lines on V. As usual, a curve $C \subset V$ is called a *line*, if the equality $(C \cdot H) = 1$ holds. In particular, a line is a smooth irreducible rational curve. We have

Proposition 1.2. *On a generic variety V there are finitely many lines through any point.*

The second and third conditions deal with linear subspaces (planes) in $\mathbb{P}$ of codimension two. Consider an arbitrary plane $P \subset \mathbb{P} = \mathbb{P}^M$ of codimension two. Generally speaking, the intersection $P \cap W$, is singular:

$$p \in \operatorname{Sing} P \cap W$$

if and only if

$$P \subset T_p W.$$

It is well known (without the assumption that the hypersurface W is generic) that the set $\operatorname{Sing} P \cap W$ is at most one-dimensional (see, for instance, [**IP, Pukh00a**]).

The assumption that W is generic makes it possible to improve this claim.

Proposition 1.3. *For a generic hypersurface W and an arbitrary plane $P \subset \mathbb{P}$ of codimension two the set* $\operatorname{Sing} P \cap W$ *is finite (or empty). In particular, the closed set $R = \sigma^{-1}(P)$ is irreducible, and the set of its singular points is at most finite.*

The third condition characterizes the singularities of the variety $\sigma^{-1}(P)$ and the singularities of its blow up on V. For a quadratic singular point, that is, a hypersurface singularity with a local equation

$$0 = w_2(u_1, \dots, u_N) + w_3(u_*) + \dots,$$

where $w_i(u_*)$ is a homogeneous polynomial of degree i, then we say that this point is of rank $\operatorname{rk} w_2$. When such a singularity is blown up, the exceptional divisor is the quadric $\{w_2 = 0\} \subset \mathbb{P}^{N-1}$ of rank $\operatorname{rk} w_2$.

Let V_P be the blow up of the subvariety $\sigma^{-1}(P)$ on V, which is irreducible by Proposition 1.3.

Proposition 1.4. *For a generic hypersurface W and an arbitrary plane $P \subset \mathbb{P}$ of codimension two:*

(i) *for* $M \geq 6$ *every singular point of the variety* V_P *is an isolated quadratic point of rank* ≥ 4,

(ii) *for* $M \geq 4$ *every singular point of the variety* V_P *is an isolated quadratic point of rank* ≥ 3,

(iii) *for* $M \geq 6$ *every singular point of the variety* $\sigma^{-1}(P)$ *is an isolated quadratic point of rank* ≥ 2.

The properties, formulated in Propositions 1.2–1.4, will be assumed to take place (sometimes we remind about this in the course of our arguments). Proof of the three propositions will be given in Section 5.

1.2. Fano fibre space over $\mathbb{P}^1$. According to Proposition 1.1, there exist a (unique) linear subspace $P \subset \mathbb{P}$ of codimension two, satisfying the estimate

$$\operatorname{mult}_R \Sigma > n,$$

where $R = \sigma^{-1}(P)$ is an irreducible variety with at most zero-dimensional singularities (Proposition 1.3). Let $\varphi \colon V^+ \to V$ be the blow up of the (possibly singular) subvariety $R = \sigma^{-1}(P)$, $E = \varphi^{-1}(R)$ the exceptional divisor.

Lemma 1.1. (i) *The variety* V^+ *is factorial and has at most finitely many isolated double points (not necessarily non-degenerate).*

(ii) *The linear projection* $\pi_{\mathbb{P}} \colon \mathbb{P} \dashrightarrow \mathbb{P}^1$ *from the plane* P *generates the regular projection*

$$\pi = \pi_{\mathbb{P}} \circ \sigma \circ \varphi \colon V^+ \to \mathbb{P}^1,$$

the general fibre of which $F_t = \pi^{-1}(t)$, $t \in \mathbb{P}^1$ *is a non-singular Fano variety of index one, and finitely many fibres have isolated double points.*

(iii) *The following equalities hold:*

$$\operatorname{Pic} V^+ = \mathbb{Z}H \oplus \mathbb{Z}E = \mathbb{Z}K^+ \oplus \mathbb{Z}F,$$

where $H = \varphi^* H$ *for simplicity of notation,* $K^+ = K_{V^+}$ *is the canonical class of the variety* V^+, F *is the class of a fibre of the projection* π, *whereas*

$$K^+ = -2H + E, \ F = H - E.$$

Proof. These claims follow directly from the definition of the blow up φ, Proposition 1.3 and the well-known fact that an isolated hypersurface singularity of a variety of dimension ≥ 4 is factorial (see [**CL**]).

Let Σ^+ be the strict transform of the system Σ on the blow up V^+ of the subvariety R.

Proposition 1.5. *The following equality holds:*

$$c_{\text{virt}}(\Sigma^+) = c(\Sigma^+, V^+).$$

Proof of Theorem 1.1. This theorem is just the union of Proposition 1.1 and Proposition 1.5. Q.E.D.

Corollary 1.1. *Assume that* $c_{\text{virt}}(\Sigma^+) = 0$. *Then the system* Σ^+ *is composed from the half-anticanonical pencil* $|H - R|$, *that is,*

$$\Sigma^+ \subset |2nF|.$$

Proof of the corollary. Assume the converse:

$$\Sigma^+ \subset |-mK^+ + lF|,$$

where $m \geq 1$. By part (iii) of Lemma 1.1,

$$m = 2n - \nu, \; l = 2\nu - 2n \geq 2,$$

so that for the threshold of canonical adjunction we get

$$c(\Sigma^+, V^+) = m.$$

Since $c_{\text{virt}}(\Sigma^+) = 0$, by Proposition 1.5 we get $m = 0$, as we claimed. Q.E.D. for the corollary.

Proof of Theorem 4.3.6. For the linear system Σ we take the strict transform with respect to χ of any linear system of the form $\lambda^*\Lambda$, where Λ is a mobile system on the base S. Applying Theorem 1.1 (or Corollary 1.1), we complete the proof. Q.E.D.

1.3. Mobile linear systems on the variety V. We start our proof of Proposition 1.5 with the well-known step: assume that the inequality

$$c_{\text{virt}}(\Sigma^+) < c(\Sigma^+, V^+) = m$$

holds. Then the pair

$$(V^+, \frac{1}{m}\Sigma^+) \tag{193}$$

is not canonical, so that the linear system Σ^+ has a maximal singularity, that is, for some birational morphism $\psi \colon \widetilde{V} \to V^+$ and irreducible exceptional divisor $E^+ \subset \widetilde{V}$ the Noether-Fano inequality holds:

$$\nu_E(\Sigma^+) > ma(E^+, V^+).$$

Lemma 1.2. *The centre of maximal singularity E^+ is contained in some fibre $F_t = \pi^{-1}(t)$, that is,*

$$B = \pi \circ \psi(E^+) = t \in \mathbb{P}^1.$$

Proof. Assume the converse: $\pi \circ \psi(E^+) = \mathbb{P}^1$. Restricting the linear system Σ^+ onto the fibre of general position $F = F_s$, we get that the pair

$$(F, \frac{1}{m}\Sigma_F)$$

is not canonical, where $\Sigma_F \subset |-mK_F|$. However, F is a smooth double space of index one and it is well known ([**Pukh89a**], see also Section 2.2), that this is impossible. Q.E.D. for the lemma.

For simplicity of notation, let $F = F_t$ be the fibre, containing the centre of singularity E^+.

Proposition 1.6. *The centre B is a singular point of the fibre F.*

Proof. Since the anticanonical degree of the divisor $D_F \in \Sigma_F$ is $2m$, and by genericity of the branch divisor the anticanonical degree of any subvariety of codimension one on F is at least 2, we get the inequality $\operatorname{codim}_F B \geq 2$, so that

$$\operatorname{codim}_{V^+} B \geq 3.$$

In the notation of Subsection 1.2, let $\Pi = \sigma \circ \varphi(F) \subset \mathbb{P}$ be the hyperplane, corresponding to the fibre F. It is easy to see that

$$\sigma_F = \sigma \circ \varphi \colon F \to \Pi = \mathbb{P}^{M-1}$$

is the double cover, branched over $W_\Pi = W \cap \Pi$: the blow up φ does not affect the divisors-elements of the pencil $|H - R|$. Now we need to consider two cases:

1) $\sigma_F(B) \not\subset W_\Pi$,

2) $\sigma_F(B) \subset W_\Pi$, but the generic point of the subvariety B is a non-singular point of the fibre F.

The case 1) is excluded by the arguments of Section 7.1. Let $o \in B$ be a point of general position,

$$\lambda \colon F^\sharp \to F$$

its blow up, $E^\sharp = \lambda^{-1}(o) \subset F^\sharp$ the exceptional divisor, $E^\sharp \cong \mathbb{P}^{M-2}$. By inversion of adjunction for a general divisor $D \in \Sigma^+$ we get: the pair

$$(F, \frac{1}{m} D_F) \tag{194}$$

is not log canonical at B, so that by Proposition 7.2.3 there is a hyperplane $\Lambda \subset E^\sharp$, satisfying the inequality

$$\operatorname{mult}_o D_F + \operatorname{mult}_\Lambda D_F^\sharp > 2m,$$

where $D_F^\sharp$ is the strict transform of the divisor D_F on $F^\sharp$. Now the arguments of Section 7.1 give a contradiction.

Consider the case 2). If $\dim B \geq 1$, then for a point $o \in B$ of general position the intersection of divisors

$$T_p W_\Pi \quad \text{and} \quad \sigma_F(D_F),$$

where $p = \sigma_F(o)$, is of codimension two (by the condition of general position, for any hyperplane $\Lambda \subset \Pi$ we get $\dim \operatorname{Sing} \Lambda \cap W = 0$, so that the tangent hyperplanes $T_p W_\Pi$, $p \in B$, form a $(\dim B)$-dimensional family). In particular, the scheme-theoretic intersection

$$(\sigma_F^{-1}(T_p W_\Pi) \circ D_F)$$

is an effective cycle of codimension two on F, of H-degree $2m$ and of multiplicity at least

$$2 \operatorname{mult}_o D_F > 2m$$

at the point o, which is impossible.

Thus it remains to consider the case when $B = o$ is a smooth point on the ramification divisor of the morphism σ_F. Since the condition of non-log canonicity of the pair (194) is linear in the divisor $D_F \in |-mK_F|$, one may assume that D_F is a prime divisor. Set $\Lambda = T_p W_\Pi$. If

$$D_F \neq \sigma_F^{-1}(\Lambda),$$

then we argue as above in the case $\dim B \geq 1$. Let us show that the equality $D_F = \sigma_F^{-1}(\Lambda)$ is impossible. It can be done by inspection of possible singularities of the intersection $W_\Pi \cap \Lambda$ for a hypersurface W of general position. There is a simpler argument: if the pair (194) is not log canonical for $D_F = \sigma_F^{-1}(\Lambda)$, then by [**Kol93**] or [**Kol95b**], the pair

$$(\Pi, \Lambda + \frac{1}{2}W_\Pi)$$

is not log canonical, either, which, in turn, implies that the pair

$$(\Lambda, \frac{1}{2}W_\Lambda)$$

is not log canonical, $W_\Lambda = (W \circ \Lambda) = W \cap \Lambda$. However, as we pointed out above, the restriction W_Λ has at most isolated double points as singularities. This contradiction proves Proposition 1.6.

Let $B = o$ be the centre of the maximal singularity E^+.

Proposition 1.7. *The point o is a singularity of the variety V^+.*

Proof. Assume the converse: the point $o \in V^+$ is non-singular. Since the pair (193) is not canonical, we get the inequality

$$\operatorname{mult}_o \Sigma^+ > m,$$

whence by Proposition 1.6 it follows that

$$\operatorname{mult}_o D_F > 2m$$

(since $o \in F$ is a singular point of the fibre). As we pointed out above, this is impossible, which proves the proposition.

1.4. The centre of the maximal singularity is a singular point of the variety V^+. We have shown above that the centre of the maximal singularity E^+ is a singular point $o \in V^+$, which we will assume from now on. Let

$$\lambda\colon V^\sharp \to V^+$$

be the blow up of the point o, $E^\sharp = \lambda^{-1}(o) \subset V^\sharp$ the exceptional divisor, which can be seen as a quadratic hypersurface in $\mathbb{P}^M$.

Recall (Proposition 1.4), that for $M \geq 6$ we may assume that for a generic hypersurface $W \subset \mathbb{P}$, arbitrary plane $P \subset \mathbb{P}$ of codimension two and any singularity $o \in V^+$ the quadric $E^\sharp$ is of rank at least 4.

Define the integer $\beta \in \mathbb{Z}_+$ by the formula

$$D^\sharp \sim \lambda^* D - \beta E^\sharp,$$

where $D \in \Sigma^+$ is a generic divisor, $D^\sharp$ its strict transform on $V^\sharp$. By Proposition 7.2.8, Proposition 1.4 implies the inequality

$$\beta > m.$$

Furthermore, the divisor

$$\lambda_F^* D_F - \beta E_F^\sharp$$

on the strict transform $F^\sharp \subset V^\sharp$ is effective (the symbols λ_F and $E_F^\sharp$ stand for the blow up of the point $o \in F$ and for the exceptional divisor $\lambda_F^{-1}(o)$, respectively).

This implies the inequality

$$\mathrm{mult}_o D_F \geq 2\beta > 2m,$$

which is impossible. Proof of Proposition 1.5 is complete. Q.E.D.

Remark 1.1. The claim of Proposition 1.5 holds for $M = 5$ as well, but the proof is a lot more difficult; see [**Pukh10**] for the details.

2. Centres of codimension two and three

In this section, we start to prove Proposition 1.1: first, we show that, except for the preimage $\sigma^{-1}(P)$, where $P \subset \mathbb{P}$ is a linear subspace of codimension two, no subvariety of codimension two can make a maximal subvariety of the system Σ (Subsections 2.1–2.4). This done, we show that the linear system Σ has no maximal singularities, the centre of which is a subvariety of codimension three.

2.1. Maximal subvarieties of codimension two. The following claim is true.

Proposition 2.1. *If an irreducible subvariety $B \subset V$ of codimension two is maximal for the mobile linear system $\Sigma \subset |2nH|$, that is, the inequality* $\mathrm{mult}_B \Sigma > n$ *holds, then $B = \sigma^{-1}(\bar{B})$, where $\bar{B} \subset \mathbb{P}$ is a linear subspace of codimension two.*

Proof. The self-intersection $Z = (D_1 \circ D_2)$, $D_i \in \Sigma$, of the linear system Σ is of H-degree $8n^2$ and contains the subvariety B with multiplicity strictly higher than n^2. Therefore, $\deg B \leq 7$. It is necessary to show that only one of these possibilities is realized: $\deg B = 2$ and, moreover, $\bar{B} = \sigma(B)$ is a $(M-2)$-plane in $\mathbb{P}$, that is, the double cover $\sigma^{-1}(\bar{B}) \to \bar{B}$ is irreducible.

Note that since $\dim V = M \geq 5$ we have $A^2V = \mathbb{Z}H^2$, so that only three possibilities occur, $B \sim H^2$ or $2H^2$ or $3H^2$. In particular, $\deg B \in \{2, 4, 6\}$.

If $B = \sigma^{-1}(\bar{B})$, then $\deg B \in \{2, 4, 6\}$. Assume that $\deg B \in \{4, 6\}$, that is, $\deg \bar{B} \in \{2, 3\}$. Let us show that these cases do not realize. Indeed, let $L \subset \mathbb{P}$ be a generic secant line of the subvariety $\bar{B} \subset \mathbb{P}$. By generality, the curve $C = \sigma^{-1}(L)$ is non-singular and irreducible, and such curves sweep out at least a divisor on V, so that $C \not\subset \mathrm{Bs}\,\Sigma$. For a general divisor $D \in \Sigma$ we get $C \not\subset D$ and $(C \cdot D) = 4n$. On the other hand, let $p_1 \neq p_2$ be the points of intersection $L \cap \bar{B}$, then (by generality) $\sigma^{-1}(p_i) = \{p_{i1}, p_{i2}\}$, $i = 1, 2$, where p_{ij} are four distinct points on B. For this reason,

$$4n = (C \cdot D) \geq \sum_{i,j} (C \cdot D)_{p_{ij}} > 4n.$$

Contradiction.

Thus, if $B = \sigma^{-1}(\bar{B})$, then $\bar{B} \subset \mathbb{P}$ is a $(M-2)$-plane, which is exactly what we need.

Starting from this moment, we assume that $\sigma^{-1}\bar{B} = B \cup B'$ breaks into two irreducible components and

$$\deg B = \deg \bar{B} \in \{2, 4, 6\}.$$

We show below that none of these cases are realized. Let us describe first the main technical tools that will be used for their exclusion.

2.2. Conics on the variety $\bar{B}$. Let $C \subset \bar{B}$ be an irreducible conic, $P = \langle C \rangle$ its linear span, a 2-plane. Assume that $C \not\subset W$, the curve $W \cap P$ is reduced and the two finite sets

$$C \cap W \quad \text{and} \quad \mathrm{Sing}(W \cap P)$$

are disjoint. Set $S = \sigma^{-1}(P)$, this is an irreducible surface with a finite set of singular points $\sigma^{-1}(\mathrm{Sing}(W \cap P))$. Let C_+ and C_- be components of the curve $\sigma^{-1}(C) = C_+ \cup C_-$, where $C_+ \subset B$, $C_- \subset B'$.

Lemma 2.1. *The surface S is contained in the base set* $\mathrm{Bs}\,\Sigma$.

Proof. Assume the converse. Then for a general divisor $D \in \Sigma$ we get $S \not\subset D$, so that $(D \circ S)$ is an effective curve on S, containing C_+ with some multiplicity

$$\nu_+ \geq \mathrm{mult}_B\,\Sigma > n$$

and C_- with some multiplicity $\nu_- \in \mathbb{Z}_+$. Let $H_S = H\,|\,_S$ be the class of a hyperplane section of S. By what we have said,

$$((2nH_S - \nu_+ C_+ - \nu_- C_-) \cdot C_\pm) \geq 0. \tag{195}$$

Note that by assumption the curves $C_\pm$ do not contain singular points of the surface S, so that the local intersection numbers $(C_+ \cdot C_-)_x$ are equal to $\frac{1}{2}(C \cdot W)_{\sigma(x)}$ and therefore

$$(C_+ \cdot C_-) = \frac{1}{2}(C \cdot W) = 2(M-1).$$

Furthermore, $C_+ + C_- \sim 2H_S$, whence we obtain

$$(C_+^2) = (C_-^2) = 2(3-M).$$

Therefore, the inequalities (195) take the form of linear inequalities

$$\begin{array}{c} 4n + 2(M-3)\nu_+ - 2(M-1)\nu_- \geq 0, \\ 4n - 2(M-1)\nu_+ + 2(M-3)\nu_- \geq 0, \end{array} \tag{196}$$

whence we get $\nu_\pm \leq n$. Contradiction. Q.E.D. for the lemma.

Corollary 2.1. *The following inequality holds:* $\deg B \geq 4$.

Proof. We have to exclude the case when $\deg B = 2$. Assume that $\deg B = 2$. Applying Lemma 2.1 to the irreducible conic $C = \bar{B} \cap P$, where $P \subset < \bar{B} >$ is a generic 2-plane in the linear span of $\bar{B}$, we get that $\sigma^{-1}(P) \subset \mathrm{Bs}\,\Sigma$. Therefore,

$$\sigma^{-1}(\langle \bar{B} \rangle) \subset \mathrm{Bs}\,\Sigma,$$

which is impossible, since $\langle \bar{B} \rangle$ is a divisor in $\mathbb{P}$. Q.E.D. for the corollary.

2.3. The secant lines of the variety $\bar{B}$. Let $C \subset \bar{B}$ be an irreducible curve, not contained in W. Let $x \in \mathbb{P}$ be a point, satisfying the following conditions of general position:

- $x \notin C$,
- for any point $p \in C \cap W$ the line $L = \langle x, p \rangle$, connecting the points x and p, intersects the hypersurface W transversally at the point p and is not a secant line of the curve C, that is, $C \cap L = \{p\}$.

Consider the cone $\Delta = \Delta(x, C)$ with the vertex at the point x and the base C. Set $S = \sigma^{-1}(\Delta)$, it is an irreducible surface. Let C_+ and C_- once again be the components of the curve $\sigma^{-1}(C) = C_+ \cup C_-$, where $C_+ \subset B$, $C_- \subset B'$. By the assumptions above, all the points of intersection of the curves C_- and C_- are smooth points of the surface S. Obviously,

$$(C_+ \cdot C_-) = \frac{1}{2}(C \cdot W) = (M-1)\deg C.$$

Furthermore, we know from Section 2.2, that on the cone Δ the curve C is numerically equivalent to the hyperplane section. Thus on the surface S,

$$C_+ + C_- \equiv H_S = \sigma^* H_\Delta,$$

where H_Δ is the hyperplane section of the cone Δ. From here we get that

$$(C_+^2) = (C_-^2) = -(M-2)\deg C.$$

The restriction $\Sigma_S = \Sigma\,|\,_S$ of the system Σ on S is a non-empty linear system of curves, containing $C_\pm$ with the multiplicity $\nu_\pm$, respectively, where $\nu_+ \geq \operatorname{mult}_B \Sigma > n$. Therefore,

$$((2nH_S - \nu_+ C_+ - \nu_- C_-) \cdot C_\pm) \geq 0,$$

which yields the system of linear inequalities

$$\begin{aligned} 2n - (M-1)\nu_+ + (M-2)\nu_- &\geq 0, \\ 2n + (M-2)\nu_+ - (M-1)\nu_- &\geq 0. \end{aligned} \tag{197}$$

From here we immediately get

Proposition 2.2. *The following estimate holds*:

$$\operatorname{mult}_{B'} \Sigma > \frac{M-3}{M-2} n \geq \frac{n}{2}. \tag{198}$$

Proof. For a general choice of the vertex x of the cone Δ we obtain

$$\nu_+ = \operatorname*{mult}_B \Sigma, \quad \nu_- = \operatorname*{mult}_{B'} \Sigma,$$

whereas the inequality

$$\nu_- > \frac{M-3}{M-2} n$$

follows directly from (197). Q.E.D. for the proposition.

Note that the estimate (198) is the stronger, the higher is M.

Using this estimate, we exclude the case $\deg B = 6$.

Proposition 2.3. *The case* $\deg B = 6$ *does not realize.*

Proof. Assume the converse: $\deg B = 6$. Let $D_1, D_2 \in \Sigma$ be general divisors, $Z = (D_1 \circ D_2)$ the self-intersection of the system Z. For the multiplicity of the subvariety B', Proposition 2.2 gives the estimate $\operatorname{mult}_{B'} \Sigma > 2n/3$. So we obtain the inequality

$$8n^2 = \deg Z \geq 6((\operatorname{mult}_B \Sigma)^2 + (\operatorname{mult}_{B'} \Sigma)^2) > 6 \cdot (1 + \frac{4}{9})n^2 = \frac{26}{3} n^2,$$

which is impossible. Contradiction. Q.E.D. for the proposition.

2.4. Three-secant lines of the variety $\bar{B}$. Thus it remains to exclude the case $\deg B = 4$. We will need another construction. Let $L \subset \mathbb{P}$ be a 3-secant line of the variety $\bar{B}$, that is, a line that intersects $\bar{B}$ at (at least) three points outside W.

Proposition 2.4. *If the curve* $\sigma^{-1}(L) = C$ *is irreducible, then* $C \subset \operatorname{Bs}\Sigma$. *If* $C = C_+ \cup C_-$ *is reducible, then at least one of the components* $C_\pm$ *is contained in* $\operatorname{Bs}\Sigma$.

Proof. Let $D \in \Sigma$ be an arbitrary divisor. The curve C intersects D at at least 6 points. The total multiplicity of D at those points is at least

$$3(\mathrm{mult}_B \Sigma + \mathrm{mult}_{B'} \Sigma) > \frac{9}{2}n,$$

whereas $C \cdot D = 4n$. Therefore $L \subset \sigma(D)$, which is what we need. Q.E.D. for the proposition.

Therefore, if the subvariety $\bar{B} \subset \mathbb{P}$ has sufficiently many 3-secant lines (more precisely, if they sweep out at least a divisor on $\mathbb{P}$), then the subvariety $B \subset V$ cannot be maximal since the linear system Σ is mobile.

Remark 2.1. The claim of Proposition 2.4 (and its proof) remain true if the line L intersects $\bar{B}$ at two distinct points outside W, and in one of them, say, $\bar{x} \in L \cap \bar{B}$, is tangent to $\bar{B}$. In that case the curve $C = \sigma^{-1}(L)$ is tangent to B and B' at the points x, x', respectively, where $\sigma^{-1}(\bar{x}) = \{x, x'\}$, $x \in B$, $x' \in B'$, and it is easy to see that the local intersection numbers satisfy the inequalities

$$(C \cdot D)_x \geq 2\,\mathrm{mult}_B \Sigma, \quad (C \cdot D)_{x'} \geq 2\,\mathrm{mult}_{B'} \Sigma,$$

which makes it possible to argue in word for word the same way as in the case of three distinct points. In the sequel, when speaking about 3-secant lines, we will include the limit case of tangency without special reservations.

As a first application of the construction of Proposition 2.4 we exclude the case $\deg B = 4$.

Proposition 2.5. *The case* $\deg B = 4$ *does not take place.*

Proof. Assume the converse. Let $P \subset \mathbb{P}$ be a generic 3-plane. For the irreducible curve $B_P = \bar{B} \cap P$ in $\mathbb{P}^3$ the four cases are possible:

1) $B_P \subset R$ is a plane curve, $R = \mathbb{P}^2$ is a plane in P;

2) $B_P = Q_1 \cap Q_2$ is a smooth elliptic curve, the intersection of quadrics Q_1 and Q_2;

3) B_P is a smooth rational curve;

4) B_P has a double point.

The case 1) does not realize, because any line $L \subset R$ is a 4-secant line. Proposition 2.4 implies that the entire surface $\sigma^{-1}(R)$ is contained in the base set $\mathrm{Bs}\,\Sigma$. This is impossible, since P is a generic 3-plane.

In the case 2) we come to a contradiction in exactly the same way as in the proof of Lemma 2.1. Namely, let Q be a generic quadric, containing the curve B_P. On the surface Q we get $B_P \sim 2H_Q$, where H_Q is the plane section. Set

$$S = \sigma^{-1}(Q), \quad \sigma^{-1}(B_P) = C_+ \cup C_-, \quad C_+ \subset B, \quad C_- \subset B',$$

so that on S we have $C_+ + C_- \sim 2H_S$, where $H_S = H\,|\,_S$ is the class of a hyperplane section. Now we argue in exactly the same way as in Lemma 2.1 and obtain the inequalities (195), which give the system of linear inequalities (196). This contradiction excludes the case 2). Note that of key importance (as in Lemma 2.1) is the fact that the curve B_P is equivalent to *two* hyperplane sections of the surface Q. In a general case, a curve can be embedded into a surface as a hyperplane section (with multiplicity one), which gives just some estimate for the multiplicity of the second component B', but does not allow to get a contradiction in one step.

Consider the case 3). Let $x \in B_P$ be a point of general position, $\pi_x\colon B_P \to \mathbb{P}^2$ the projection from the point x. The image $\pi_x(B_P) \subset \mathbb{P}^2$ is a rational cubic curve with a double point. Therefore, the curve B_P has a 3-secant line, passing through the point x. Since P is a 3-plane and $x \in B_P$ is a general point, we apply Proposition 2.4 and obtain a contradiction.

Consider the case 4). The curve B_P has a unique double point. This implies that the variety $\bar{B}$ contains a $(M-3)$-plane Π of double points. Let $L \subset \Pi$ be a generic line, $\Lambda \supset L$ a generic 3-plane, containing L. Now the curve $B_\Lambda = B \cap \Lambda$ is a quartic in $\mathbb{P}^3$, containing the line L with multiplicity 2. Therefore,

$$B_\Lambda = C_\Lambda + 2L,$$

where C_Λ is a (in the general case irreducible) conic. The variety $\bar{B}$ is swept out by the conics C_Λ. Now we apply Lemma 2.1 and obtain a contradiction.

Q.E.D. for Proposition 2.5.

2.5. Exclusion of maximal centres of codimension three. Maximal subvarieties of codimension two being now completely studied, let us prove that the linear system Σ has no maximal singularities, the centre of which is a subvariety of codimension three on V.

Recall that the linear system Σ has a maximal singularity, the centre of which is an irreducible subvariety $B \subset V$. We have already proved that if Σ has no maximal subvariety of the form $\sigma^{-1}(P)$, where $P \subset \mathbb{P}$ is a linear subspace of codimension two, then Σ has no maximal subvarieties of codimension two at all. Therefore we may assume that $\operatorname{codim} B \geq 3$.

Proposition 2.6. *The subvariety B is of codimension ≥ 4.*

Proof. Assume the converse. By Proposition 2.1, $\operatorname{codim} B = 3$. Since $M \geq 5$, we may assume that $\deg B \geq 2$.

Note that the morphism

$$\sigma|_B\colon B \to \sigma(B) = \bar{B}$$

is birational. Indeed, let $Z = (D_1 \circ D_2)$ be the self-intersection of the system Σ, then

$$\operatorname{mult}_B Z > 4n^2,$$

whence it follows that if $\sigma|_B$ is a double cover, then

$$\operatorname{mult}_{\bar{B}} \sigma_* Z > 8n^2,$$

however, $\sigma_* Z$ is an effective cycle of codimension two on $\mathbb{P}$ of degree $8n^2$. We get a contradiction.

Therefore, $\deg \bar{B} = \deg_H B \geq 2$ (since the branch divisor W does not contain linear subspaces of codimension three). Let $p, q \in \bar{B}$ be points of general position, $L \subset \mathbb{P}$ the line, connecting these points, $\Pi \supset L$ a generic (two-dimensional) plane, $\Lambda = \sigma^{-1}(\Pi)$ an irreducible surface on V. If $L \not\subset \operatorname{Supp} \sigma_* Z$, then the intersection $\Pi \cap \operatorname{Supp} \sigma_* Z$, and therefore, also the intersection $\Lambda \cap \operatorname{Supp} Z$, is zero-dimensional, so that we get

$$8n^2 = (\Lambda \cdot Z) \geq \sum_{x \in \sigma^{-1}(L) \cap B} (\Lambda \cdot Z)_x$$
$$\geq \sum_{x \in \sigma^{-1}(L) \cap B} \operatorname{mult}_x Z > 8n^2,$$

a contradiction. Therefore, $L \subset \operatorname{Supp} \sigma_* Z$.

Let $Q \subset \mathbb{P}$ be the irreducible subvariety, swept out by all secant lines of the variety $\bar{B}$. By what we have proved, $\operatorname{codim} Q = 2$, so that Q is a subspace of codimension two and $\bar{B} \subset Q$ is some hypersurface.

Now let us write

$$Z = a\sigma^{-1}(Q) + Z^\sharp,$$

where $Z^\sharp$ does not contain the subvariety $\sigma^{-1}(Q)$ as a component and $a \geq 1$. The cycle Z satisfies the linear inequality

$$2 \operatorname{mult}_B Z > \deg Z.$$

It is easy to see that any effective cycle of codimension two, satisfying this inequality, contains the subvariety $\sigma^{-1}(Q)$ as a component: as above,

$$\deg Z = (\Lambda \cdot Z) \geq \sum_{x \in \sigma^{-1}(L) \cap B} \operatorname{mult}_B Z > \deg Z$$

for every secant line L of the variety $\bar{B}$, which is not contained in the support of the cycle $\sigma_* Z$ (and a generic plane $\Pi \supset L$). However,

$$\operatorname{mult}_B \sigma^{-1}(Q) = 1 \quad \text{and} \quad \deg \sigma^{-1}(Q) = 2$$

(recall that for a general hypersurface W the intersection $Q \cap W$ has at most zero-dimensional singularity, so that $\sigma^{-1}(Q)$ is an irreducible set), whence it follows that the cycle $Z^\sharp$ satisfies the inequality

$$2 \operatorname{mult}_B Z^\sharp > \deg Z^\sharp$$

and therefore contains the subvariety $\sigma^{-1}(Q)$ as a component. Contradiction.

Q.E.D. for Proposition 2.6.

3. Counting multiplicities

In this section we give a stronger version of the technique of counting multiplicities for the self-intersection of a mobile linear system. We obtain the result that, together with the $8n^2$-inequality, the technical basis for the exclusion of maximal singularities is formed, the centre of which is of codimension ≥ 4. The notations in this section are independent of the other parts of this chapter.

3.1. Set up of the problem. Let $o \in X$ be a germ of a smooth three-dimensional variety, $\varphi \colon \widetilde{X} \to X$ a birational morphism, $E \subset \widetilde{X}$ an irreducible exceptional divisor over the point o, that is, $\varphi(E) = o$. Consider the resolution of the discrete valuation ν_E, that is, the sequence of blow ups

$$\varphi_{i,i-1} : X_i \to X_{i-1},$$

$i = 1, \ldots, N$, where $X_0 = X$, $\varphi_{i,i-1}$ blows up an irreducible subvariety $B_{i-1} \subset X_{i-1}$ (a point or a curve),

$$E_i = \varphi_{i,i-1}^{-1}(B_{i-1}) \subset X_i$$

is the exceptional divisor, where B_i is uniquely defined by the conditions $B_0 = o$ and for $i = 1, \ldots, N-1$,

$$B_i = \text{centre}(E, X_i),$$

and, finally, the geometric discrete valuations

$$\nu_E \quad \text{and} \quad \nu_{E_N}$$

of the field of rational functions of the variety X coincide. Geometrically this means that the birational map

$$\varphi_{N,0}^{-1} \circ \varphi \colon \widetilde{X} \dashrightarrow X_N$$

is biregular at the generic point of the divisor E and maps E onto E_N. Here

$$\varphi_{N,0} = \varphi_{1,0} \circ \cdots \circ \varphi_{N,N-1} \colon X_N \to X_0,$$

and, more generally, for $i > j$ set

$$\varphi_{i,j} = \varphi_{j+1,j} \circ \cdots \circ \varphi_{i,i-1} \colon X_i \to X_j.$$

The strict transform of an irreducible subvariety (by linearity, also of an effective cycle) $Y \subset X_j$ on X_i we denote, as usual, by adding the upper index i: that is, we write Y^i.

Assume that for $i = 1, \ldots, L \leq N$ the centres B_{i-1} of blow ups are points, for $i \geq L+1$ they are curves. Let Γ be the graph of the constructed resolution, that is, an oriented graph with the vertices $1, \ldots, N$, where an oriented edge (arrow) joins i and j for $i > j$ (notation: $i \to j$), if and only if

$$B_{i-1} \subset E_j^{i-1}.$$

In particular, by construction always $i + 1 \to i$.

Let us describe the obvious combinatorial properties of the graph Γ.

Lemma 3.1. *Let $i < j < k$ be three distinct vertices. If $k \to i$, then $j \to i$.*

Proof. By definition, $k \to i$ means that $B_{k-1} \subset E_i^{k-1}$. By construction of the resolution of singularities, we get

$$\varphi_{k-1,j-1}(B_{k-1}) = B_{j-1}$$

(the centres of blow ups with higher numbers cover the centres of previous blow ups) and, also, obviously

$$\varphi_{k-1,j-1}(E_i^{k-1}) = E_i^{j-1}.$$

From here the lemma follows directly. Q.E.D.

Definition 3.1. We say that the vertex i of the graph Γ is of *class* $e \geq 1$ (notation: $\varepsilon(i) = e$), if precisely e arrows come out of it, that is,

$$\sharp\{j \mid i \to j\} = e.$$

We say, furthermore, that the graph Γ is of class $e \geq 1$, if for each vertex i we have $\varepsilon(i) \leq e$.

For instance, a graph of class 1 is a chain:

$$1 \longleftarrow 2 \longleftarrow \ldots \longleftarrow N$$

(no other arrows but $i+1 \to i$). The graph of a sequence of blow ups of points on a non-singular surface is of class 2.

Lemma 3.2. *The graph of the resolution of the valuation ν_E is of class 3. If for some vertex i we have $\varepsilon(i) = 3$, then $i \le L$, that is, B_{i-1} is a point.*

Proof. By definition,

$$B_{i-1} \subset E(i) = \bigcup_{\{j|i\to j\}} E_j^{i-1}$$

and, moreover, at the general point B_{i-1} is a smooth variety and $E(i)$ is a normal crossings divisor, each component of which contains B_{i-1}. Q.E.D. for the lemma.

Remark 3.1. Word for word the same arguments show that in the case of arbitrary dimension $\dim X$ the graph of the resolution of any valuation is of class at most $\dim X$ and if

$$\sharp\{j \mid i \to j\} = a,$$

then $\operatorname{codim} B_{i-1} \ge a$.

Let Σ be a germ of a mobile (that is, free from fixed components) linear system on X, Σ^i its strict transform on X_i,

$$\nu_i = \operatorname{mult}_{B_{i-1}} \Sigma^{i-1},$$

so that for a general divisor $D \in \Sigma$ we have

$$D^i = D - \sum_{j=1}^{i} \nu_j E_j,$$

where we write D instead of $\varphi_{i,0}^* D$ and similarly for the exceptional divisors E_i.

Consider a pair of general divisors $D_1, D_2 \in \Sigma$ and construct the self-intersections of linear systems Σ^i, which are (not uniquely determined) effective 1-cycles

$$Z_i = (D_1^i \circ D_2^i)$$

on X_i. These cycles admit the natural decomposition

$$\begin{array}{rcl} Z_1 & = & Z_0^1 + Z_{1,1}, \\ Z_2 & = & Z_1^2 + Z_{2,2} = Z_0^2 + Z_{1,2} + Z_{2,2}, \\ & \dots & \\ Z_i & = & Z_{i-1}^i + Z_{i,i} = Z_0^i + Z_{1,i} + \dots + Z_{i,i}, \end{array}$$

where $Z_{a,i} = (Z_{a,i-1})^i = \dots = Z_{a,a}^i$, $i = 1, \dots, L$.

Recall (Section 2.2) that a function $a\colon \{1, \dots, L\} \to \mathbb{Z}_+$ is said to be compatible with the structure of the graph Γ, if the following inequalities hold:

$$a(i) \ge \sum_{j \to i} a(j).$$

By construction, $B_L \subset E_L \cong \mathbb{P}^2$ is a plane curve. Set $\beta_L = \deg B_L$ to be its degree in $\mathbb{P}^2$ and for an arbitrary $i \ge L+1$,

$$\beta_i = \beta_L \deg[\varphi_{i,L}|_{B_i} \colon B_i \to B_L]. \tag{199}$$

By Proposition 2.2.4, for any function $a(\cdot)$, compatible with the graph structure, the inequality

$$\sum_{i=1}^{L} a(i)m_i \geq \sum_{i=1}^{L} a(i)\nu_i^2 + a(L) \sum_{i=L+1}^{N} \beta_i \nu_i^2 \tag{200}$$

holds, where $m_i = \mathrm{mult}_{B_{i-1}} Z_0^{i-1}$, $i = 1, \ldots, L$.

The aim of this section is to prove a stronger estimate that includes (200) as a particular case.

Definition 3.2. A vertex $i \in \{4, \ldots, L\}$ of the graph Γ is said to be *complex*, if precisely three arrows come out of this vertex, that is, for three distinct vertices $i_1 < i_2 < i_3$ we have

$$i \to i_1, \ i \to i_2, \ i \to i_3.$$

Note that in the notation of Definition 3.2, by Lemma 3.1, we always have

$$i_2 \to i_1, \quad i_3 \to i_1, \quad \text{and} \quad i_3 \to i_2.$$

If in the graph Γ there are no complex vertices, then it is of class ≤ 2.

Definition 3.3. A *simplification* of the graph Γ is the oriented graph Γ^* of class ≤ 2 with the set of vertices

$$1, \ldots, L,$$

the arrows in which join the vertices in accordance with the following rule: if

$$\sharp\{j \mid i \to j\} \leq 2,$$

then $i \to j$ in Γ^* if and only if $i \to j$ in Γ; however, if i is a complex vertex, then in the notation of Definition 3.2 in Γ^*, there are two arrows coming out of the complex vertex i,

$$i \to i_2 \quad \text{and} \quad i \to i_3,$$

that is, the arrow $i \to i_1$ is deleted.

Thus the graph Γ^* is obtained from Γ by means of deleting some arrows (and the vertices that correspond to the blow ups of curves).

Proposition 3.1. *For any function $a\colon \{1, \ldots, L\} \to \mathbb{Z}_+$, compatible with the structure of the graph Γ^*, the inequality* (200) *holds.*

3.2. Proof of the improved inequality. If the function $a(\cdot)$ is compatible with the structure of Γ, then, in addition, it is compatible with Γ^*, so that Proposition 3.1 implies Proposition 2.2.4. Following the general scheme of the proof of the inequality (200) in Chapter 2, set for $i = 1, \ldots, L$,

$$d_i = \deg Z_{i,i},$$

where $Z_{i,i} \subset E_i \cong \mathbb{P}^2$ is a plane curve and for $i < j \leq L$,

$$m_{i,j} = \mathrm{mult}_{B_{j-1}} Z_{i,j-1}.$$

We obtain the following system of equalities

$$\begin{array}{rcl} \nu_1^2 + d_1 & = & m_1, \\ \nu_2^2 + d_2 & = & m_2 + m_{1,2}, \\ & \cdots & \\ \nu_i^2 + d_i & = & m_i + m_{1,i} + \cdots + m_{i-1,i}, \\ & \cdots & \\ \nu_L^2 + d_L & = & m_L + m_{1,L} + \cdots + m_{L-1,L}. \end{array} \tag{201}$$

Also, we get the obvious inequality

$$d_L \geq \sum_{i=L+1}^{N} \beta_i \nu_i^2,$$

where the numbers β_i are defined by the formula (199). Let us multiply the i-th equality in (201) by $a(i)$ and put together the resulting equalities. In the left-hand side we get

$$\sum_{i=1}^{L} a(i)\nu_i^2 + \sum_{i=1}^{L} a(i)d_i.$$

In the right-hand side we get

$$\sum_{i=1}^{L} a(i)m_i + \sum_{i=1}^{L-1} \left(\sum_{j=i+1}^{L} a(j)m_{i,j} \right).$$

Thus Proposition 3.1 follows immediately from the following claim.

Lemma 3.3. *For any $i = 1, \ldots, L-1$ the following inequality holds:*

$$a(i)d_i \geq \sum_{j=i+1}^{L} a(j)m_{i,j}. \tag{202}$$

Proof. Up to this moment our argument repeated word for word the corresponding arguments in Section 2.2. Recall that if the function $a(\cdot)$ is compatible with the structure of the graph Γ (not just Γ^*), then, taking into account that the inequality $m_{i,j} > 0$ is possible only if $j \to i$, and that always

$$d_i \geq m_{i,j},$$

we obtain (202) from the inequality $a(i) \geq \sum_{j \to i} a(j)$, this is how Proposition 2.2.4 was proved. However, this is not sufficient for the proof of Proposition 3.1, since the function $a(\cdot)$ is compatible with the structure of the graph Γ^* only, and the latter has, generally speaking, less arrows than Γ.

To prove (202), recall, first of all, that the integer-valued functions

$$d_i = \deg Z_{i,i}, \quad m_{i,j} = \operatorname{mult}_{B_{j-1}} Z_{i,j-1}$$

are linear functions of effective 1-cycles $Z_{i,i}$ on the exceptional plane $E_i \cong \mathbb{P}^2$. Since the inequality (202) is also linear, the claim of Lemma 3.3 follows from a simpler fact.

Lemma 3.4. *For any irreducible curve $C \subset E_i$ the following inequality holds:*

$$a(i) \deg C \geq \sum_{j=i+1}^{L} a(j) \operatorname{mult}_{B_{j-1}} C^{j-1}. \tag{203}$$

Proof. Set $d = \deg C$,

$$\mu_j = \mathrm{mult}_{B_{j-1}} C^{j-1}, \quad j = i+1, \dots, L.$$

As we pointed out above, if $\mu_j > 0$, then $j \to i$, so that it is necessary to prove the inequality

$$a(i)d \geq \sum_{j \to i} a(j)\mu_j.$$

This inequality is a claim about singularities of plane curves. Let us consider two cases:

1) $C \subset E_i$ is a line in $E_i \cong \mathbb{P}^2$,

2) C is a curve of degree $d \geq 2$.

In the case 1) define the integer $k \geq 1$ by the condition

$$\{j \mid B_{j-1} \in C^{j-1}\} = \{i+1, \dots, i+k\}. \tag{204}$$

In order to distinguish between the arrows in the graphs Γ and Γ^*, we write $a \overset{*}{\to} b$, if the vertices a, b are joined by an arrow in Γ^*, leaving the usual arrow for Γ.

The following fact is of crucial importance.

Lemma 3.5. *For each* e, $1 \leq e \leq k$, *we have*

$$i + e \overset{*}{\to} i.$$

Proof. Assume that $(i+e)$ is a complex vertex of the graph Γ (otherwise $i + e \overset{*}{\to} i$ by definition). Recall that the simplification procedure removes, of the three arrows, coming out of $i+e$, the one that goes to the *lowest* vertex. Therefore, we may assume that $e \geq 3$. However, the points

$$B_i, B_{i+1}, B_{i+2}, \dots, B_{i+k-1}$$

lie on the strict transform of the smooth curve C, and therefore the subgraph

$$i+1 \longleftarrow i+2 \longleftarrow i+3 \longleftarrow \cdots \longleftarrow 1+k$$

is a chain, that is, between the vertices $i+1, \dots, i+k$ in the original graph Γ there are no arrows, except for the consecutive ones. In any case there are two arrows coming out of the vertex $i+e$:

$$i + e \to i + e - 1 \quad \text{and} \quad i + e \to i.$$

What has been said implies that if a third arrow comes out the vertex $i+e$, that is, $i + e \to j$, then inevitably,

$$j \leq i - 1.$$

It is this arrow that the simplification procedure deletes. Therefore, the arrow $i + e \to i$ will not be deleted. Q.E.D. for the lemma.

Let us come back to the case 1). The inequality (203) takes the form of the estimate

$$a(i) \geq \sum_{j=i+1}^{i+k} a(j). \tag{205}$$

By the lemma we have just proved, (205) is true, because the function $a(\cdot)$ is compatible with the structure of the graph Γ^*. Q.E.D. for Lemma 3.4 in the case 1).

Let us consider the case 2). Again let us define $k \geq 1$ by the condition (204). If $k = 1$, then there is nothing to prove, since

$$i + 1 \stackrel{*}{\to} i$$

and $\mu_j \leq d$ for any j. For this reason we assume that $k \geq 2$.

Lemma 3.6. *The inequality $d \geq \mu_1 + \mu_2$ holds.*

Proof. Let $L \subset E_i$ be the line, passing through the point B_i in the direction of the infinitely near point $B_{i+1} \in E_i^{i+1}$. By assumption, $C \neq L$. Then for the intersection number on the surface E_i^{i+2} we get

$$0 \leq (C^{i+2} \cdot L^{i+2}) = d - \mu_1 - \mu_2,$$

which is what we need. Q.E.D. for the lemma.

Let Γ_C be the subgraph of the graph Γ with the vertices $\{i+1, \ldots, i+k\}$ (and the same arrows), Γ_C^* the subgraph of the graph Γ^* with the same set of vertices. By definition of the simplification procedure, the arrow

$$i + a \to i + b, \quad a > b \leq 1,$$

cannot be deleted, because in Γ there is an arrow that goes to a lower vertex:

$$i + a \to i. \tag{206}$$

This implies that $\Gamma_C = \Gamma_C^*$. Furthermore, the arrow (206) is deleted by the simplification procedure if and only if $i + a$ is a complex vertex and the arrow (206) is the lowest one. In the language of the graph Γ_C this means that the vertex $i + a$ is of class 2 (as a vertex of that graph). The following claim is obvious.

Lemma 3.7. *The integer-valued function*

$$i + a \longmapsto \mu_{i+a} \in \mathbb{Z}_+$$

is compatible with the structure of the graph Γ_C.

Proof. This is a well-known property of multiplicities of a curve at infinitely near points (on a non-singular surface). Q.E.D.

By what has been said, the proof of the inequality (203) in the case 2) is reduced to the following combinatorial fact. Let Δ be a subgraph of class 2 with the set of vertices $\{1, \ldots, k\}$ and $\varepsilon_\Delta(\cdot) \in \{0, 1, 2\}$ the function of the class of a vertex ($\varepsilon_\Delta(1) = 0$). Let $\mu(\cdot)$ and $a(\cdot)$ be $\mathbb{Z}_+$-valued functions, compatible with the structure of the graph Δ.

Lemma 3.8. *The following inequality holds:*

$$(\mu(1) + \mu(2)) \sum_{\varepsilon_\Delta(j) \leq 1} a(j) \geq \sum_{j=1}^{k} \mu(j) a(j). \tag{207}$$

Proof is given by induction on the number of vertices $k \geq 2$. If $k = 2$, then $\varepsilon_\Delta(1) = 0$, $\varepsilon_\Delta(2) = 1$ and the inequality (207) is of the form

$$(\mu(1) + \mu(2))(a(1) + a(2)) \geq \mu(1)a(1) + \mu(2)a(2),$$

so that there is nothing to prove.

Assume that $k \geq 3$ and the vertices 3 and 1 are not joined by an arrow: $3 \not\to 1$, that is, $\varepsilon_\Delta(3) = 1$. In that case we apply the induction hypothesis to the graph Δ_1 with the vertices $\{2, \dots, k\}$ and the same arrows as in Δ: for that graph the inequality (207) takes the form

$$(\mu(2) + \mu(3)) \sum_{\varepsilon_\Delta(j)=1} a(j) \geq \sum_{j=2}^{k} \mu(j)a(j),$$

whence, taking into account the inequality $\mu(1) \geq \mu(2) \geq \mu(3)$, we obtain the required inequality (207) for Δ.

Assume that $k \geq 3$ and there are arrows

$$3 \to 1, \dots, 2 + l \to 1,$$

where $l \geq 1$. In that case let us denote the left-hand side of (207) as

$$(\mu(1) + \mu(2)) \left(a(1) + a(2) + \sum_{\substack{j \geq l+3, \\ \varepsilon_\Delta(j) = 1}} a(j) \right),$$

and the right-hand side of (207) as

$$\mu(1)a(1) + \mu(2)a(2) + \sum_{j=3}^{l+2} \mu(j)a(j) + \sum_{j=l+3}^{k} \mu(j)a(j).$$

Since the functions $\mu(\cdot)$ and $a(\cdot)$ are compatible with the structure of the graph, we get the inequality

$$\mu(2)a(1) \geq \mu(2)(a(2) + a(3) + \dots + a(l + 2))$$

and the symmetric inequality for $\mu(1)a(2)$. Applying the induction hypothesis to the subgraph Δ_{l+3} with the vertices $\{l + 3, \dots, k\}$, we complete the proof of the lemma.

Q.E.D. for Proposition 3.1.

3.3. Counting multiplicities of the self-intersection for a non-log canonical singularity. Let $o \in X$ be a non-singular three-dimensional germ, Σ a mobile linear system, such that

$$\operatorname{mult}_o \Sigma \leq 2n,$$

but the point o is an isolated centre of non-log canonical singularities of the pair

$$\left(X, \frac{1}{n}\Sigma \right), \tag{208}$$

where $n > 0$ is some number. Let $Z = (D_1 \circ D_2)$ be the self-intersection of the system Σ (an effective 1-cycle on X). Let

$$\varphi_{i,i-1} \colon X_i \to X_{i-1},$$

$i = 1, \dots, K$, be the sequence of blow ups of the centres of some non-log canonical singularity of the pair (208), $X_0 = X$, $\varphi_{i,i-1}$ blows up an irreducible subvariety

$B_{i-1} \subset X_{i-1}$, a point or a curve, $B_0 = o$,

$$E_i = \varphi_{i,i-1}^{-1}(B_{i-1}) \subset X_i$$

is the exceptional divisor, finally, $E_K \subset X_K$ realizes a non-log canonical singularity of the pair (208), that is, the log Noether-Fano inequality

$$\sum_{i=1}^{K} p_{Ki}\nu_i > n\left(\sum_{i=1}^{K} p_{Ki}\delta_i + 1\right) \tag{209}$$

holds, where $\delta_i = \operatorname{codim} B_{i-1} - 1$, p_{K_i} is the number of paths from E_K to E_i and, as usual,

$$\nu_i = \operatorname{mult}_{B_{i-1}} \Sigma^{i-1} \leq 2n,$$

Σ^i is the strict transform of the mobile system Σ on X_i. Set

$$\{1, \ldots, L\} = \{j \mid \dim B_{j-1} = 0\}.$$

Since by assumption $\nu_i \leq 2n$, (209) implies that $K \geq L + 1$, that is, among the centres of blow ups there is at least one curve.

Denote by the symbol Γ the oriented graph of the sequence of blow ups $\varphi_{i,i-1}$, $i = 1, \ldots, K$, and by the symbol Γ^* its simplification, a graph of class ≤ 2. The number of paths from the vertex i to the vertex j, $i > j$, in the graph Γ^* we denote by the symbol p_{ij}^*. By definition, $p_{ii}^* = 1$. Set also,

$$m_i = \operatorname{mult}_{B_{i-1}} Z^{i-1},$$

$i = 1, \ldots, L$. The following fact improves the classical $4n^2$-inequality for the case of a non-*log* canonical singularity of the pair (208).

Proposition 3.2. *The following estimate holds:*

$$\sum_{i=1}^{L} p_{Li}^* m_i > 4n^2\left(\sum_{i=1}^{L} p_{Li}^* + 1\right).$$

Proof. Set

$$N = \min\left\{e \,\middle|\, \sum_{i=1}^{e} p_{ei}(\nu_i - \delta_i n) > 0\right\},$$

that is, E_N is the non-canonical singularity of the pair (208) with the minimal number. In particular, the pair (208) is canonical at $E_1, \ldots, E_{N-1}$. It follows from the inequality $\nu_1 \leq 2n$ (see Section 4), that the segment of the graph Γ with the vertices

$$L-1,\ L,\ L+1,\ \ldots,\ N$$

is a chain. Define the number a, $0 \leq a \leq 1$, by the equality

$$a = \frac{1}{n}\sum_{i=1}^{N-1} p_{Ni}(\delta_i n - \nu_i)$$

(the numbers of paths p_{N_i} and $p_{N-1,i}$ coincide by what was said above). If $a = 1$, then it is easy to see that

$$\nu_1 = \cdots = \nu_N = 2n,$$

whence the claim of the proposition follows directly. For that reason, we assume that $a < 1$.

First assume that $N = K$. In that case the technique of counting multiplicities together with the inequality (209) gives:

$$\sum_{i=1}^{L} p^*_{Li} m_i > \frac{(2\Sigma_0^* + \Sigma_1^* + 1)^2}{\Sigma_0^* + \Sigma_1^*} n^2 = 4(\Sigma_0^* + 1)n^2 + \frac{(\Sigma_1^* - 1)^2}{\Sigma_0^* + \Sigma_1^*} n^2,$$

where

$$\Sigma^*_{2-\alpha} = \sum_{\delta_i = \alpha} p^*_{Ni}, \quad \alpha = 1, 2,$$

which was required (note that the log Noether-Fano inequality can get only stronger when we replace p_{Ki} by p^*_{Ki}, since those coefficients are changed only for $i = 1, \dots, L-1$ and $\nu_i \leq 2n$ for these values of i, this is a well-known fact).

Thus for $N = K$ Proposition 3.2 holds.

Assume that $K \geq N + 1$.

Lemma 3.9. *The pair*

$$(X_{N-1}, \frac{1}{n}\Sigma^{N-1} - aE_{N-1})$$

is not log canonical at the curve $B_{N-1} \subset E_{N-1}$.

Proof. Taking into account the standard properties of the numbers p_{ij}, rewrite the inequality (209) in the form

$$\left(\sum_{\alpha \to N-1} p_{K\alpha}\right)\left(\sum_{i=1}^{N-1} p_{Ni}(\nu_i - \delta_i n)\right) + \sum_{i=N}^{K} p_{Ki}(\nu_i - \delta_i n) > n$$

(obviously, for $i \leq N-1$ the equality $p_{Ni} = p_{N-1,i}$ holds). Now the claim of the lemma follows directly from the definition of the number a and the fact that $N \to N-1$. Q.E.D. for the lemma.

Now setting

$$\Sigma_0^* = \sum_{i=1}^{L} p^*_{Li}, \quad \Sigma_1^* = \sum_{i=L+1}^{N-1} p^*_{Ni},$$

by Propositions 2.4.1 and 3.1 we get

$$\sum_{i=1}^{L} p^*_{Li} m_i > \sum_{i=1}^{N-1} p^*_{Ni} \nu_i^2 + 4(1+a)n^2,$$

whence, taking into account the Noether-Fano inequality, we get

$$(210) \quad \sum_{i=1}^{L} p^*_{Li} m_i > \left[\frac{(2\Sigma_0^* + \Sigma_1^* - a)^2}{\Sigma_0^* + \Sigma_1^*} + 4(1+a)\right] n^2 = 4(\Sigma_0^* + 1)n^2 + \frac{(\Sigma_1^* + a)^2}{\Sigma_0^* + \Sigma_1^*} n^2.$$

Q.E.D. for Proposition 3.2.

4. Infinitely near singularities

In this section we show that under the assumption that the system Σ has no maximal subvariety of the form $\sigma^{-1}(P)$, where $P \subset \mathbb{P}$ is a linear subspace of codimension two, we prove that Σ has no maximal singularities with the centre B of codimension ≥ 4. Given what was shown in Section 2, it means that the linear system Σ has no maximal singularities at all. This contradiction completes the proof of Proposition 1.1.

4.1. The centre of the singularity is not contained in the ramification divisor. By Propositions 2.1 and 2.6 we may assume that the centre B of the maximal singularity of the linear system Σ is of codimension ≥ 4 (and the same is true for any other maximal singularity). Let $P \subset \mathbb{P}$ be a generic linear subspace of dimension $\operatorname{codim} B$ and $o \in \sigma^{-1}(P) \cap B$ is some point. Set $V_P = \sigma^{-1}(P)$.

Consider first the case when $\sigma(B) \not\subset W$. In this case $p = \sigma(o) \notin W$. The variety V_P is smooth,

$$\sigma_P = \sigma|_{V_P} \colon V_P \to P$$

is the double cover, branched over $W_P = W \cap P$,

$$\operatorname{Pic} V_P = \mathbb{Z} H_P,$$

where $H_P = H|_{V_P}$. Let Σ_P be the restriction of the system Σ onto V_P. This is a mobile linear system and the pair

$$(V_P, \frac{1}{n}\Sigma_P)$$

is not log canonical, where the point o is an isolated centre of a non-log canonical singularity. Let

$$\varphi \colon V_P^+ \to V_P$$

be the blow up of the point o, $E = \varphi^{-1}(o)$ the exceptional divisor. Set

$$Z_P = (D_1 \circ D_2)$$

to be the self-intersection of the linear system Σ_P and Z_P^+ its strict transform on V_P^+. By Theorem 2.4.1, for some plane $\Pi \subset E$ of codimension two the inequality

$$\operatorname{mult}_o Z_P + \operatorname{mult}_\Pi Z_P^+ > 8n^2 \tag{211}$$

holds. Now let

$$\varphi_P \colon P^+ \to P$$

be the blow up of the point $p = \sigma(o)$ and $E_P = \varphi_P^{-1}(p)$ the exceptional divisor, E identifies naturally with E_P. Let $\Lambda \subset P$ be the unique plane of codimension two, containing the point p and cutting out Π on $E_P = E$:

$$\Lambda^+ \cap E_P = \Pi,$$

where $\Lambda^+ \subset P^+$ is the strict transform. The subvariety $Q = \sigma_P^{-1}(\Lambda) \subset V_P$ is irreducible, of codimension two (with respect to V_P) and, moreover,

$$\deg Q = 2, \quad \operatorname{mult}_o Q = \operatorname{mult}_\Pi Q^+ = 1, \tag{212}$$

where $Q^+ \subset V_P^+$ is the strict transform. Since the cycle Z_P satisfies the inequality

$$\operatorname{mult}_o Z_P + \operatorname{mult}_\Pi Z_P^+ > \deg Z_P = 8n^2$$

and writing

$$Z_P = aQ + Z_P^\sharp,$$

where $a \in \mathbb{Z}_+$ and $Z_P^\sharp$ does not contain Q as a component, we obtain

$$\operatorname{mult}_o Z_P^\sharp + \operatorname{mult}_\Pi (Z_P^\sharp)^+ > \deg Z_P^\sharp,$$

$(Z_P^\sharp)^+$ is the strict transform. Finally, let R be the σ-preimage of a generic hyperplane in P, containing the point p and cutting out Π on $E_P = E$, that is,

$$\sigma(R)^+ \cap E_P = \Pi.$$

The divisor R contains no component of the effective cycle $Z_P^\sharp$, so that for the scheme-theoretic intersection

$$Z_R^\sharp = (Z_P^\sharp \circ R)$$

we obtain the inequality

$$\mathrm{mult}_o Z_R^\sharp \geq \mathrm{mult}_o Z_P^\sharp + \mathrm{mult}_\Pi (Z_P^\sharp)^+ > \deg Z_R^\sharp,$$

which is impossible. This contradiction proves Proposition 1.1 in the case when $\sigma(B) \not\subset W$.

4.2. The centre of the singularity is contained in the ramification divisor: the simple case. Consider, finally, the last case when $\sigma(B) \subset W$. Once again, we work on the variety $V_P = \sigma^{-1}(P)$, the linear system Σ_P is mobile and the point o is an isolated centre of log maximal singularity of this linear system. Set $p = \sigma(o)$. For the blow ups

$$\varphi\colon V_P^+ \to V_P \quad \text{and} \quad \varphi\colon P^+ \to P$$

of the points o and p, respectively, with the exceptional divisors

$$E = \varphi^{-1}(o) \quad \text{and} \quad E_P = \varphi_P^{-1}(p),$$

the double cover $\sigma_P\colon V_P \to P$ does not extend to an isomorphism of the exceptional divisors E and E_P. Let $T_P = T_p W_P$ be the tangent hyperplane to the branch divisor $W_P = W \cap P$ at the point p, $\mathbb{T}_P$ the corresponding hyperplane in E_P. It is easy to see that there exist a hyperplane $\mathbb{T} \subset E$ and a point $\xi \in E \backslash \mathbb{T}$ such that σ_P induces an isomorphism of $\mathbb{T}$ and $\mathbb{T}_P$ and the rational map

$$\sigma_E\colon E \dashrightarrow E_P$$

is the composition of the projection $\mathrm{pr}_\xi\colon E \dashrightarrow \mathbb{T}$ from the point ξ and the isomorphism $\mathbb{T} \cong \mathbb{T}_P$. In particular, $\sigma_E(E) = \mathbb{T}_P$ (all these facts are easy to check in suitable local coordinates $z_1, \dots, z_k$ at the point p on P, in which V_P is given by the local equation $y^2 = z_1$).

Let, as above, Z_P and Z_P^+ be the self-intersection of the linear system Σ_P and its strict transform on V_P^+, respectively. Let $\Pi \subset E$ be the plane of codimension two, satisfying the inequality (211),

$$\Pi_P = \sigma_E(\Pi) \subset \mathbb{T}_P \subset E_P$$

the image of the plane Π. Obviously, Π_P is a linear subspace in $\mathbb{T}_P$ of codimension 1 or 2. In the latter case for a generic hyperplane $R \ni p$, $R \subset P$, such that $R^+ \supset \Pi_P$, we get: none of the components of the effective cycle $(\sigma_P)_* Z_P$ of codimension two are contained in R. By the inequality (211) we get

$$\mathrm{mult}_o(\sigma^{-1}(R) \circ Z_P) \geq \mathrm{mult}_o Z_P + \mathrm{mult}_\Pi Z_P^+$$
$$> 8n^2 = \deg Z_P = \deg(\sigma^{-1}(R) \circ Z_P),$$

which is impossible. Therefore, Π_P is a hyperplane in $\mathbb{T}_P$. Let $\Lambda \subset P$ be the unique plane of codimension two, such that $\Lambda \ni p$ and $\Lambda^+ \cap E_P = \Pi_P$. The subvariety $Q = \sigma_P^{-1}(\Lambda)$ is irreducible. If Q^+ does not contain Π, then we get

$$\mathrm{mult}_o Q = 2, \quad \mathrm{mult}_\Pi Q^+ = 0,$$

so that $\mathrm{mult}_o Q + \mathrm{mult}_\Pi Q^+ = \deg Q$ and we may argue as above. Recall that by the conditions of general position (and taking into account the genericity of the subspace P), the point $o \in Q$ is an isolated quadratic singularity. If the rank of that singularity is at least 3, then the quadric $Q^+ \cap E$ is irreducible (and reduced), so that $\Pi \not\subset Q^+$. An easy dimension count (similar to the one carried out below in Section 5) shows that for $M \geq 8$ for a sufficiently general hypersurface W, a general subspace P and a general subspace $\Lambda \subset P$, $\Lambda \ni p$ of codimension two in P, where $p \in P \cap W$ is an arbitrary point, the rank of the corresponding quadratic singularity $o \in Q$ is at least 3.

However, for $M \leq 7$ the possibility $\Pi \subset Q^+$ takes place for any hypersurface W. Now we cannot argue in the same way as in the case when the point o does not lie on the ramification divisor: let $\Lambda \subset P$ be the unique plane of codimension two, such that $\Lambda \ni p$ and $\Lambda^+ \cap E_P = \Pi_P$. The subvariety $Q = \sigma_P^{-1}(\Lambda)$ is irreducible, however,

$$\operatorname{mult}_o Q = 2, \quad \operatorname{mult}_\Pi Q^+ = 1,$$

so that the arguments, similar to the case when $\sigma(o) \notin W$, do not work. To exclude this case, we need the improved technique of counting multiplicities (developed in Section 3).

4.3. The centre of the singularity is contained in the ramification divisor: the hard case. The following claim is true.

Lemma 4.1. *The inequality* $\nu = \operatorname{mult}_o \Sigma \leq 2n$ *holds.*

Proof. Consider the divisor $T = \sigma^{-1}(T_p W)$, where $T_p W \subset \mathbb{P}$ is the tangent hyperplane. The system Σ is mobile, so that the effective cycle $(D \circ T)$ is well defined, where $D \in \Sigma$ is a general divisor. Now we get

$$2\nu \leq \operatorname{mult}_o (D \circ T) \leq \deg (D \circ T) = \deg D = 4n,$$

as we claimed. Q.E.D.

Let $\Delta \subset P$, $\Delta \ni p$, be a generic 3-plane, so that $V_\Delta = \sigma^{-1}(\Delta)$ is a smooth variety, $\sigma_\Delta \colon V_\Delta \to \Delta$ a double cover, Σ_Δ the restriction of the system Σ onto V_Δ. The pair

$$(V_\Delta, \frac{1}{n}\Sigma_\Delta) \tag{213}$$

is not log canonical at the point o, and, moreover, o is an isolated centre of non-log canonical singularities of this pair.

Set $C = Q \cap V_\Delta = \sigma^{-1}(L)$, where $L = \Delta \cap \Lambda$ is the line, passing through the point p and tangent to W at that point (the plane Λ was defined in the end of the previous subsection). Set also

$$y = \Pi \cap V_\Delta^+,$$

where V_Δ^+ is the strict transform of V_Δ on V^+, that is,

$$\varphi_\Delta = \varphi|_{V_\Delta^+} \colon V_\Delta^+ \to V_\Delta$$

is the blow up of the point o with the exceptional plane E_Δ, $y = \Pi \cap E_\Delta$. There is a non-log canonical singularity of the pair (213), the centre of which on V_Δ^+ is the point y.

Consider the self-intersection

$$Z_\Delta = (D_1 \circ D_2) = Z|_{V_\Delta}$$

of the mobile linear system Σ_Δ and write

$$Z_\Delta = bC + Z_1,$$

where $b \in \mathbb{Z}_+$, the effective 1-cycle Z_1 does not contain the curve C as a component and for this reason satisfies the inequality

$$\operatorname{mult}_o Z_1 + \operatorname{mult}_y Z_1^+ \leq \deg Z_1 = 8n^2 - 2b. \tag{214}$$

Assume now that at the point o the curve C has two distinct branches:

$$C^+ \cap E_\Delta = \{y, y^*\},$$

where $y, y^* \in E_\Delta$ are distinct points. This assumption is justified for $M \geq 6$ by the conditions of general position, since for any subspace $\mathcal{U} \subset \mathbb{P}$ of codimension two, $\mathcal{U} \ni p$, the quadratic point $o \in \sigma^{-1}(\mathcal{U})$ is of rank at least two (see Proposition 1.4(iii)), so that the same is true for the quadratic point $o \in \sigma^{-1}(L) = C$, as well, since $L = \mathcal{U} \cap \Delta$, where Δ is a generic 3-plane, containing the point p.

In the notation of Subsection 3.3 let

$$\varphi_{i,i-1} \colon X_i \to X_{i-1},$$

$i = 1, \ldots, N$, $X_0 = V_\Delta$, be the resolution of the non-log canonical singularity, the centre of which on $X_1 = V_\Delta^+$ is the point $B_1 = y$. Set

$$\{1, \ldots, k\} = \{i \,|\, 1 \leq i \leq L, B_{i-1} \in C^{i-1}\}.$$

By the assumption about the branches of the curve C, we get $k \geq 2$, and, moreover, the subgraph with the vertices $1, \ldots, k$ is a chain.

Note that $b \geq 1$: otherwise the inequality

$$\operatorname{mult}_o Z_\Delta + \operatorname{mult}_y Z_\Delta^+ \leq \deg Z_\Delta = 8n^2$$

holds and one can argue word for word the same way as for $\sigma(o) \notin W$.

We have

$$p^*_{L1} = \cdots = p^*_{L,k-1}.$$

Furthermore, set

$$\mu_i = \operatorname{mult}_{B_{i-1}} Z_1^{i-1}$$

for $i = 1, \ldots, L$. By the inequality (214), we get the estimate

$$\begin{aligned} \sum_{i=1}^{L} p^*_{Li} m_i &= b\left(\sum_{i=1}^{k} p^*_{Li} + p^*_{L1}\right) + \sum_{i=1}^{L} p^*_{Li}\mu_i \\ &\leq b\left(\sum_{i=1}^{k} p^*_{Li} + p^*_{L1}\right) + \tfrac{1}{2}\deg Z_1 \sum_{i=1}^{L} p^*_{Li}. \end{aligned} \tag{215}$$

Lemma 4.2. *The following estimate holds:*

$$p^*_{L1} = \cdots = p^*_{L,k-1} \leq 1 + \sum_{i=k+1}^{L} p^*_{Li}.$$

Proof. For $L \geq i > j \geq 1$ denote by the symbol $\mathcal{P}^*_{i,j}$ the set of paths in the graph Γ^* going from the vertex i to the vertex j. We say that an arrow $i \overset{*}{\to} j$ is a *jump*, if $i \geq j + 2$. A path $\pi \in \mathcal{P}^*_{i,j}$ is said to be *simple*, if it contains no jumps, that is, it goes subsequently through all vertices

$$i \overset{*}{\to} i-1 \overset{*}{\to} \cdots \overset{*}{\to} a \overset{*}{\to} a-1 \overset{*}{\to} \cdots \overset{*}{\to} j.$$

It is obvious that each set $\mathcal{P}^*_{i,j}$ contains precisely one simple path. Denote it by the symbol σ_{ij}.

If the set $\mathcal{P}^*_{i,j} \setminus \{\sigma_{ij}\}$ is non-empty, then every path $\pi \in \mathcal{P}^*_{i,j} \setminus \{\sigma_{ij}\}$ has at least one jump. Let

$$l(\pi) \stackrel{*}{\to} q(\pi)$$

be the *last* jump in the path π (that is, the jump from the vertex $l(\pi)$ with the *smallest* number). Since the graph Γ^* is of class ≤ 2, the jump $a \stackrel{*}{\to} b$ is uniquely determined by the vertex a (since there are precisely two arrows coming out of the vertex a, and one of them is the arrow $a \stackrel{*}{\to} a-1$). Therefore, the path π is uniquely determined after the vertex $l(\pi)$: first, there is the uniquely determined jump to the vertex $q(\pi)$, and after that vertex (if $q(\pi) \neq j$) the path π is simple.

Denote by the symbol $\mathcal{J}^*_{i,j} \subset \{j+2, \ldots, i\}$ the set of such indices l, where $j+2 \leq l \leq i$, that from the vertex l there is a jump $l \stackrel{*}{\to} a \geq j$. It follows from what was said that, associating to every path $\pi \in \mathcal{P}^*_{i,j} \setminus \{\sigma_{ij}\}$ the vertex $l(\pi)$ of the last jump, we get a one-to-one correspondence between the sets

$$\mathcal{P}^*_{i,j} \setminus \{\sigma_{ij}\} \quad \text{and} \quad \coprod_{l \in \mathcal{J}^*_{i,j}} \mathcal{P}^*_{i,l}$$

(where the symbol $\coprod$ means the disjoint union). Therefore,

$$p^*_{ij} = 1 + \sum_{l \in \mathcal{J}^*_{i,j}} p^*_{il} \leq 1 + \sum_{l \geq j+2} p^*_{il},$$

since for $l \in \mathcal{J}^*_{i,j}$ in any case $j+2 \leq l$. Q.E.D. for the lemma.

Note that the argument above gives a new proof of Lemma 2.2.7 and, therefore, of the $4n^2$-inequality.

By Lemma 4.2, the right-hand side of the inequality (215) is bounded from above by the number

$$b + (b + \frac{1}{2} \deg Z_1) \sum_{i=1}^{L} p^*_{Li} = b + 4n^2 \sum_{i=1}^{L} p^*_{Li}.$$

Now Proposition 3.2 implies the estimate

$$b > 4n^2,$$

which is impossible. As we assume that $M \geq 6$, this contradiction completes the proof of Proposition 1.1 (for $M = 5$ more work is needed, see [**Pukh10**]). Q.E.D.

5. Generic double spaces

In this section we prove Propositions 1.2–1.4 on the geometric properties of double spaces of general position.

5.1. Lines on the variety V. Let us prove Proposition 1.2. The non-trivial part of that claim is that through every point there are at most finitely many lines; the fact that any (not necessarily generic) double space of index two is swept out by lines, is almost obvious. It is easy to see that the image $L = \sigma(C)$ of a line $C \subset V$ on $\mathbb{P}$ is a line in the usual sense and

$$\sigma|_C \colon C \to L \subset \mathbb{P}$$

is an isomorphism. Thus there are two possible cases: either $L \not\subset W$, so that $\sigma^{-1}(L) = C \cup C^*$ is a part of smooth rational curves (permuted by the Galois involution of the double cover σ), or $L \subset W$ is contained entirely in the branch divisor, that is, $\sigma^{-1}(L) = C$. The converse is also true: if a line $L \subset \mathbb{P}$ is such that the curve $\sigma^{-1}(L)$ is reducible or $L \subset W$, then $\sigma^{-1}(L)$ consists of two or one lines on V, respectively. An easy dimension count shows that on a generic hypersurface in $\mathbb{P}$ of degree $2(M-1)$ there are no lines, so that the second option does not take place. Furthermore, the double cover $\sigma^{-1}(L) \to L$ is reducible if and only if the divisor $W\,|\,_L$ on $L = \mathbb{P}^1$ is divisible by 2, that is,

$$\frac{1}{2}(W|_L) \in \operatorname{Div} L$$

is an integral divisor. Thus Proposition 1.2 follows immediately from the following fact.

Proposition 5.1. *For a generic smooth hypersurface $W \subset \mathbb{P}$ of degree $2(M-1)$ through every point $x \in \mathbb{P}$ there are finitely many lines L such that $W|_L \in 2\operatorname{Div} L$.*

Proof. Let us denote by the symbol $\mathcal{P}_k(\mathbb{P}^l)$ the space of homogeneous polynomials of degree k on the projective space $\mathbb{P}^l$ (that is, $H^0(\mathbb{P}^l, \mathcal{O}_{\mathbb{P}^l}(k))$), considered as an affine algebraic variety of dimension $\binom{k+l}{l}$. Let

$$\begin{array}{rccc} \operatorname{sq}\colon & \mathcal{P}_k(\mathbb{P}^l) & \to & \mathcal{P}_{2k}(\mathbb{P}^l), \\ \operatorname{sq}\colon & f & \mapsto & f^2 \end{array}$$

be the map of taking the square. Its image

$$\operatorname{sq}(\mathcal{P}_k(\mathbb{P}^l)) \subset \mathcal{P}_{2k}(\mathbb{P}^l)$$

will be denoted by the symbol $[\mathcal{P}_k(\mathbb{P}^l)]^2$. Consider the space of pairs

$$\Pi = \mathbb{P} \times \mathcal{P}_{2(M-1)}(\mathbb{P})$$

and set $\Pi(x) = \{x\} \times \mathcal{P}_{2(M-1)}(\mathbb{P})$ for an arbitrary point $x \in \mathbb{P}$. Set

$$Y(x) \subset \Pi(x)$$

to be the closed algebraic subset of pairs (x, F), $F \in \mathcal{P}_{2(M-1)}(\mathbb{P})$, defined by the condition

(+) the set of lines $L \subset \mathbb{P}$, $L \ni x$, for which $F\,|\,_L \in [\mathcal{P}_{M-1}(L)]^2$, is of positive dimension.

It is easy to see that the closure

$$\overline{\bigcup_{x \in \mathbb{P}} Y(x)} \subset \Pi$$

is a closed algebraic subset of dimension $\leq M + \dim Y(x)$. Therefore, Proposition 5.1, in turn, is implied by the following fact.

Proposition 5.2. *The codimension of the closed set $Y(x)$ in $\Pi(x) \cong \mathcal{P}_{2(M-1)}(\mathbb{P})$ is at least $M+1$.*

In fact, as we will see from the proof, a much stronger estimate for the codimension of the set $Y(x)$ holds. In particular, the claim of Proposition 5.1 remains true for double spaces of index two with elementary singularities (quadratic points). However, we do not need it here.

Proof of Proposition 5.2. Let $z_1, \dots, z_M$ be a system of affine coordinates on $\mathbb{P}$ with the origin at the point $x = (0, \dots, 0)$. We write the polynomial $F \in \mathcal{P}_{2(M-1)}(\mathbb{P})$ in the form

$$F = q_0 + q_1(z_1, \dots, z_M) + \cdots + q_{2(M-1)}(z_1, \dots, z_M),$$

where $q_i(z_*)$ is a homogeneous polynomial of degree i. The line $L \ni x$ corresponds to a set of homogeneous coordinates

$$(a_1 : \cdots : a_M) \in \mathbb{P}^{M-1},$$

$L = \{t(a_1, \dots, a_M) \,|\, t \in \mathbb{C}\}$. Obviously, $F|_L \in [\mathcal{P}_{M-1}(L)]^2$ if and only if the polynomial

$$q_0 + tq_1(a_*) + \cdots + t^{2(M-1)} q_{2(M-1)}(a_*) \in \mathbb{C}[t]$$

is a full square in $\mathbb{C}[t]$.

For each $k = 0, 1, \dots, 2(M-1)$ we define the set $Y_k(x) \subset \Pi(x)$, by the condition $+_k$ there exists an irreducible closed subset $Z \subset \mathbb{P}^{M-1}$ of positive dimension such that for a general line $L \in Z$ we have $F_L \in [\mathcal{P}_{M-1}(L)]^2$ and, moreover, $F|_L$ has a zero of order k at the point $x \in L$.

(Recall that we identify the points $\mathbb{P}^{M-1}$ with the lines in $\mathbb{P}$, passing through the point x.) Obviously, for an odd $k \notin 2\mathbb{Z}$ we have $Y_k = \emptyset$ and

$$Y(x) = \bigcap_{i=0}^{M-1} Y_{2i}(x).$$

Therefore, it is sufficient to prove the estimate of Proposition 5.2 for each of the (constructive) sets $Y_{2i}(x)$, $i = 0, \dots, M-1$. Consider first the set $Y_0(X)$ (corresponding to the lines on V, passing through the point outside the branch divisor). For $F \in Y_0(x)$ we have $q_0 \neq 0$ and we may assume that $q_0 = 1$.

Lemma 5.1. *For any $m \geq 1$ there exists a set of quasi-homogeneous polynomials,*

$$A_{m,i}(s_1, \dots, s_m) \in \mathbb{Q}[s_1, \dots, s_m],$$

of degree $\deg A_{m,i} = i \in \{m+1, \dots, 2m\}$, *where the weight of the variable s_j is* $\mathrm{wt}(s_j) = j$, *such that the polynomial*

$$1 + b_1 t + \cdots + b_{2m} t^{2m}$$

is a full square in $\mathbb{Q}[t]$ for $b_1, \dots, b_{2m} \in \mathbb{C}$ if and only if the following system of equalities is satisfied:

$$b_i = A_{m,i}(b_1, \dots, b_m),$$

$i = m+1, \dots, 2m.$

Proof. Consider the equality

$$1 + s_1 t + \cdots + s_{2m} t^{2m} = (1 + r_1 t + \cdots + r_m t^m)^2.$$

Equating the coefficients at the same powers of t, we find r_i as polynomials in $s_1, \dots, s_i$ for $i \leq m$ (with coefficients in $\mathbb{Z}[\frac{1}{2}]$). The equality of coefficients at $t^{m+1}, \dots, t^{2m}$, gives the required system of equations. Q.E.D.

By the lemma, the restriction of F onto the line $\{t(a_*)\}$ is a full square if and only if the system of equations

$$q_i(a_*) = A_{M-1,i}(q_1(a_*), \dots, q_{M-1}(a_*)), \tag{216}$$

$i = M, \dots, 2(M-1)$, is satisfied. This is a system of $(M-1)$ polynomial homogeneous equations in $(a_1 : \cdots : a_M)$ of degrees $M, \dots, 2(M-1)$, respectively (which, in particular, implies immediately that through every point $x \in V$ there are at least two lines, and in the case of general position through $x \in V$ there are

$$2 \cdot (M \cdot (M+1) \cdot \dots \cdot 2(M-1)) = 2\frac{(2M-2)!}{(M-1)!}$$

lines). As can be seen from (216), the coefficients of the right-hand side depend polynomially on the coefficients of the polynomials $q_1, \dots, q_{M-1}$. Therefore, $Y_0(x)$ consists of such polynomials $F \in \mathcal{P}_{2(M-1)}(\mathbb{P})$, for which the system (216) defines an algebraic set of positive dimension.

Now the codimension $\operatorname{codim} Y_0(x)$ can be estimated by the method used in Section 3.1. Assuming the polynomials $q_1, \dots, q_{M-1}$ to be fixed, we obtain the equality

$$\operatorname{codim} Y_0(x) = \operatorname{codim} Y_0^*(x),$$

where the closed set $Y_0^* \subset \mathcal{P}_M(\mathbb{P}^{M-1}) \times \dots \times \mathcal{P}_{2(M-1)}(\mathbb{P}^{M-1})$ (which consists of sets $q_M^*, \dots, q_{2(M-1)}^*$ of homogeneous polynomials of the corresponding degrees) is defined by the condition: the system of equations

$$q_M^* = \dots = q_{2(M-1)}^* = 0$$

has a positive-dimensional set of solutions. Repeating the proof of Lemma 3.1.3, define the subsets $Y_{0,j}^* \subset Y_0^*$ for $j = M, \dots, 2(M-1)$, fixing the first "incorrect" codimension:

$$Y_{0,j}^* = \{(q_M^*, \dots, q_{2(M-1)}^*) \mid \operatorname{codim}\{q_M^* = \dots = q_j^* = 0\} = j - M\},$$

so that

$$Y_0^* = \coprod_{j=M}^{2(M-1)} Y_{0,j}^*,$$

where the symbol $\coprod$ stands for a disjoint union (for instance, $Y_{0,M}^*$ consists of the tuples (q_*^*) with $q_M^* \equiv 0$). For $(q_M^*, \dots, q_{2(M-1)}^*) \in Y_{0,j}^*$ there is an irreducible component

$$B \subset \{q_M^* = \dots = q_{j-1}^* = 0\}$$

of codimension precisely $j - M$, on which q_j^* vanishes identically. Let

$$\pi \colon B \to \mathbb{P}^{\dim B} \subset \mathbb{P}^{M-1}$$

be a generic linear projection onto a generic $\dim B$-dimensional plane. Since the π-pullback of a non-zero homogeneous polynomial on $\mathbb{P}^{\dim B}$ does not vanish on B identically, we get the estimate

$$\operatorname{codim} Y_{0,j}^* \geq \dim \mathcal{P}_j(\mathbb{P}^{2M-1-j}) = \binom{2M-1}{j}$$

(since $\dim B = 2M - 1 - j$, $j \in \{M, \dots, 2M-2\}$), so that

$$\operatorname{codim} Y_0^* \geq \min\left\{ \binom{2M-1}{j} \,\middle|\, j = M, \dots, 2M-2 \right\} = 2M - 1.$$

Thus $\operatorname{codim} Y_0(x) \geq 2M - 1 \geq M + 1$, as required.

Let us consider now the problem of estimating the codimension of the set $Y_k(x)$, $k = 2e \geq 2$. For $F \in Y_k(x)$ there exists a set $Z_F \subset \mathbb{P}^{M-1}$ of positive dimension, on which the following polynomials vanish identically,

$$q_0, q_1(z_*), \dots, q_{k-1}(z_*),$$

and for a point of general position $(a_1 : \dots : a_M) \in Z_F$ the polynomial

$$t^k q_k(a_*) + \dots + t^{2(M-1)} q_{2(M-1)}(a_*) \in \mathbb{C}[t]$$

is a full square and, moreover, $q_k(a_*) \neq 0$. Applying Lemma 5.1, we obtain the system of equalities

$$\frac{q_i(a_*)}{q_k(a_*)} = A_{M-e-1,i}\left(\frac{q_{k+1}(a_*)}{q_k(a_*)}, \dots, \frac{q_{M+e-1}(a_*)}{q_k(a_*)}\right),$$

$i = M-e, \dots, 2(M-e-1)$, or, after multiplying by $q_k(a_*)^i$,

$$q_i(a_*) q_k(a_*)^{i-1} = A^+_{M-e-1,i}(q_k(a_*), q_{k+1}(a_*), \dots, q_{M+e-1}(a_*)),$$

$i = M-e, \dots, 2(M-e-1)$, where $A^+(\cdot)$ is the appropriately modified polynomial. We obtain a system of $M+e-1$ homogeneous equations on $\mathbb{P}^{M-1}$, and the set $Y_k(x)$ consists of those polynomials F, for which that system has a set of solutions of positive dimension, on which q_k does not vanish identically. Here q_0 is a non-zero constant by definition of the set $Y_k(x)$ for $k = 2e \geq 2$.

Now we argue as above: We assume the polynomials $q_k, \dots, q_{M+e-1}$ to be fixed, so that the codimension of the set $Y_k(x)$ is the codimension of the subset

$$Y_k^* \subset \mathbb{C} \times \mathcal{P}_1(\mathbb{P}^{M-1}) \times \dots \times \mathcal{P}_{k-1}(\mathbb{P}^{M-1}) \times \mathcal{P}_{M+e}(\mathbb{P}^{M-1}) \times \dots \times \mathcal{P}_{2(M-1)}(\mathbb{P}^{M-1}),$$

defined by the condition: the sequence

$$(q_0, q_1, \dots, q_{k-1}, q_{M+e}, \dots, q_{2(M-1)}) \in Y_k^*$$

if and only if $q_0 = 0$ and the system of equations

$$q_1 = \dots = q_{k-1} = q_{M+e} = \dots = q_{2(M-1)} = 0$$

defines an algebraic set that has a component of positive dimension, on which $q_k \not\equiv 0$. Now we may forget about the latter condition.

Now we estimate the codimension of the set Y_k^* by the method of Section 3.1 in precisely the same way, as it was done above for $k = 0$: We fix the first "incorrect" codimension, when the next polynomial q_i in the sequence that was written above vanishes on an irreducible component, defined by the previous equations (which is of "correct" codimension). We get the worst estimate at the first step: the condition $q_1 \equiv 0$ (together with the condition $q_0 = 0$) gives the codimension

$$\operatorname{codim} Y_2^* = M + 1,$$

and this estimate is optimal. Indeed, $q_1 \equiv 0$ means that the branch divisor is singular at the point x, and then through this point there is a one-dimensional family of lines. In all other cases the estimate for $\operatorname{codim} Y_k^*$ is considerably stronger (we omit the elementary computations). Q.E.D. for Proposition 1.2.

5.2. Isolated singular points. The proof of Proposition 1.3 is elementary and we just point out its main steps. Assume that for some subspace $P \subset \mathbb{P}$ of codimension two the intersection $P \cap W$ has a whole curve C of singular points. It is convenient to consider the pair $p \in C$, where p is an arbitrary point, so that in any case

$$P \subset T_p W.$$

There is a $(2M-3)$-dimensional family of pairs $(p, P \ni p)$, satisfying this condition. It is sufficient to show that the number of independent conditions, which are imposed on the (non-homogeneous) polynomial

$$f|_P = f(z_1, \ldots, z_{M-2})$$

of degree $2M-2$ by the condition that the hypersurface $\{f|_P = 0\} = W \cap P$ contains a curve C of singular points, passing through $p = (0, \ldots, 0)$, is at least $2M-2$. There are three possible cases:

— C is a line,

— C is a plane curve, $C \subset \Lambda \subset P$, where Λ is some 2-plane,

— the linear span of the curve C is a k-plane, where $k \geq 3$.

In the first case one can compute the number of independent conditions precisely, this is an elementary exercise.

In the second case the plane curve $\{f|_\Lambda = 0\}$ has an irreducible component C of degree ≥ 2 and multiplicity ≥ 2, which gives an estimate from below for the number of independent conditions (which is essentially stronger than we need).

In the third case on the curve C we choose $3(M-1)$ points in general position (neither three lie on a line and neither four lie in the same plane). It is easy to check (considering hypersurfaces that are unions of singular quadrics), that being singular at these points imposes on f independent conditions, which completes the proof of Proposition 1.3. (In fact, the codimension of the set of hypersurfaces with a whole curve of singular points is much higher, but we do not need that.) The details are left to the reader. Q.E.D.

5.3. The rank of quadratic singularities. Let us prove Proposition 1.4. We will show the claim (i). The claims (ii) and (iii) are proved in a similar way. It is easy to check that the planes $P \subset \mathbb{P}$ of codimension two that are tangent to W at at least one point (that is, $\operatorname{Sing} P \cap W \neq \emptyset$) form a $(2M-3)$-dimensional family. So it is sufficient to prove the following fact.

Let $P \subset \mathbb{P}$ be a fixed plane of codimension two, $p \in P$ a fixed point,

$$\mathcal{W} = \mathbb{P}(H^0(\mathbb{P}, \mathcal{O}_{\mathbb{P}}(2M-2)))$$

the space of hypersurfaces of degree $2M-2$. Let us define the subset $\mathcal{W}_P \subset \mathcal{W}$ by the conditions:

- a hypersurface $W \in \mathcal{W}_P$ is non-singular at the point p,
- the tangent hyperplane T_pW contains P,
- the point p is an isolated singular point of the intersection $P \cap W$.

For $W \in \mathcal{W}_P$ set $\sigma_W \colon V_W \to \mathbb{P}$ to be the double cover, branched over W, $o = \sigma_W^{-1}(p)$ is a singular point, $R = \sigma_W^{-1}(P)$, and

$$\varphi \colon V_W^+ \to V_W$$

the blow up of the subvariety R. On the variety V_W^+ there is a unique singular point $o^+ \in \varphi^{-1}(o)$. Let us define the closed subset $Y \subset \mathcal{W}_P$ by the condition that for $W \in Y$ the rank of the quadratic singularity $o^+ \in V_W^+$ is at most 3. Now part (i) of Proposition 1.4 follows immediately from the estimate

$$\operatorname{codim}(Y \subset \mathcal{W}_P) \geq 2M-2 \tag{217}$$

for $M \geq 6$.

The proof of the inequality (217) is obtained by simple local computations which we will just describe. Let $(z_1, \dots, z_M)$ be affine coordinates at the point p, where the plane P is defined by the system of equations $z_1 = z_2 = 0$, and the tangent hyperplane to W is $z_1 = 0$. The local equation of the double cover V_W at the point $o = \sigma_W^{-1}(p)$ is of the form

$$y^2 = z_1 + q_2(z_1, \dots, z_M) + q_3(z_*) + \dots ,$$

and the local equation of the blow up V_W^+ at the point o^+ is of the form

$$u^2 = u_1 u_2 + q_2(0, u_2, \dots, u_M) + \dots .$$

This implies that the condition that the rank of the quadratic point o^+ is at most $M + 1 - k$, imposes

$$\frac{k(k+1)}{2}$$

independent conditions on the coefficients of the equation of the hypersurface W. If $M + 1 - k \leq 3$, then we get at least

$$\frac{(M-2)(M-1)}{2}$$

independent conditions, which for $M \geq 6$ is strictly higher than $2M - 3$. Q.E.D. part (i) of Proposition 1.4. The claims (ii) and (iii) are shown in a similar way.

Notes and references

This paper is based entirely on the paper [**Pukh10**]. Up to this day, there were only very few papers describing the birational type of Fano varieties of index $r \geq 2$. Of course, quite a lot of work has been done on *particular* birational maps between Fano varieties of higher index; for instance, there is a huge literature on birational transformations of projective spaces $\mathbb{P}^m$, $m \geq 3$, but we mean the work providing some *global and complete* information.

Fano himself pioneered the study of varieties of index ≥ 2, trying to describe the group of birational self-maps and prove non-rationality of three-dimensional cubics [**Fano3**]. However, his ideas were much ahead of the techniques of his time and the problem was much too complicated for the methods that were available to him. Only after the success of the paper of Iskovskikh and Manin [**IM**] it became possible to consider the cubics and other Fano varieties of higher index. It was natural to try to apply the new technique of the method of maximal singularities to those varieties and such attempts were immediately made: in [**I80**] the basic definitions and some constructions were given for varieties of an arbitrary index $r \geq 1$ and some progress was made towards a description of the birational type of the Veronese double cone of dimension three. In [**Kh84**] that description was announced to be completed (in particular, the Veronese double cone was claimed to be non-rational), however, the proof was faulty and a correct proof, hard and lengthy, was presented only 20 years later in [**Grin03b, Grin04**]. Since a Veronese double cone admits a two-dimensional family of structures of a del Pezzo fibration over $\mathbb{P}^1$, it was simply not possible to describe its birational geometry without being able to work with del Pezzo fibrations, in other words, before the corresponding technique was developed in [**Pukh98a**].

Apart from the Veronese double cone, attempts were made to apply the method of maximal singularities to Fano three-fold double spaces of index two (Khashin) and even the cubic three-fold hypersurfaces (Tregub). The techniques available today are much stronger and more efficient and versatile, and it is not clear whether these two classes of three-folds are within reach (although there are grounds for cautious optimism); what could be done in the 80s with only very primitive tools available? Nevertheless, a description of the group of birational self-maps of the three-fold double spaces of index two (corresponding to the value $M = 3$ in the notations of this chapter) was announced as a completed result at ICM'83 in Warsaw [**I83**]. The list of generators of the group $\mathrm{Bir}\, V$ was produced by Khashin in his PhD thesis but no publication followed. (As we mentioned in Chapters 2 and 6, a lot of work on birational geometry of three-folds that was done in Moscow in the 80s did not produce complete results and for that reason remained unpublished, only very few short accounts like [**Kh86**] appeared.) See [**Grin04, Grin06**] and Chapter 6 for a discussion of the current state of this problem.

Thus, up to [**Pukh10**], the Veronese double cone of dimension three remained the only Fano variety of index two, birational geometry which was completely studied. A series of remarkable results were obtained by means of other methods: non-rationality of the three-dimensional cubic was proved by Clemens and Griffiths in [**CG**] (see also [**Tyu72, Tyu79**]), non-rationality of "very general" Fano hypersurfaces of arbitrary dimension and of index two and higher was proved by Kollár (see [**Kol95a**]), non-rationality of double spaces $\mathbb{P}^3$ of index two follows from the fact that they admit no structures of a conic bundle, as Tikhomirov showed in [**Tikh81, Tikh83, Tikh86**]. These are just three examples; we do not give a complete list of those results, since the methods used in the above mentioned papers are very far from the method of maximal singularities.

Bibliography

[AlgS] Algebraic surfaces. By the members of the seminar of I.R. Shafarevich. I.R. Shafarevich, ed., Proc. Steklov Math. Inst. **75**. 1965. English transl. by AMS, 1965. 281 pp.

[Al87] Alexeev V.A., On conditions for the rationality of three-folds with a pencil of del Pezzo surfaces of degree 4, Math. Notes **41** (1987), 408-411.

[Al93] Alexeev V. Two two-dimensional terminations. Duke Math. J. **69** (1993), 527-545.

[ArMum] Artin M., Mumford D., Some elementary examples of unirational varieties which are not rational, Proc. London Math. Soc. **25** (1972), 75-95.

[BaBeau00] Bayle L. and Beauville A., Birational involutions of $\mathbb{P}^2$. Asian J. Math. **4** (2000), No. 1, 11-17.

[Bar] Bardelli F., Polarized mixed Hodge structures: On irrationality of threefolds via degeneration. Ann. Mat. Pura et Appl. **137** (1984), 287-369.

[Beau77] Beauville A., Variétés de Prym et jacobiennes intermédiares. Ann. Sci. École Norm. Sup. **10** (1977), No. 3, 309-391.

[Beau82] Beauville A., Les singularités du diviseur Θ de la jacobienne intermédiare de l'hypersurface cubique dans $\mathbb{P}^4$. In: Lecture Notes in Math. **947** (1982), 190-208.

[Beau08] Beauville A., p-elementary subgroups of the Cremona group. J. Algebra **314** (2007), No. 2, 553-564.

[Beau11] Beauville A., Non-rationality of the symmetric sextic Fano three-fold. arXiv:1102.1255 [math.AG], 3 pp.

[BeBl04] Beauville A. and Blanc, J. On Cremona transformations of prime order. C.R. Math. Acad. Sci. **339** (2004), No. 4, 257-259.

[BCTSSD] Beauville A., Colliot-Thélène J.-L., Sansuc J.-J. and Swinnerton-Dyer P., Variétés stablement rationnelles non rationnelles. Ann. of Math. **121** (1985), No. 2, 283-318.

[BeS] Beheshti R. and Starr J. M., Rational surfaces in index-one Fano hypersurfaces, arXiv:math.AG/0601710, 13 pp.

[BCHM] Birkar C., Cascini P., Hacon C.D. and McKernan J., Existence of minimal models for varieties of log general type. J. Amer. Math. Soc. **23** (2010), No. 2, 405-468.

[Bi07] Birkar C., Ascending chain condition for log canonical thresholds and termination of log flips. Duke Math. J. **136** (2007), 173-180.

[Bl07] Blanc J., Finite abelian subgroups of the Cremona group of the plane. C. R. Math. Acad. Sci. Paris **344** (2007), No. 1, 21-26.

[Bl10] Blanc J., Groupes de Cremona, connexité et simplicité. Ann. Sci. Éc. Norm. Supér. **43** (2010), No. 2, 357-364.

[Bl11] Blanc J., Elements and cyclic subgroups of finite order of the Cremona group. Comment. Math. Helv. **86** (2011), No. 2, 469-497.

[BCZ] Brown G., Corti A. and Zucconi F. Birational geometry of 3-fold Mori fibre spaces. In: The Fano conference. Univ. Torino, Turin. 2004. 235-275.

[CL] Call F. and Lyubeznik G., A simple proof of Grothendieck's theorem on the parafactoriality of local rings, Contemp. Math. **159** (1994), 15-18.

[Cam92] Campana F., Connexité rationnelle des variétés de Fano. Ann. Sci. Éc. Norm. Supér., 1992, IV. Sér. **25**, 539-545.

[CPP] Campana F., Peternell Th. and Pukhlikov A. V., Generalized Tsen theorem and rationally connected Fano fibrations. Sbornik: Mathematics. **193** (2002), No. 9-10, 1443-1468.

[CaL10] Cantat S. and Lamy S., Normal subgroups in the Cremona group. arXiv: math1007.0895, 53 p.

[Cas] Castravet A.-M., Examples of Fano varieties of index one that are not birationally rigid, Proc. Amer. Math. Soc. **135** (2007), No. 12, 3783-3788.

[Ch00] Cheltsov I. A., On a smooth four-dimensional quintic. Sbornik: Mathematics. **191** (2000), No. 9-10, 1399-1419.

[Ch00b] Cheltsov I.A., Log pairs on hypersurfaces of degree N in $\mathbb{P}^N$, Math. Notes. **68** (2000), No.1-2, 113-119.

[Ch01] Cheltsov I. A., Log canonical thresholds on hypersurfaces. Sbornik: Mathematics. **192** (2001), No. 8, 155-172.

[Ch03] Cheltsov I. A., Non-rationality of a four-dimensional smooth complete intersection of a quadric and a quartic, not containing a plane, Sbornik: Mathematics, **194** (2003), 1679-1699.

[Ch04a] Cheltsov I. A., Conic bundles with a large discriminant. Izvestiya: Mathematics **68** (2004), No. 2, 429-434.

[Ch04b] Cheltsov I. A., A double space with a double line. Sbornik: Mathematics **195** (2004), No. 9-10, 1503-1544.

[Ch05a] Cheltsov I. A., Nonexistence of elliptic structures on general Fano complete intersections of index one. Moscow Univ. Math. Bull. **60** (2005), No. 3, 30-33.

[Ch05b] Cheltsov I. A., Birationally rigid del Pezzo fibrations. Manuscripta Math. **116** (2005), No. 4, 385-396.

[Ch05c] Cheltsov I. A., Birationally rigid Fano varieties. Russian Math. Surveys. **60** (2005), No. 5, 875-965.

[Ch06a] Cheltsov I., Double cubics and double quartics, Math. Z. **253** (2006), no. 1, 75-86.

[Ch06b] Cheltsov I. A., Local inequalities and the birational superrigidity of Fano varieties. Izvestiya: Mathematics. **70** (2006), no. 3, 605-639.

[Ch07a] Cheltsov I. A., On nodal sextic fivefold. Math. Nachr. **280** (2007), No. 12, 1344-1353.

[Ch07b] Cheltsov I. A., Elliptic structures on weighted three-dimensional Fano hypersurfaces. Izvestiya: Mathematics **71** (2007), No. 4, 765-862.

[Ch08] Cheltsov I. A., Fano varieties with many self-maps, Adv. Math. **217** (2008), No. 1, 97-124.

[ChG] Cheltsov I. A. and Grinenko M. M., Birational rigidity is not an open property. 2006. arXiv:math/0612159, 26 pp.

[ChK] Cheltsov I. and Karzhemanov I., Halphen pencils on quartic threefolds. Adv. Math. **223** (2010), No. 2, 594-618.

[ChP06] Cheltsov I. A. and Park J., Weighted Fano threefold hypersurfaces, Crelle J. reine angew. Math. **600** (2006), 81-116.

[ChP09] Cheltsov I. and Park J., Halphen pencils on weighted Fano threefold hypersurfaces. Cent. Eur. J. Math. **7** (2009), No. 1, 1-45.

[ChPW07] Cheltsov I., Park J. and Won J., Log canonical thresholds of certain Fano hypersurfaces. 2007. arXiv:math/0706.075, 32 pp.

[Clem74] Clemens H., Applications of the theory of Prym varieties. Proc. Int. Cong. Math. (Vancouver, 1974). Vol. 1, 415-421.

[Clem82] Clemens H., Degeneration techniques in the study of threefolds. Lecture Notes in Math. **947**, 93-154.

[Clem83] Clemens H., Double solids. Adv. in Math. **47** (1983), No. 2, 107-230.

[CG] Clemens H. and Griffiths Ph., The intermediate Jacobian of the cubic threefold. Ann. of Math. **95** (1972), 281-356.

[CKM] Clemens H., Kollár J. and Mori Sh., Higher-dimensional complex geometry. Asterisque **166** (1988).

[Co95] Corti A., Factoring birational maps of threefolds after Sarkisov. J. Algebraic Geom. **4** (1995), no. 2, 223-254.

[Co00] Corti A., Singularities of linear systems and 3-fold birational geometry, in "Explicit Birational Geometry of Threefolds", London Mathematical Society Lecture Note Series **281** (2000), Cambridge University Press, 259-312.

[CoMe] Corti A. and Mella M., Birational geometry of terminal quartic 3-folds. I. Amer. J. Math. **126** (2004), No. 4, 739-761.

[CPR] Corti A., Pukhlikov A. and Reid M., Fano 3-fold hypersurfaces. In: "Explicit Birational Geometry of Threefolds", London Mathematical Society Lecture Note Series **281** (2000), Cambridge University Press, 175-258.

[FEM03] de Fernex T., Ein L. and Mustaţă M., Bounds for log canonical thresholds with applications to birational rigidity. Math. Res. Lett. **10** (2003), No. 2-3, 219-236.

[dF04] de Fernex T., On planar Cremona maps of prime order. Nagoya Math. J. **174** (2004), 1-28.

[dF06] de Fernex T., Birationally rigid hypersurfaces. arXiv: math/0604213, 23 pp.

[FEM11] de Fernex T., Ein L. and Mustaţă M., Log canonical thresholds on varieties with bounded singularities. Classification of Algebraic Varieties, 221-257, EMS , Zürich, 2011.

[Dol66] Dolgachev I. V., Rational surfaces with a pencil of elliptic curves. Math. USSR Izvestiya **30** (1966), 1073-1100.

[Dol83] Dolgachev I. V., Weyl groups and Cremona transformations. In: Proc. Symp. Pure Math. **40**, AMS, 1983.

[Dol09] Dolgachev I. V., On elements of order p^s in the plane Cremona group over a field of characteristic p. Proc. Steklov Inst. Math. **264** (2009), No. 1, 48-55.

[Dol10] Dolgachev I. V., Finite subgroups of the plane Cremona group. Adv. Stud. Pure Math. **60**, Math. Soc. Japan, 2010.

[DI09a] Dolgachev I. V. and Iskovskikh V. A., On elements of prime order in the plane Cremona group over a perfect field. Int. Math. Res. Not. 2009, No. 18, 3467-3485.

[DI09b] Dolgachev I. V. and Iskovskikh V. A., Finite subgroups of the plane Cremona group. In: Algebra, Arithmetic and Geometry: in honor of Yu. I. Manin., Vol. I, 443-548.

[EM08] Ein L. and Mustata M., Log canonical thresholds on smooth varieties: the ascending chain condition, arXiv:0811.4444 [math.AG]. 7 pp.

[EP] Eckl Th. and Pukhlikov A.V., On the locus of non-rigid hypersurfaces, arXiv:1210.3715 [math.AG]. 17 pp.

[Fano1] Fano G., Sopra alcune varieta algebriche a tre dimensioni aventi tutti i generi nulli, Atti Acc. Torino **43** (1908), 973-977.

[Fano2] Fano G., Osservazioni sopra alcune varieta non razionali aventi tutti i generi nulli, Atti Acc. Torino **50** (1915), 1067-1072.

[Fano3] Fano G., Nuove ricerche sulle varieta algebriche a tre dimensioni a curve-sezioni canoniche, Comm. Rent. Ac. Sci. **11** (1947), 635-720.

[Ful] Fulton W., Intersection Theory, Springer-Verlag, 1984.

[FL10] Furter J.-Ph. and Lamy S. Normal subgroup generated by a plane polynomial automorphism. Transf. Groups **15** (2010), No. 3, 577-610.

[Giz] Gizatullin M. Kh., Defining relations for the Cremona group of the plane. Izv. Akad. Nauk SSSR Ser. Mat. **46** (1982), No. 5, 909-970, 1134.

[GHS] Graber T., Harris J. and Starr J., Families of rationally connected varieties. J. Amer. Math. Soc. **16** (2002), no. 1, 57-67.

[GrH] Griffiths Ph., Harris J., Principles of algebraic geometry. Wiley-Interscience. New York, 1978.

[Grin98] Grinenko M. M., Birational automorphisms of a three-dimensional double cone. Sbornik: Mathematics **189** (1998), no. 7-8, 991-1007.

[Grin00] Grinenko M. M., Birational properties of pencils of del Pezzo surfaces of degrees 1 and 2. Sbornik: Mathematics. **191** (2000), no. 5-6, 633-653.

[Grin03a] Grinenko M. M., Birational properties of pencils of del Pezzo surfaces of degrees 1 and 2. II. Sbornik: Mathematics. **194** (2003), no. 5-6, 669-696.

[Grin03b] Grinenko M. M., On the double cone over the Veronese surface. Izvestiya: Mathematics **67** (2003), No. 3, 421-438.

[Grin04] Grinenko M. M., Mori structures on Fano threefold of index 2 and degree 1. Proc. Steklov Inst. Math. **246** (2004, no. 3), 103-128.

[Grin06] Grinenko M. M., Fibrations into del Pezzo surfaces, Russian Math. Surveys. **61** (2006), no. 2, 255-300.

[Grin11] Grinenko M. M., Three-dimensional quartics with a plane. Sbornik: Mathematics **202** (2011), No. 1-2, 207-241.

[HM09] Hacon C. D. and McKernan J., The Sarkisov program, arXiv:0905.0946 [math.AG]

[HM10] Hacon C. D. and McKernan J., Existence of minimal modesl for varieties of log general type. II. J. Amer. Math. Soc. **23** (2010), No. 2, 469-490.

[Hal] Halphen M., Sur les curbes planes du sixiéme degré à neuf points doubles, Bull. Soc. Math. France **X** (1881), 162-172.

[Hart] Hartshorne R., Algebraic geometry. Graduate Texts in Mathematics, **52**. Springer-Verlag, New York, Heidelberg, 1977.

[HasTsch] Hassett B., Tschinkel Yu., Weak approximation over function fields. Invent. Math. **163** (2006), 171-190.

[Hi] Hironaka H., Resolution of singularities of an algebraic variety over a field of characteristic zero. I, II. Ann. of Math. **79** (1964), 109-203, 205-326.

[Hud] Hudson H., Cremona transformations in plane and space. Cambridge University Press, 1927.

[Hul] Hulek, K. Elementary algebraic geometry. Student Mathematical Library, **20**. American Mathematical Society, Providence, RI, 2003.

[I67] Iskovskikh V. A., Rational surfaces with a pencil of rational curves. Math. Sbornik **74** (1967), 608-638.

[I70] Iskovskikh V. A., Rational surfaces with a pencil of rational curves and with a positive square of canonical class. Math. Sbornik **83** (1970), 90-119.

[I77a] Iskovskikh V. A., Birational automorphisms of three-dimensional algebraic varieties. Dokl. Akad. Nauk SSSR **234** (1977), No. 4, 743-745.

[I77b] Iskovskikh V. A., Birational automorphisms of the Fano 3-fold V_6^3. Dokl. Akad. Nauk SSSR **235** (1977), No. 3, 509-511.

[I80] Iskovskikh V. A., Birational automorphisms of three-dimensional algebraic varieties, J. Soviet Math. **13** (1980), 815-868.

[I83] Iskovskih V. A., Algebraic threefolds with special regard to the problem of rationality. Proc. of Intern. Cong. of Math. 1983. V. 1-2, 733-746.

[I87] Iskovskikh V. A., On the rationality problem for conic bundles. Duke Math. J. **54** (1987), No. 2, 271-294.

[I91a] Iskovskikh V. A., A simple proof of a theorem of Gizatullin. Proc. Steklov Inst. Math. (1991), 127-133.

[I91b] Iskovskikh V. A., On the rationality problem for conic bundles. Math. USSR-Sbornik **72** (1992), No. 1, 105-111.

[IT94] Iskovskikh V. A. and Tregub S. L., Relations in the two-dimensional Cremona group over a non-closed field. Proc. Steklov Inst. Math. **207** (1995), No. 6, 111-135.

[I95] Iskovskikh V. A., On the rationality problem for three-dimensional algebraic varieties fibred into del Pezzo surfaces. Proc. Steklov Inst. Math. **208** (1995), 128-138.

[I96a] Iskovskikh V. A., On a rationality criterion for conic bundles. Sbornik: Mathematics **187** (1996), No. 7, 1021-1038.

[I96b] Iskovskikh V. A., Factorization of birational mappings of rational surfaces from the viewpoint of Mori theory. Russian Math. Surveys, **51** (1996), No. 4, 585-652.

[I01] Iskovskikh V. A., Birational rigidity of Fano hypersurfaces in the framework of Mori theory. Russian Math. Surveys. **56** (2001), no. 2, 207-291.

[IM] Iskovskikh V. A. and Manin Yu. I., Three-dimensional quartics and counterexamples to the Lüroth problem, Math. USSR Sb. **86** (1971), no. 1, 140-166.

[IPr] Iskovskikh V. A. and Prokhorov Yu. G., Fano varieties. Encyclopaedia Math. Sci. Vol. 47. Springer, Berlin, 1999.

[IP] Iskovskikh V. A. and Pukhlikov A. V., Birational automorphisms of multi-dimensional algebraic varieties, J. Math. Sci. **82** (1996), 3528-3613.

[ISh] Iskovskikh V. A. and Shokurov V. V., Birational models and flips. Russian Math. Surveys. **60** (2005), No. 1, 27-94.

[JPR05] Jahnke P., Peternell Th. and Radloff I., Threefolds with big and nef anticanonical bundles. I. Math. Ann. **333** (2005), No. 3, 569-631.

[JPR11] Jahnke P., Peternell Th. and Radloff I., Threefolds with big and nef anticanonical bundles. II. Cent. Eur. J. Math. **9** (2011), No. 3, 449-488.

[Ka96] Kawamata Yu., Divisorial contractions to 3-dimensional terminal quotient singularities. In: Higher-Dimensional Complex Varieties, de Gruyter, Berlin, 1996, 241-246.

[KMM] Kawamata Yu., Matsuda K. and Matsuki K., Introduction to the minimal model program. Algebraic Geometry, Sendai, 1985, 283-360, Adv. Stud. Pure Math. **10**, North-Holland, Amsterdam, 1987.

[Kh84] Khashin S. I., Birational automorphisms of the Veronese double cone of dimension three. Moscow Univ. Math. Bull. 1984, no. 1, 13-16.

[Kh86] Khashin S. I., Birational automorphisms of Fano varieties of index 1 of degree 10. In: Algebra, Logic and Number Theory. (In Russian) Moscow State Univ., Moscow, 1986, 84-86.

[KolMM] Kollár J., Miyaoka Y. and Mori S., Rationally connected varieties, J. Alg. Geom. **1** (1992), 429-448.

[Kol93] Kollár J., et al., Flips and Abundance for Algebraic Threefolds, Asterisque **211**, 1993.

[Kol95a] Kollár J., Nonrational hypersurfaces, J. Amer. Math. Soc. **8** (1995), 241-249.

[Kol95b] Kollár J., Singularities of pairs, in: Algebraic Geometry, Santa Cruz 1995, 221-286.

[Kol96] Kollár J., Rational curves on algebraic varieties. Springer-Verlag, Berlin, 1996.

[Kol00] Kollár J., Nonrational covers of $\mathbb{C}P^m \times \mathbb{C}P^n$. In: "Explicit Birational Geometry of Threefolds", London Mathematical Society Lecture Note Series **281** (2000), Cambridge University Press, 51-71.

[Kol07] Kollár J., Lectures on resolution of singularities. Annals of Mathematics Studies. **166** Princeton Univ. Press, Princeton, 2007.

[Kol09] Kollár J., Birational rigidity of Fano varieties and field extensions. Proc. Steklov Math. Inst. **264** (2009), No. 1, 96-101.

[Kul86] Kulikov, Vik. S., Minimal objects of algebraic spaces. Izv. Akad. Nauk SSSR Ser. Mat. **50** (1986), no. 4, 726–740, 878.

[Kul06] Kulikov V.S., A remark on the non-rationality problem for generic cubic fourfolds, arXiv:math.AG/0610963.

[M66] Manin Yu. I., Rational surfaces over perfect fields. Publ. Math. IHES. **30** (1966), 55-113.

[M67] Manin Yu. I., Rational surfaces over perfect fields. II. Mat. Sb. **72** (1967), 161-192.

[M72] Manin Yu. I., Cubic forms. Algebra, geometry, arithmetic. Russian original — Moscow, Nauka, 1972. English translation, second edition. North-Holland Mathematical Library, **4.** North-Holland Publishing Co., Amsterdam, 1986.

[Me04] Mella M., Birational geometry of quartic 3-folds. II. Math. Ann. **330** (2004), No. 1, 107-126.

[MoPr08a] Mori Sh. and Prokhorov Yu., On $\mathbb{Q}$-conic bundles. Publ. Res. Math. Sci. **44** (2008), No. 2, 315-369.

[MoPr08b] Mori Sh. and Prokhorov Yu., On $\mathbb{Q}$-conic bundles. II. Publ. Res. Math. Sci. **44** (2008), No. 3, 955-971.

[MoPr09] Mori Sh. and Prokhorov Yu., On $\mathbb{Q}$-conic bundles. III. Publ. Res. Math. Sci. **45** (2009), No. 3, 787-810.

[Mul] Mullany R., Fano double spaces with a big singular locus, Math. Notes **87** (2010), No. 3, 444-448.

[Mum] Mumford D., Algebraic geometry. I. Complex projective varieties. Grundlehren der Mathematischen Wissenschaften, 221. Springer-Verlag, Berlin, New York, 1981.

[N] Noether M., Über Flächen welche Schaaren rationaler Curven besitzen, Math. Ann. **3** (1871), 161-227.

[Pukh87] Pukhlikov A. V., Birational isomorphisms of four-dimensional quintics, Invent. Math. **87** (1987), 303-329.

[Pukh89a] Pukhlikov A. V., Birational automorphisms of a double space and a double quadric, Math. USSR Izv. **32** (1989), 233-243.

[Pukh89b] Pukhlikov A. V., Maximal singularities on the Fano variety V_6^3. Moscow Univ. Math. Bull. **44** (1989), no. 2, 70-75.

[Pukh89c] Pukhlikov A. V., Birational automorphisms of a three-dimensional quartic with an elementary singularity, Math. USSR Sb. **63** (1989), 457-482.

[Pukh95] Pukhlikov A. V., A note on the theorem of V.A. Iskovskikh and Yu. I. Manin on the three-dimensional quartic, Proc. Steklov Math. Inst. **208** (1995), 244-254.

[Pukh95b] Pukhlikov A. V., Rational curves with prescribed branches, Moscow Univ. Math. Bull. **50** (1995), No. 1, 20-22.

[Pukh98a] Pukhlikov A. V., Birational automorphisms of three-dimensional algebraic varieties with a pencil of del Pezzo surfaces, Izvestiya: Mathematics **62** (1998), no. 1, 115-155.

[Pukh98b] Pukhlikov A. V., Birational automorphisms of Fano hypersurfaces, Invent. Math. **134** (1998), no. 2, 401-426.

[Pukh00a] Pukhlikov A. V., Fiberwise birational correspondences. Math. Notes. **68** (2000), No. 1-2, 102-112.

[Pukh00b] Pukhlikov A. V., Birational automorphisms of algebraic varieties with a pencil of double quadrics. Math. Notes. **67** (2000), No. 1-2, 192-199.

[Pukh00c] Pukhlikov A. V., Essentials of the method of maximal singularities, in "Explicit Birational Geometry of Threefolds", London Mathematical Society Lecture Note Series **281** (2000), Cambridge University Press, 73-100.

[Pukh00d] Pukhlikov A. V., Birationally rigid Fano fibrations, Izvestiya: Mathematics **64** (2000), No. 3, 563-581.

[Pukh00e] Pukhlikov A. V., Birationally rigid Fano double hypersurfaces. Sbornik: Mathematics **191** (2000), No. 6, 883-908.

[Pukh01] Pukhlikov A. V., Birationally rigid Fano complete intersections, Crelle J. für die reine und angew. Math. **541** (2001), 55-79.

[Pukh02a] Pukhlikov A. V., Birationally rigid Fano hypersurfaces with isolated singularities, Sbornik: Mathematics **193** (2002), No. 3, 445-471.

[Pukh02b] Pukhlikov A. V., Birationally rigid Fano hypersurfaces, Izvestiya: Mathematics. **66** (2002), No. 6, 1243-1269.

[Pukh03] Pukhlikov A. V., Birationally rigid iterated Fano double covers. Izvestiya: Mathematics. **67** (2003), No. 3, 555-596.

[Pukh04a] Pukhlikov A. V., Birationally rigid varieties with a pencil of Fano double covers. I, Sbornik: Mathematics. **195** (2004), No. 7, 1039-1071.

[Pukh04b] Pukhlikov A. V., Birationally rigid varieties with a pencil of Fano double covers. II. Sbornik: Mathematics. **195** (2004), No. 11, 1665-1702.

[Pukh04c] Pukhlikov A. V., Birationally rigid Fano varieties. In: The Fano conference, Univ. Torino, Turin, 2004, 659-681.

[Pukh05] Pukhlikov A. V., Birational geometry of Fano direct products, Izvestiya: Mathematics, **69** (2005), no. 6, 1225-1255.

[Pukh06a] Pukhlikov A. V., Birationally rigid varieties with a pencil of Fano double covers. III, Sbornik: Mathematics. **197** (2006), No. 3-4, 335-368.

[Pukh06b] Pukhlikov A. V., Birational geometry of algebraic varieties with a pencil of Fano complete intersections, Manuscripta Mathematica. **121** (2006), 491-526.

[Pukh07a] Pukhlikov A. V., Explicit examples of birationally rigid Fano varieties. Mosc. Math. J. **7** (2007), No. 3, 543-560.

[Pukh07b] Pukhlikov A. V., Birationally rigid varieties. I. Fano varieties. Russian Math. Surveys. **62** (2007), No. 5, 857-942.

[Pukh08a] Pukhlikov A. V., Birational geometry of Fano double covers. Sbornik: Mathematics. **199** (2008), No. 8, 1225-1250.

[Pukh08b] Pukhlikov A. V., On the $8n^2$-inequality. 2008. arXiv:0811.0183, 8 pp.

[Pukh09a] Pukhlikov A. V., Birational geometry of algebraic varieties with a pencil of Fano cyclic covers. Pure and Appl. Math. Quart. **5** (2009), No. 2, 641-700.

[Pukh09b] Pukhlikov A. V., Birational geometry of singular Fano varieties. Proc. Steklov Inst. Math. **264** (2009), 159-177.

[Pukh10] Pukhlikov A. V., Birational geometry of Fano double spaces of index two. Izvestiya: Mathematics, **74** (2010), No. 5, 925 - 991.

[Pukh10b] Pukhlikov A. V., Birationally rigid varieties. II. Fano fiber spaces. Russian Math. Surveys **65** (2010), No. 6, 1083-1171.

[Pukh11] Pukhlikov A. V., Bitationally rigid Fano complete intersections. II. arXiv: 1110.2052 [math.AG], 10 pp. To appear in Crelle J. reine angew. Math.

[Pukh12a] Pukhlikov A. V., On the multiplicity of solutions of a system of algebraic equations, Proc. Steklov Inst. Math. **276** (2012), 234-249.

[Pukh12b] Pukhlikov A. V., K-trivial structures on Fano complete intersections, Math. Notes **91** (2012), No. 4, 568-574.

[Pukh12c] Pukhlikov A. V., Birationally rigid complete intersections of quadrics and cubics, arXiv: 1205.1998 [math.AG], 55 pp.

[R87] Reid M., Young person's guide to canonical singularities. Algebraic Geometry, Bowdoin, 1985, 345-414. Proc. Symp. Pure Math. **46**, Part 1, AMS, Providence, 1987.

[R91] Reid M., Birational geometry of 3-folds according to Sarkisov. Warwick Preprint, 1991.

[R00] Reid M., Twenty-five years of 3-folds — an old person's view. In: "Explicit Birational Geometry of Threefolds", London Mathematical Society Lecture Note Series **281** (2000), Cambridge University Press, 313-343.

[Ry06] Ryder D., Classification of elliptic and $K3$ fibrations birational to some $\mathbb{Q}$-Fano 3-folds, J. Math. Sci. Univ. Tokyo **13** (2006), No. 1, 13-42.

[Ry09] Ryder D., The curve exclusion theorem for elliptic and $K3$ fibrations birational to Fano 3-fold hypersurfaces, Proc. Edinb. Math. Soc. **52** (2009), No. 1, 189-210.

[S80] Sarkisov V. G., Birational automorphisms of conic bundles. Izv. Akad. Nauk SSSR Ser. Mat. **44** (1980), no. 4, 918-945, 974.

[S82] Sarkisov V. G., On conic bundle structures. Izv. Akad. Nauk SSSR Ser. Mat. **46** (1982), no. 2, 371-408, 432.

[S89] Sarkisov V. G., Birational maps of standard $\mathbb{Q}$-Fano fibrations, Preprint, Kurchatov Institute of Atomic Energy, 1989.

[Shaf] Shafarevich I. R., Basic algebraic geometry. 1. Varieties in projective space. Second edition. Translated from the 1988 Russian edition. Springer-Verlag, Berlin, 1994.

[Shr08] Shramov K. A., On the birational rigidity and $\mathbb{Q}$-factoriality of a singular double covering of a quadric with branching over a divisor of degree 4. Math. Notes **84** (2008), No. 1-2, 280-289.

[Shr08b] Shramov K. A., Birational automorphisms of nodal quartic threefolds. Preprint, arXiv:0803.4348, 32 p.

[Sh93] Shokurov V. V., Three-dimensional log flips. Izvestiya: Mathematics. **40** (1993), no. 1, 95-202.

[ShCh11] Shokurov V. V., and Choi S .R., Geography of log models: theory and applications. Cent. Eur. J. Math. **9** (2011), No. 3, 489-534.

[Sob00] Sobolev I. V., On the birational rigidity of a series of multidimensional Fano fibrations. Math. Notes **68** (2000), No. 3-4, 392-396.

[Sob01] Sobolev I. V., On a series of birationally rigid varieties with a pencil of Fano hypersurfaces. Sbornik: Mathematics. **192** (2001), No. 9-10, 1543-1551.

[Sob02] Sobolev I. V., Birational automorphisms of a class of varieties fibred by cubic surfaces. Izvestiya: Mathematics. **66** (2002), No. 1, 201-222.

[Tikh81] Tikhomirov A. S., The intermediate Jacobian of the double covering of P^3 branched at a quartic. Math. USSR-Izvestiya. **17** (1981), no. 3, 523-566.

[Tikh83] Tikhomirov A. S., Singularities of the theta-divisor of the intermediate Jacobian of a double cover of $\mathbb{P}^3$ of index two. Math. USSR-Izvestiya. **21** (1983), no. 2, 355-373.

[Tikh86] Tikhomirov A. S., Letter to the editors of the journal "Izvestiya AN SSSR. Seriya Matematicheskaya", Math. USSR-Izvestiya. **27** (1986), no. 1, 201.

[Tyu72] Tyurin A. N., Five lectures on threefolds. Russian Math. Surveys. **27** (1972), No. 5, 3-50.

[Tyu75a] Tyurin A. N., The geometry of the Poincaré divisor of a Prym variety. Math. USSR Izvestiya **39** (1975), No. 5, 1003-1043.

[Tyu75b] Tyurin A. N., The intersection of quadrics. Russian Math. Surveys **30** (1975), No. 6, 51-99.

[Tyu78] Tyurin A. N., Corrections to the paper: "The geometry of the Poincaré divisor of a Prym variety" (Math. USSR Izvestiya **39** (1975), No. 5, 1003-1043). Math. USSR Izvestiya **42** (1978), No. 2, 468.

[Tyu79] Tyurin A. N., Intermediate Jacobian of three-dimensional varieties. J. Soviet Math. **13** (1980), No. 6, 707-745.

Index

Selected Published Titles in This Series

190 **Aleksandr Pukhlikov,** Birationally Rigid Varieties, 2013
189 **Alberto Elduque and Mikhail Kochetov,** Gradings on Simple Algebras, 2013
188 **David Lannes,** The Water Waves Problem, 2013
187 **Nassif Ghoussoub and Amir Moradifam,** Functional Inequalities: New Perspectives and New Applications, 2013
186 **Gregory Berkolaiko and Peter Kuchment,** Introduction to Quantum Graphs, 2013
185 **Patrick Iglesias-Zemmour,** Diffeology, 2013
184 **Frederick W. Gehring and Kari Hag,** The Ubiquitous Quasidisk, 2012
183 **Gershon Kresin and Vladimir Maz'ya,** Maximum Principles and Sharp Constants for Solutions of Elliptic and Parabolic Systems, 2012
182 **Neil A. Watson,** Introduction to Heat Potential Theory, 2012
181 **Graham J. Leuschke and Roger Wiegand,** Cohen-Macaulay Representations, 2012
180 **Martin W. Liebeck and Gary M. Seitz,** Unipotent and Nilpotent Classes in Simple Algebraic Groups and Lie Algebras, 2012
179 **Stephen D. Smith,** Subgroup complexes, 2011
178 **Helmut Brass and Knut Petras,** Quadrature Theory, 2011
177 **Alexei Myasnikov, Vladimir Shpilrain, and Alexander Ushakov,** Non-commutative Cryptography and Complexity of Group-theoretic Problems, 2011
176 **Peter E. Kloeden and Martin Rasmussen,** Nonautonomous Dynamical Systems, 2011
175 **Warwick de Launey and Dane Flannery,** Algebraic Design Theory, 2011
174 **Lawrence S. Levy and J. Chris Robson,** Hereditary Noetherian Prime Rings and Idealizers, 2011
173 **Sariel Har-Peled,** Geometric Approximation Algorithms, 2011
172 **Michael Aschbacher, Richard Lyons, Stephen D. Smith, and Ronald Solomon,** The Classification of Finite Simple Groups, 2011
171 **Leonid Pastur and Mariya Shcherbina,** Eigenvalue Distribution of Large Random Matrices, 2011
170 **Kevin Costello,** Renormalization and Effective Field Theory, 2011
169 **Robert R. Bruner and J. P. C. Greenlees,** Connective Real K-Theory of Finite Groups, 2010
168 **Michiel Hazewinkel, Nadiya Gubareni, and V. V. Kirichenko,** Algebras, Rings and Modules, 2010
167 **Michael Gekhtman, Michael Shapiro, and Alek Vainshtein,** Cluster Algebras and Poisson Geometry, 2010
166 **Kyung Bai Lee and Frank Raymond,** Seifert Fiberings, 2010
165 **Fuensanta Andreu-Vaillo, José M. Mazón, Julio D. Rossi, and J. Julián Toledo-Melero,** Nonlocal Diffusion Problems, 2010
164 **Vladimir I. Bogachev,** Differentiable Measures and the Malliavin Calculus, 2010
163 **Bennett Chow, Sun-Chin Chu, David Glickenstein, Christine Guenther, James Isenberg, Tom Ivey, Dan Knopf, Peng Lu, Feng Luo, and Lei Ni,** The Ricci Flow: Techniques and Applications: Part III: Geometric-Analytic Aspects, 2010
162 **Vladimir Maz′ya and Jürgen Rossmann,** Elliptic Equations in Polyhedral Domains, 2010
161 **Kanishka Perera, Ravi P. Agarwal, and Donal O'Regan,** Morse Theoretic Aspects of p-Laplacian Type Operators, 2010
160 **Alexander S. Kechris,** Global Aspects of Ergodic Group Actions, 2010

For a complete list of titles in this series, visit the
AMS Bookstore at **www.ams.org/bookstore/survseries/**.